AA003129

American Institute of Chemical Engineers

23rd Center for Chemical Process Safety International Conference

Topical Conference at the
2008 AIChE Spring National Meeting

April 6-10, 2008
New Orleans, Louisiana, USA

Printed from e-media with permission by:

Curran Associates, Inc.
57 Morehouse Lane
Red Hook, NY 12571
www.proceedings.com

ISBN: 978-1-60560-210-3

Some format issues inherent in the e-media version may also appear in this print version.

ISBN: 978-1-60560-210-3

Copyright (2008) by the American Institute of Chemical Engineers.
All rights reserved.

For permission requests, please contact the American Institute of Chemical Engineers at the address below.

American Institute of Chemical Engineers
Proceedings
Three Park Avenue
New York, NY 10016-5991
Phone: 212-591-8100

www.aiche.org

American Institute of Chemical Engineers

23rd Center for Chemical Process Safety International Conference
2008

TABLE OF CONTENTS

Three Years On From Texas City

Michael P. Broadribb
BP Refining Safety & Operations Excellence
Sunbury UK

Abstract

Following the tragic Texas City incident in March 2005, the executive leadership of BP reviewed the fundamentals of the company's approach to process safety and operational excellence. This evaluation has led to significant changes designed to improve safety and operations, not just at Texas City refinery, but across the BP Group worldwide.

For instance, U.S. Refining has already made progress with respect to the Baker Panel recommendations, which can be grouped under 4 main headings:

- Leadership
- Integrated and comprehensive process safety management system
- Process safety knowledge and expertise
- Process safety culture

The changes go beyond adoption of the Baker Panel's recommendations, and the paper will also discuss other issues, such as competencies, practices/procedures, metrics, and auditing, and it will highlight some wider messages for the industry.

Lessons Learned while developing and implementing Company Process Safety Metrics

John Herber, Senior Specialist, Corporate Safety and Industrial Hygiene, 3M Company

Abstract

In 2004 3M Company developed an internal PSM metric with the intent of improving PSM performance and reducing serious fires in hazardous processes. This metric uses internal PSM requirements as the framework for a self assessment. Goals are established and achievement is published along with other EHS performance indicators. A paper on this topic was presented at the 2005 CCPS Conference.

In addition, the 3M internal chemical manufacturing group has developed an activity-based metric using key PSM activities. Activities such as Process Hazard Analyses PHA's and Management of Change (MOC) are used as positive indicators while past due action items and incidents detract from the index.

This paper will present how these metrics have positively impacted the PSM systems since 2004 and reduced the measured risk to the company. It will outline the lessons learned while developing and implementing a company-wide PSM metrics system.

Author & Presenter – John Herber, Senior Specialist, Corporate Safety and Industrial Hygiene - 3M Company, St. Paul Minnesota

3M is a $23 billion diversified technology company that manufactures thousands of products ranging from Scotch® tapes and Post-it ® Notes to flexible electronic circuits, industrial chemicals and specialty films. Headquartered in St. Paul, Minnesota, the company has operations in more than 60 countries and serves customers in nearly 200 countries. This diversity in processes, products, business units, countries and cultures creates a challenge to consistently manage process risk.

This paper will review how 3M uses hazard and risk assessments to screen locations for inclusion in its PSM programs, how the internal metrics are applied, the results over three years of implementation and the lessons learned during the development, and implementation of a company-wide PSM metric system.

History of PSM in 3M

Process Safety Management (PSM) started in 3M in the late 1980s and was focused on conducting Process Hazard Analyses (PHA's). After the OSHA PSM regulation was published in 1992, 3M issued a PSM Policy. It set 3M's threshold quantities (TQs) for some materials at lower levels than the OSHA limits and covered some processes that historically had experienced higher rates of relevant incidents at 3M.

When the U.S. Environmental Protection Agency (US EPA) enacted the Risk Management Program, (RMP) in 1999, 3M expanded the original PSM Policy into the Process Hazard Management (PHM) Policy include TQs for other global PSM-like regulations (e.g., Seveso II) and EPA RMP. The PHM Policy requires all locations to:

- Understand the hazards of the processes operating at a location,
- Define the hazards and risks posed by the processes, and
- Implement an appropriate management system to manage the hazards.

In addition, 3M facilities having corporate or government-regulated hazardous chemicals in excess of the threshold quantities must comply with applicable government regulations or the 3M Process Safety Management requirements if they are more stringent.

This policy is supported by the 3M Process Hazard Management Standard, which provides the guidance on complying with the policy and directs the user to the 3M PSM Manual, which provides element-by-element guidance on implementing PSM.

While implementation of this policy was global, there was no consistent global auditing system to review this approach or the effectiveness of local systems. Without consistent oversight, full compliance with the 3M PHM Policy could not be determined.

In 2002, 3M developed the EHS Management System, which integrated corporate-wide performance expectations for 20 to 40 key metrics across all aspects of Environmental, Health and Safety (EHS) activity. Some of the metrics are traditional, i.e. Injury and Illness Rate, and number of fires. Others measure progress toward internal goals such as waste reduction and Industrial Hygiene (IH) survey completion. These score cards are updated quarterly and the results are reported to management. Emphasis on Six Sigma and scorecard-style reporting made this an efficient and effective tool to report EHS performance and to evoke action.

In 2004, a metric specifically for Process Safety Management was added to the EHS Management System. The balance of this paper will explain the details of implementing the system and the effects it has had on our PSM programs.

The need for a PSM Metric

The PSM metric was intended to close two gaps: 1) lack of accurate process risk information for global operations and 2) lack of a consistent method to measure PSM systems. Several typical metrics were initially considered - audit and PHA action item closure rates, number of incidents and fires, etc. Each of these metrics had major drawbacks that kept the measure from accurately reflecting PSM program effectiveness.

After several iterations of metric ideas, the task team struck upon the obligation to comply with all applicable local regulations and 3M internal policies. Once the team focused on the PHM policy, it became clear that the objective was to get each 3M location to implement a hazard management system that was proper for the hazard and risk level at the respective location. This redirected the metric effort to: 1) determine which locations need PSM; 2) help locations determine their need and apply the correct management system; and 3) measure the progress toward a highly functional PSM system. The team developed the following three-step process for the metric.

Phase 1 - Location Identification

With 3M's broad array of manufacturing processes and locations in many countries, it was necessary to develop a quick, but accurate way to sort the locations definitely NOT needing PSM, from those who definitely DID, and a few "maybes". A questionnaire-style location screening tool was deployed by e-mail. It asked locations to describe types of processes they operated. It sorted the number of locations of interest from 160 to 110. This rough screen has been revised and validated by the EHS managers around the globe. As each step was developed, the potential failure mechanisms were reviewed, and if the problems were critical to the success of the metric, modifications were made.

Phase 2 - Risk Screening
The second phase of the metric needed to verify the output of the screening and quantify the risk levels for the remaining 110 locations. Since 1999, 3M had been using an internally-developed risk tool, Quantitative Process Risk Screening Tool or QPRST. The tool is based on a risk assessment tool developed by Eli Lilly [1] combined with the Dow's Fire and Explosion Index Hazard Classification Guide [2]. A poster session on this tool was presented at the 2002 CCPS international conference. [3] It uses information on the chemicals, quantities, reaction data, controls, safeguards and the incident history for a given process to determine the risk class (I to V).

Initially, each location was required to select their 10 most hazardous processes and perform a QPRST analysis on them. The highest risk class process at a location determines the risk class of that location. This was based on two assumptions: a location will employ one process hazard management system and the PSM management system must be sufficiently robust to address the highest risk process on-site. Also, the plan called for locations to review additional hazardous processes with each ensuing year until they have reviewed all of their processes; this keeps the risk profile accurate.

QPRST deployment required a concerted effort to properly train locations in the use of the tool. Locations operating similar processes were encouraged to compare data. Since the frequency of releases is a factor, locations operating similar processes may calculate different risk classes for comparable processes.

The PHM Standard outlines how QPRST risk classes are used to determine the level of PSM system required. A summary of the requirements are shown below:

Risk Class I - These processes handle small quantities of hazardous material. For locations determined to be a Risk Class I, the use of 3M's Global Safety & Health Plan (GSHP) is recommended. This system covers the full spectrum of safety and health concerns including ergonomics, electrical safety, industrial hygiene, etc.

Risk Class II – These processes may use PSM-covered materials, they may be below OSHA and 3M threshold quantities. For these locations, in addition to the GSHP, the recommended management system is PSM Basic, which is based on the EPA RMP Program 2 guidance.

Risk Class III - Processes in this risk class approach or possibly exceed the OSHA TQs. This is the level where 3M requires the use of PSM. For locations outside the US, compliance with local process safety regulations is required. In the absence of any PSM-like regulation, the 3M PHM standard defines the management system.

Risk Class IV – Generally, processes in this class are above OSHA TQs or have potential of serious consequence. Full PSM and the 3M GSHP are employed, but in addition, locations are required to perform a consequence analysis, and use additional semi-quantitative methods to better define the risk presented. The results of these studies are reported to Division and EHS management outlining the hazards that create the higher risk level, the potential consequences of a serious event to the location, the community and the business, and what can be done to reduce the risk level.

Risk Class V – This is the highest risk class defined by QPRST. Again, all of the requirements for Risk Class IV processes must be met (GSHP, PSM, etc.) For these processes, 3M also requires a full Quantitative Risk Analysis (QRA); the results of the QRAs must be communicated to Division and EHS management.

Locations with processes only in the lowest risk class (I) were excused from further work in the metric since in our system these processes are appropriately managed using our general safety system. This further reduced the number of locations of interest from 110 to 82.

Phase 3 – Management System Self Assessment

Finally, the metric needed a method to measure the quality of the PSM management systems being employed. The concern was how to accurately measure PSM performance. In 2001, Gary York of Rhodia presented a paper entitled "Facilitated Self-Assessment – An Approach to Measuring PSM Systems, [4] at the CCPS International Conference and Workshop. It outlined a management system assessment used by Rhodia worldwide. 3M was given permission to use the Rhodia self-assessment tool as the basis for the PHM management system assessment.

The Rhodia tool also employs a second set of questions which are considered exemplary practices and beyond compliance. This feature was not included in the 3M program thus far.

3M modified this tool to use 3M terminology, changed the mandatory questions to match the 3M PSM auditing protocol, added a weighted score for Program and Implementation, and selected segments that are unique to PSM and not duplicated in other 3M Safety Management systems; those are: Process Safety Information (PSI), Process Hazard Analysis (PHA), Management Of Change (MOC) , Pre Startup Safety Review (PSSR), Operating Procedures, Mechanical Integrity (MI), Training, Incident Investigation, and Audits.
The scoring for each element is given a number of points based on how the PSM elements are weighted in Rhodia's self assessment and the "ProSmart: Performance Measurements of Process Safety Management and Measurement Software"[5], developed by CCPS. Inside each element the points are divided among the questions with the implementation questions having 3 times the value of the program questions.

Rhodia's process uses a facilitator to review the plants' input and develop the scores. This requires a trained EHS person from the corporate group to lead each location through the self assessment. With the short timeline, the large number of locations and limited corporate resources, this was not feasible for the 3M rollout. Since 3M locations had been using self assessment tools for safety and environmental issues, we felt they could accurately reply to the questions. We employed the 3M internal TV network and a number of netmeetings to train people on the intent of each question and to explain the metric expectations. We also referenced corresponding sections of the Corporate PSM manual, our internal guidance on PSM, as an element-by-element reference.

The final step in the development of the PSM Metric was to set expected level of self-assessment performance for each risk class. This step involved reaching a consensus on which elements are the most important in reducing overall risk and what rate of improvement can be reasonably achieved. For locations required to have full PSM compliance, the target score was obvious –

100%. The expectation for Risk Class II locations or those who recently began developing PSM programs was set at 85%.

Also, a 5 year plan for expectation was developed so the long term targets could be communicated to the users.

Implementation

2004 was the first year the rollout year for the metric. The focus was on all manufacturing locations globally. Laboratories and pilot plants were initially exempted until the applicability of the metric for those operations could be determined.

While these expectations would be elementary for locations presently using PSM, it would allow those new to PSM or who have significant PSM gaps to establish the required tools at a rate that encourages sustainability. Each metric phase was approached as a quarterly goal. The locations completed the location screening in the first quarter of the year, the QPRST analysis in the second quarter and the self assessment in the third quarter. The self assessment only reported the results of the first group of elements – Process Safety Information, Process Hazard Analysis, Management of Change & Pre Startup Safety Review-with no targets or goals applied to the first year.

The second year location screening and risk analysis expectations were unchanged. Location either submitted more QPRSTs or updated existing ones if needed. The self assessment shifted focus to the remaining elements: Operating Procedures, Training, Mechanical Integrity, incident Investigation and Audits. Again, no scoring targets were set, but the data was available to all EHS personnel for planning necessary improvements.

The reception at the locations was positive. The locations especially appreciated the ramp-up of effort and the graduated PSM implementation requirements based on risk class.

Progression of the Metric Program

This metric is now an on-going process. Location Screening is conducted every year to assure the risk picture reflects changes in existing operations as well as inclusion of locations new to 3M through merger and acquisitions or capital growth. We have rolled the location screening questions into the GSHP self assessment, so that all locations globally answer the screening even if PSM is not implemented there. This placed the PSM metric on the same schedule of events as other EHS Scorecard metrics. Quarterly conference calls are conducted with the North American, Asian, European, and South American regions to provide latest summaries, upcoming activities and invite consultation on PSM implementation.

The QPRST risk results are also reviewed or renewed every year. Processes and equipment changes could alter the risk conditions; those changes could increase or decrease the risk level for that process. Also, some locations may have a high turnover of hazardous processes. We want to keep the risk profile data as accurate as possible.

The Self Assessment is also required annually with quarterly updates expected from those below metric targets. A plan was developed for improvement of Management System Self Assessment scores across the implementation years. The proposed targets for these goals are shown in the graph below.

3M has an acquisition integration procedure that requires a number of assessments shortly after the deal is complete. Some are business related, but a thorough EHS assessment is conducted and used to develop a risk-based program for closing gaps. Newly acquired locations are phased into the EHS Scorecard system over the course of their first year in 3M. This allows them time to develop or improve their systems before being measured.

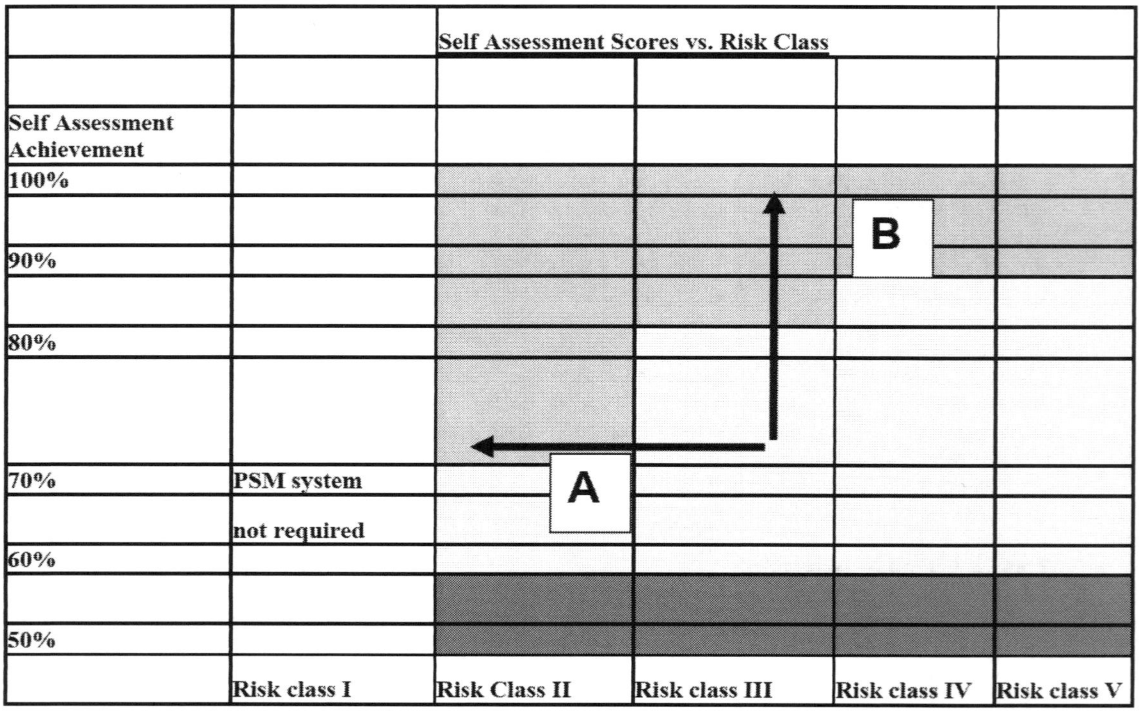

Locations have two options for meeting the expectations of the metric. The first and most desirable is to change the process conditions to reduce the risk level. This is shown in the diagram above by arrow A. Depending on the process, this may not be an option. The second improvement option is to improve PSM implementation as measured by the MSSA score and shown above by arrow B.

The initial response to this was somewhat negative, since in our culture, plants have little control on major process changes. The intent was to open a dialog inside the business about higher risk operations and make a case for process changes that could reduce risk. The metric was used successfully to justify capital improvements that also reduced risk. Pilot plants are now included in the location screening, however most do not have processes that require the rigor of PSM.

Over the last 3 years, we have been "working the system" of the metric goals established for 2004-2008. Here is a synopsis of the achievements:

1. Reduction of the number of serious fires in PSM-covered locations by 60%
2. A clearer picture of hazardous processes globally.
3. Improved communications among the network of PSM practitioners
4. A method to track PSM improvement during Merger & Acquisition integration
5. More confidence that 3M meets or exceeds its general duty to operate safe processes globally.
6. The decision to omit pilot plants and laboratories from this metric has proven to be a sound one. The existing EHS guidance for these operations provides sufficient protection.
7. Calibrated self assessments during PSM audits as they are conducted at covered locations

Other metric systems

While our team was developing this PSM metric, our counterparts in Ergonomics were on a parallel path. We became collaborators to share approaches and methods. Ironically, their metric had the same basic approach of screening the operations for ergonomic hazards, sorting the hazards into risk "buckets", and using an established tool to reduce the ergonomic risk. Their metric measures the completion of the risk assessments and implementation of corrective measures. They too have seen a measurable reduction of ergonomic–related injuries and a corresponding reduction in ergo-related workers comp claims.

3M's chemical manufacturing division had developed an activity-based metric to chart to "quality" of the PSM programs at their manufacturing locations. This metric combines data on positive PSM activities such as MOC & PHA action item closure, with negative data such and overdue action items, Mechanical Integrity Inspections and contractor incidents. This metric is a great indicator of PSM activity, but had 2 aspects that made it less desirable than the PSM Metric, when applied on a corporate basis. First, it did not address the issues of basic PSM implementation such as collection of Process Safety Information, quality of Process Hazard Analyses, etc. Secondly, it required a lot of activity tracking that would be onerous to many locations without the well-established database systems that the Chemical division uses.

Many times, the need to "normalize" the metric derails the development or implementation. Normalizing is trying to adjust for location difference, as is done with the OSHA Injury and Illness rate – normalizing injuries per 200,000 worker hours. Part of the success of the PSM metric was that there were no event-driven aspects (number of fires, injuries, etc) to normalize. Regardless of the size of a location, they all need a PSM system rigorous enough to manage the risks at that location.

Some basic assumptions to consider when developing a metric are:

1. People are willing to work toward a well-understood goal.
2. All users are going to be honest and objective
3. There will be some mistakes by both metric developers and users
4. Metrics may cause embarrassment to some.
5. You cannot predict or control how your metric data will be used.
6. Internal metrics and those available to the public can have different goals or targets.

Lessons learned while developing the PSM metric:

1. Define the long term goal, purpose and/or intended outcome of the metric before developing the measurement system. For 3M that was to reduce the number of serious fires through improved PSM Implementation.
2. Start small, build from a solid foundation and build on small successes. Ramp toward the intended goals.
3. Look for possible negative outcomes from developing metric goals. As seen from experience with Injury & Illness reduction goals, setting a goal may cause undue pressure to meet the goal with little regard for the longer term impact or accuracy of the results
4. Garner high level support for the concept. High level support ensures that management is prepared to make the commitment of resources to accomplish your metric goals.
5. Build a solid team to develop and review metric proposals, and include IT people. The team needs to represent the using community, but not include every user. For us, integrating the metric into the existing EHS management system and scorecard required involvement of our IT group. IT pointed out some efficiency they could provide to the long term success.
6. Make the metric fair to all participants – do not penalize based on size, complexity, history or other parameters that the locations can't control.
7. Determine if existing data or tools can be used to jump-start the metric. If pertinent data is available in other systems, find a way to use it vs. asking for it in both metrics.
8. Design some flexibility into the early phases of implementation, then later ramp up the expectations
9. You won't get it completely right the first time. If you start from a solid base and keep the users informed of where the metric is going, they will be more willing to point out possible pitfalls as seen from their perspective.
10. There will always be a few obstinate users to any metric system; use them to assist system development and review.
11. Metrics are best used for large improvement opportunities and may need revision to get the last 10%

Lessons for implementing metrics:

1. Automate data entry, collection and analysis as much as possible. This is for the benefit of both those who input the data and those we need to create the summary reports.
2. Use a broad variety of users to pilot the program before rollout. A common mistake is to take the proposed system to the "power-users". Test it with the lower skilled users who will find more failure modes for you.
3. Pilot metric questions and guidance to several geographic areas to ensure they are clearly worded.
4. Use real examples to demonstrate the metric system, so the users can relate them to their processes and systems.
5. Evaluate potential failure modes of the measurement system and what can be done to prevent future problems.
6. Review the proposed system with other groups who have successful metric systems in your area or company
7. Avoid the urge to have locations "compete", but instead, encourage collaboration (Steal shamelessly; Share openly)

Verification after implementation

1. Does the metric accurately reflect the condition being measured? If not, find ways to adapt it.
2. Is there progress toward the long term goal? Realize that there may be some initial decline before actual progress is made.
3. Are there opportunities to make the system more user-friendly? Address all user problems quickly, but sustainably.
4. Develop reports that use the metric data and recognize the successes; be accurate but tactful when reporting poor performance.
5. Understand why some locations are not meeting the metric goals and provide "safe" coaching opportunities for them.
6. Publish guidance to all users based on reported metric system problems – let them suggest solutions and improvements
7. Publish how other users are using the data to achieve objectives outside the primary goal such as capital justification for better or safer systems.
8. Use the development team as the jury to verify accurate & consistent evaluation of metric results.

Conclusions

Developing internal PSM metrics for a company can be challenging enough, and the challenges increase as the operations become more diverse. Start with a small team to develop a "straw man" proposal that includes major goals and expected benefits. Use this proposal to garner management support for the concept. Use a broader cross-functional team to develop the detailed goal(s), and measurement systems. Pilot the draft concepts to a broader audience to work out inconsistencies in wording and scoring. Prior to release, revisit the metric proposal with management to confirm their agreement and commitment. Prepare the using community by frequent communication before the metric is launched. Plan for a large technical support effort to answer the inevitable questions and calibrate the users. Review all results to ensure accuracy before publication; inaccurate results will torpedo the effectiveness of a metric. Decide what level of performance constitutes a concern and communicate to management based on that level.

Corporate metric results can be easily changed by closing a very competent plant or acquiring one with a history of poor PSM performance. However, in most cases, the performance of a given plant should continue to improve over time. In all cases, provide balanced feedback on a location's progress. Some of the locations who have struggled in the past may become the best mentors for other locations who need improvement.

While the PSM Metric journey has been a challenge, it has raised the awareness of PSM across 3M company and improved our understanding of our process risks. It has given us a measurement system to apply to assess locations new to 3M or new to PSM. It has provided accurate data on the performance levels of all PSM covered locations which can be used in discussions with locations as well as corporate management.

Looking forward, the basic operation of the metric will not change radically unless there are significant changes to PSM expectations. These changes may be initiated by countries other than the US, which will enhance the global impact of the metric. With the rapport built through development and implementation of the metric, the corporate group has a better chance of learning about global regulatory changes and how they could impact our operations and improve PSM implementation.

Bibliography

1. Steven J. Schultz, William R. Fassel, Michael P. Roesner, and William M. Walasinski, "Process Hazard Screening – A Method for Identifying Batch Pharmaceutical Processes with the Greatest Risk for Catastrophic Accident", Proceedings of the CCPS International Process Safety Management Conference and Workshop, September 1993, Pages 475-495, AIChE, New York, 1993

2. Dow's Fire and Explosion Index Hazard Classification Guide, 7th Ed., AIChE, New York, 1994

3. Arlyn H. Poppen and Robert E. Hollenbeck, *3M Quantitative Process Risk Screening Tool,* Proceedings of the CCPS International Process Safety Management Conference and Workshop, October 2002, AIChE, New York, 2002. and Proceedings of the Mary Kay O'Connor Process Safety Symposium, October 2002, Mary Kay O'Connor Process Safety Center, Texas A&M University, College Station, TX, 2002.

4. Gary York, Rhodia Inc. Facilitated Self-Assessment – An Approach to measuring PSM Systems, Proceedings of the Center for Chemical Process Safety International Conference and Workshop: Making Process Safety Pay; The Business Case, AIChE, New York 2001

5. "ProSmart: Performance Measurements of Process Safety Management and Measurement Software", ISBN 0-8169-0813-3, AIChE, New York, 2001

" STRUCTURAL CAPITAL A UNIQUE MANAGEMENT TOOL FOR ACHIEVING HIGHER LEVEL IN QUALITY AND SAFETY "

Topic related to: Knowledge Portal enhancing quality, yield & product performance

P. N. Trivedi

TITLE

" STRUCTURAL CAPITAL A UNIQUE MANAGEMENT TOOL FOR ACHIEVING HIGHER LEVEL IN QUALITY AND SAFETY "

Introduction:

Objectives of Paper:

1. Apply the Structural Capital framework to develop an integrated solution which will improve the process quality, productivity & operating performance
2. Evaluate the impact and use of technology in performance improvement.
3. Managing & utilizing of Process Safety Information, therefore applying the database for implementation & best practice of Process Safety Management.

Reliance is a rapidly growing organization in multi-locations. As a part of growth plan, experienced employees move to new projects at other locations and relatively less experienced employees, after formal training and validation look after day-to-day plant operations at our manufacturing locations. Reliance is ISO 9000/14000/, OHSAS-18000 certified organization. Each product quality is in line with international requirement of the customers.

Leadership focuses on empowering employees through continual learning to manage the Business processes effectively and efficiently. To achieve this, the effectiveness of our old on-line Quality & Learning management system was regularly evaluated and improved. But, the overall effectiveness was still at level 2 (learning). This necessitated quality & training function to transform its role from a Service Provider to a Strategic Business Partner. The top management of company felt that, it should sustain it's continual improved performance for more than coming 100 years.

The main reason for low effectiveness of the Quality & Learning system was that it was not capturing the organic (evolving) expert knowledge or success and failure learning. While the juniors had the scriptural knowledge gained from technology manuals, SOP's, Check sheets they lacked the organic knowledge gained through experience.

A Structural Capital (SC) was developed by line function, central technical/engineering, quality & training professionals to bridge this gap. SC represents the knowledge assets that will complement human capital to create value. It ensures that appropriate knowledge is made available to empower right people at right time. By integrating with earlier knowledge management efforts, the effectiveness of quality & learning systems improved to level 5. Return on investment is 112%.

The success of this practice in one Plant encouraged us to replicate in other manufacturing Plants. Successful integration, utilization and business result has demonstrated that organizational learning in a conductive environment can create win-win for all stakeholders.

Structural Capital (SC):

In a rapidly growing multi-location organization, there is a need to relocate experienced employees for commissioning new projects, turn around jobs and improving efficiencies of existing plant operations at short notice leaving day-to-day operations to relatively less experienced employees.

Top Management focuses on empowering employees through continual learning to manage the processes effectively and efficiently. The integrated quality management system is implemented to carry out effective internal auditing system, management review. The voice of customer is captured & addressed in integrated management system for continual improvement & for the customer satisfaction. On an average, each and every employee spends about 64 hours of training per annum and our annual quality management system & training expenses comes to approximately 5 % of our payroll expenses. Some of the identified employees are groomed as **Specialists** in their disciplines as a part of the Learning initiative. In spite of the regular improvements made, the effectiveness of the quality/ learning system was still at level 2 (learning) and in a segmented manner at level 3 (Behavior) two years ago. On analyzing, it was found that the main reason for this low effectiveness was that there was a gap in transferring learning from training context to the workplace and its consequent impact on business results. It was noted that the quality/learning system was not capturing the organic (evolving) knowledge within the organization and outside from those in similar business. The success and failure learning was also not captured in a structured manner. As a result, the Top Management recognized the need for the quality/training function to expand its role from being a **Service Provider** to a ***Strategic Business Partner***.

To implement this transformation, the topmost executive of the organization, a firm believer of quality/learning organization was put in-charge. A core team representing professionals from Production, Maintenance, Quality Assurance, Training, Information Technology and Technical Services functions assisted him. The team consisting of employees drawn from top, middle and junior management levels identified the actions to bridge the gaps in the Learning system, by making it more application oriented in the workplace.

The purpose of developing Structural Capital was to capture not only the scriptural information documented in technology/vendor manuals but also the knowledge generated in the minds of experienced personnel as they experience success and failure learning. Less experienced personnel across multiple locations capture this knowledge. It empowers them take informed decisions to solve problems related to process upsets and equipment/systems failures, improvement in the effectiveness and efficiency of processes and systems, reducing wastage and improving product quality & safety. This has led to enhanced productivity of people and assets, increased profits, a culture of knowledge sharing, reduced cycle time for problem solving, reduced problems, enhanced training effectiveness, effective decision making and increased employee loyalty due to greater empowerment.

The goals of this practice which have remained same since its inception are

- To empower the relatively less experienced employees with the right kind of knowledge at the right time to enable them operate and maintain the existing plant operations at optimum levels and
- Integrate knowledge management efforts across the organization and improve the effectiveness of the learning system.

This practice was successfully designed, implemented, effectively utilized and its business impact measured in one of our world-scale plant 18 months ago. The effectiveness of the Learning Management System in that Plant improved to Level 5 (Return on Investment) from the earlier Level 2.Encouraged by the good results, the practice was replicated in 4 more World-scale Plants in the same location thereafter and the results have been uniformly encouraging. The Phase II of Implementation in 2 more world-scale plants in the same location will be completed by May 2008. The Top Management has planned to replicate this practice in other locations also without making any changes. The Top Management has planned to replicate this practice in other locations also without making any changes.

1300 employees (72 % of total employees in the respective plants) currently utilize this practice. They include employees at all levels - the plant operatives, engineers and managers. Developing this practice required commitment of resources in terms of money, people and infrastructure. Developing of SC framework including the portal was done in a structured manner by forming various teams. The teams were involved in needs assessment, design, implementation and evaluation of the practice.

State-of-the-art high-speed servers were specifically procured for this purpose. These servers are loaded with IBM applications for various applications; Lotus notes utilities, domini.doc software for library management system. These servers are connected with users desktop/PC on fiber optic network to ensure best speed possible for access. Integration of various servers used for different applications, Distribution Control System, which controls & monitors manufacturing plant operation

This practice is complimentary to the other knowledge management KM initiatives and it is well aligned to other Learning management systems. The exhibit below illustrates the interaction between SC and training, learning and performance improvement systems.

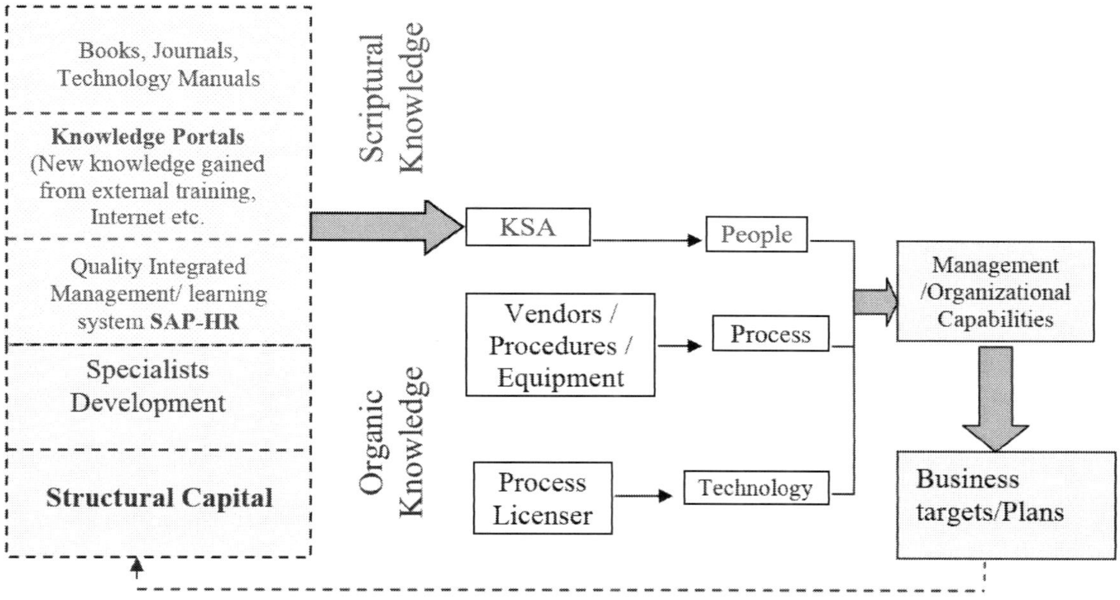

Needs assessment was done by group of new engineers, 15 junior level executives, 5 middle level executives, and 3 senior level executives from 7 plants along with one junior and one middle level executive from training & safety function. One person was nominated as overall coordinator for this project. Training, Technical Services along with Information Technology (IT) professionals, did design of the Framework. The same people who made the Needs assessment did implementation. Users did evaluation of the Framework from a cross-section of the organization at all levels (including 50 plant operatives (junior level) in various plants, 25 engineers (junior level), 10 managers (Middle level) and two senior level executives in the plants and technical support services). The cost incurred towards development of SC framework (hardware and software) was US $ 300,000. While all Plants at our location share the hardware cost, other Plant-specific costs are booked to individual Plant account. The return on investment works out to 112 %

It is an excellent case of quality & training functions transforming it into a strategic business partner driving organizational results through creation of a conducive knowledge-sharing environment for application-focused learning. It is unique in terms of its design for continued relevance of learning to meet the changing business needs. This is done by capturing the knowledge of experts at various levels while experiencing success and failure learning and making it available to the knowledge seeker at the right time. Implementation of this practice has brought about a cultural change in the organization. The mindsets of the people have changed from knowledge hoarding to knowledge sharing and reuse. It has paved the way for application focused organizational learning systems developed by the employees and for the employees.

Major activities of the Project:

1. Formation of Team: Team comprising of sectional head of each discipline (Production/ Mechanical / Instrumentation /Electrical /Quality control Technical services / Central Engineering Services / IT) was formed.
2. Visit To Companies Like Du-Pont, INFOSYS & GE
3. Identification Of The Server, Identification of The Software / suitable IT platform
4. Designing The Portal (Designing of Taxonomy of database, Designing of Home Page, Uploading all Soft Copies To The Portal)
5. Digitization Of All Hard Copies : (A4/A3 Sheets: 115000 nos. & Drawings A4/A3/A2/A1/A0 : 8500)
6. Uploading of Soft copies of documentation
7. Networking of various application servers, Distributed Control Systems, different portals
6. Training to all engineers, new graduate engineers & Technicians / Operators and their Validation
7. Estimation of benefits (Quality, yield & Productivity) & Forward Plan

Evaluation strategies & methods of benefits calculations:

The effectiveness of this practice is evaluated through a structured mechanism using a multi-pronged approach. The evaluation is done viz-a viz goals of this practice. The goals of this practice as envisaged initially were

a. Empower the relatively less experienced employees with the success and failure learning of the experts.
b. Integrate the knowledge & integrated quality management efforts across the entire organization.
c. Improve the effectiveness of the learning & quality system by giving an opportunity to the employees acquire knowledge which they can apply in their workplace and which has a positive impact on the business results. This gives ultimately the Return on Investment.
d. Reduce cycle time for relocation of experienced personnel and induction of new personnel.

Some of the factors that have been considered for evaluation of this practice are

- Increase in production levels without corresponding increase in Human resources or equipment capacity.
- Reduction in equipment failures or improvement in reliability
- Number of hits in identified portals that are the vehicles for Knowledge sharing.
- Reduction in lead-time for accessing documents those were in for the Seniors.
- Reduction in cycle time for problem solving or any improvement resulted by knowledge sharing.
- Reduction in lead-time for exchanging information with customers.
- Reduction in lead- time required for knowledge seekers to acquire the requisite knowledge.
- Reduction in lead-time required for an employee to take independent charge of a new position after his/her predecessor has moved away to a new position or a new assignment.
- Reduction in cycle time required for induction of a fresh employee into the organization.
- Efficient use database for the use of Process Safety Management team.

These factors have a direct implication in improving the bottom line or profits. Other factors include enhanced involvement of employees in organizational improvement activities, reduced employee turnover, improved interpersonal relationships and consequent teamwork, and enhanced open communication and lower absenteeism.

Financial costs of this practice are calculated based on the cost of development of SC framework including the portal, cost to maintain the framework, cost of uploading the knowledge in electronic form and man power cost. The return on investment is 112%.

Learning's from designing and implementing this practice:

Lessons learned from designing and implementing this practice is as follows:

a) SC can be successfully implemented by not only creating the tools, but more importantly a conducive environment within the organization.
b) The relevance of learning framework needs to be continuously evaluated to ensure that it is aligned with the changing business needs. Creation and sustenance of effective learning & quality systems requires commitment from top management, sincere involvement of associates and alertness to changing business needs by the training professionals.
c) Knowledge generated in the minds of people needs to be captured and shared with all concerned on an ongoing basis. Its integration with the learning system will definitely improve training & quality system effectiveness resulting in enhanced workplace place performance and bottom line.
d) Integration of isolated knowledge management efforts will add value to Business by providing complete learning solutions to all concerned employees for continuous improvement.
e) Knowledge sharing needs to be appropriately rewarded and recognized in order to create a culture that promotes the same. Technology alone cannot sustain knowledge sharing systems.

This practice is already replicated within the 5 world scale plants. A plan has been drawn to replicate this practice in 2 more world scale plants in the same location and 70 world scale plants in 12 other locations as the benefits that have resulted can be compounded several times. We would be willing to share the road map for implementation of SC with any organization in the spirit of truly being a world-class learning organization.

Benefits:

The information, which supported the fact that the organization has achieved the results of implementing the new practice of learning & integrated quality management system are given below.
The new practice was implemented since January 2006 and the comparative figures are given for the periods as follows:

- **Before'** implementing the practice: March 2005.
- During '**Launch'** of the practice: January 2006
- '**After'** implementing the practice: March 2007

The Business results of implementing the new practice, which have a direct and long-term implication in improving the bottom line or profits of the organization, are as follows:

a) Increased Employee Productivity

Effective decision-making and job planning, together with faster problem solving, due to easy and timely availability of expert knowledge, has resulted in improved Employee productivity. People productivity (MT of product/ Employee) has increased by **13.8 %** in March 2007, as compared to 9.3 % in March 2006. The site has recorded the highest production in all its plants during 2006-07.

Before	Launch	% Increase	After	% Increase
430	470	**9.3%**	535	**13.8%**

b) Reduced Conversion cost

Enhanced participation in various organizational improvement activities has resulted in reduction of conversion cost in all the plants. Conversion cost (US $/ MT of product) was reduced in the largest plant, by **4.0 %** in March 2007, as compared to 0.6 % in March 2006.

Before	Launch	% Reduction	After	% Reduction
140.7	139.8	**0.6%**	134.3	**4.0%**

c) Reduced Utility consumption

Due to reasons mentioned in (b), specific consumptions of Power and Steam in plant P1 have reduced by **6.4%** and **34.2 %** respectively in March 2007, as compared to 1.1% and 7.3 % respectively in March 2006.

Parameter	Before	Launch	% Reduction	After	% Reduction
Power (KWH/MT)	440	435	**1.1%**	407	**6.4%**
Steam (MT/ MT)	0.905	0.839	**7.3%**	0.552	**34.2%**

d) Improved Equipment Reliability/ Productivity

After the implementation of the practice, the time required to locate and retrieve the emergency spares of the equipment reduced by 60-70 %. On an average, emergency spares can be located in 15 minutes, with the help of information available online. This helped in reducing the cycle time of any problem solving considerably. In plant P1, number of equipment failures have reduced by **10.2 %.**

Equipment failures	Before	After	% Reduction
	1022	917	**10.2%**

The maintenance cost has been reduced in Plant P1. (in million US $)

Before	Launch	% increase	After	% reduction
9.95	13.26	**33.3%**	11.32	**14.6 %**

e) Reduced Inventory cost

Inventory (Million US $) in the plants have been reduced by **10.4 %,** in March 2007.

Before	After	% Reduction
29.75	26.64	**10.4 %**

f) Increased Contribution / Profit

The improvement in the processes and equipment reliability and reduction in various costs as a result of changes in people behavior has improved the contribution/ profit for the organization. Contribution (Million US $) of the site has improved by **7%** in March 2007, as compared to 5.8 % reduction in March 2006

Before	Launch	% Reduction	After	% Increase
199.7	188.2	**5.8%**	201.4	**7.0%**

g) Improved Quality, Yield and First time on-specification products (without rework/recycle)

The easy and timely availability of expert knowledge and enhanced customer focus have resulted in improved yield and first time on-specification products. In plant P2, both the yield and first time on specification products have improved, as compared to those in the previous 2 years:

Parameter	Before	Launch	% Reduction	After	% Increase
Yield (%)	96.89	96.53	**0.40**	96.97	**0.45**
First time on Spec.%)	92.38	91.81	**0.60**	92.90	**1.20**

h) Enhanced Involvement in Organizational improvement

Due to enhanced sense of belongingness to the organization, there has been active participation at all working levels in the form of Suggestions for improvement. No. of suggestions has increased by **180 %** in March 2007, as compared to 81.6 % increase in March 2006:

Before	Launch	% Increase	After	% Increase
87	158	**81.6%**	443	**180.0%**

i) Reduced Employee unauthorized absenteeism

Employee Absenteeism has got reduced from 0.5 % in March 2006, to 0.4 % in March 2007 – a 20**%** reduction in unauthorized absenteeism.

Before	After	% Reduction
0.5 %	0.4 %	**20 %**

j) Product quality and enhanced Customer Satisfaction

The quality report of raw materials, intermediates and products are made available on LIMS (Laboratory Information Management System) under SC. This has resulted in 30% reduction in time taken for Process operators to communicate with Laboratory Chemists, enabling them make timely decisions on maintaining or changing process parameters, required to achieve desired quality of product. The use of LIMS has reduced the recycling of one product by **20%,** during start-up of the plant. Product quality can also be viewed by the internal customers through a software titled IP 21, available on SC, which collects instantaneous data from D.C.S. (Distributed Control System) in the Control Rooms.

No. of **Customer Complaints** in plant P2 has come down from 45 in March 2006, to 40 in March 2007 – a reduction of **11%.**

Before	After	% Reduction
45	40	11.0%

k) **Reduced waste generation**

100 % availability of documentation online has helped in reducing requirements of hard copies and hence the paper waste has got reduced by **25 %.** This has also resulted in **20 %**

reduction in hard disc space requirements for all the employees' computers. Presently, computer hard disc space is optimally utilized, after the implementation of SC.

Waste generation in the plant P2 has come down by **30.1 %,** in March 2007.

Before	After	% Reduction
59.8	41.8	30.1%

l) **Reduced cycle time for problem solving**

The availability of 100% of documents online has enabled an employee to locate documents (manuals, standard operating procedures, handbooks, standards) at any location in the organization in **70 %** less time, as compared to the time required before implementing the practice. This in turn has reduced the cycle time for any problem solving substantially.

m) **Reduced time required for Shutdown Planning**

SC enabled employees know previous Shutdown experiences in the plant, shutdown budgets, manpower requirements, safety performance, contractor details and their performances. This has reduced the time required for retrieving the history of previous shutdown reports by **50%** and overall cycle time in shutdown planning by **70 %.**

n) **Reduced time required for induction of new employees**

In our multi-location organization, where experienced employees are re-located at regular intervals, the organic (evolving) knowledge of experts has been made available to new employees through SC. This enabled new employee get inducted to the organizational practices within a shorter time, as compared to the time required before the new practice was in place. There has been a **66 %** reduction in time required for induction of new employees.

o) **Reduced cycle time for Start-up**

The cycle time for start-up operations for the entire location was reduced from 30 days to 12 days – **reduction by 60 %**

p) **Benefits in Process Safety Management (PSM)**

Structural Capital provides several benefits for effective implementation of Process Safety Management. Here we can see how the PSM elements are integrated SC & the organization gets the benefits.

1. Managing Process Safety Information
With uploading technology information to the portal & regularly updating the same Reliance is directly managing the Process Safety Information, which is one of the important elements in PSM.

2. Employee Participation
The portal is accessible to all employees; hence Process Safety Information is available to all employees. Indirectly Employee Participation is practiced efficiently. This is one of the very important criteria for effective management of PSM

3. Process Hazard Analysis
Process Safety Information is used to the teams for those who are conducting of Process Hazard Analysis (PHA). The database is useful for baseline PHA as well as revalidation cases.

4. Management of Technology Changes
During plant modification, alteration, addition/ deletion of equipments etc. it requires extensive study as well as assesses the risk before doing the same. Plant history & technology information is vital for those to evaluate & approve it. This information's are easily available in SC. Thus is saves the time for evaluation team.

5. Investigating & analyzing of Process Safety Incidents
Process Safety Incidents are to be analyzed, to find out the root cause of the incidents & implementation of recommendation to avoid any future recurrence of such events. For analyzing the incidents requires past data as well as basic technology information & operating practices. The SC database is the ideal platform to access the information & keeping the record.

6. Operating Procedures
Plant operating procedure is one of the vital information for those directly operating & managing the process. The field operator needs day-to-day information for running the plant safely & efficiently. These manuals are stored in SC, hence can be easily retrieved by them as & when required.

7. Mechanical Integrity & Quality Assurance
Maintaining plant assets is very much important to run the plant safely. It also reduces the

downtime & available for more production hence increasing the productivity. Original plant equipment design data, & subsequent keeping & updating records in SC helps for efficient management of MIQA program for PSM.

Conclusion:

Following objectives were achieved by implementing Structural Capital project.

aa) Applied the Structural Capital framework to develop an integrated solution which will improve the process quality & operating performance
bb) Evaluate the impact and use of technology in performance improvement.
cc) Increase in Company's repository
dd) Integration of various activities, portals, systems enabled helped management & company for over all improvement of yield, quality & training effectiveness
ee) Effective implementation of PSM elements.

: Results given under benefits section is clearly indicates the success story/achievement of the SC project.

Human Factors Programs within Contra Costa County

2008 CCPS Conference on Human Factors

Michael Dossey
Contra Costa Health Services
Hazardous Materials Programs
Martinez, California
Email: mdossey@hsd.cccounty.us

ABSTRACT

Starting in 1999, regulated industrial facilities in Contra Costa County, California were required to develop Human Factors Programs as part of the local Industrial Safety Ordinance (ISO). Human Factors Programs within the County are designed to be a thorough assessment of the interactions between equipment, work processes, procedures, practices, work environment, and management systems as they relate to health and safety. Committees were established at each facility to oversee the development, training, and execution of the site-specific Programs. A primary area of focus in the ongoing implementation of an effective Human Factors Program is periodic assessment and resolution of latent conditions in the health and safety arena. This paper describes the steps that facilities in Contra Costa County have taken to develop comprehensive site specific Human Factors Programs to continuously improve the work environment by identifying and minimizing situations that could lead to incidents or accidents.

1. BACKGROUND

Contra Costa Health Services (CCHS) is the local Certified Unified Program Agency (CUPA) for Contra Costa County in the San Francisco Bay Area. Due to a large number of significant incidents between 1989 and 1999, the County Board of Supervisors was urged by the local community to develop additional oversight requirements for local facilities that handle extremely hazardous materials. This resulted in the development of the Industrial Safety Ordinance (ISO) in 1998, and supplemental revisions in 2000 and 2006. The ISO requires the regulated facilities within its jurisdiction to develop and implement programs to assess the following:

- Human Factors *(topic of this paper)*
- Inherently Safer Systems
- Root Cause Analysis / Incident Investigation

- Safety Culture Assessment
- Security Vulnerability Assessment
- Safety Plan Submittal

CCHS developed a Safety Program Guidance Document to assist local industry in complying with the ISO requirements. This guidance document includes criteria to assist in the development of the Human Factors Program and is further described within the following sections of this paper. This document is also available online at the Contra Costa Health Services website: http://www.cchealth.org/groups/hazmat/industrial_safety_ordinance_guidance.php

2. WRITTEN HUMAN FACTORS PROGRAM

The Human Factors Program is developed to evaluate and resolve active failures and latent conditions within four primary subject areas, or interfaces between subject areas:

- Individuals (e.g., motivation, physical being, emotional states)
- Task being conducted, including the procedures for the activity or task (e.g., routine, non-routine, written, practice, formal, informal)
- Physical environment (e.g., equipment, work surrounding) or workplace
- Management or organization (e.g., communication, reward and discipline system)

To accomplish the above, the written Human Factors Program should describe the how, when, who, what, and where personnel will assess their workplace for hidden problems (i.e., latent conditions) and other studies to ensure changes in the facility's organizational structure do not diminish capabilities in operations, maintenance, safety, and emergency response. The following safety program elements must be enhanced/developed to incorporate Human Factors Assessments:

- Evaluation and Minimization of Latent Conditions
- Process Hazard Analysis
- Incident Investigations
- Training
- Operating Procedures
- Maintenance Safe Work Practices and Maintenance Procedures
- Management of Organizational Change

The written Human Factors Program must include employee participation in its development and ongoing implementation. To achieve this, employees were sent to train-the-trainer classes to learn concepts on human factors, human systems, human error, latent conditions, active failures, etc. These concepts were then customized and incorporated into the facility's safety programs. A number of different training sessions were then developed for different audiences at the facility: basic awareness training, overall training and specialized training. Refresher training is performed at least every three years for basic awareness and overall training to cover the major points of the facility's Human Factors Program. Usually specialized training is just in time training performed for individuals preparing to evaluate latent conditions.

3. DEFINITIONS

This section defines select terms used within Contra Costa County.

3.1 Human Factors

"A discipline concerned with designing machines, operations, and work environments so that they match human capabilities, limitations, and needs."[1] Human Factors can be further referred to as "environmental, organizational, and job factors, and human and individual characteristics which influence behavior at work in a way which can affect health and safety."[2]

3.2 Human Error

Human Error is defined as "departure from acceptable or desirable practice on the part of an individual that can result in unacceptable or undesirable results".[3]

3.3 Human Systems

To effectively evaluate Human Factors, you need to look into the Human Systems that make up the workplace, which are defined as: "the systems, such as written and unwritten policies, procedures, and practices, in effect to minimize the existence / persistence of latent conditions at the stationary source. It also includes the broad area of safety culture of a stationary source to the extent that it influences the actions of individuals or groups of individuals." (Section 450-8.014(s) of Contra Costa County Ordinance Code)

3.4 Latent Conditions

Latent Conditions are situations that exist in all systems that for the most part lie unrecognized (e.g., a missing step in an operating procedure, unlabeled valves). In many situations, human errors can occur when a mistake (i.e., active failure) is made when combined with an unrecognized latent condition to result in an incident. Latent Conditions can arise due to decisions made throughout the organization (e.g., marketing personnel, designers, managers) and outside of the organization (e.g., regulating agencies). The same Latent Condition may contribute to a number of different accidents.

---

1. Lorenzo, Don K., *A Manager's Guide to Reducing Human Errors*, Washington DC, Chemical Manufactures Association, 1990.
2. *Reducing Error and Influencing Behavior*, UK, Health and Safety Executive, HSG48, 1999.
3. Bea, Robert, Holdsworth, Rodger, and Smith, Charles, *Human and Organization Factors in the Safety of Offshore Platforms*, a paper presented at the International Workshop on Human Factors in Offshore Operations, 1996.

3.5 Active Failures

Active failures are mistakes in actions taken or not taken by people at the human–system interface such as operators and maintenance personnel. Active failures can be further described by the types of error mechanisms: slips and lapses; mistakes; and violations. Slips involve the correct intent to do something but failure occurs in implementation (e.g., operator intended to isolate pump #1, but isolated pump #2 instead). Lapses involve error in memory recall, such as, skipping a step or task that was supposed to be performed, or losing one's place in an assignment. Mistakes involve when the action taken was intended but the intention was wrong, which could be caused by inappropriately applying rules, procedures, or reasoning. Violations are errors when an intended action is made that deliberately ignores known operations rules, restrictions, or procedures (excluding sabotage).

4. LATENT CONDITIONS CHECKLISTS

One of the main components of the Human Factors Program in Contra Costa County is the evaluation and minimization of Latent Conditions. Eliminating Latent Conditions will make incidents and accidents less likely to occur or at least reduce the severity if a mistake is still made. To assist with this evaluation, CCHS included a Latent Conditions Checklist (LCC) within its Safety Program Guidance Document. The LCC contains approximately 250 questions that cover the following broad topics:

- Individual
  - Experience / Knowledge
  - Stress / Fatigue / Substance Abuse
  - Shiftwork
- Activity / Task
  - Procedures
  - Practices
  - Conflicts Between Practice and Procedure
- Physical Environment / Workplace
  - Process Design and Labeling
  - Control Room / Panel Design
  - Hardware
  - Safeguards
  - Work Environment
- Organization / Management
  - Communications
  - Training
  - Staffing / Overtime
  - Worker Selection
  - Climate / Culture
  - Management System

4.1 Example LCC Questions

Table 1 below provides example LCC questions excerpted from Contra Costa County's LCC.

Table 1 – Example Latent Conditions Checklist Questions

Checklist Questions	Y/N/NA	Justification / Examples
1. INDIVIDUAL:		
Experience/Knowledge:		
1.1 Do employees remain in each unit for a sufficient amount of time to develop the experience and knowledge base necessary to safely operate the unit and respond to emergencies?		
1.2 Is there experience available for each of the different Safety Programs, including Human Factors?		
1.8 Are there jobs that are beyond employees' physical limits or safe physical limits (e.g., carrying equipment, that requires both hands, up stairs that are poorly lighted at night? [vii]		
Shiftwork:		
1.12 Is the length of a shift during startup and turnaround appropriate given the degree of alertness required and potential for operator fatigue? [i]		
1.13 Are shift turnover periods sufficient to adequately communicate plant operating conditions from off-shift to on-shift personnel? [i]		
1.16 Are job turnover communications between shifts adequate?		
2. ACTIVITY/TASK:		
Procedures:		
2.2 Does the facility ensure that only current, approved versions of procedures are available?		
2.11 Is the procedure difficult to use?		
2.17 Does a different person subsequently make an independent check that mandatory procedures have been carried out? [iii]		
2.21 Do Cautions, Warning, and Notes stand out from procedure steps? [iv]		
2.35 Is equipment and instrumentation clearly labeled and are the equipment and instrument tag numbers used in the procedures? [v]		
3. PHYSICAL ENVIRONMENT/WORKPLACE		
Work Environment:		
3.64 Is the lighting adequate in the unit [*consider local instrument panels, battery or plot limit valve manifold locations, equipment and valves requiring operation during emergency conditions, etc.*]? [i]		
3.65 Is the emergency lighting (light fixtures on the emergency power circuit) adequate in the unit? [i]		
3.69 Are the control building air conditioning and pressurization adequate to prevent intrusion of toxics, flammables, or corrosive		

Checklist Questions	Y/N/NA	Justification / Examples
contaminants (if applicable)? [i]		
3.71 Can the HVAC system in the control room be isolated quickly from outside air? [v]		
4. ORGANIZATION/MANAGEMENT		
Staffing/Overtime:		
4.27 Are staff levels (both size and experience) sufficient to handle routine and non-routine duties that can be reasonably expected to occur during a shift? [i]		
4.29 Is there backup assistance when an operator emergency responder must respond to an emergency? [ii]		
4.32 Are restrictions applied to employee overtime (e.g., employees are not allowed to work consecutive 12-hr. shifts)?		
Climate/Culture:		
4.44 Do workers feel that unsafe operations or maintenance activities can be shutdown without fear of retaliation? [vi]		
4.47 Do the rewards and penalties for safety compare to those for production performance (i.e., equal rewards for safety performance and production). For example, do employees receive a company hat for safety rewards and substantial monetary rewards for production? [i]		
4.50 Does senior management promote the understanding of the different Safety Program elements, including Human Factors?		
4.51 Are managers held accountable for their health and safety record? [i]		

[i] EQE PHA Checklist: Human Factors, developed for and used by Chevron and Tosco (1996)
[ii] MRC Human Factors Review Checklist (1994)
[iii] Primatech Human Factors Checklist (1994)
[iv] Guide to Reducing Human Error In Process Operation, Short Version (1985)
[v] CCPS, Guidelines for Writing Effective Operating and Maintenance Procedures (1996)
[vi] U.S. Department of Energy workbook for Conducting Accident Investigations, Revision 1, November 21, 1999
[vii] RRS Engineering Human Factors Checklist (1999)

4.2 Guidance on Using a Latent Conditions Checklist

Depending on the facility and the individual(s), applying a checklist properly can be challenging. To make sure that the LCCs are properly completed and any issues identified are properly addressed, formal programs and procedures must be developed to provide oversight and direction. Having key individuals to champion their facility's Human Factors Programs is invaluable. The following are examples of some of the guidelines facilities should adhere to in Contra Costa County when applying their LCCs:

- Individual LCC questions are designed to prompt further discussion of an issue since each question may only be an indicator

- If an issue is found, solutions or remedies are required to address the big picture to resolve a program deficiency (if there is one) instead of just fixing the symptoms or the problem at hand
- Individuals asking the LCC questions need to be trained not only in the concepts of human factors but also to understand that the intent of the question is not to identify errors but rather to identify and rectify situations that could cause an error to be made
- Personnel applying the checklist must be adequately trained to: understand the specific reason for each question, understand the relative importance of different questions, and understand the degree to which items fail to meet the criteria
- A formal "feedback" loop must be developed to ensure that those who developed the recommendation agree that the resolution was appropriate
- A formal tracking mechanism must be developed to ensure that recommendations are resolved in a timely fashion

5. PERFORMING HUMAN FACTORS ASSESSMENTS

5.1 Latent Conditions Checklist Assessments

The regulated facilities in Contra Costa County use LCCs to evaluate the required four primary subject areas, as identified in Section 2, for latent conditions: Individuals, Tasks, Workplace, and Organization. Most facilities formed committees to evaluate the relevance of each of the County's LCC questions and developed customized checklists to be used to evaluate individual Human Factors Program areas within the facilities. In doing so, individual checklists were developed for use during: the development and review of operating procedures, PHA studies, and incident investigations. Many of the same LCC questions appear on multiple LCCs due to the thought-provoking nature of the questions and to make sure that concepts are captured in all areas of a facility's work process. In some cases, facilities found a number of LCC questions (e.g., staffing levels) difficult to address during the review of an individual operating procedure or unit Process Hazard Analysis (PHA), so these selected questions were placed into a Facility-Wide or Management System checklist to be addressed globally. Most facilities added questions, modified or expanded the checklists to account for specific areas they wanted to analyze or those areas that may not have been fully captured by the County's standard LCC questions. Some of the questions were reworded to better reflect the organizational structure and culture of the facility and to make the LCC questions more user friendly to the individual sites. Many questions are worded such that a "Yes" answer does not always imply a favorable condition to minimize the tendency of quickly going through the questions. Very few questions from the County LCC were dropped, but if they were, written rationale is maintained to justify why the question is not applicable.

When using the individual LCCs, facilities use assessment teams that include an operator, a maintenance employee, an employee representative, engineering and/or safety staff and management. This team concept is used to capitalize on the knowledge and experience of a cross-functional group instead of relying on only those individuals directly responsible or involved with the topic. The team concept also allows all individuals involved to get a sense of contribution or ownership. Assessments performed are documented along with items found that need improvement. In general, the team performing the assessment is allowed to decide whether an issue is acceptable or the condition(s) needs to be addressed. If an issue needs to be addressed, facilities are required to track the issue to resolution. Facilities are required to have a process for management to review and sign-off on the completed assessments to verify they were performed adequately and action items have been assigned.

Facilities perform their assessments in a variety of ways. Some use teams to visit one operating area at a time and assess the procedures, work environment, control stations, field equipment, panels, practices, etc. Some facilities performed site-wide studies to review all areas at the site for certain topics like operating procedures, field lighting, equipment labeling, staffing for operations and emergency response. Some facilities use small groups (e.g., 1-3 people) and some use larger teams (e.g., up to 8) to conduct these assessments.

Latent Conditions identified within PHA sessions are allowed to follow the normal PHA recommendation development process for identifying likelihood, severity, and ranking. For the regulated facilities, the ISO requires that PHA recommendations must be resolved within one year unless a turnaround is required. Resolution of PHA recommendations is subject to evaluation of Inherently Safer Systems as well.

The frequency of performing Latent Conditions assessments varies depending on where they are being assessed. Latent Conditions must be evaluated with every PHA and PHA revalidation, every new operating/emergency procedure, and every major incident investigation that meets certain criteria. Facility-Wide or Management System LCCs need to be repeated at least every five years.

5.2 Management of Organizational Change Assessments

To adequately assess this primary subject area identified in Section 2 of this paper, Management or Organization, a process needs to be developed to make sure the proper organizational structure is maintained when confronted with changes that inevitably occur. The County's ISO requires that Management of Organizational Change (MOOC) reviews be conducted prior to making changes in staffing in operations, maintenance, safety, or emergency response. The MOOC requirements apply to reductions in the number of positions or number of personnel; substantive increase in the duties; or changes in responsibilities. The intent of this requirement is to make sure that the duties and responsibilities for the positions to be changed are effectively captured and re-assigned to others without overburdening the positions that remain. MOOC requirements apply to changes in permanent staffing levels/reorganization in these areas, and apply to contractors in permanent positions in operations and maintenance as well. Staffing changes that last longer than 90 days are considered permanent. Prior to conducting the management of organizational change, the facility must have current and accurate job function descriptions for each position under consideration. To perform the MOOC, a few facilities have customized a Latent Conditions Checklist to assist with this evaluation. Many facilities have used a process similar to that referenced in the County Guidance Document developed by the Chemical Manufacturing Association (now the American Chemistry Council).

Facilities are required to evaluate organization or staffing changes with a team, or Change Team. This team develops the technical basis for the change, as well as the purpose, scope, and schedule of the change, and determines the potentially affected positions. The Change Team assesses the impact of the organizational change on responsibilities and safety and health (e.g., demands on remaining operations staff, emergency response capabilities) before the change proceeds. Potential impacts to the safety and health of others, or increased duties/responsibilities are identified, and suggested action items to mitigate the impacts are developed when deemed necessary. One of the key factors in this program is to require that employees affected by a change are informed of, and trained in, the change prior to the change occurring.

To date, facilities have used dedicated Change Teams to assess proposed organizational changes that meet the ISO requirements. These Change Teams typically are comprised of 2-3 persons; one individual responsible for the facility's MOOC program, and an operator/crafts person or union representative. This core team jointly performs all aspects of the MOOC review, including affected employee interviews and developing recommendations.

Most facilities when confronted with a proposed change start by defining the technical basis for the change (e.g., purpose, scope, and schedule). The next step is typically performing a preliminary screening assessment to determine the positions/responsibilities most likely impacted by the change. Facilities need to define the existing situation (e.g., program requirements, personnel duties/tasks/responsibilities) before they can assess the impact on safety and health. CCHS notes that very few facilities have thoroughly developed job descriptions for every position, and hence, many rely on this phase of the MOOC assessment to develop detailed job descriptions for the impacted positions. Once the existing situation is known and the reason for the change defined, safety and health impacts of the proposed change are then evaluated.

Occasionally, the Change Team identifies additional positions that may be impacted by the change as they go more in depth to assess the impacts. When this occurs, the Team must re-define the existing situation to make sure that they fully understand the responsibilities of the positions to be impacted or changed before they can proceed with assessing impacts to safety and health. One of the main components of assessing safety and health impacts is to identify action items or requirements that must be met to mitigate negative consequences. There have been instances where unique responsibilities or duties of the position were not fully recognized until the MOOC assessment, and some instances where the Change Team recommended against proceeding with the proposed change.

6. ASSESSMENT RESULTS

6.1 LCC Results

After beginning the Latent Conditions assessments, most facilities realized that their operating procedures needed to be improved to address human factor concerns. As a result, most of the ISO regulated facilities within the County decided to overhaul all of their operating and emergency procedures. In most cases, a phased-in approach was used to overhaul the operating procedures within the different process units over a five-year period. Procedure templates were developed that eliminated many existing Latent Conditions (e.g., compound and unclear instructions, responsibilities identified, inconsistencies between process units, lack of procedures). Experienced operators were involved with the drafting of new and review of existing procedures.

Most facilities realized that developing "program fixes" to correct identified Latent Conditions was problematic. For example, just because a facility completed upgrading field lighting site-wide as a result of a site-wide lighting study does not mean that everybody is satisfied. An operator or maintenance person may still identify specific lighting issues, especially if they were involved with a Latent Condition assessment and their particular lighting concern was not addressed to their satisfaction. Other areas found to need improvement are consistently communicating corrections back to those who identified the latent condition, ensuring that management properly reviews and signs off that latent condition checklists were properly applied, continue to strive for program fixes, and having a represented employee participate in completing the checklist or in the evaluation of the latent conditions.

Additional positive actions taken include:

- Facilities perform procedural PHA on critical procedures (e.g., startup, emergency shutdown) to verify the steps were accurately and clearly presented, in the proper order, and contain all the necessary information
- Manual valves were relocated, chain pulls were added, or they were converted to motor operated valves to improve access and ease of use
- New procedures were written, redundant or overlapping procedures eliminated, and existing procedures were made clearer
- Consistent procedure format and content between process units assist operators to acclimate quicker if they relocate to another unit
- Equipment identifiers, such as tags/labels, were mounted on equipment. Equipment labels matched the terms used within the operating procedures
- Numerous lighting studies were performed that resulted in additional lighting installations
- Select field instrumentation readouts (e.g., pressure or temperature gauges) were routed to control room board operator panels
- Walkways were enhanced (e.g., non-slip surfaces, improved handrails/steps)
- Platforms were built to eliminate the use of step stools or excessive reaching
- Alarm management studies were performed and protocols were implemented
- Represented employees obtained feedback from their suggestions, and actually saw improvement being made in the units for situations they

identified. This typically results in additional employee participation as well as improved moral

- Incident investigation teams continued to drill down into the issues and did not stop until management system failures were identified and corrected. Simply identifying the need for better communication or retraining was generally no longer good enough
- Human systems were being considered as causal factors (e.g., direct cause, contributing causes, root causes) within incident investigations
- Reporting and investigation of near misses improved
- All facility personnel, from field operator to senior management received training on the concepts and benefits of maintaining an effective Human Factors program

6.2 Management of Organizational Change Results

All ISO regulated facilities have developed programs to manage organizational changes. A few facilities have embraced the MOOC concepts and perform reviews when not specifically required (e.g., direct replacement of Plant Manager, elimination of positions not subject to the ISO). Documentation of the MOOC assessments has been inconsistent. Some documentation was found to be missing during CCHS audits even though organizational changes were made. One facility decided to completely bypass the MOOC assessment requirement, so they could try to keep impending sweeping organizational changes a secret until the day of the change. On the other side of the spectrum, another facility performs a preliminary assessment for all organizational changes to determine if qualifying changes will be made. For this facility, if an MOOC is not required, a Negative Declaration form is filled out and the basis for no MOOC being required is documented. One of the most troubling concerns we have heard is from the represented workforce that indicates management continues to make reductions in personnel even when the MOOC assessment does not support the change. CCHS has not found sufficient supporting information to document these claims, but it continues to be a topic of keen interest. CCHS has also observed proposed organizational changes rejected, possibly based in part on the results of the MOOC assessment (e.g., the level of effort needed to implement the MOOC recommendations outweigh the benefits of the organizational change).

CCHS has noted a number of positive outcomes from MOOC evaluations at the ISO regulated facilities:

- Organizational changes are performed after thorough consideration of how operations, maintenance, safety, and emergency response activities were to be maintained or improved
- Job descriptions are written to clarify the assigned duties of personnel
- Important tasks are maintained to ensure that there was no decrease in coverage
- Positions/tasks affected by a proposed change are ranked depending on the extent of the health and safety impacts expected. Those positions/tasks that ranked highest receive the most impact reduction effort

- Many motor operated valves were installed to reduce demand on field operators during emergencies
- New procedures were written, or existing procedures revised, to identify individual responsibilities
- Additional information was routed to control board operator panels (e.g., redundant temperature or pressure readings) instead of requiring field operator visuals
- Additional personnel were trained to operate process units and some trained to write effective operating procedures
- Select emergency procedures received in-depth timeline sequencing analysis to map all required tasks to eliminate conflicts where simultaneous tasks were required

7. NEXT STEPS

Similar to maintaining Process Safety Management programs, regulated facilities need to consistently evaluate their Human Factors program to ensure it remains fresh and viable and does not become stagnant. Regulated facilities are currently being asked to evaluate their Maintenance Procedures and Maintenance Safe Work Practices for Latent Conditions by June 30, 2011. CCHS anticipates that facilities will overhaul these procedures and practices with similar thoroughness as their operating and emergency procedures.

The Principle and Practice of the HSE Management in SINOPEC

Mu Shanjun
Deputy Director, SINOPEC Safety Engineering Institute;
Qingdao, China,
Email: mushanjun@vip.sina.com
Xu Gang
Director, SINOPEC HSE Department,
Beijing, China
Email: xug@SINOPEC.com.cn
Zhang Mingxia
Engineer, SINOPEC Safety Engineering Institute
Qingdao, China,
Email: zmxhse@163.com

ABSTRACT

China Petroleum & Chemical Corporation (hereafter referred to as "SINOPEC") is one of the largest integrated energy and chemical companies in China which rank 17 on the World Top 500 by Fortune in 2007. SINOPEC always insist on the principle of "Human Oriented" and "Focus on Behavior of Employees" since developed and implemented HSE management system in 2001. "Preventing accident and avoiding injury" are the common objective and commitment of all personnel in SINOPEC from chairman to operator. SINOPEC supported by Qingdao safety engineering institute which is the SINOPEC HSE research and development base has established several HSE regulations and procedures and used a series of methods and technique to monitor the behavior of the employees and prevent the accident. Taking these approaches, SINOPEC has achieved obvious performance improvement that the number of the major accidents decreased from 69 in 2001 to 30 in 2006.

1. General Introduction of HSE management system frame in SINOPEC

1.1 Introduction of SINOPEC

China Petroleum & Chemical Corporation (hereafter referred to as "SINOPEC") is an integrated energy and chemical company. The scope of its business mainly covers oil and gas exploration, development, production and marketing; oil refining; production and sales of petrochemicals, chemical fibers, chemical fertilizers and other chemical products; storage and pipeline transportation of crude oil and natural gas; import, export and import/export agency business of crude oil, natural gas, refined oil products, petrochemicals, chemicals, and other commodities and technologies; research, development and application of technology and information. It is China's largest producer and supplier of oil products and major petrochemical products. It is rank 17 on the World Top 500 by Fortune in 2007.

1.2 Introduction of HSE management system in SINOPEC

SINOPEC has formally declared to implement the HSE management system on April 4, 2001, which includes one system, four criteria and five manuals. The system is HSE management system which is concerned with the following ten constituents:

（1） Leadership Commitment, Policy, Objectives and Responsibilities

The strong leadership and linear commitment provided by the senior management of SINOPEC are the basis of successful implementation for HSE management system. SINOPEC has the specific commitment and the documented policy and objectives which are guiding ideology and principle to ensure achieve excellent HSE performance.

HSE policy of SINOPEC is "safety first, priority on prevention; comprehensive administration; improving environment, protecting health; scientific management and sustainable development". HSE objectives of SINOPEC are "to pursue no accidents occurred, not to damage personal heath, not to contaminate the environment and to create the excellent HSE performance".

（2） Organization, Responsibilities, Resources and Document Control

In order to ensure the effective implementation of the HSE management system, SINOPEC and all levels directly under enterprises of SINOPEC assign necessary human resources, material resources and financial resources, and define the management responsibilities for each department or employee.

（3） Risk Assessment and Hidden Hazards Control

Risk assessment which is the basis of all HSE elements is the core in development and implementation of HSE management system of SINOPEC. All levels directly under SINOPEC identify hazards and assess risks regularly to decide the degree and scope of influence of the major hazards, so that the corporation can adopt the efficient prevention measures for reducing risk to a level as low as reasonably practicable.

（4） Management of Contractors and Suppliers

SINOPEC manage the contractors and suppliers in several aspects, include selection of contractor, audit of contractor's qualification, preparation before construction, operation process control and performance evaluation.

（5） Design and Construction of Installations

Facilities newly-building, rebuilding and expanding comply with the "Three Simultaneities" principles strictly and with the relative regulations and standards to design, purchase, install and test.

（6） Operation and Maintenance

The subsidiaries in SINOPEC should control the production installations, equipments, hazardous materials, special technical process and hazardous operation environment efficiently to improve the safety and reliability of operation by taken effective management system and methods.

(7) Management of Change and Emergency Response

All subsidiaries in SINOPEC shall control the permanent and temporary changes of personnel, operation process, operation procedure, technique to avoid the hazards and effect on environment and health. For emergency response management, they analyze and research the production system totally, systematically and carefully, then establish the prevention measures and emergency response planning and exercise to ensure the accidents could be control and reduce losses.

(8) Inspection, Review and Supervision

The subsidiaries should establish the regular inspect and supervise system to achieve the HSE management policy.

(9) Accident Management and Prevention

Accident reporting, investigating and treating procedures are established to investigate and decide the reasons of accidents or near-miss that happened, and establish the relevant corrective and preventing measures to avoid the accidents happen again.

(10) Auditing, Reviewing and Continuous Improvement

All subsidiaries audit and review the HSE management system regularly to ensure the management system is effective.

The four criteria are HSE management criterions for oilfields, refining and petrochemicals, sales and constructing enterprises. The five manuals are the HSE implementation programming manuals for oilfields, refining and petrochemicals, sales and constructing enterprises as well as functional departments.

1.3 HSE Performance of SINOPEC

SINOPEC has achieved obvious performance improvement by implementing HSE management system since 2001.In contrast to the sales increased year after year, the number of the major accidents decreased from 69 in 2001 to 30 in 2006 and the exhaust and COD emission decreased every year, which is shown in figure 1 and figure 2.

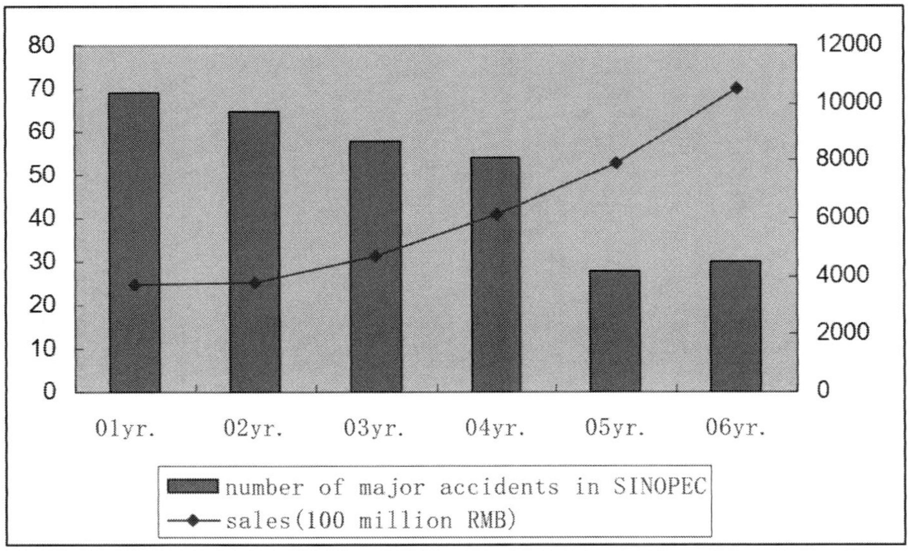

Figure 1 safety performance improvement

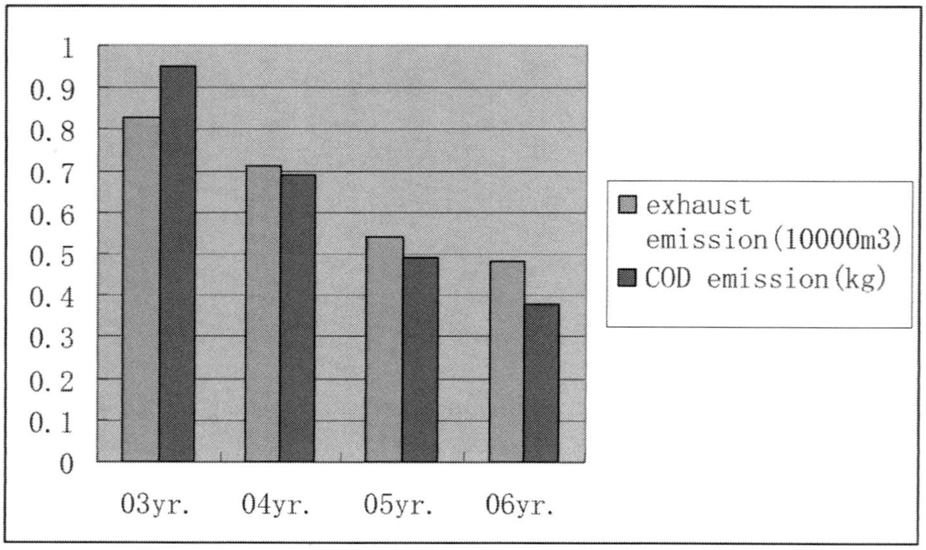

Figure 2 environment performance improvements

2 The Principle and Practice of the HSE Management in SINOPEC

SINOPEC always insist on the principle of "Human Oriented" and "Focus on Behavior of Employees" since developed and implemented HSE management system. "Preventing accident and avoiding injury" are the common objective and commitment of all personnel in SINOPEC from chairman to operator, who all believes operation safety is the basis and premise for company to gain the economic benefit. SINOPEC supported by Qingdao safety engineering institute which is the SINOPEC HSE research and development base has established several HSE regulations and procedures and used a series of methods and technique to monitor the behavior of the employees and prevent the accident to achieve the prominent performance, shown in figure 3.

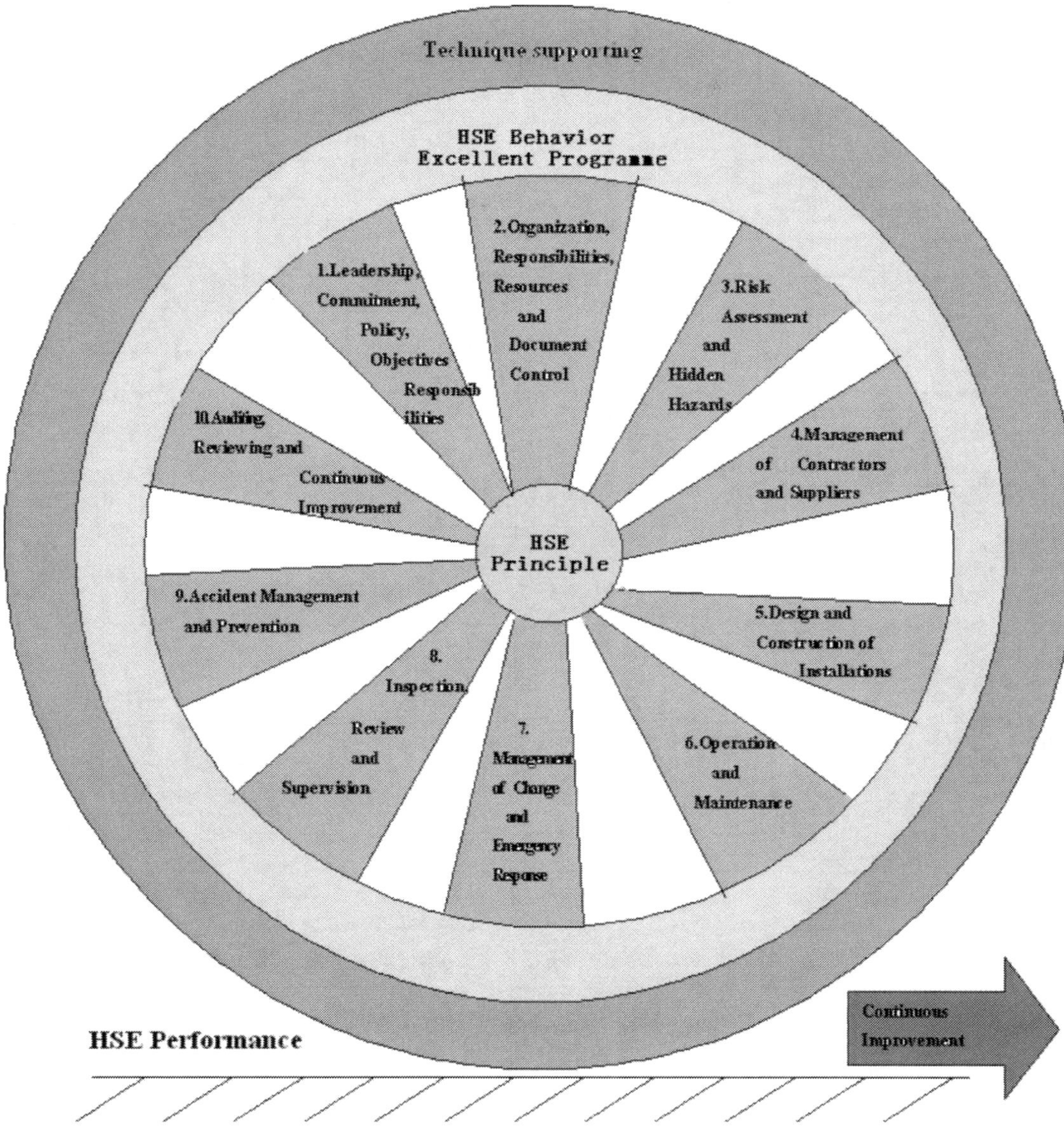

Figure 3 wheel shown SINOPEC HSE management principle

SINOPEC stick to the principle of "Human Oriented" in the following aspects:

1) Create the environment of Zero Accident—— "All accidents can be avoiding"

SINOPEC adopt the methods that integrate the active management with passive management, which is the risk management that integrates prevention before accident with accident management, shown in figure 4.

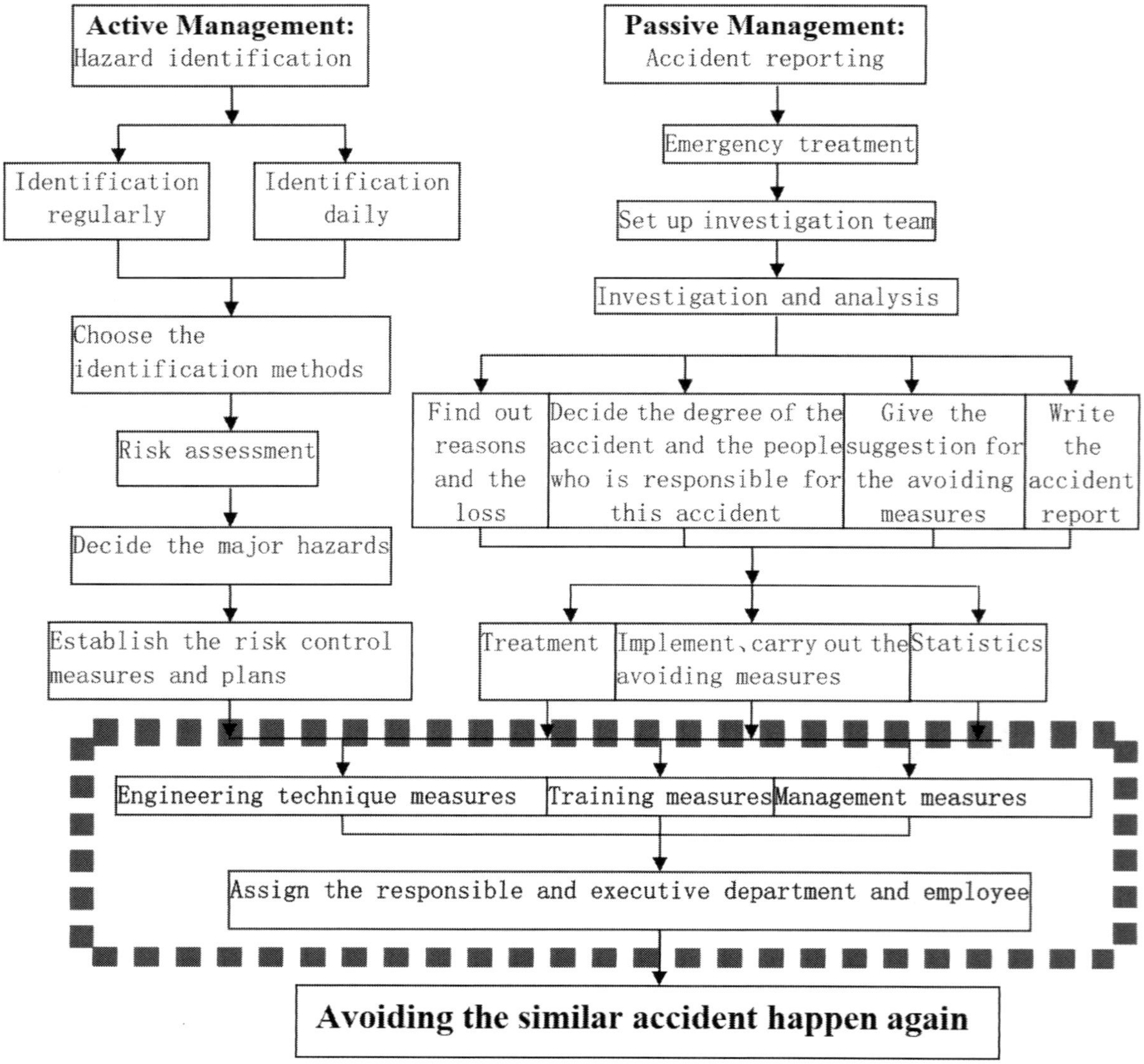

1 Figure 4 the active management and passive management

SINOPEC has established a strict management procedure from accident classification, report, investigation, treatment to statistics, and implemented the "Four No-Passing Elements" which is accident management principle to prevent the similar accidents happen again.

(1) It is no-passing that the reasons of accidents are not found out: SINOPEC requires that all levels units should analyze the reasons of accidents first when the accidents were investigated, find the truth of accidents, avoid rushing to conclusions before the main reasons discovered, and also could not take the minor reasons for the main reasons, could not give up until find the truth of accidents happened, and make sure of the relationships between all kinds of elements.

(2) It is no-passing that the people responsible for the accident are not punished: This is the corresponding embodiment for responsibility system; SINOPEC requires that the accident principal should be punished seriously according to the relevant laws and rules.

(3) It is no-passing that most of employees are not learned from the accident: SINOPEC requires that all levels units should make most of staff and accident principal learned from the accident and understood the reasons and hazards caused the accident when the accident was investigated.

(4) It is no-passing that the prevention measures are not followed: SINOPEC requires that all levels units should propose measures to prevent the similar accidents happened again while investigate the accident and implement the corrective measures.

2） "Line Responsibility"

According to the principle "Who is responsible for production, then for HSE" and "Line Responsibility", the leading member of SINOPEC is the first principal of HSE protection in group, and assumes overall responsibilities. HSE department is responsible for managing and supervising HSE affairs of entire group and the management department for oilfields, refining and petrochemicals, sales and constructing are respectively responsible for the HSE affairs of subsidiary companies in SINOPEC.

All levels leading member of enterprises directly under the SINOPEC is the first principal of HSE protection in enterprise. The other leaders, functional departments and staff are responsible for safety production in their work. All levels enterprise sign the "HSE Responsibility Paper" with the headquarters of SINOPEC every year； all directly under enterprises sign the "Objective Responsibility Paper for Safety Production" level upon level; all workgroup sign the "Responsibility Paper of Safety Production" to follow the responsible mechanism level upon level.

3） Focus on Behavior of Employees

According to the developmental tendency of enterprise safety management, the technical improvement and the equipment update shall provide a balance between accidents and trend on injuries finally as the time passed by. After this, the function of technical improvement and the equipment update is not evident to reducing accidents and injuries, so the enterprise safety management will establish on the basis of management of safety behavior of people. According to statistics, the proportion of accidents caused by unsafe behavior factors is 65% inside the petrochemical industries of china, which is far exceed to the international standard 15.6%, so that the management of safety behavior of employees is still the key project concerned by SINOPEC.

SINOPEC has established many management regulations, and adopted the effective methods to prevent the unsafe behavior. SINOPEC has issued "The Ban of SINOPEC for Safety Operation" to control the HSE behavior of local units and employees. The ban includes ten items, which is the high pressure line for safety production of SINOPEC, the guide line for staff behavior, the ideologically alertness line for staff and the guarantee for lives. SINOPEC implement the operation safety observation to put the ban into effect. By operation safety inspecting, employee finds and reports the unsafe behavior, unsafe conditions and near-misses in workplace in time, and records the HSE observation information. Then analyses the record information regularly and concludes the satisfied operation measures or behavior specifications to provide the support for modifying operation regulations and management plans, etc. to control and avoid the potential hazards.

4) The scientific and effective method is the key to avoid the accident

Setting up the HSE culture, SINOPEC go through the four stages, in which there is different method used to promote the HSE quality of the employee and prevent the accident:

(1) The first stage: To improve the HSE quality of employees

The HSE quality is achieved by people from various measures and channels in the long term study. It is the internal psychological character with spirit of humanity and science from comprehensive understanding of HSE knowledge. To cultivate and improve HSE quality for employees is also a gradual learning process. In this stage, the enterprise HSE culture is relatively underdeveloped. There are few employees obtain the safety knowledge from outside, and most of them safeguard personal safety from their instinct. Therefore, in this stage, SINOPEC adopt the effective measures to strengthen the HSE knowledge training for employees, especially the safety operation regulations training for operators, so as to reducing accidents.

(2) The second stage: Putting HSE responsibility into effect

In this stage, the effective measures to avoid accidents is establishing various regulations, including rewards and punishment regulations, and require the all levels employees make commitments safety responsibilities. SINOPEC has established a full set of regulations for HSE management. The "work permit system" is implemented for jobs of high risks. Serious management procedures are compiled for work application, review and approval, and implementation, includes "Safety Management Regulations for Hot work", "Safety Management Regulations for Work in confined space", "Radiography work Management Regulations", "Safety Management Regulations for Construction", "Safety Management Regulations for Work at height", "Safety Management Regulations for Lifting Operation", "Safety Management Regulations for hot work", and so on. The safety consciousness of employees has an obvious improvement than the last stage when the regulations and systems have been established. But the safety consciousness in this level is passive and controlled because employees lack subjective initiative to engage in HSE management, and the safety consciousness of employees also has being the passive state as "made me safety".

(3)The third stage: Make the safety behaviour a habit

In this stage, the safety consciousness of employees has the active impetus, from "made me safety" of the last stage to "I need safety". The behavior of employees changed greatly. There is a well communication atmosphere among employees, between employees and their supervisors. The HSE culture can be breathe into employees and employees could communicate experience in safety operation with each other. On the other hand, SINOPEC not only has better regulations and systems, but also transfer the modern safety philosophy to employees by effective and various approaches. The safety consciousness of employees is improving and their requirements about the safety working situation are also steadily improving in this time. They will report actively the hazards in time, in order to solve them quickly. So the safety consciousness of employees has the active exhibition in this time.

(4)The fourth stage: HSE Culture Construction of Enterprises

HSE culture is the values, activity habit and activity guide line which formed from safety production practice by employees. The target of HSE culture established by SINOPEC is to promote the HSE management to the rank of culture management and to establish the long-term mechanism for providing the forceful spiritual dynamics for safety production. In this stage, the employees of SINOPEC take full advantage of emotion and attention on the existing HSE knowledge to find the potential hazards accurately and take effective measures to protect themselves and their workmates. The safety consciousness of employees has new changes from "I need safety" of last stage to "I can safety". The HSE qualities of employees have been combined with the HSE culture of enterprises.

5) "There is only starting point, no terminal point in safety operation"

SINOPEC have consistently followed the principles as "the production must be kept safe, the safety is the guarantee for production". Although the production often changed according to the market and has the difference the right order of priority, but the safety operation should be implemented all the time. There is only starting point, no terminal point in safety operation. The safety operation should be maintained permanently to ensure the safety production. Therefore, SINOPEC has established perfect HSE inspection and supervision mechanism formed by the review of headquarters, directly under enterprise, plant, workgroup, safety administrator, etc., to ensure the continual HSE management improvement of SINOPEC. The review of group headquarters put emphasis on examination for the ideology of leadership, management and responsibility to keep the safety management operation systematization, standardization and regularity. The self-review of local enterprises is the basic in HSE management, and is the most practical and effective measures. The emphasis of self-review for local units is to find the potential hazards in all workshop, organize and implement the daily examination of workgroup, weekly examination of plant, quarterly examination of the second class unit and the examination for every six months of enterprise, establish the closed-loop management system of HSE review.

3 The role of the SINOPEC Qingdao Safety Engineering Institute for SINOPEC HSE management

While SINOPEC set up the HSE culture, it needs the professional institute given technique support. SINOPEC Qingdao Safety Engineering Institute (SSEI) was founded in 1979, as the comprehensive safety research institute directly under SINOPEC. Through the developing of more than 20 years, SSEI becomes an important researching base in the technological innovation, technical progress, application development of Health, Safety & Environment (HSE) in petroleum and petrochemical industries in China. It is also the leading technical support organization for the safe management, hazards prevention and control, emergency response of hazardous chemicals.

SSEI is the school for HSE training, center for HSE technique service and base for HSE research and development in SINOPEC. The general goal of SSEI is based on the key safety work within SINOPEC, following the principles of close to enterprise, close to production, close to market, emphasizing on application development of HSE technology, providing technical guidance for the decision-making of work safety in SINOPEC, providing technical support for the safe production, environment protection and occupational health within SINOPEC enterprises. Based on the work safety situation in petroleum industries and practical needs of enterprises, SSEI develops main research in the following fields:

1) Technical Research on Long-term Safe Operation of Process plants.

According to SINOPEC development strategies, centered on the plant safe operation to improve SINOPEC economic profit, emphases are put on the processing the high sulfur and acid crude oil equipment corrosion and mechanical integrity researches for the oil deterioration, progressing the anti-corrosion technology, developing the advanced reliability technique results, to improve the inherent safety.

2) Technical Research on Chemical Emergency Response

Mainly develops research in prevention, monitoring and control technology in chemical emergency response, which plays the leading role domestically. The national emergency response hotline was established on January 1st of 1998 servicing 24 hours a day and 7 days a week, which provides nearly 1000 times of consultation in each year.

3) Research on Typical Chemical Reaction in Petroleum Production

Mainly studies the fire and explosion mechanism for typical chemicals and related prevention and control technology, to avoid and reduce the occurrence of calamitous accidents in production link and other links.

4) Research on Risk Assessment Technology for Petroleum Installations and Construction Projects

Mainly develops research in risk assessment and control technology for large-scale tank farms, long-distance pipelines, as well as petroleum installations, to guarantee the safe and stable operation.

5) Research on HSE Management and Safety Training

Deal with the developing of HSE management system, drafting of standards, consultation to enterprise, HSE supervision and safety training, to provide technical support for the establishment of SINOPEC HSE management system.

6) Technical Research on Electrostatic and Lightning Protection

Mainly studies the sets of technology of electrostatic and lightning prevention for large-scale oil depots and pipe lines in order to prevent the fires and explosions in the fields caused by electrostatic and lightning. The electrostatic laboratory is high-class and the lightning laboratory is of advanced measure instruments and surge protector.

7) Research on Monitoring and Testing Technology for Workplace Safety

Mainly studies and develops monitoring and testing technology of hazardous factors in workplace and safety monitoring instrument, provides more powerful technology support for standardizing management of monitoring and testing and relative technical standards of SINOPEC.

8) Technical Research on Occupational Hazard Prevention and Control.

Deal with the occupational hazard control in petrochemical work place, health assessment methods and occupational disease prevention and control methods, to fulfill the demands of SINOPEC in chemical, noise and other hazards prevention and management.

Process Industry Tool for Operator Action Analysis

Katharina Löwe, Diah Indriani Widiputri

Technische Universität Berlin, Institute of Process and Plant Technology
Berlin, Germany
diah.i.widiputri@tu-berlin.de

Heinz-Josef Löher

Bayer CropScience AG
Frankfurt a. M., Germany

ABSTRACT

Despite the increasing number of accidents in process industry caused by human error, there are still too few efforts done to improve safety of operation. This is mainly caused by people's tendency to overestimate human's capability in interacting with their working environment and moreover by the unavailability of reliable analysis methods. A new computer-based method for elaborate human factors analysis on existing plants, Process Industry Tool for Operator Action Analysis (PITOPA), is presented in this paper, which enables the improvement in design and operation to increase system's safety.

Human factors consideration during design of a process plant holds the key in realizing the attempts to minimize accidents caused by human error in process industry. Thus, an approach to incorporate HF into design phase is developed in this paper, along with an adaptation of PITOPA into $PITOPA_{DESIGN}$ which can help achieving a safer operation as early as possible with minimized cost implications.

Key words: human factors, process safety, task analysis, operator action, process plant design

1. INTRODUCTION

During the latest decades human error has been accused to have bigger contribution in causing major accidents in process industry. Studies have shown that around 80% of accidents happened in process industry were caused by human failures, particularly in highly automated systems [9]. Nevertheless, there have been too few efforts done in discovering the underlying causes of such errors. The fact that human beings are easily affected and therefore can be highly influenced by surrounding conditions is often ignored and most people tend to overestimate human's capability in interacting with their working environment. This condition is moreover propped by the unavailability of systematic analysis methods. The widely implemented general safety analyses do not include human factors (HF) consideration systematically. Moreover, in spite of the fact that some HF approaches have been developed during the latest time, there is still a lack of adequate qualitative HF analysis method, since the existing ones focus more on symptoms rather than the root causes of human failures [9].

Optimizing process safety by reducing the likelihood of human errors should be addressed by improving the designs of equipment, operations, procedures and working environment in a way that is compatible with human beings capabilities and limitations. Unrealistic demands on operators must be recognized and therefore, the system must be configured correspondingly. The new method for operator action analysis [2] provides a way to identify and evaluate operator workload during task performance, as well as the consequences following executed errors in both normal and abnormal operation. In addition to it, an evaluation on how immense all performance influencing factors (PIFs) are affecting human during task performance is conducted. By identifying the most influencing factors on human during a certain task, improvement potentials can be revealed and operation safety can thus be maximized.

A new computer-based method for HF analysis, Process Industry Tool for Operator Action Analysis (PITOPA) is presented in this paper. This method provides a more subtle way in performing an overall human factors analysis on existing plants. Since the attempts to lessen the probable consequences of human error and to minimize risks must actually be started during design, an approach to incorporate HF consideration into design of a process plant is developed. Consequently, an enhancement of the developed computerized method is conceptualized for an application during design, configuring $\text{PITOPA}_{\text{DESIGN}}$.

2. HF FUNDAMENTALS

Human factors can be defined as the interface between scientific knowledge of human, facilities and procedural or managerial system regarding all activities in different stages of a process, which can lead either to a better or poorer system efficiency, safety and reliability. An analysis on HF should be carried out with the aim to optimize system's qualification and ability to be used comfortably, safely and effectively by human beings. The performance of such analysis can be time and effort consuming in accordance with the increasing complexity of the plant under assessment, if not carried out systematically.

Several techniques have been developed for HF analysis purposes which can be differentiated between the quantitative and the qualitative ones. The quantitative HF analysis methods are taking human beings as one part of the system which either functions or fails [2]. They do not pay more attention to the underlying causes of an error occurrence. For that reason, qualitative methods are considered to be more liable to provide an adequate way for HF analysis.

A very essential technique to be performed initially during a HF analysis is the task analysis. Task analysis is referred to as systematic and structured attempts to identify and examine tasks performed by operators as they interact with systems [6]. The performance of task analysis can be very tedious and therefore should only focus on human related tasks, especially those with perceived criticality [6,7]. By exploiting the available plant documents and technical information such as process flow diagrams, P&IDs, plant layout, safety and near-miss reports, as well as operating and working instructions, task analysis can be initiated.

Most of the existing safety analysis methods are not including HF considerations. Nevertheless, several methods were developed in process industry to analyse and observe the operator actions performed during operation. Unfortunately, these operator action analysis techniques are not widely implemented, since they have several drawbacks to be emphasized; they do not provide information about the distance between equipment that must be operated simultaneously, some of the methods do not give prediction of consequences that could arise following an error occurrence and they do not represent both normal and abnormal conditions concurrently. On that account, a new method to analyse operator actions was developed [2].

Human beings are very easily influenced by the surroundings during their work. All factors that affect human during task performance are referred to as performance influencing factors (PIFs). Since these factors influence human in different ways and immensities, an evaluation must be done in order to discover factors with greatest influence on human performance. Such evaluation can help recognizing system's strengths and weaknesses from human factors' point of view. Since the influencing factors can differ broadly depending on the kind of activities performed by the operator, e.g. in nuclear, aerospace and chemical industries, these factors must be identified at the first place. Based on several guidelines, regulations and standards, an outline of assessment areas to be examined in assessing the quality of human and organisational factors in a process industry were defined [3,4].

As the occurrence of human errors is mainly caused by inadequacy in design and management system of a plant, taking HF considerations into account only during operation cannot be viewed as an optimal attempt in avoiding and minimizing the number of human errors in process industry [8]. A survey conducted in the scope of a previous work [4] indicates that most respondents begin the consideration of HF during the detail engineering phase. Even this is already too late. Reducing human error should be addressed by inherently configuring the process, the plant design, the operability and the working environment to best suit human capabilities and their limitations. This involves identifying early in the design phase the unrealistic demands on operators and maintainers by system characteristics. This will reduce the likeliness of later changes and can result in cost reduction during operation.

HF activities should be started as the earliest technical design has begun. Information concerning the process type, chemical to be utilized, draft of plant layout, and process flow diagrams should be available. It is also essential to obtain lessons learnt from similar plants in operation in form of reports and plant documents.

3. NEW APPROACH FOR QUALITATIVE HF ANALYSIS

Due to the lack of adequate methods, HF analysis is not widely performed in process industry. Several HF techniques developed in process industry were examined regarding their practical applicability and were unfortunately proved to be unsuitable for an elaborate HF analysis [2]. Therefore, a new approach was developed which provides a systematic way in performing a qualitative HF analysis in order to achieve a safer operation and decrease the number of human failure events.

3.1 Systematic Method for Operator Action Analysis

This new method for operator action analysis provides a way in evaluating the consequences of incorrect or omitted actions during task performance in both normal and abnormal operation. For this purpose, distance between components in the system which have to be operated simultaneously must be evaluated, since distance description can help determining operator workload during task performance. Identified tasks are to be decomposed into subtasks and for each of which, information concerning location, time frame, possible errors and their consequences should be collected. The evaluation is described in diagrams (Figure 1 and 2) where each step is represented using a numbered rectangle, demonstrating the task sequence horizontally. The x-axis of the diagram represents time span, while the y-axis shows the distance between equipment that should be operated simultaneously. The horizontal lines parallel to x-axis indicate the grade levels, on which each task has to be performed. The method is explained in more detail in [2].

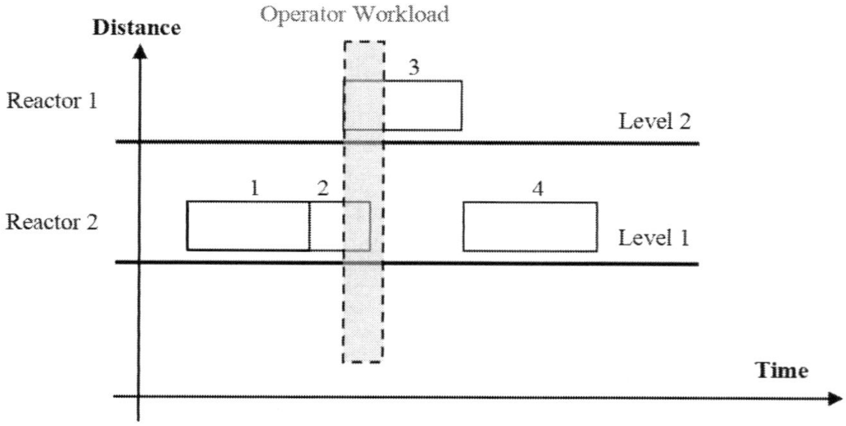

Fig. 1 *Operator Action Analysis Diagram for Normal Operation*

During an abnormal condition, operator who has to overcome with a disturbance while at the same time has to keep performing normal tasks could accumulate a certain workload. This can affect their performance and can result in errors. Consequences of such error can be analysed to be able to arrange the required countermeasures. During performance of a countermeasure further error can also occur and should as well be anticipated. All relevant consequences and the required counteractions can be demonstrated in the same diagram (Figure 2). Checklists were developed to help performing this method [1,2].

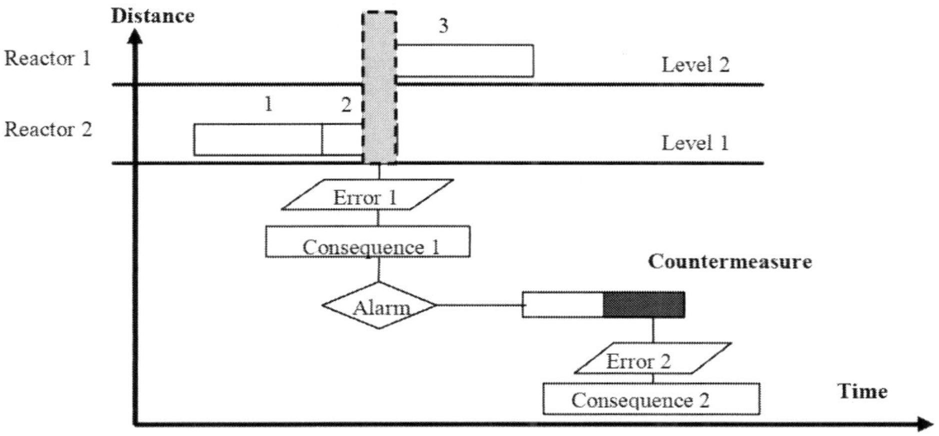

Fig. 2 *Operator Action Analysis Diagram for Abnormal Operation*

By using this new method for operator action analysis, different plants of BayerCropscience AG, Germany were examined. The method enables the identification of each subtask and all possible errors during task performance, while allowing the consequences of executed incorrect actions to be anticipated. Results of this method indicate necessary modifications of the existing system's design and management in order to improve HF quality of the plant. This may include changes in e.g. plant lay-outs, design of procedures and instructions, or improvement on technical equipment condition that can lead to a safer and more reliable plant operation.

3.2 Evaluation of Performance Influencing Factors (PIFs)

Working environment should be configured in such a way which best suits human beings during their work. Due to the complex interaction between various factors within a man-machine system, it becomes very complicated to settle on factors which could offer an improvement to the whole system. For that reason, a new method for evaluation of various performance influencing factors (PIFs) was developed, which allows the determination of the most influencing factors on operators during a certain task performance. This enables the recognition of improvement potentials within the system in order to maximize process safety. A classification of all influencing factors was done initially.

PIFs can be classified into three HF groups; technical facilities, management system and human beings (Table 1). The factors within each group are further decomposed into attributes. The influence of one HF group is determined by the influence of each PIF within it. Likewise, the influence of each PIF is determined by influence of each attribute contributing to it. The determination of PIFs level of influence is carried out through the implementation of Analytical Hierarchy Process (AHP), a technique which has been widely implemented in multi criteria decision making [12]. In order to evaluate and compare the level of influence of each factor, a weighting system was developed (Table

1), which shows how important one factor can contribute in influencing the whole system, compared to other factors. The development of this weighting system involved a European-wide industrial survey [5,11].

Although each factor has a certain potential in influencing human performance, those factors can have different level of influences for different situations, depends on the specific task. Hence, the condition of each HF aspect must be rated for every particular task carried out by the operator. The rating process is assisted by a questionnaire that lists all related questions concerning the attributes with a rating scale between 1 and 10, where 1 represents a minor influence and the score 10 the largest.

Table 1 *PIFs Classification and Relative Weighting System*

PIF	Relative Weight	Attributes	Attribute Relative Weight	Global Weight
Equipment	0.073	Technical System	0.750	0.018
		Process Safety Condition	0.250	0.006
System Interface	0.296	Display	0.404	0.040
		Communication System	0.453	0.045
		Feed-back and Alarm	0.143	0.014
Environment	0.272	Lighting	0.230	0.021
		Temperature	0.257	0.023
		Vibration and Noise	0.203	0.018
		Air Quality	0.310	0.028
Workplace Design	0.358	Lay-out	0.486	0.058
		Accessibility	0.514	0.061
Skill & Knowledge	0.429	Type of Task	0.316	0.045
		Qualification & Experience	0.684	0.098
Manual & Physical Handling	0.429	Physical Load	0.528	0.075
		Tools	0.472	0.067
Stress Level	0.143	Concentration Demand	0.401	0.019
		Monotony	0.229	0.011
		Health Hazards	0.370	0.018
Job Design	0.310	Task Frequency	0.167	0.017
		Job Description	0.833	0.086
Line Management & Instructions	0.390	Line of Responsibilities	0.343	0.045
		Procedures	0.314	0.041
		Supervision	0.343	0.045
Information	0.300	Labels and Signs	0.192	0.019
		Communication	0.532	0.053
		Documentation	0.276	0.028

3.3 Process Industry Tool for Operator Action Analysis (PITOPA)

The difficulty in performing an elaborate HF analysis increases with the increasing complexity of the process plant under assessment. Beside the unavailability of a systematic HF approach, the enormous size of information collected during plant investigation can even more confound the whole evaluation. Information concerning the process should be maintained in such a way that enables its reutilization to take place effortlessly. For this reason, a computer-based method Process Industry Tool for Operator Action Analysis (PITOPA) was developed. This new method supports the performance of an extensive HF analysis while maintaining the huge amount of data collected during assessment in an associated database system.

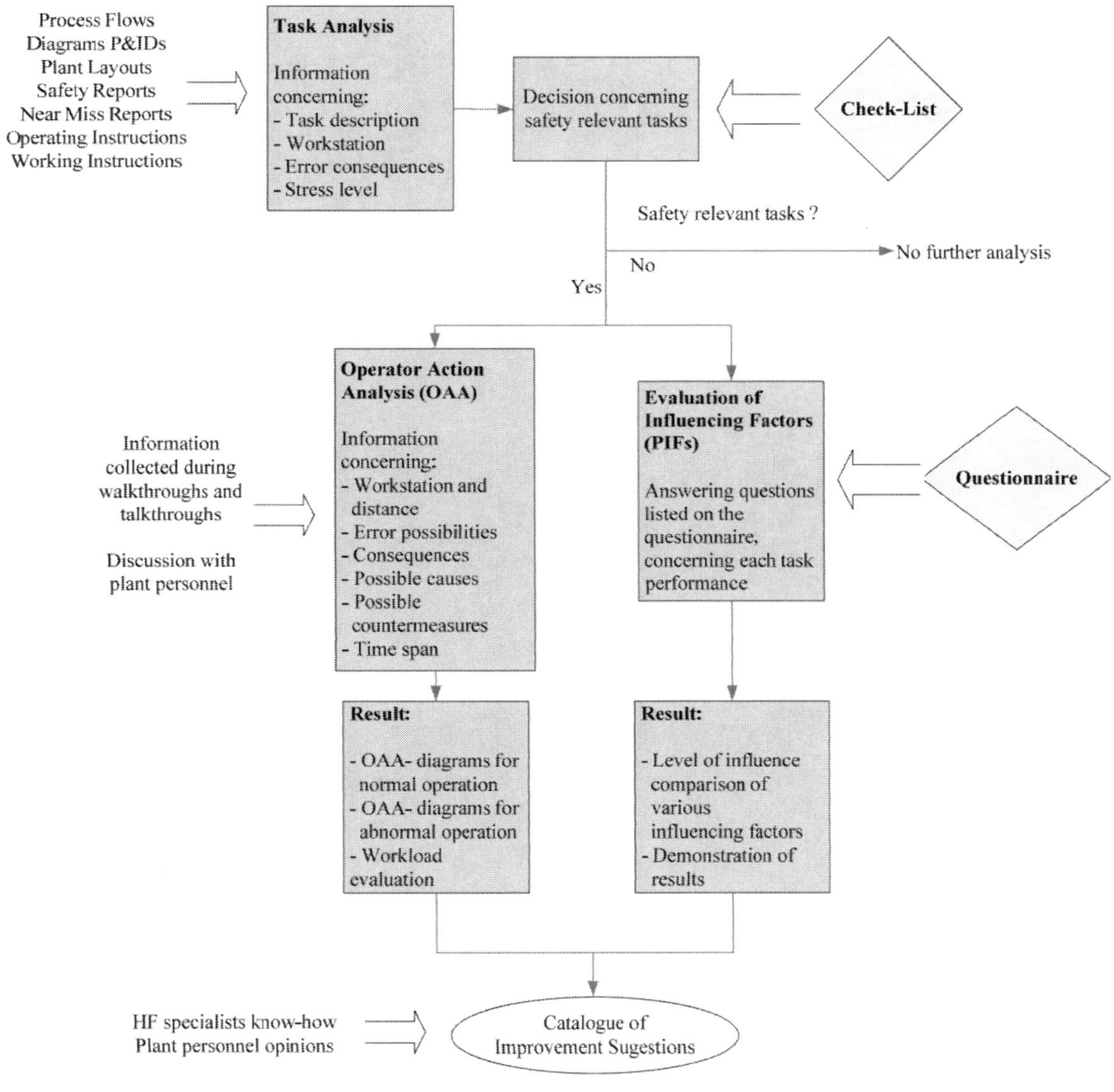

Fig. 3 *New Approach for HF Analysis*

PITOPA is based on the new HF analysis approach shown in Figure 3, which comprises the earlier discussed methods, operator action analysis (Chapter 3.1) and PIFs evaluation (Chapter 3.2). An initial step of the approach is a plant investigation, where necessary information concerning the process is gained. This step requires up to circa three weeks for complex process plants, and obliges supports in the form of available plant documents. By employing a checklist for task analysis [1,2], safety relevant tasks are determined. The next steps of this analysis will only be carried out on safety relevant tasks. Ultimately, suggestions on how the system should be improved to better support operators in performing their work are presented in form of catalogue. These can require some design modifications and improvements of the working circumstances to be done.

The structure of PITOPA is shown in Figure 4. Some results obtained from the application of PITOPA are:

- A collection of data and information about the process under assessment in form of a computerized database.
- Identification and evaluation of operator workload demonstrated by operator action analysis diagrams for both normal and abnormal operating conditions.
- Consequences recognition of possible executed operator errors.
- Determination and comparison of PIFs influence level (Figure 5) to reveal improvement potentials within the system.
- Suggestions for improvement on plant design and working condition in the form of catalogue.

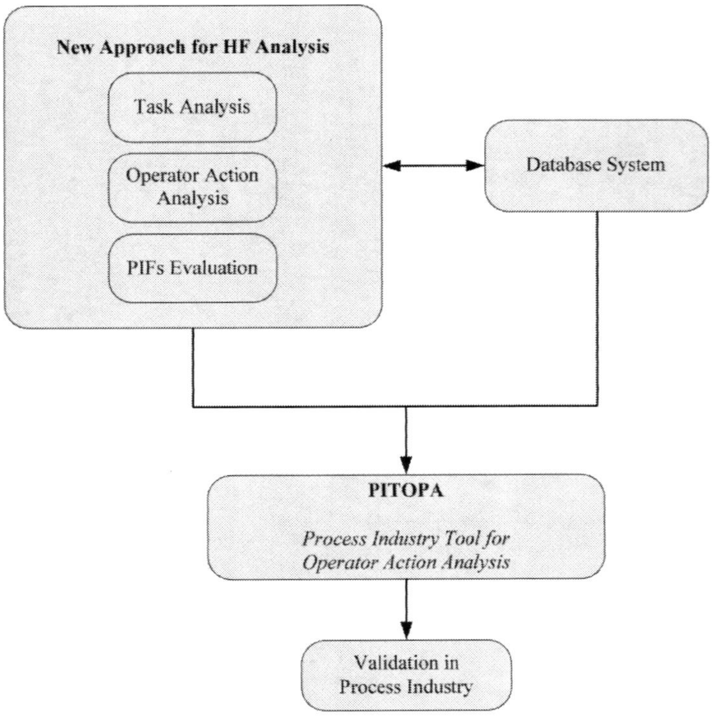

Fig. 4 *Development of PITOPA*

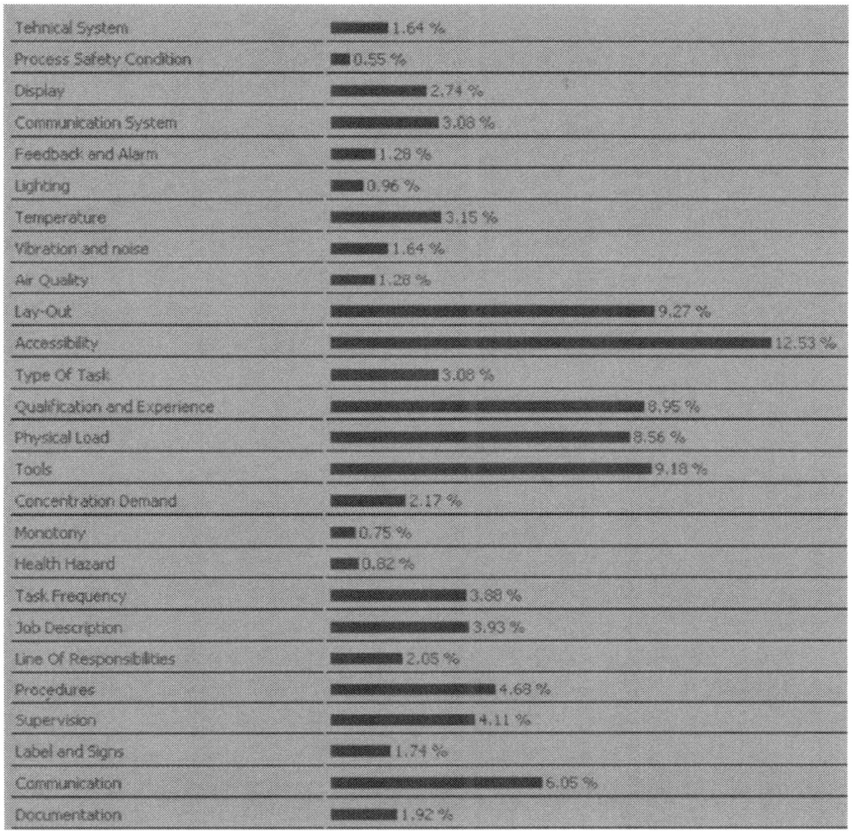

Fig. 5 *Level of influence comparison between various PIFs*

PITOPA has been validated in several plants of BayerCropscience AG, Germany and was proved to be very practical in its application. One of the results of PITOPA is delivered in form of influence level comparison between various PIFs (figure 5), which indicates the parts of the system, on which modifications must be carried out. As an illustration, from the comparison it can be recognized that during the corresponding task performance, the factors accessibility, plant lay-out and additional tools give highest influence on operators. This shows that some actions must be done regarding to these factors, in order to better support operators. On the other hand, since environmental aspects such as lighting, vibration and air quality have lower level of influence; modifications on these factors are not required.

Based on the influence level comparison and results of the evaluation on operator workload, improvement suggestions are offered in a form of catalogue. This supports the realization of a safer operation through an optimization of HF quality of the system. Performing a HF assessment as general was proved to be beneficial, as industry can take delivery of feed-back concerning their current process condition. Moreover, operators' motivation can as well be increased, as they become conscious that their needs and requirements are being paid attention to.

5. NEW APPROACH FOR HF ANALYSIS DURING DESIGN

It has been acknowledged that configuring a system that meets human beings requirements should be addressed by considering HF early in plant design, in order to avoid later modifications during operation. This enables on one hand the realization of inherently safer design, as it leads to improvement of operation safety while minimizing later potential costs. However, performing a HF analysis on a plant that is not yet existed is even more complicated than on an existing one. Hence, a new approach to provide a means to consider and analyse HF aspects during design phase is conceptualized, which is referred to as $\text{PITOPA}_{\text{DESIGN}}$.

Process plant design can be divided into 3 phases, the conceptual design, the basic engineering and the detailed engineering phase. HF should be considered in every design phase. The performance of the new method (Figure 6) starts during conceptual design with an implementation of functional analysis and functions allocation. Functional analysis and functions allocation have the aim to identify different activities required to achieve the intended purpose of the system and to later decide, which tasks are to be performed manually, automatically or a combination of both (semi-automatically) [10]. Both techniques should be performed iteratively in connection with the development of technical plant design. Changes made to technical plans require these analyses to be performed over accordingly.

Subsequent to the performance of functional analysis and functions allocation, task analysis is to be completed as the next step. This is a very critical step in design process, during which all information concerning all tasks and operator needs to complete them are collected and analysed [6]. Task analysis has also the responsibility to define all subtasks and steps necessary to complete specific tasks. Information to be collected during task analysis includes identification of task location, task description, possible errors and their consequences.

Analysing a particular task performance and operator requirements depends entirely on the available facilities and system's specifications. Since at this point, it is still impossible to examine the exact system's arrangement; task analysis must be carried out in relation to the ongoing HF-oriented system design. This is one step of the method, in which considerations are made in ensuring that the system design incorporates HF design criteria [7]. Task analysis and HF-oriented system design must be performed through the whole design process. Both techniques are therefore differentiated for every design phase with increasing particularity.

Human factors concerns should be addressed mostly during basic and detailed engineering phase to avoid modifications at later stages. However, a preliminary HF design should be done prior to it, early in the conceptual design phase [7]. During this phase, the first obvious attempts are made to reduce potential sources of human errors [10]. In the preliminary HF design, considerations concerning selection of geographical plant location, preference of process types and decisions on subject of the utilized chemicals are made. These aspects can have further influence on the design of working procedures, safety equipment, labelling and signage, and other components within the system, which is carried out in the next stages. Simultaneously in conceptual design, task analysis is to be performed. Outputs of the preliminary HF design can help in analyzing the task performance, for instance in identifying locations where human errors are most likely to take place. Such event can occur in form of contacts with hazardous chemicals or overheated equipment.

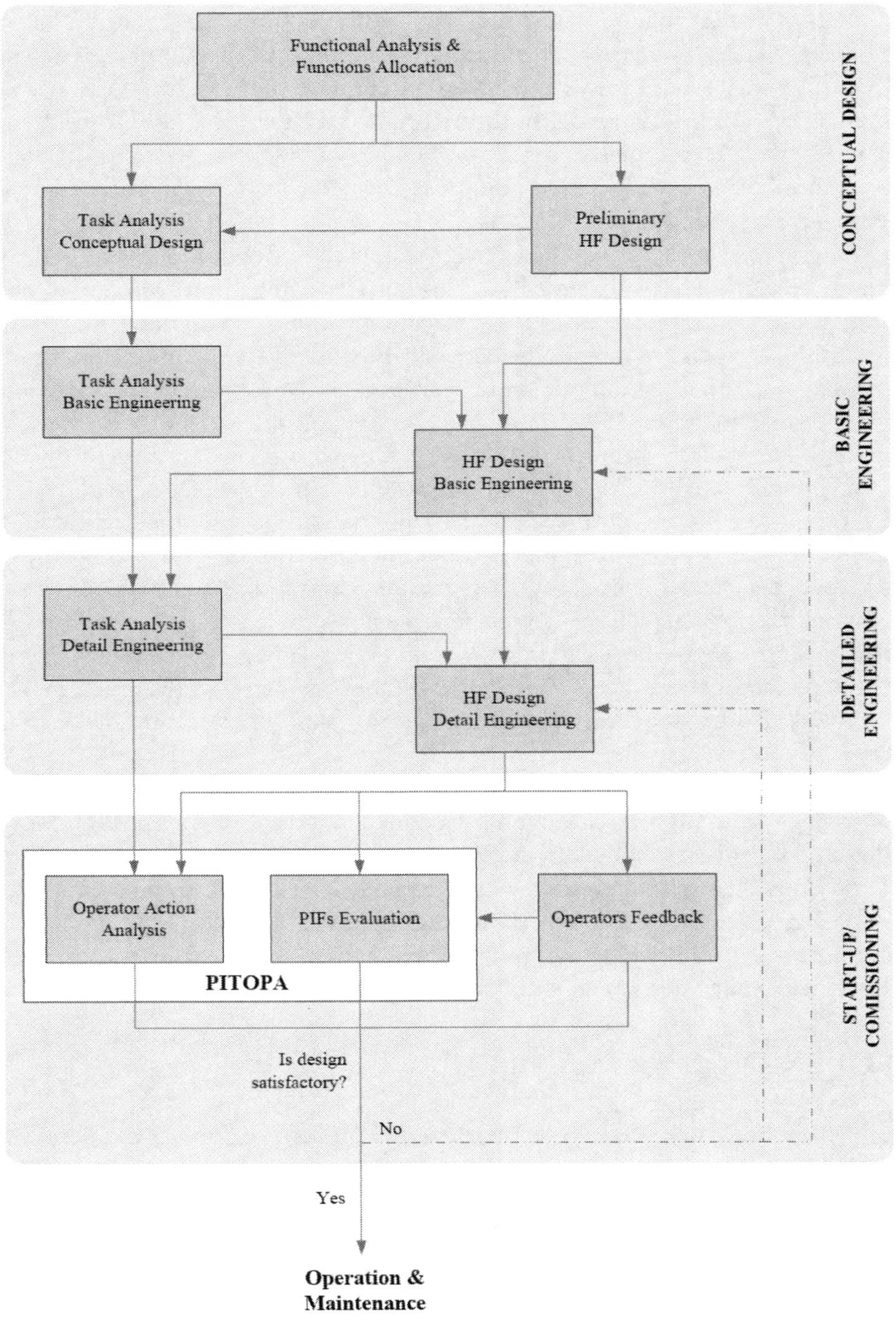

Fig. 6 *Structure of PITOPA*$_{DESIGN}$

In basic engineering, task analysis has the responsibility to define steps needed to complete all tasks. Operator requirements can be better acknowledged after a closer examination on each subtask. The result can then be used as an input for HF-oriented system design during this phase.

Similarly in detailed engineering, output of task analysis will be made use in HF design. In this design phase, task analysis will provide all required information concerning every task to be performed during operation, in order to design a system that meets HF design criteria. Therefore, in detailed engineering design phase, opinions and suggestions of HF specialists are of high importance, concerning how to best situate the installation so that it optimally supports human beings in performing their work. Guidelines, handbooks of HF and ergonomics, as well as software procedures can be employed to support them.

In making sure that the developed design meets the intended purpose, an evaluation is essential to be carried out subsequently. This activity can be done directly after detailed engineering phase is completed or as the design comes into the commissioning and start-up stage. Problems found at this point can still be solved with minor cost implications [10]. Design evaluation must include an observation concerning how adequate it supports human in performing their work, prediction of possible errors, as well as their consequences, so that it would be possible to recognize and identify necessary modifications. This can be done through an application of PITOPA and by exploiting operator's opinion concerning the design. This would be the most prudent way in verifying that the constructed design suits operators' capability without putting irrational demands on them. Unsatisfactory evaluation results can require some modifications to be done, which can bring the whole design process back to basic or detailed engineering phase. The design can be brought into operation only after an acceptable satisfactory result of this evaluation has been achieved. HF analysis through the implementation of PITOPA should also be frequently performed during plant operation in order to maintain the system's HF quality.

A realization of the new approach as an iterative computer program is being undertaken. The program will have an associated database system that maintains all information and considerations taken during design. This new program will automatically decide if the constructed design reaches the expected HF quality, will then suggest the points in earlier design phases where modifications are still required, and will run iteratively until a satisfactory design is obtained. Moreover, the computer program will support the performance of general safety analyses during plant design.

6. CONCLUSION

A new computer-based HF method, Process Industry Tool for Operator Action Analysis (PITOPA) is introduced in this paper. The application of this method delivers results which give suggestions concerning the necessary system improvements in order to optimize safety of operation. A validation in process industry has proved PITOPA to be very practical in its application and have reliable results, since it enables the identification of operator workload in both normal and abnormal operating condition and determines the most influencing factors on human beings during task performance. This way, improvement potentials within the system can be revealed, so that changes in design and management system can be done specifically.

HF considerations must already be incorporated early in the design phase in order to prevent later changes during operation. Therefore, a new approach for HF analysis during design was developed, as an enhancement of PITOPA into $PITOPA_{DESIGN}$. With the new approach, configuring a safer operation can be achieved while cost implication is avoided. Efforts are being made to bring the developed approach into realization as an iterative computer program and to afterwards validate it in process industry.

7. REFERENCES

[1] Dalijono, T., Castro, J., Löwe, K., Löher, H.-J., *Reducing Human Error by Improvement of Design and Organisation,* Process Safety and Environmental Protection 84(B3), pp 191 – 199, Trans IChemE, Part B, 2006.

[2] Dalijono, T., Löwe, K., Löher, H.-J., *Development and Verification of a New Approach for Operator Action Analysis,* Process Safety and Environmental Protection, 83(B4), pp 331-337, Trans IChemE, Part B, 2005.

[3] Kariuki, S. G., Löwe, K., *Increasing Human Reliability in the Chemical Process Industry Using Human Factors Techniques,* Process Safety and Environmental Protection, 84(B3), pp 200 – 207, Trans IChemE, Part B, May 2006.

[4] Kariuki, S. G.; Löwe, K., *Integrating Human Factors into Process Hazard Analysis*, Reliability Engineering & System Safety, Volume 92, Issue 12, pp 1764-1773, Elsevier, 2007.

[5] Kariuki, S. G.; Mayor Vadillo, E.; Löwe, K.: *Human Factors Assessment Technique for the Chemical Process Industry,* Chemical Engineering & Technology, 30, No. 10, pp 1319-1327, 2007.

[6] Kirwan, B., Ainsworth, L. K., *A Guide to Task Analysis,* ISBN 0-7484-0057-5, Great Britain: Taylor and Francis Ltd., 2001.

[7] Löwe, K., Kariuki, S. G., *Methods for Incorporating Human Factors during Design Phase,* Loss Prevention and Safety Promotion in the Process Industries, Loss Prevention, pp 5205 – 5215, 2004.

[8] Löwe, K., Kariuki, S. G., Porcsalmy, L., Fröhlich, B., *Development and Validation of a Human Factors Engineering Guideline for Process Industries,* Loss Prevention Bulletin 182, IChemE, pp 9 – 14, 2005.

[9] Löwe, K.; Kariuki, S. G., *Quantification of Human Factors for Quantitative Risk Analysis,* Conference Proceedings, 3rd Global Congress on Process Safety, ISBN 978-0-816910-14-4, AIChE, 2007.

[10] McCafferty, D. B., *Successful System Design Through Integrating Engineering and Human Factors,* Process Safety Progress Vol. 14, No. 2, pp 147 – 151, 1995.

[11] PRISM, Homepage of the EU-project PRISM: www.prism-network.org.

[12] Saaty, Thomas L., *The Analytic Hierarchy Process,* ISBN 0-7-054371-2, New York: McGrawHill, 1980.

Making sure that lessons from past major incidents are known and kept alive.

Frederic Gil CEng IFireE
Process Safety & Fire Engineering Advisor
BP Refining Safety & Operations Excellence
Sunbury UK

ABSTRACT

In discussions with young graduates or even experienced technicians it has become clear that many had not heard of Flixborough, Feyzin or Piper Alpha, and if they had, few of them received enough information to be able to transfer lessons to their current activities. BP initiated a project in early 2004 to develop a process safety teaching aid in the form of a booklet containing analyses of non-process and process industries major accidents against their Integrity Management program. BP has offered this booklet to CCPS for wider publication under the title "Incidents that define Process Safety – ISBN 978-0-470-12204-4". To complement this document, video animations are being developed to help disseminate widely and keep alive the main lessons from major incidents in a simple, user friendly way.

1. Context

In a survey[0] done in 2006 by the Mary Kay O'Connor Process Safety Center of nearly two hundred Universities across the United States, to establish the status of the Process Safety in the chemical engineering curriculums of American Universities, it was found that:

- 11% of the Universities that answered offer process safety as part of the core curriculum,
- 13% offer a process safety course as an option,
- 3% have plans to introduce process safety in their engineering curriculums in the near future.

In discussions with young graduates or even experienced technicians it has become clear that many had not heard of Flixborough, Bhopal or Piper Alpha, and if they had, few of them received enough information to be able to transfer lessons to their current activities.

It is well known that, to be effective, training has to be constantly refreshed and repeated in different ways. Material used to illustrate Process Safety course was often found to lack consistent incident and lesson descriptions, to have limited access to data for older or foreign incidents and to be in a non user-friendly format.

Jesse C. Ducommun was Vice-President, Manufacturing and a Director of American Oil Company in 1961 (joined in 1929) and Vice-President American Petroleum Institute in 1964. While working for Amoco, he conceived, inspired and co-authored the unique series of booklets on process safety in the 1960s that was then updated and complemented after 2002 by BP Refining Process Safety Community of Practices. In the preface, he stated:

> *"It should not be necessary for each generation to rediscover principles of process safety which the generation before discovered. We must learn from the experience of others rather than learn the hard way. We must pass on to the next generation a record of what we have learned."*

2. Step 1: developing a user-friendly booklet on past major incidents [2]

We searched long and hard for a book that could illustrate the different elements of Process Safety with actual incident cases. In the end, we found that we had to write our own: BP initiated a project in early 2004 to develop a process safety teaching aid in the form of a 264 pages booklet containing analyses of non-process and process industries accidents against their Integrity Management program. Titled "Integrity Management: Learning from past major industrial incidents", the first edition was published internally in early 2005, with a second edition containing lessons from BP Texas City March 2005 accident in 2006. BP has offered this booklet to CCPS for wider publication[1]. The objective of this booklet was to disseminate more widely the main lessons from major incidents from wherever they originated using a simple, user friendly, integrity management tool. It therefore consolidates and archives concise information on representative incidents that are relevant today so readers can apply the resulting safety principles to their current operations.

Included in the booklet are a number of major process incidents that have been influential in shaping process safety, such as Flixborough, Bhopal, Phillips Pasadena, and Piper Alpha, that are used to illustrate synergies and because they are important in their own right. It also incorporates events from other industries that have implications for the chemical industry, such as the NASA Challenger Disaster, Chernobyl, and references for additional study. Events are described in detail so readers get a fundamental understanding of the root cause(s), the consequences, the lessons learned, and actions that can prevent a recurrence.

The scenarios used in the booklet were chosen against a number of topic headings:

- Blind operations
- Design
- External causes
- Inspection & maintenance
- Knowledge and training
- Lack of HAZID (hazard identification)
- Management of Change
- Not learning from near misses
- Operating practices
- Permit to Work
- Emergency Response
- Human Factors

As the booklet also contains non-process accidents (such as Titanic, Columbia or Concorde), some may wonder why use these when there is plenty of fertile ground within our own oil and petrochemical industry. There are a number of reasons:

- Most people have heard of Titanic, Chernobyl, the NASA accidents (Challenger and Columbia), Concorde and other aviation accidents through the press and media.
- Oil and petrochemical companies exist in a very similar environment to other commercial ventures, in that:
    - They operate in a highly competitive business environment
    - They have hierarchal management structures
    - They employ skilled people who have their own career progression aspirations, which can impact on an individual's behavior
    - They are strictly regulated.
    - They involve similar technologies or practices (for example board operators in a refinery control room and air traffic controllers in a control tower both have to work remotely and can have issues mentally visualizing what is going on; replacing critical parts on an aircraft engine or a refinery compressor has to be done with similar rigorous procedures).
- It is very easy for individuals reading accidents reports to focus on technical details, whereas the majority of root causes relate to softer, cultural and human behavior issues. Lessons learned from an incident on a crude oil distillation unit (CDU), for example, may receive a technical review at another refinery which leads to a conclusion that a similar accident could not occur on their CDU as it has a different equipment arrangement or process conditions. In most cases, the root causes are likely to be around human factors and management protocols. Lessons from non-process accidents do not have a technical equivalence and readers are forced to consider the full range of lessons learned.
- Some industrial sites, or locations, can build up a negative reputation for process safety if there is a history of accidents and incidents, and others may assume that they are better than the site suffering the incident, and that the same thing could not happen to them. Lessons from accidents such as Titanic and the aviation industry can excite the imagination, leading to the learning of valuable lessons that remain in the minds of those who had studied them.

Presented to raise process safety awareness, to help readers learn from previous incidents, and to supplement established initiatives and materials, this book has proved to be a valuable reference for engineers and technicians involved in the design and operation of chemical and petroleum processing facilities, as well as managers and decision makers in these industries. It is also an enlightening supplement to many chemical engineering courses.

3. Step 2: developing process safety video animations on past major incidents

Taking this one step further, BP Refining Process Safety Community Of Practice initiated a project in 2007 to develop a series of short process safety animations on major past industrial accidents. For ease to use, these animations were developed as Flash-Player files (see example of screen grabs below), embedded into PowerPoint presentations.

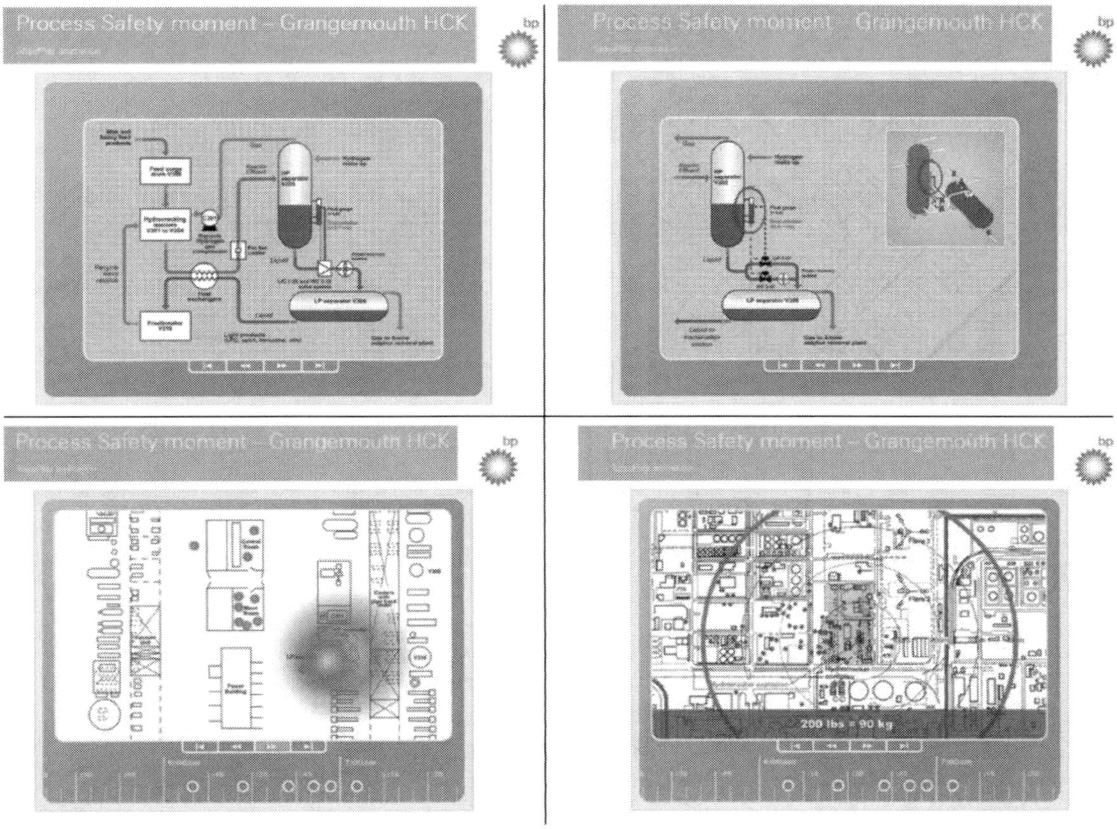

The Flash-Player files were developed from a script using available public domain information on each incident (some examples of references used in the BP booklet and animations are given in reference [3]) and shortly describe the plant and process involved, the conditions leading to the incident, and the direct consequences of the event. A mix of animated 2D, 3D drawings, pictures and video plus a simple voice-over running commentary are combined to attract the audience attention and point out the main issues. All animations have been kept below a maximum of five minutes long, with an average of three to four minutes. Therefore, some simplification of the incident circumstances may have been necessary and this has to be understood by the user who is given references to read more should the need arise.

Each animation has been developed in two versions: the first one is continuously running from start to finish for use as presentation during Process Safety moments, while the second one can be stopped at any time and also stops at the end of key sections to allow discussion during exercise or training sessions.

The PowerPoint files contain:

- the main lessons learned from the incident,
- the main causes and contributing factors,
- a graphic representation of the layers of protection that failed (see example below),

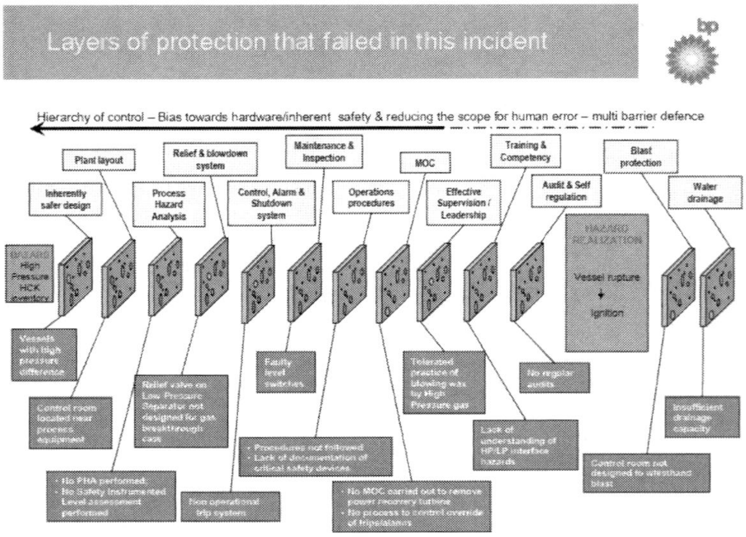

- specific exercises on the main causes of each incident (for example Management Of Change, Shift Handover, HAZOP study, etc.: see examples in table 1),
- references to read more on the incident and the specific topic(s) selected,
- additional information that may be useful to run the exercises (i.e. P&ID, PFD).

PowerPoint files allow local users to adapt the message to their plant and to translate the content to local language. A few animations have been translated in German, Dutch and Spanish to test feasibility but the English voice-over text is also provided: this allows the animation to be used with the sound off and a running commentary to be made in any language by the presenter.
For all incidents the following PowerPoint files with animation embedded are developed:

- Process Safety Moment
- Exercise 1: 'can you identify the measures of protection that failed and the ones that worked as intended?'
- Exercise 2 is specific to each incident (see table 1 below).
- Exercise 3: 'Do we have similar risks here and what are the prevention/mitigation measures in place?'
- Q&A

Table 1: animations developed or in preparation:

Incident	Exercise 2 selected theme
LPG release and BLEVE - Feyzin LPG storage BLEVE, 1966	Procedure to drain water from LPG vessels
Overfilling of flare system - Milford Haven FCCU, 1994	Overfilling process vessels
Major platform fire - Piper Alpha platform, 1988	Shift handover
High Pressure / Low Pressure incident - Grangemouth Hydrocracker, 1987	Inherently Safer Design, Process Hazard Analysis and Risk Assessment
Major plant explosion - Flixborough, 1974	Management Of Change process
Multiple tank fire - Singapore Refinery, 1988	Compliance with Industry standards
Jetty and ship fire – Bantry Bay, 1979	Use of Recognized And Generally Accepted Good Engineering Practices (RAGAGEP)
Ferry capsize - Herald of Free Enterprise 1987	Operating Procedures
Crude tower fatal fire - Lavera, 1994	Control Of Work
Hydrocracker thermal excursion and fire - Martinez, 1997	Instrumented Safety Systems
Hydroformer explosion and tank farm giant fire - Whiting, 1955	Hazard Identification

Three animations were released as pilots at the end of 2007 on the Company's Intranet for the plants use. Monitoring of the visits on the Intranet site, demands for translation of the material in different languages and feedback from users have proved that these were very quickly growing popular. These have already proved to be flexible and have been used for starting management meetings, in training sessions, as preamble to HAZOP or Management of Change studies, as site-wide TV broadcast, in meetings with contractors, etc.

Making sure that lessons from past major incidents are known and kept alive.

4. Conclusion and way forward

Both the booklet and the Process Safety animations have proved very successful: the booklet had to be reprinted 3 times to cover internal demands; and monitoring of the visits on the Intranet site, demands for translation of the material in different languages and feedback from users have proved that the animations were very quickly growing popular.

BP has offered this booklet to CCPS for wider publication: CCPS and Wiley Editions will publish it in 2008 under a new title "Incidents that define Process Safety"[1] – ISBN 978-0-470-12204-4.

A BP internal program has been set up to release the Process Safety animations and PowerPoint files at a frequency of one per month and to carry out a users survey after one year.

After reviewing the material developed by BP, the IPSG (International Process Safety Group) has agreed to fund two Process Safety animation developments in 2008, and ways to make these available to Academia for Chemical Engineers education will be explored.

BP and the Institution of Chemical Engineers (UK) are also exploring ways to make the Process Safety animations available to the Industry, as a complement to the BP Process Safety booklet series already published by the ICHEME [4].

5. REFERENCES

0. Mary Kay O'Connor, *Process Safety Curriculum in US Universities,* Texas, 2006

1. BP, *Incidents that define Process Safety,* ISBN 978-0-470-12204-4, New York: American Institute of Chemical Engineers, Center for Chemical Process Safety, 2008.

2. John Atherton & Frederic Gil, *Can we still use learning's from past major incidents in non-process industries?*, IChemE "Hazards XX" symposium, Manchester, UK, 2008.

3. Some of the selected reading used to draft the various sections of the booklet or for the animations:

- Nuclear Industry:

- Three Mile Island: - US Nuclear Regulatory Commission Fact sheet on the Accident at Three Mile Island, from: http://www.nrc.gov
- "United States President's Commission on the Accident at Three Mile Island." Vol. 9. U.S. Government Printing Office: Washington, D.C., 1979.
- Hodgson, Peter. "Chernobyl After Five Years". Contemporary Review. (July, 1992)
- Lewis, H.W.. "The Accident at the Chernobyl Nuclear Power Plant and Its Consequences". Environment. (November, 1986).

Space and aircraft incidents:

- Challenger - NASA web-site: http://history.nasa.gov/sts51l.html: The Challenger Decision to Launch, Diane Vaughan, University of Chicago Press, ISBN: 0-226-85176-1
- MD83/Shorts 330 crash at Paris Charles de Gaulle airport: Bureau Enquêtes-Accidents Report "Accident on 25 May 2000at Paris Charles de Gaulle (95) to aircraft F-GHED operated by Air Liberté and G-SSWN operated by Streamline Aviation.

- "US summary comments on draft final report of aircraft accident Flash Airlines flight 604, Boeing 737–300, SU-ZCF January 3, 2004, Red Sea near Sharm El-Sheikh, Egypt", National Transportation Safety Board.
- "Factual Report of Investigation of Accident, Flash Airlines flight 604, January 3, 2004, Boeing 737–300, SU-ZCF, Red Sea off Sharm El-Sheikh, Egypt", Egyptian Ministry of Civil Aviation.
- TWA 800 - National Transportation Safety Board Report "In-flight Break up Over The Atlantic Ocean Trans World Airlines Flight 800 Boeing 747-131, N93119 Near East Moriches, New York July 17, 1996" NTSB/AAR-00/03 Adopted August 23, 2000
- Hindenburg - Flight 100 years of Aviation, R.G. Grant, Smithsonian National Air and Space Museum/Duxford Imperial War Museum, ISBN 0-7513037323
- Columbia Accident Investigation Board – Volume 1, Report published by the CAIB on 26 August 2003 and subsequently printed and distributed by NASA and the Government Printing Office, Washington DC. Available on Board web-site: http://www.caib.us/
- Concorde - Bureau Enquêtes-Accidents Preliminary Report "Accident on 25 July 2000at "La Patte d'Oie"in Gonesse (95), to the Concorde, registered F-BTSC, operated by Air France – English Version
- R101: The Airship Disaster 1930, 1931, The Stationary Office, London, ISBN: 0 11 702407 4

Shipping incidents :

- The Last Log of the Titanic, David G. Brown, McGraw-Hill, 2001, ISBN 0-07-136447-1
- Disasters at Sea, Capt. Richard A. Cahil, Century, 1990, ISBN: 0-7126-3814
- Bantry Bay - "Disaster at Whiddy Island, Bantry, Co. Cork", Report of the Tribunal of Inquiry, published by the Stationary Office, Dublin
- "Report of the enquiry into the sinking of the "Erika" off the coast of Brittany on 12 December 1999" - Permanent Commission of Enquiry into Accidents at sea (Commission Permanente d'Enquetes sur les Evenements de Mer – CPEM)
- "Grounding of the US tankship Exxon Valdez on Bligh Reef, Prince William Sound near Valdez, Alaska. March 24, 1989" Marine Accident Report PB90-916405 NTSB/MAR-90/04, National Transportation Safety Board.
- "Fire on HMS Glasgow 23 September 1976", Health & Safety Executive/ HM Factory Inspectorate, HMSO, 1978, ISBN: 0 11 883075 9
- K-Boats, Steam Powered Submarines in World War 1, Dan Everitt, Airlife Publishing Ltd, ISBN 1-84037-057-2
- MV Herald of Free Enterprise. Report of Court No. 8074 - Formal Investigation (Hon. Mr Justice Sheen - Wreck Commissioner), July 29th 1987. London, Her Majesty's Stationery Office, 1987, ISBN 0 11 550828 7

Petrochemical Industry:

- Tosco Hydrocracker explosion & fire - US Environmental Protection Agency Chemical Accident Investigation Report: EPA 550-R-98-009, November 1998
- Mexico City - Lees, F.P., 'Loss Prevention in the Process Industries – Hazard Identification, Assessment and Control', Volume 3, Appendix 4, Butterworth Heinemann, ISBN 0 7506 1547 8, 1996.
- Marsh and McLennan, 'Large Property Damage Losses in the Hydrocarbon-Chemical Industries a thirty-year Review', 16th Edition, Marsh and McLennan Protection Consultants, 1995.
- "Analysis of the LPG Disaster in Mexico City", C.M. Pietersen, TNO, Apeldoorn, Netherlands.
- Tuprus - US Geological Survey Circular 1193, Implications for Earthquake Risk Reduction in the United States from the Kocaeli, Turkey, Earthquake of August 17, 1999
- Bhopal: - "TED Case Studies: Bhopal Disaster," Trade and the Environment Database, 1996. Ashford, N.A.; Gobbell, J.G.; Lachman, J.; Matiesen, M.; Minzner, A.; Stone, R.
- Kalelkar, A.S.; Little, Arthur D., "Investigation of Large-Magnitude Incidents: Bhopal as a case study," Chemical Engineers Conference on Preventing Major Chemical Accidents, London, England, May 1988;
- Marathon HF release - UK Health & Safety Executive Hazardous Installations Directorate "Release of Hydrofluoric Acid from Marathon Petroleum Refinery, Texas, USA. 30th October 1987; website links directly to H&SE's Lifting Procedures guidance. http://www.hse.gov.uk/comah/sragtech/casemarathon87.htm

- Health and Safety Executive, 'The explosion and fires at the Texaco Refinery, Milford Haven, 24 July 1994: A report of the investigation by the Health and Safety Executive into the explosion and fires on the Pembroke Cracking Company Plant at the Texaco Refinery, Milford Haven on 24 July 1994, ISBN 0 7176 1413 1, 1997.
- Total la Mede - "Etude de l'Accident de la Raffinerie "TOTAL" a la Mede", Dr Ir. P.Michaelis, Risk Manager Total Raffinage Distribution, presented at a meeting of GESIP at Tours on 12 June 1996 (in French).
- Conoco Phillips - Health and Safety Executive « Public report of the fire and explosion at the Conoco Philips Humber refinery on 16 April 2001"
- Feyzin - (French) Ministère chargé de l'environnement – DPPR/ SEI/ BARPI – Inspection des installations classées : Septembre 2005
- P 36 Accident analysis, ANC/DPC Inquiry Commission Report, Agência Nacional do Petróleo/ Diretoria de Portos e Costas, July 2001. (Brazilian National Petroleum Agency and Directorate of Port and Coasts joint report).
- Esso Longford - Report of the Royal Commission into the accident at Esso Longford.
- Lessons from Longford, Andrew Hopkins, CCH Australia Ltd., ISBN 1 86468 422 4
- The Fires and Explosion at BP Oil (Grangemouth) Refinery Ltd, a report of the investigations by the Health and Safety Executive into the fire and explosion at Grangemouth and Dalmeny, Scotland, 13 March, 22 March and 11 June 1987, HSE Books 1989, ISBN: 0 1188 5493 3
- The Flixborough Disaster, Report of the Court of Inquiry, HMSO, 1975, ISBN 011 361075 0
- BP Texas City – web site http://www.bp.com/genericarticle.do?categoryId=9005029&contentId=7015905
- CSB Investigation Report No. 2005-04-I-TX, March 2007 "Refinery Explosion And Fire, Bp Texas City, Texas March 23, 2005" from http://www.csb.gov
- Motiva - US Chemical Safety and Hazard Investigation Report – Refinery Incident - Motiva Enterprises LLC, Delaware City Refinery, Delaware City, Delaware, July 17 2001; issue date October 2002.
- Phillips 66 Company Houston Chemical Complex Explosion and Fire, Implications for Safety and Health in the Petrochemical Industry, A Report to the President by US Department of Labor Occupational Safety and Health Administration, April 1990.
- Public Inquiry into the Piper Alpha Disaster, Volumes 1 & 2, November 1990, HMSO Publications Centre ISBN 0-10-113102-X
- Skikda: "The Incident at the Skikda Plant: Description and Preliminary Conclusions," LNG14, Session 1, March 21, 2004, DOHA-Qatar, Sonatrach.

<u>Ammonium nitrate:</u>

- Accident on the 21st of September 2001 at a factory belonging to the Grande Paroisse Company in Toulouse, Report Of The General Inspectorate For The Environment Ministry For Regional Development And The Environment (France) , 24 October 2001
- Damages of the Toulouse Disaster 21st September 2001, Nicolas Dechy and Yvon Mouilleau, 11th International Symposium Loss Prevention 2004, Prague.

<u>Dust explosions:</u>

- "Courrières 10 mars 1906 : la terrible catastrophe" by Bruno Vouters (2006). Lille: Editions La Voix du Nord. ISBN 2-84393-100-2
- US Chemical Safety Board report 2003-07-I-NC Sept. 2004 (West plant)
- US Chemical Safety Board report 2003-09-I-KY February 2005 (CTA Acoustics plant)
- INERIS report "Explosion d'un silo de céréales, Blaye (33)", July 1998
- "Agricultural Dust Explosions in 1997 in the USA", February 27, 1998, University of Kansas State, Department of Grain.

<u>Reactive chemicals:</u>

- "BPS Inc., West Helena Arkansas" EPA/OSHA chemical accident investigation report, EPA 550-R99-003, April 1999.
- "Napp Technologies Inc., Lodi New Jersey" EPA/OSHA chemical accident investigation report, EPA 550-R97-002, October 1997.

- "Morton International Inc.", Chemical Safety Board investigation report 1998-06-I-NJ.
- "The explosion at Concept Sciences: hazards of Hydroxilamine", Chemical Safety Board case study 1999-13-C-PA, March 2002.

Warehouses:

- "Sherwin-Williams Paint Warehouse Fire, Dayton, Ohio (May 27, 1987) – With Supplement on Sandoz Chemical Plant Fire, Basel, Switzerland," United States Fire Administration, Federal Emergency Management Agency, Technical Report Series, 1987.
- "Flammable Liquid Warehouse Fire, Dayton, Ohio, May 27, 1987," NFPA Fire Investigations Report.

4. BP Process Safety series, IChemE:

• Hazards of Water,	2004 Edition	9780852954669
• Hazards of Air & Oxygen,	2004 Edition	9780852954676
• Hazards of Steam,	2004 Edition	9780852954683
• Safe Furnace & Boiler Firing,	2005 Edition	9780852954720
• Hazards of Trapped Pressure & Vacuum,	2005 Edition	9780852954751
• Safe Handling of Light Ends,	2005 Edition	9780852954782
• Confined Space Entry,	2005 Edition	9780852954799
• Hotel Fire Safety,	2005 Edition	9780852954805
• Hazardous Substances in Refineries,	2005 Edition	9780852954829
• Safe Ups and Downs for Process Units,	2006 Edition	9780852955024
• Hazards of Electricity & Static Electricity,	2006 Edition	9780852955031
• Liquid Hydrocarbon Tank Fires,	2006 Edition	9780852955048
• Hazards of Nitrogen & Catalyst Handling,	2006 Edition	9780852955079
• Safe Tank Farms & (Un)loading Operations,	2006 Edition	9780852955093
• Control of Work,	2007 Edition	9780852955147
• LNG Fire Protection and Emergency Response	2007 Edition	9780852955154
• Hazards Of Oil Refining Distillation Units	2008 Edition	9780852955222

Implementing Process Safety - Kuwait Petroleum Corp

Ujwal Ritwik *CRSP, QEP, REM, CPSMA*
Kuwait National Petroleum Co.
Kuwait
u.ritwik@knpc.com

Sami Al-Yakoot
Kuwait Petroleum Corporation
Kuwait
s.yakoot@kpc.com.kw

Fadhel Al-Ali
Kuwait Petroleum Corporation
Kuwait
f.ali@knpc.com.kw

ABSTRACT

Kuwait Petroleum Corporation (KPC) is the holding company for a host of subsidiary companies engaged in upstream and downstream businesses in oil and gas industry mainly in Kuwait. Refineries, petrochemicals and part of exploration and production facilities have high risks pertaining to process safety. In want of explicit legal requirements, it was left to subsidiaries to adopt and negotiate process safety issues. Several high visibility incidents, at the turn of century, forced KPC and subsidiary companies to look beyond optional compliance to some vague 'industry standards.' All operators understood the fact that process safety was more than looking good, very business survival was at stake.

A spurt of initiatives followed, starting with OSHA PSM standards 29 CFR 1910.119, API RP 750, adopting specifics of Du Pont, BP, Chevron and like. The recently published Baker Panel Report recommendations and CCPS Risk Based Process Safety are the latest in the list of wishes, rendering the implementation increasingly confused. It has been a real challenge for the practitioners and managers what to do, where to start and will it really make difference to the bottom line.

In proportion to its risks, company responsible for operating three refineries, Kuwait National Petroleum Co. (KNPC), a subsidiary of KPC, has been pouring much on process safety. The efforts have been immense, although diffused with some veritable pockets of excellence. Other subsidiaries, one operating petrochemicals plants and the other engaged in exploration and production, too are catching up. As predictable, the results of the past six plus years have been mixed. KPC is working actively to implement minimum common requirements on process safety among all its subsidiaries.

1. KPC PROFILE

Kuwait Petroleum Corporation (KPC), one of the top ten oil energy corporations, is the state-owned conglomerate responsible for Kuwait's hydrocarbon interests throughout the world. It supplies oil and gas by exploring for, producing, refining, transporting and marketing these natural resources both at home and internationally. Established in 1980 in order to bring together all state-owned elements of the Kuwait oil sector under one corporate umbrella, today, KPC overseas a fully-integrated industry with operations spanning six continents. With total assets of more than 11 billion Kuwaiti dinars (US$37 billion) and annual revenues of almost 9 billion Kuwaiti dinars (US$28 billion), KPC is a major contributor to the revenues of the State of Kuwait.

Kuwait Petroleum Corporation operates through a series of specialized subsidiaries in Kuwait and across the world, with activities encompassing all aspects of the hydrocarbon industry. All of these subsidiaries are fully owned by KPC.

1.1 Subsidiaries

Under KPC, Kuwait Oil Company (KOC) operates for exploration and production of oil and gas in the State of Kuwait, for the responsible management of Kuwait's domestic reservoirs, and export of crude oil. It has recently executed projects to attain substantial production capacity increase.

Kuwait National Petroleum Company (KNPC) operates three modern petroleum refineries at Shuaiba, Mina Al-Ahmadi and Mina Abdulla. Major part of its products, gasoline, naphtha, kerosene, diesel, is exported. KNPC also locally markets the refined petroleum products and manages and supplies several local petrol station network.

Petroleum Industries Company (PIC) produces liquid ammonia, urea fertilizer, sulfur and polypropylene. It also holds a 45% interest in the Kuwait based Equate Petrochemicals Co., which produces ethylene, polypropylene, and ethylene glycol.

Other subsidiaries include Kuwait Oil Tanker Co. (KOTC) which operates a fleet of modern crude, product and gas carriers; Kuwait Aviation Fuelling Co. (KAFCO) to service aircrafts; Kuwait Foreign Petroleum Exploration Company (KUFPEC) manages the KPC's foreign upstream activities with interests in oil and gas fields world wide; Kuwait Petroleum International (KPI) directs refining and marketing throughout Europe and operates wholesale, retail and aviation fuel sales network in Asia; Kuwait Gulf Oil Co. (KGOC) controls and manages the Kuwait share of natural resources of offshore and onshore divided area between Kuwait and Saudi Arabia; Oil Sector Service Co. (OSCO) undertakes the responsibility of providing security and support services for KPC and its subsidiaries; and Oil Development Company (ODC) is formed to develop oil fields in Northern Kuwait and increase their production.

2. HISTORY OF OIL IN KUWAIT

In 1938, oil was first discovered in Kuwait by Kuwait Oil Company (KOC), a London-based joint venture of the Anglo-Persian Oil Company (now BP) and Gulf Oil (now Chevron Corporation). On June 30, 1946, His Highness Sheikh Ahmed Al-Jaber Al-Sabah, turned a silver wheel to commence the flow of the first Kuwait oil exports.

Over the next three decades, extensive developments occurred both in the upstream and downstream elements of the industry. The nationalization process started in the year 1974 was completed in 1979 marked a key turning point for Kuwait's oil industry. On January 27, 1980, Kuwait Petroleum Corporation was formed, which brought together all elements of the industry under one holding company. Throughout the 1980's, vigorous expansion of refineries was executed.

3. REGULATORY REGIME

3.1 Health & Safety

Historically Kuwait's health and safety laws are derivative of United Kingdom's age old Factories Act. Employers are obliged to take precautions to protect their employees against physical hazards and occupational diseases work.

Accidents: If an employee is injured at work, the employer must report the matter to the local police station and the Ministry of Social Affairs and labor (MSA&L). The injured employee has the right to treatment, at the employer's expense

Compensation: An employee has the right to compensation for work-related injuries without having to prove that the employer was at fault,

3.2 Concerned Statute

In Kuwait, the enforcement of occupational safety and health is entrusted to the Ministry of Social Affairs and Labor. Occupational safety and health policies are also implemented by other ministries and governmental bodies such as the Ministry of Public Health, Ministry of Commerce and Industry, and the Fire Department. Legislation is covered by the Labor Law (38) 1964; Act (28) 1965 which protects workers in the petroleum industry and the Ministerial Decision (57) 1982 for the protection against benzol; Ministerial Decision (114) 1996; Ministerial Decision (17) 1973; Ministerial Decision (56) 1982 for machinery guarding and Ministerial Decision (45) 1979 on chemical safety.

As an oil producer, Kuwait is concerned with chemical safety and has legislation that addresses preliminary and periodic medical examination of petroleum workers. There are also provisions aimed at the protection of workers from exposure to chemicals and at emergency plans for handling chemical leakages and industrial fires. Such plans include specific measures to handle chlorine, hydrogen and ammonia leakage in major refining and petrochemical areas such as the Shuaiba region.

Whilst inter-ministerial coordination in chemical safety exists between the Ministries of Labor and Health and the Shuaiba industrial authority, occupational safety and health is coordinated at the national level in the form of joint committees between several ministries.

Establishment of Environment Public Authority (1995-96) and promulgation of its by-laws (2001) brought in much needed standards pertaining to protection of work and external environment. Environmental impact assessment of new projects is covered under these.

Kuwait Law 56 of 1996 - PAI Article 29 Clause 11- Specifies adequate and protective measures for the safety of public utilities and properties, and supervising industrial safety and protection against fire in relation to all existing installations within its jurisdiction.

The prevention of major industrial accidents in the petroleum sector is the responsibility of a special unit within the Ministry of Labor: Article 9 of the Ministerial Decision (14) 1996 requires the notification of major accidents within 24 hours of their occurrence.

Not all the above laws and by-laws, however, cover process safety requirements comparable to US OSHA PSM 29CFR1910.119 or Seveso Directives of European Union.

3.3 Regulations for the Conservation of Petroleum Resources

The 1989 Petroleum Conservation Regulations provide for the implementation of Law No. 19 of 1973, setting authorization and reporting requirements along with health, safety and environment standards according to which upstream and downstream petroleum activities must be conducted. Issued initially in 1975 and based on the then current Canadian Petroleum Conservation Regulations, Technical Affairs Dept in the MOE is assigned the task of implementing the Regulations.

A few years back MOE has revised the subject regulations. A steering committee comprising of representatives from the oil sector was formed. An international consultant "M/S IHS Energy" was hired and a revised document was prepared and submitted to the Ministry of Energy in 2005. It is waiting approval.

3.4 Process Safety

Traditionally oil and gas companies in Kuwait have been buying technology from US, UK and Europe who as a good practice follow their respective country legislation on personnel and process safety. With prevalence of US contracts, after mid nineteen eighties, the engineering standards have generally followed NFPA, API, ANSI, ASME and other US standard making bodies. More focused process safety issues began getting addressed in late eighties. US OSHA PSM was noticed at KNPC in 1992, when localized implementation of some identified issues began.

Elevating a checklist based review of modifications a formal PHA administration was established in 1996. Later he same year the first structured 'management system' implementation was undertaken at KNPC in 1996. With the help of consulting arm of then Conoco-DuPont a 'Process for Continuous Safety Improvement' (PCSI) was initiated. It was top-management supported and saw several key elements implemented. The program, had 14 elements, grouped in four divisions, many named same as OSHA PSM elements. After a few months, the initial euphoria waned and barring a few elements, the system generally remained on paper and topic of discussions at conferences and seminars. Nonetheless, a beginning was made and pockets of initiatives were taken by some well-intended people around the company. Mechanical Integrity feature, which was already a strength, was joined by new elements such as Contractor, Incident Investigations, Audit and Metrics in the list of success stories.

In the corporation as a whole, the focus was still on personnel safety and major hazard risk analysis was not structured and unified.

4. CATASTROPHIC INCIDENTS

At the turn of the century, two catastrophic incidents at KNPC and one at KOC shook the very foundation of Kuwaiti industrial safety culture. Following is the recount of these two and some more major industrial catastrophic incidents Kuwait has seen in the last 25 years.

4.1 Shuaiba Refinery, the starting point of events

On 16 June 2000, an accident occurred described then as the worst accident to take place in the petroleum sector occurred at Shuaiba Refinery, leading to the death of two employees because of inhaling the poisonous hydrogen sulphide gas. While inspecting the unit they fell from the top of a tower of 21 meters height. The accident occurred at the hydrocracking unit, which have been out of service since the Iraqi invasion. Following the accident, the minister of oil then, Sheikh Saud Al Nasser Al Sabah, formed an investigation committee.

4.2 Mina Al-Ahmadi Refinery accident

Shortly after Shuaiba Refinery accident, Mina Al-Ahmadi Refinery accident took place which led to the death of five persons and 50 others were injured. An explosion occurred on Tuesday morning 25June, 2000 when employees were attempting to isolate a leak on a condensate line between the NGL plant and the refinery. Three crude units were damaged and two reformers producing gasoline were destroyed, with an estimated loss of $400 millions. The fire was extinguished approximately nine hours after the initial explosion. Initial investigation into the loss indicates a lack of inspection and maintenance of the condensate line.

4.3 Al Roudhatain explosion

On January 31, 2002, an explosion and fire at an oil-gathering center near Kuwait's northern Raudhatain oil field killed four workers, while cutting the field's oil and gas output sharply. The explosion, which reportedly was caused by a leak at a major oil pipeline, knocked out three critical gathering centers (the 280,000-bbl/d GC-15, 120,000-bbl/d GC-23, and 200,000-bbl/d GC-25), an electrical substation, and a natural gas booster station (BS-130). The accident also caused injury to 9 workers and 39 firefighters of the oil sector and general fire fighting.

4.4 Other Major Incidents

Time and again, Kuwait's safety record was put under the spotlight. There were occurrences of 11 accidents of different sizes and human and material effects within 24 months from February 2000 to February 2002 at the rate of one accident every two months, which is a high rate compared to other oil countries.

Among major incidents before 2000 were -

- LPG storage arson fire in Mina Al-Ahmadi 1986
- Tank farm fire in Shuaiba refinery 1987
- Leak in Mina Al-Ahmadi pipeline polluting several kilometers of coastline 1998

4.5 Tank Farm Fire 1981

The history of catastrophic incidents in Kuwait also records an oft referred tank farm fire. On August 20, 1981 Shuaiba refinery tank farm fire destroyed eight tanks and damaged several others resulting in an estimated loss of $70 millions. It appears to have originated at a pump manifold within the common dike serving six 160,000-barrel floating roof tanks containing petrochemical grade naphtha. Naphtha was being pumped into one of the tanks when the initial explosion and fire occurred. About one-half hour into the fire, the seal of the first tank caught fire. This spread progressively, eventually involving all the six tanks in the group. In spite of strong fire fighting attack that lasted several days flame front and radiated heat caused the fire to spread into an adjoining row of five more tanks. Water curtains set up between the tank groups and nearby process units at the refinery and petrochemical plant proved quite effective.

4.6 Kuwaiti oil fires

No list of major man made catastrophes is complete without mention of one of the biggest act of arson mankind has suffered in the recent times. In February 1991, nearly 800 oil wells were set ablaze by the retreating Iraqi army and the fires were not fully extinguished until November 6, 1991. The fires consumed an estimated six million barrels of oil daily. The resulting fires burned out of control because of the dangers of sending in firefighting crews. Land mines had been placed in areas around the oil wells, and a military cleaning of the areas was necessary before the fires could be put out. Eventually, privately contracted crews extinguished the fires. By that time, however, the fires had burned for months, causing widespread pollution. The byproducts of the petroleum burn caused pollution to the soil and air, and the oil fires have been linked in the popular imagination with what was later called Gulf War Syndrome. Whether the oil fires, by chemical attack, or other causes have caused this syndrome has not been determined, and the long-term environmental effects of the fires have yet to be fully understood.

5. INVESTIGATIONS AND IMPROVEMENTS

5.1 KNPC

Following the incidents at KNPC in June of 2000, as expected, a number of investigations were carried out. Significant among them were company's internal investigation, ministerial investigation with the help of international experts, a parliamentary inquiry and an inquisition by Exxon experts into the methods and systems at KNPC. Spate of recommendations followed. The company laid down a consolidated program for implementing the recommendations. From the 58 recommendations 203 action items emanated the majority of which were implemented at a cost of around 91 million Kuwaiti dinar.

After the accident, KNPC acted upon a number of recommendations the most significant of which was forming the higher permanent committee for safety and making the role of the refineries managers and senior officials more effective by granting them administrative, financial and control authorities. The safety, health and environment department was restructured to escort the new systems, laws and programs that were adopted. New equipment were purchased which comprised the latest safety systems which the safety, health and environment department trained the company staff on using them.

KNPC also embarked upon a concerted effort to build a new safety, health and environment management system (SHEMS). Developed with some support from the consulting company DNV the new system was a step further from the earlier PCSI. Again consisting of 14 elements, though somewhat different, a great emphasis was placed on the program implementation.

5.2 KOC

Like its sister concern KNPC, Kuwait Oil Company, KOC, following Al Roudatain incident, adopted a new HSE policy. It updated its Health, Safety & Environment Management System document and carried out comprehensive program for its implementation. On 3 March 2003, the Company issued HSEMS guide along with the HSE policy, Quality Cycle Model, HSE Commitments and HSE System Characteristics. The Company established twelve HSE System elements and eighty five expectations to develop and implement the HSEMS. These elements integrate strategic HSE management into Company's operations.

5.3 PIC

Liquid ammonia is a material of great process safety concern to the Petroleum Industries Company. Even without getting hurt by any major accident, as a proactive measure, it initiated implementing a Du Pont based safety improvement program in 1999. It covered personnel safety and behavioral safety aspects vigorously. The core of the program was Du Pont's Process Safety Management, an adaptation of US OSHA requirements. By 2002, the program was successfully implemented.

5.4 Governmental Reaction

On the governmental radar, safety of petroleum installations got the highest priority. People at high places realized extraordinary situation demanded extraordinary measures. The very continuance of operations, safety of population and employees and credibility of the industry was at stake. The then acting minister of oil, issued instructions to the chairmen of the boards of KPC subsidiaries, for the necessity of addressing concern to health, safety and maintenance and placing them on top of the agenda of all board meetings. This required also to submit a periodical report to the minister on the works accomplished and recommendations adopted.

6. KPC HSE MANAGEMENT SYSTEM

6.1 Initiation

In order to provide directions to and coordinate efforts of its subsidiary companies KPC in late nineties had began developing a core team on safety, health and environment. By 2002, it had a full time deputy managing director and one manager for health, safety and environment. Soon due to the under mentioned reasons it realized the need to cover the legislation vacuum -

- Recent accident history,
- Tragic and unnecessary loss of lives,
- Asset and business losses in hundreds of millions of dollars,
- Lack of effective, comprehensive and integrated HSE management system, and
- Corporate priority to improve process safety performance.

It commissioned DuPont to carry out the task. It was a structured project approach beginning with a survey of systems and processes at the subsidiary companies in 2003. The visiting consultants shared their observations with the respective subsidiaries but the main thing they brought back with them was data on subsidiaries' status and individual needs. What followed was writing of fifty-nine standards and guidelines covering various areas of health, safety and environment. This was named KPC Health, Safety and Environment Management System (HSE MS).

Figure1: MS model

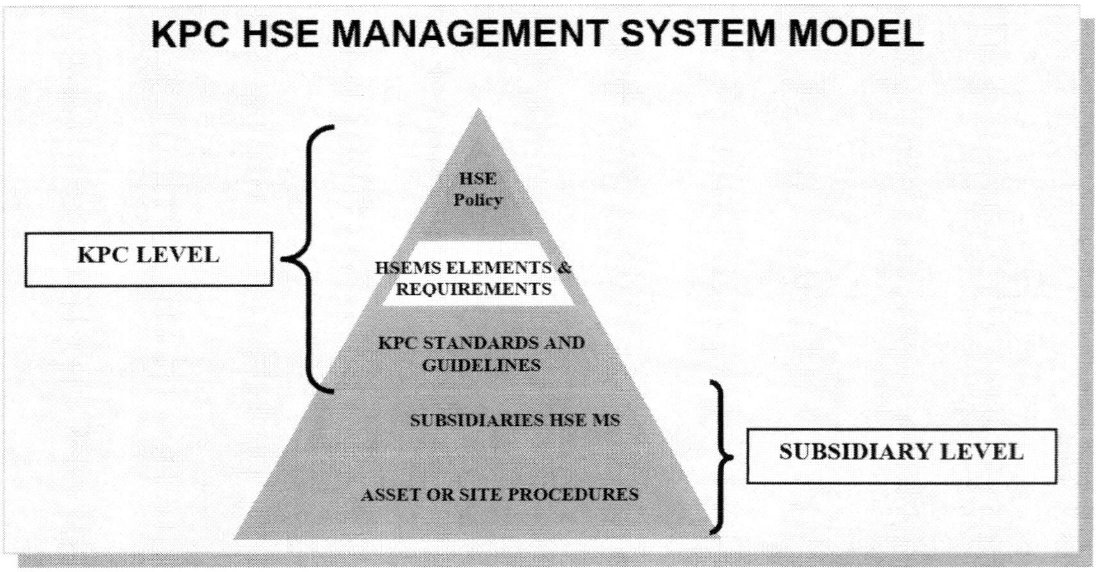

The KPC top management made its commitment visible by –

- Close follow up of the project
- Financial support
- Additional manpower
- Increased training

6.2 Elements of KPC HSE MS

The new KPC Health, Safety and Environment Management System HSE MS has following sixteen elements. KPC pursued new strategic direction on HSE front to comply with local HSE standards and regulations and application of most appropriate and best international HSE standards. As is evident it has heavily borrowed from OSHA PSM 29CFR1910.119, and other similar industry standards such as API.

Management Commitment and Responsibilities
Integrated HSE Organisation
Challenging HSE Goals and Objectives
High Standards of Performance
Continuous Training and Performance
Effective Two-way Communication & Motivation
Behaviour and System Audit
Incident Prevention through Effective Investigations
Management of Change
Managing Contractors
Mechanical Integrity and Quality Assurance
Hazard Identification and Risk Management

Emergency Preparedness and Response
Environmental Management
Occupational Health Management
Legal Compliance and Other Requirements

The 16 elements are grouped in subject areas such as Corporate Polices, Process Safety, Environmental Management, Occupational Health, Incident Management, Emergency Response etc. Altogether 59 documents are written to support the system.

Figure 2: KPC HSE MS Document System

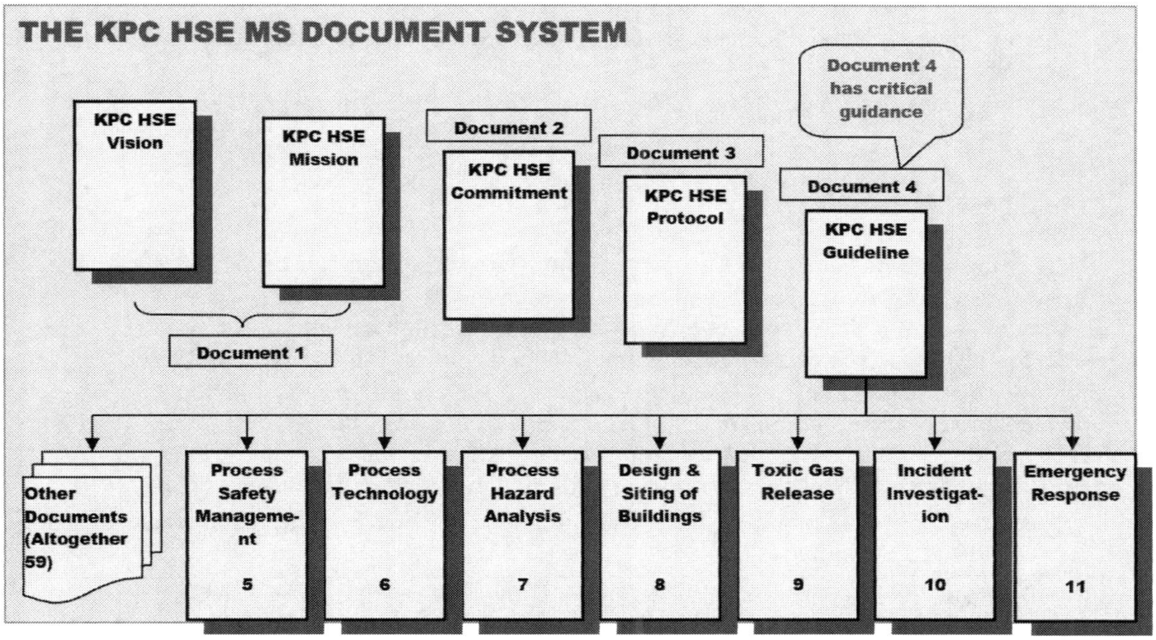

6.3 KPC HSE MS Working Groups

Two Working Groups were formed to accomplish the HSE MS Project:

- HSE MS Project Steering Committee
- HSE MS Project Implementation Committee

Similarly, two Working Groups are formed to oversee the implementation of HSE MS at the Subsidiaries:

- HSE MS Implementation - Steering Committee
- HSE MS Implementation in Subsidiaries -Implementation Committee

6.4 Project Challenges

- To deal with Subsidiaries having varied interests like Exploration & Drilling, Refining, Petrochemical, Tanker Operations, inside Kuwait and overseas operations with interests in extraction & drilling, refining etc.
- Garnering buy-in and support from the various subsidiaries for the project
- Approving the standards after catering to all the comments and suggestions from the Subsidiaries
- Taking the Subsidiaries on board explaining the tangible and intangible benefits a Corporate HSE MS has on the Subsidiaries
- To deal with some subsidiaries that had their own HSEMS Systems in place
- The time challenges and resource constraints
- Coaching and training that has to be done to enhance the skills of staff and employees
- Convincing the top and line management on their safety accountabilities and responsibilities
- Creating a new dynamic network and team-working processes inside KPC organization

6.5 Implementation Status

After the standards and guidelines were formulated the next challenge, far more difficult, was to implement them at Subsidiaries. Each Subsidiary is a full fledged company in its own right, having its own board, financially independent, profit making, and having history of safety management practices and culture. While KNPC and KOC were only a few years in their new safety management system, PIC had completed its implementation of the program. On the other extreme of the spectrum were KOTC, KAFCO, and KGOC, without a system altogether or just started as a company. KPI and KUFPEC with international operations were still to fathom implications. As of date

- KPC HSE MS Standards have been approved by the KPC Management and MDs of Subsidiaries
- Implementation signposting (gap identification) has been done at most of the subsidiaries
- HSE MS Implementation status audit is being done
- Implementation Sub-committees have been formed on
    - Process Safety Management
    - Environment
    - Occupational Health

6.6 Future Tasks for the HSE MS Project at KPC

The future tasks for completing the implementation of the KPC HSE MS are:

- Sub-committee tasks for implementation
- Implementation tasks by Subsidiaries
- Preparing the audit protocol
- Initial audit by KPC Team
- Final implementation audit by third party

7. PROCESS SAFETY MANAGEMENT

KPC process safety management program is directed toward preventing serious, process-related chemical and petrochemical incidents that might affect plant personnel, off-site communities, or the environment, or result in significant loss of property or business. It involves the application of systems and controls to chemical and petrochemical manufacturing processes in a way that hazards are identified, understood, and controlled so that process-related injuries and incidents are prevented.

As understood at KPC four key steps involve the implementation and sustainability of an effective PSM program:

Establish a safety culture.
Provide management leadership and commitment.
Implement a comprehensive PSM program.
Achieve operational excellence through operational discipline.

7.1 High Hazard Processes (HHP)

The KPC PSM guideline document (Document 5) describes the principles and essential features of PSM. In want of any local regulation or description of hazard threshold, the guideline classifies refineries, petrochemicals and petroleum exploration and processing as High Hazard Processes (HHP).

As defined at KPC HHP is any activity manufacturing, handling, storing, or using hazardous substances that, when they are released, ignited, or intentionally combined, can result in death or irreversible human health effects, significant property or environmental damage, or off-site impacts due to acute toxicity, flammability, explosiveness, corrosiveness, thermal instability, latent heat or compression. Local regulations may add additional considerations for processes in this category.

Examples of higher-hazard substances include large quantities of pressurized fuel gases, or flammables, when their use poses a potential for causing any of the effects listed above. Exclusions: Excluded from coverage under the KPC standard are simple enclosed transportation, distribution, and material handling generally associated with warehousing activities.

Application to HHPs

Each site operating an HHP is required to develop, a PSM program suited to its specific parameters. Such guideline / standard should embrace all the principles and essential features of corporate KPC PSM.

7.2 Operational discipline

A major requirement in the PSM program is maintaining a high degree of operational discipline (OD). OD translates the PSM paper program, knowledge of the process hazards, operating conditions, and layers of protection into tangible, real performance results to eliminate incidents and protect employees, the environment and the community.

Some elements or characteristics that are essential to achieving and maintaining a strong sense of OD within the PSM program for a site or Subsidiary are:

Leadership by example
Sufficient and capable resources
Employee involvement
Active lines of communication
Strong teamwork
Pride in the organization

Sites are required to implement and maintain programs and systems for HHPs to achieve and maintain a high degree of operational discipline for all PSM elements in a manner that supports business and operating objectives.

7.3 Support resources

The PSM program requires support structure both at KPC and Subsidiaries levels.

KPC 'Process Safety, Fire protection & Emergency Response Department' - to consult on PSM issues and train, in conjunction with the KPC Training Centre, personnel in Process Hazard Analysis and release modeling.

KPC Legal Department - to answer questions regarding regulatory intent, interpretations, and enforcement activities, and to assist Subsidiaries and sites undergoing regulatory audits.

KPC Risk Management Department - to assist in developing comprehensive assessments of business risk and relative risk.

The Subsidiary PSM function maintains staff and consulting resources to help resolve issues related to mandatory regulations, standards, and policies.

Each Subsidiary maintains a PSM coordinator to assist sites in PSM activities where appropriate, and to represent the Subsidiary on KPC's PSM Leadership Team.

7.4 KPC PSM Guidelines and Standards

The core of KPC PSM is built over OSHA PSM 29CFR1910.119 kernel. However, there is an extra emphasis on developing internal competency. Taken from the books of Du Pont an interesting inclusion is 'Management of Subtle Changes' which is in addition to classical OSHA element of 'Management of Change'.
The PSM group of documents are comprised of following documents.

- Doc. 4.9 – HSEMS Guideline: Management of Change
- Doc. 4.11– HSEMS Guideline: Mechanical Integrity & Quality Assurance
- Doc. 5 – Process Safety Management
- Doc. 6 – Process Technology
- Doc. 7 – Process Hazard Analysis
- Doc. 8 – Design & Siting of Buildings
- Doc. 9 – Protection of Personnel from Toxic Gas Release in Emergencies
- Doc 10 - Incident Prevention through Effective Investigations
- Doc 11- Emergency Preparedness and Response

- Doc. 28 – Operating Discipline
- Doc. 31 – Identification and Color Coding of Pipes

Several other of the 59 documents, at least in parts, feed into the requirements of the PSM program.

8. KPC PSM DOCUMENT

8.1 Main Document

The main document (KPC HSE MS Document 5) describes the PSM implementation requirements in detail. This guideline provides information to help management implement procedures to control the hazards associated with processing to levels that are consistent with the KPC commitment to Health, Safety and the Environment (KPC HSE Document 2). The procedures are intended to protect personnel from serious injury and prevent significant environmental harm, property damage, and business losses in support of the principles outlined in other KPC HSE Documents.

The task of process safety management (PSM) implementation is complex because it crosses over many functional areas of an organization, including engineering, manufacturing, construction, maintenance, and training. An integrated, systematic approach to implementing process safety elements across these functions in a consistent manner is important to manage changes and sustain strong PSM-related performance.

In the KPC PSM, the fourteen elements are there again, with some differences.

A.1 Process technology
A.2 Process hazards analysis (PHA)
A.3 Operating procedures and safe work practices
A.4 Management of change technology
A.5 Personnel training and performance
A.6 Contractor safety and performance
A.7 Management of change personnel
A.8 Incident investigation and communication
A.9 Emergency planning and response
A.10 Auditing
A.11 Quality assurance (QA)
A.12 Mechanical integrity (MI)
A.13 Pre-start-up safety review
A.14 Management of "subtle" changes

8.2 Principles and essential features of PSM Elements

The intent of the Guideline is to act as a methodology that potentially will lead to excellence in PSM. It is essential to remember the inter-connectivity of these elements and the advantages and insights that come from correctly applying this process in a controlled repetitive manner.

Technology

1. Process technology (PT)

The process technology package provides a description of the chemical or petrochemical process or operation. It provides the foundation for identifying and understanding the hazards involved, which are the first steps in the PSM effort. The process technology package consists of three following parts:

- Hazards of materials
- Process design basis
- Equipment design basis

2. Process hazards analysis (PHA)

The guideline prescribes use of structured process hazard analysis (PHA) to identify, evaluate, and develop methods to control significant hazards associated with HHPs. These hazards generally represent the potential for fires, explosions, and/or the release of toxic materials. PHAs must use an organized, methodical study approach; seek to achieve a multidiscipline consensus on hazard control; and document results for future use in follow-up, emergency planning, and training of personnel involved in operating and maintaining the process.

3. Operating procedures and safe work practices

The operating procedure provides a clear understanding of the detailed operating parameters and limits for safe operation for those who are operating the process. It also clearly explains the safety, health, and environmental consequences of operation outside process limits and describes the steps to be taken to correct and/or avoid deviations. Safe work practices provide a carefully planned system of procedures and/or permits involving checks and authorizations prior to performing non-routine work in process areas.

4. Management of change - technology

Changes to documented process technology (e.g., hazards of materials, equipment design basis, and process design basis) potentially invalidate prior hazard assessments. Accordingly, all changes to the documented process technology are subjected to the same rigorous review that is applied to new processes.

Personnel

5. Personnel training and performance

Site-specific training programs are an essential element of PSM throughout each segment of the manufacturing facility. This training provides all operations and engineering personnel with the appropriate information, tools and understanding to conduct tasks in alignment with the operating procedures and process technology. There is also emphasis on demonstrated performance.

6. Contractor safety and performance

It is essential that all tasks be completed safely, in accordance with established procedures and/or safe work practices, and consistent with the principles and essential features of PSM-whether the tasks be completed by site employees or by contract employees.

7. Management of change - personnel

Recognizing that people are the essential ingredient inter-woven through all the elements of PSM, it is important to maintain a minimum level of staffing.

(1) Specific direct process experience

(2) Knowledge and skill in managing process safety.

It is essential that personnel changes at all levels be controlled to pre-established criteria so that minimum levels of experience and knowledge are maintained.

8. Incident investigation and communication

Serious incidents and serious potential incidents (including near-miss occurrences) recur unless key factors are identified and corrected. Thorough and persistent reporting and investigation of all serious and serious potential incidents, coupled with actions directed at prevention of recurrence, continually improve safety.

9. Emergency planning and response

In-depth planning for potential emergencies is essential for an effective response by site personnel working in close conjunction with supporting community emergency response organizations. The important products of these efforts are

(1) Mitigation of the impact on personnel, environment, and facilities

(2) Prompt control of the emergency.

10. Auditing

Audits are important tools in the establishment, measurement, maintenance, and continuous improvement of process safety performance. Audits compare performance vs. established standards. Proper auditing provides a process that nourishes and reinforces correct systems, behavior and work practices while identifying opportunities for improvement.

<u>Facilities</u>

11. Quality assurance (QA)

The QA effort bridges the gap between design specifications and the initial installation.

QA efforts help verify that process equipment meets the following standards:

- Fabricated in accordance with design specification
- Delivered to the proper location
- Assembled and installed properly

12. Mechanical integrity (MI)

The MI element covers the life of the facility from the initial installation through dismantlement. MI focuses on maintaining and continuously improving a system's integrity to contain hazardous substances throughout the life of the facility. It addresses such topics as the following:

- Maintenance procedures
- Training and performance of maintenance personnel
- Quality control procedures
- Equipment tests and inspections, including predictive and preventive maintenance
- Repairs and changes
- Reliability engineering
- Documentation

The predictive and preventive maintenance topics are important to maintain reliable and incident free operation. Such programs help prevent premature failure and maintain

13. Pre-start-up safety review

The pre-start-up safety review (PSSR) provides a final checkpoint for new and modified equipment to confirm that all appropriate elements of process safety management have been addressed satisfactorily and the facility is safe to start up.

14. Management of "subtle" changes

Subtle changes in the field can lead to and have led to catastrophic events. All changes, including those within the documented process technology that are not "replacement-in-kind," should receive appropriate review and authorization.

8.3 Special considerations

Though specifically applicable to high hazard processes, KPC PSM also intends to cover certain other activities.

Warehouses

Warehouses are a good example of where a PSM program, within the framework of KPC HSE MS Doc 5, may be specifically tailored to a storage application. For example, an extensive process technology (PT) is not expected for a warehouse, nor are process and instrument diagrams (P&ID), vessel calculations, or process limits. However, important PT and other PSM elements for a warehouse include the following:

Hazards of materials
Facility and storage design basis
Equipment design basis
Emergency planning and response
Mechanical integrity

Highly toxic materials

Sites that handle highly toxic materials, for example ammonia, should pay particular attention to concepts like inherently safer design, emergency containment and personnel alert and protection systems.

9. IMPLEMENTATION ISSUES

9.1 Management responsibilities

For effective implementation of the system, specific roles and responsibilities are defined. Line management in KPC, Subsidiaries and functions have the responsibility to implement the document.

The guideline requires KPC as a corporate entity, its CEO and the Board to devote time to clearly demonstrate the value that an effective PSM program brings to the business success and help ensure proper resources are committed to PSM as a needed technical discipline throughout KPC and its Subsidiaries

Technology and Skills Management

Under the KPC HSE Department a dedicated department is formed with a 'Process Safety, Fire protection & Emergency Response Department' Manager. He leads a corporate team at KPC to provide consultation and guidance on a range of PSM elements.

With regard to PHAs, the department is assigned to make the following services available:

- Process hazards analysis resourcing and documentation
- Process hazards analysis training
- Facility siting evaluations
- Consequence analyses
- Process risk analyses
- Human factors analyses

Other PSM-related services and consultation assistance required from the department include - explosion and fire analysis, on-site incident investigations, hazardous service equipment design guidance and regulatory and code compliance assistance, Publish and maintain KPC PSM standards, provide training in the principles and essential features of PSM, provide PSM auditor training and maintain a list of qualified PSM lead auditors, and maintaining content and structure of the KPC PSM intranet website .

The Process Safety, Fire protection & Emergency Response Department is further responsible for formation and supporting across KPC and subsidiaries

Process Safety Technology Leadership Team

Process Hazards Analysis Technology Leadership Team, and

Consequence Analysis Technology Leadership Team.

PSM Networking

The KPC HSE Department and the Process Safety, Fire protection & Emergency Response Department Manager are required to provide leadership and foster the PSM networking within and among the Subsidiaries.

9.2 Subsidiary Responsibilities

Subsidiary leadership and the technology function have the responsibility to assign adequate resources to PSM. There are widely differing structures within the Subsidiaries. The following structures have been interpreted as indicative of structures known to work. Where other restraints are priorities, the structures could be modified to suit but the responsibilities must be discharged.

9.2.1 Subsidiary Managing Director

The Subsidiary Managing Director must ensure proper resources are committed to

- Demonstrate overall management leadership and commitment.
- Ensure that adequate attention is paid to PSM during change over activities (e.g., forming joint ventures, making acquisitions, contract operations/toll manufacturing, long-term standby processes, and process phase-out).

9.2.2 Demonstrate overall management leadership and commitment

Management leadership and commitment are basic and essential to successful management of process safety. Effective management leadership shall include the following activities:

- Establish Subsidiary PSM policies and guidelines Commit resources
- Establish clear accountability for performance against specific process safety goals
- Verify compliance with PSM policies and guidelines and implement appropriate corrective actions
- Participate personally in activities that visibly demonstrate a commitment to PSM.
- Evaluate the need for and establish where justified any special PSM technology committees

9.2.3 Designate a PSM Coordinator for the Subsidiary

Each Subsidiary is required to designate a PSM coordinator to -

- Assist line managers in fulfilling their PSM responsibilities
- Monitor operating unit PSM programs
- Serve on the KPC PSM Leadership Team.

9.2.4 Audit Subsidiary PSM programs

Subsidiary management, audit site PSM programs to help ensure they are complete, current, and of good quality.

9.2.5 Co-ordinate information exchange

Each Subsidiary need to establish procedures to appropriately disseminate information among sites within the Subsidiary as well as to other Subsidiaries.

9.2.6 Co-ordination in transfer of PSM responsibilities

Where two or more Subsidiaries are involved in a product line (e.g., manufacturer and distributor or supplier and consumer), each should share responsibility for PSM. While one Subsidiary (usually the producer) may possess most of the process know-how, the other Subsidiary (consumer, distributor at the manufacturing location) is responsible for making relevant process technology available and for properly managing process hazards.

9.2.7 Evaluate and apply PSM during changes in Subsidiary operations

Guidelines are designed to focus adequate attention on PSM during business-related decisions and activities that may result in substantive changes to Subsidiary manufacturing operations and processes (e.g., forming joint ventures, acquisitions, technology transactions, contract operations/tolling, long-term standby processes, and process operations phase-out).

As part of the due diligence process for any venture, the KPC and/or the applicable Subsidiary shall conduct and document an assessment of process safety hazards, liabilities and risks, and overall PSM program status.

For ventures in which KPC is not considering ownership of a majority interest, the KPC and/or the applicable Subsidiary should encourage the operating entity to adopt PSM programs and protection measures comparable to those described in this guideline.

9.2.8 Contract operations/toll manufacturing

Chemical and petrochemical operations involving hazardous substances are at times contracted out to non-KPC organizations. Contractors are required to establish and follow PSM procedures and practices that provide protection equivalent to that provided KPC and subsidiary employees doing similar work.

9.2.9 Long-term standby processes

Long-term standby, mothballing, of process equipment that has handled hazardous substances offers many opportunities for knowingly and unknowingly compromising the mechanical integrity of the process facilities. Prior to long-term standby, a PHA must be conducted to address safety issues.

9.2.10 Process phase-out

To protect against degradation of process safety, Subsidiary management shall ensure that knowledgeable personnel are retained throughout the phase-out and that all appropriate PSM activities are carried out until the process is fully terminated.

A shutdown procedure, analogous to start-up procedures commonly developed for new facilities, should be developed to provide adequate attention to all of the unique conditions that may develop during the phase-out (e.g., equipment decontamination, inventory monitoring and disposal, and customer service during the transition to new suppliers).

9.2.11 Designate a site PSM committee

Sites handling hazardous materials or operating hazardous processes must establish a PSM subcommittee as part of the site's Central Safety (HSE) Committee. The following activities shall be the primary responsibilities of this sub-committee:

- Document each process covered by this guidelines standards.
- Develop, document, issue, and update site PSM standards
- Provide training
- Audit the line organization for compliance with site PSM guidelines.
- Network with other sites, Subsidiaries and KPC to effectively share and leverage PSM expertise and knowledge.

9.2.12 Subsidiary PSM Competency Centre

Each Subsidiary is required to have a PSM Competency Centre, is responsible for the following:

- Chairing the Subsidiary PSM Leadership Team where justified
- Facilitating process safety technology leadership teams (if needed in addition to KPC teams)
- Publishing and maintaining Subsidiary PSM standards
- Conducting quality assurance (QA) assessments of PSM programs
- Collect and analyze internal and external PSM metrics for trends, key learnings, and appropriate corrective actions

9.2.13 Subsidiary Engineering

On projects or other work for which it is responsible, the Subsidiary's Engineering Department function must provide leadership for the incorporation of process safety concepts in the design and construction of new or modified facilities, with involvement and participation by the operating unit.

9.3 KPC PSM Committee Progress

A specially constituted PSM Sub Committee was the first of the sub committees to start working for implementation of KPC HSEMS in April 2007.

- A clear Terms of Reference (TOR) is agreed
- A high level roadmap is agreed (given below)
- Five PSM documents have been reviewed, separately and jointly by the committee members

Figure 3: High Level Road Map of KPC PSM Committee

HIGH LEVEL ROAD MAP OF PSM IMPLEMENTATION

Phase 1 (to complete by June 2008)
Review of KPC HSE MS PSM Documents
Identification of resources and training for PSM Committee members
Prioritizing PS requirements and identifying monitoring measures
Gap analysis by Subsidiaries
Review of Subsidiary gap analysis

Phase two (to complete by October 2008)
Action plan by Subsidiaries
Review Subsidiaries action plan

Phase 3 (to begin by March 2009)
PSM implementation by Subsidiaries
Auditing Subsidiary PSM
KPC PSM system review
Propose PSM enhancement measures (ongoing)

10. NEW ADVANCES

The new guidelines from Centre for Chemical Process Safety (CCPS) on Risk Based Process Safety was released in 2007, almost the same time the Baker Panel investigating the BP refineries in US released its report. The refining company KNPC is on way to implement some of these new advances. As a result

- New recommended practices on stacks and temporary shelters (API 752) has been adopted as a company engineering standard,
- A new reporting process on 'Process Safety Near Miss' has been established,
- An awareness program has been kicked off to emphasize process safety as compared to personnel safety,
- A Process Safety Culture Survey has been initiated to gauge the values and beliefs among the company employees, and
- Discussions at various levels are underway to addressed Baker Panel report (its findings, recommendations, etc.) and Risk Based PSM approach

10.1 Current PS culture

There never has been any effort so far to measure process safety culture at KPC or any subsidiary level. Barring a few pockets, the score is not expected to be high, at least for the present. More controllable aspect of process safety such as Mechanical Integrity is a universal high scorer. However, when it comes to soft issues such as management involvement or employee participation things become sluggish.

10.3 Efforts by other bodies – ASSE etc.

There are not many non-petroleum companies or agencies working on process safety issue in the region. There are capable individuals, but they do not make a system. Education and research institutes, NGOs or even private consulting bodies do not find it motivating enough to get their feet wet beyond ankle.

Local Chapter of American Society of Safety Engineers has been doing some yeoman job in spreading awareness by arranging seminars, training courses and discussion forums.

11. CONCLUSION

Most process incidents result from errors or conditions that can ultimately be traced to breakdowns in management control. Examples are inadequate understanding of process technology, incomplete or obsolete operating or emergency procedures, unauthorized or inadequately designed equipment modifications, inadequate inspection or maintenance programs, inadequate job knowledge and/or training, inadequate supervision and failure to communicate essential process technology. KPC PSM focuses on providing sufficient controls and/or redundancies to avoid a set of conditions that can lead to process incidents. While this guideline standard focuses on chemical and petrochemical processes, most of the elements can be applied to any operation in an appropriate manner and result in improvement in the safety of the operation.

A complete implementation of KPC HSE MS and including PSM is planned over the next three years. The KPC PSM journey has started and it's challenging.

12. REFERENCES

1. Kuwait Petroleum Corporation Health Safety & Environment Management System (KPC HSE MS) Documents
2. *KPC Profile,* KPC Internet website. http://www.kpc.com.kw/index2.htm
3. Process Safety Management of Highly Hazardous Chemicals (29 CFR 1910.119), U.S. Occupational Safety and Health Administration, February 24, 1992, www.osha.gov
4. *International Petroleum Accidents ... Have They Transformed into a Phenomenon* 2004-7-1 Oil Magazine, Quarterly published by the Ministry of Oil, Govt. of Kuwait, July 2007. http://www.moo.gov.kw/magazine/en/index.asp?More=yes&NewsID=305&mode=0&day=9&page=3
5. The Report of the B.P. U.S. Refineries Independent Safety Review Panel, January 2007 (Baker Panel Report),
6. *Risk Based Process Safety Management*, New York: American Institute of Chemical Engineers, Center for Chemical Process Safety, 2007.
7. American Institute of Chemical Engineers (AIChE), Centre for Chemical Process Safety (CCPS) Technical Guidance and Publications
8. *EU (Seveso II) Directive 96/82/EEC.* [Seveso II is often interpreted at Country/National level, e.g. BRZO 1999 for use in the Netherlands]

Minimizing the use and hazard of solvents during process design

Julie B. Manley, ACS GCI Pharmaceutical Roundtable, 1155 Sixteenth St, NW, Washington, DC 20036 USA, 202-872-6102, gcipr@acs.org.

Inherently safer design and green chemistry& engineering have a common philosophy. That is to minimize or eliminate the hazard through the design of the process. The incorporation of green chemistry and engineering principles in the design of the process not only minimizes environmental and safety liability, but it also has been demonstrated to increase process efficiency with an impact on the financial bottom line. Many pharmaceutical companies have been working together in partnership with the ACS Green Chemistry Institute to facilitate the implementation of green chemistry and engineering in the global pharmaceutical industry. They recognize that the incorporation of these principles is essential for a sustainable business and environment.

The ACS GCI Pharmaceutical Roundtable has gathered benchmarking data indicating that approximately 50% of the process mass intensity of active pharmaceutical ingredients (API) is related to solvent use. The findings support other published results [C. Jimenez-Gonzalez *et.al.*, Int. J. LCA, ***9(2)***, 114 (2004)] indicating that most of the waste (c.a. 80%) of API is solvent related with the remaining being solids. These findings suggest a focus on the use of solvents in the design of the process will minimize environmental and safety hazards. While recognizing the important role of solvents, the Roundtable is working to identify mechanisms to reduce the use of solvent while also investing resources to improve our understanding of less hazardous alternatives to many commonly used solvents. The Roundtable is also considering how the life cycle impact of solvents can influence choice. Examples of inherently safer design and green chemistry & engineering principles will be demonstrated with an emphasis on the opportunities to minimize the use of solvents and to identify less hazardous alternatives when used.

DYNAMIC MODEL OF PROCESS SAFETY MANAGEMENT

INDICATOR OF PROCESS RISK MANAGEMENT

Author: Américo Diniz Carvalho Neto – Brasil

One of the great challenges of the current Chemical and Petrochemical industry is to have an efficient and effective way of knowing, measure and manage their risks of process in order to minimize the possibility of causing accidents to people, the environment and their assets.
Knowing, measure and control risks means having a proper management of their process, ensuring that issues that may initiate an unwanted event as an accident of process with impacts on human life, in active or in the environment have reduced their likelihood.
What we see now is that there are different systems being created, composed of elements of management, standards and guidelines that according to its authors, if implemented, these businesses will be safe, and the process accidents will not occur.
In practice, we observed that catastrophic accidents have occurred in organizations that have, according to their corporate reports, excellent management systems deployed and that indicators of management are monitored to ensure that the system is deployed.
In fact what we try to show is that not only enough have a HSE management system, because sometimes it is not able to ensure that process accidents do not occur.
There are also two major dimensions that may also contribute to a culture of prevention of accidents in the company is targeting decisions, attitudes and behaviors focused on the reduction of losses through the maintenance of facilities and reliable barrier of protection with their integrity guaranteed.
The proposal is to integrate the program of assessment of the management system we call the creation of the map of risk scenarios of the process.
This technology enables the identification and management of the areas of process scenarios with low risk, medium or high, depending on risk studies performed.
Identifies which scenarios and studies, recommendations of the studies of risk, which preventive barriers that minimize the probability of occurrence of the scenario, which reactive barriers that minimize the consequence of it. With the map of risk of process we integrate the security matters of processes between the areas of maintenance, operation, engineering and contractors. The integration makes the issue of process safety can be seen in a more systemic, helping to guide organizations on how they should act in equipment, instruments and systems, based on the criticality of these scenarios with regard to the risks of procedures to which they are Related, seeking to ensure that the scenarios of risk of process are managed. You should also define how empowering teams, select contractors based on the levels of risk and exposure to the risks of these people in the plant.

We know that in any management system, the process of continuous improvement is achieved through a program of audits. With the methodology proposed the integration of systemic audit, based on the model of management and audit focused on the map of scenarios based risk process that will ensure effectively the non occurrence of accidents at the process from the increased reliability and integrity of reactive and preventive barriers and the scenarios of risk.

The frequency and interaction between these audits based on the level of implementation of the systemic model and culture of the organization, taking up the premise that the existence of itself does not eliminate systemic evaluation of the necessity of the audit focused on barriers and the scenarios of medium to high risk of the plant.

Finally, with the results of the audit focused on the barriers and systemic audit can be an indicator that relates create and continually evaluates the level of efficiency of the management system of process safety of the plant with regard to the objective of ensuring systems, equipment and tools reliable and that can contribute to the reduction of accidents through no occurrence of failures.

Below follows a matrix of risk scenarios, one of the plants, where we can see that based on studies of risk have made 1289 scenarios assessed, where 14 scenarios are medium risk and low risk scenarios 1275. This matrix reflects the situation of scenarios considering that the safeguards evaluated during the study and contribute to the classification of the same are OK with their integrity, or are operating if an event initiator and they have to act avoiding the accident. In practice what we see is that if the process safety management system that is responsible for ensuring the integrity and reliability of the barriers is not effective, can occur changes in the situation of a matrix of risk scenarios, making scenarios that are classified as medium risk, can be given time with high risk, exposing the facilities, people and the environment, according to the change of frequency of occurrence of the non-availability of the scenario or the barriers of protection or mitigation of the risk of setting process. This imbalance between the possible situation envisaged scenarios of risk process, which has been identified in studies of risk and the actual situation that are the scenarios that call for dynamic model of process safety, which determined scenario that had been classified as low or average and has been treated / managed as such, can be as high risk according to the non-availability of or operational barriers of protection of the same.

MATRIZ DE RISCO		Freqüência					Total
		Improvável	Remoto	Ocasional	Provável	Frequente	
Severidade	Catastrófico	0	4	0	0	0	4
	Crítico	208	308	10	0	0	526
	Moderada	104	197	58	0	0	359
	Baixa	29	279	54	38	0	400
Total		341	788	122	38	0	1289

Tabela 1 (Matriz de cenários de risco de processo de uma planta)

Figure 1 below graphically represents the scenarios risk process of the matrix above explained in their respective areas of process. Thus we can see from simple way which areas in the plant that have low risk, medium or high and thus ensure a differentiated management with respect to technology, people and facilities depending on the risk in their area of process.

This product, showed in the fig.1 below, represent the map of scenarios risks of the process plant, is very important during the audit of the barriers of protection that we performed to evaluate the process of managing security process.

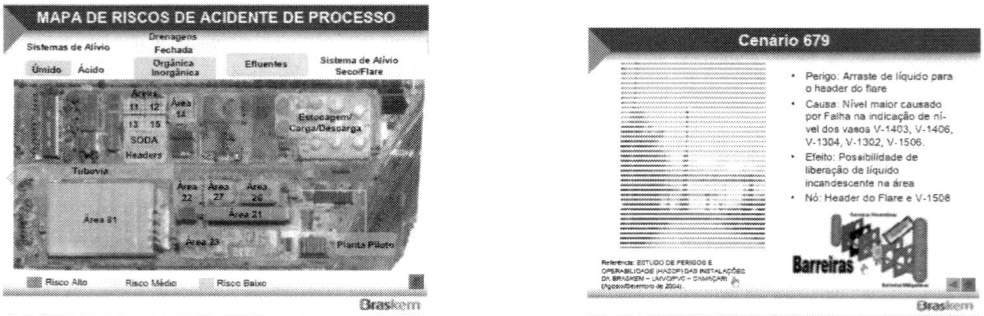

Figure 1 (Map of scenarios Risk process of the plant)

Our challenge is to seek a way that through process safety audits will evaluate the system and while the situation of the scenarios of risk management process as to the integrity of the systems and protections, as do the people of operation and maintenance reflect how has played its role in the search for safety of the plant.

This experience has allowed us give people an opportunity for reflection and incorporation of concepts and practices, evaluating and acting on a conscious and based on prior knowledge of the risks of the process and criticality and importance of barriers to protect certain scenario at risk.

Below, on fig. 2 follows the model chart that rep the process safety management audits where propose the holding of two types of audit. The systemic audit, that use the requirements based on OSHA 29 CFR 1910, which depend on the maturity of implementation and management of the system in the enterprise, we can manage on a different frequency of the completion of the audits, and the other type of audit is based careful in the assessment of scenarios for risk analysis more critical of the plant and how this scenario and the barriers of protection of the same are being managed.

A joint analysis of the results of both audits the systemic and the focused on the scenarios barriers allow a more effective in the control and management of the possible loss of process and also allows the areas operation, maintenance and engineering have a vision of how each of these areas is with respect to its role of contributing to the security of processes. In addition to allowing also have a systemic view of the importance of their work for the safety and reliability of the entire production process, equipment and systems involved.

As a long-term vision, with the maturation of the process safety system in the company, it is our expectation that there will be improvement in the reliability and integrity of barriers to protect the scenarios of risk and may well be the planned audit focused on barriers of protection with a frequency less because the actual management system PSM should ensure more effective way of the integrity and reliability of systems, equipment and protections.

The systemic audit, process safety management audit,have redefined its frequency depending on the maturity of the same plant, ensuring an audit at least every three years. These two audits performed together and understanding that the results are not static, or any slide in the process safety management directly affects the integrity and reliability and operational scenarios of risk is what we call dynamic model of process safety.
The impairment of integrity and degradation of the barriers of protection makes the process scenarios more vulnerable, in the case of the occurrence of any event initiator that cause a process accident.

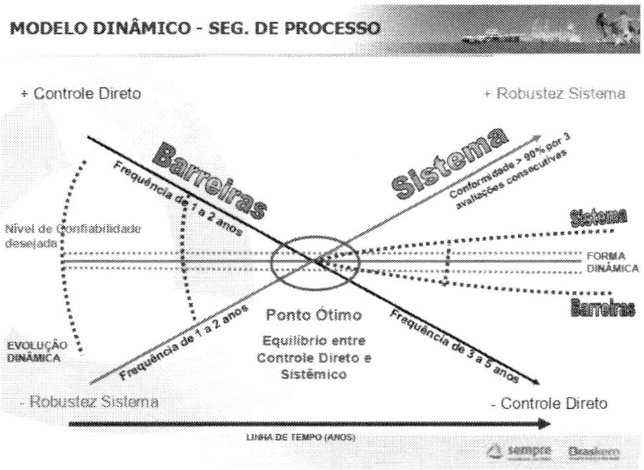

Figure 2

The experience we have acquired in a combination of the two audits in plants has allowed us to improve process safety not only on the systemic level, but has helped to improve the process of classification, maintenance and inspection of equipment based on scenarios risk process and the importance of such equipment and instruments in the risk scenario.
We have also observed an improvement in the perception of risk and the understanding of how each area, maintenance, operation and engineering can add greater value to the issues of process safety management in the plant.

Key Performance Indicators (KPIs): Consequence-Based Metric for the Assessment of Inherent Safety

Alessandro Tugnoli, Valerio Cozzani

Dipartimento di Ingegneria Chimica, Mineraria e delle tecnologie Ambientali
Alma Mater Studiorum - Università di Bologna
Bologna, Italy
alessandro.tugnoli@mail.ing.unibo.it, valerio.cozzani@unibo.it

Gabriele Landucci

Dipartimento di Ingegneria Chimica, Chimica Industriale e Scienza dei Materiali
Università di Pisa
Pisa, Italy
gabriele.landucci@ing.unipi.it

ABSTRACT

The practice of inherent safety in process development and design asks for reliable and systematic assessment tools. Despite some valuable procedures have been proposed in the literature, they are commonly based on general scoring techniques and involve subjective judgment in large extent. Hence relevant biases may be introduced in the analysis, resulting in unreliable outcomes. In the present study, a novel quantitative approach developed for the inherent safety assessment of process flow diagrams in early design stages is presented. The output of the approach is a metric that quantifies the inherent safety fingerprint of the process scheme by a set of Key Performance Indicators. Physical parameters are used for the quantification of the hazard deriving from materials, process conditions and equipment characteristics. The assessment resorts to the identification and modeling of credible accident consequences on humans and equipment. The adoption of tangible parameters based on consequence modeling yields a clear and sound picture of the inherent safety performance of a design option. This was demonstrated by the application to several case studies and by the comparison with other literature approaches. In the following an example concerning hydrogen production and storage is presented.

1. INTRODUCTION

Current strategies for the improvement of the safety performance of the process industry require the implementation of the inherent safety concept [1-3]. However the practice of inherent safety in process development and design is presently limited or non-systematic. The lack of reliable tools for inherent safety assessment is a key issue in this context [2]. Despite some valuable assessment procedures have been proposed in the literature, they are commonly based on general scoring techniques and involve subjective judgment by a large extent. Hence, relevant biases may be introduced in the analysis, resulting in unreliable outcomes [4,5].

In the present study, a novel quantitative approach developed for inherent safety assessment of process flow sheets is presented. The output of the approach is a metric which quantifies the inherent safety fingerprint of the process scheme. The metric is based on a set of Key Performance Indicators (KPIs). The analysis is particularly suitable as a decision supporting tool to be applied in the early PFD design. Physical parameters are used for a solid quantification of the hazard from materials, process conditions and equipment characteristics. Heuristic rules for scenario definition allow identifying the credible expected accidents. Actual consequence modeling provides quantification of the hazards through standard damage levels for humans and equipment targets.

The described method allows identifying in a physically sound manner the inherent safety performance of a design option. This is demonstrated by the application to several case studies, where the expected hazard profiles were correctly identified. An example of case study will be provided in the following. The comparison with previous scoring approaches reported in the literature on the one hand confirms the validity of the results and on the other clearly shows the limits in the application of the latter [6]. The KPIs allow easily taking into account the hazards coming from auxiliary equipment (e.g. pumps, compressors, heat exchangers, etc.), that are often overlooked by other assessment methods, but appear to be significant in several practical cases. Moreover the open and physically based framework of the KPIs method allows managing any specific case by the best available model, thus providing reliable outputs. Finally, the method is low time-demanding, since current availability of simulation software and computer power speed-up the onerous step of consequence modeling.

2. DEFINITION AND ASSESSMENT OF THE KPIs

The present method is aimed to the definition and calculation of inherent safety quantitative key performance indicators (KPIs) of a given process flow diagram. The resulting inherent safety fingerprint may be used for both the improvement of the inherent safety of the process and the comparison with alternative process options. Hence the methodology is particularly suitable for the use as a decision support tool in early stages of design for inherent safety (i.e. conceptual and basic design of a plant).

The flow diagram of the method is shown in figure 1. Table 1 reports the required input and the output provided by the method. The analysis of the table clearly evidences that a low level of detail is required for the input, confirming the applicability of the method also during early stages of basic design.

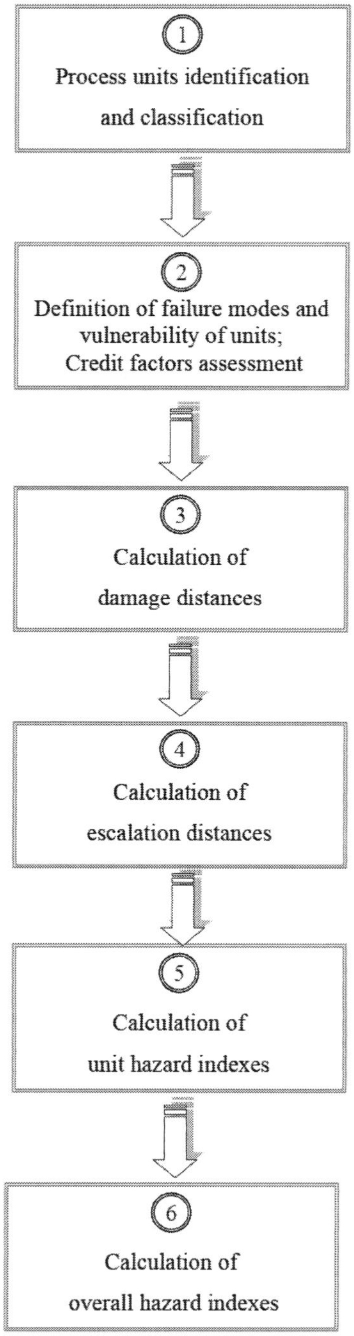

Figure 1: Flow diagram of the method used for the calculation of the inherent safety Key Performance Indicators.

INPUT	
1) Definition of substances and operative conditions in each unit of the process	
2) Quantification of flows in main process lines and preliminary pipe sizing	
3) Preliminary design of equipment units	
4) Estimation of inventories in the equipment units of each process alternative	
OUTPUT	
UPI_k	Unit potential hazard index (m^2)
UHI_k	Unit inherent hazard index (m^2/a)
PI	Overall potential hazard index (m^2)
HI	Overall inherent hazard index (m^2/a)
UPD_k	Domino unit potential hazard index (m^2)
UHD_k	Domino unit inherent hazard index (m^2/a)
PD	Domino overall potential hazard index (m^2)
HD	Domino overall inherent hazard index (m^2/a)

Table 1: Required input for the application of the proposed method and Inherent Safety KPIs used in the present study.

As shown in figure 1, the first step of the methodology is the identification of process units (k) to be considered in each alternative scheme, which are classified on the basis of their structural and geometrical features. In the second step, different failure modes related to loss of containment (LOC) are associated to each unit on the basis of literature data analysis [7-8]. Table 4 reports an example of standard set of LOCs proposed for some categories of units. A credit factor ($cf_{i,k}$) is assessed for each LOC (i), on the basis of the expected release and failure frequency data reported for standard technologies in several publications [7,9]. A simplified Failure Mode and Effect Analysis (FMEA) was applied to obtain data for non-standard equipment [6].

In step 3 of the procedure, the consequence analysis of each possible scenario following the LOC is performed. Standard event trees are used to identify the scenarios, while loss intensities and consequences are calculated by conventional literature models [10,11]. The threshold values for damage to humans reported in table 2 and derived from technical standards are used to calculate a conventional damage distance for each scenario ($d_{i,j,k}$). A unit hazard matrix is thus obtained ($m_{i,j,k}$, for the k-th unit), having a number of rows equal to all the credible LOC events (i) and a number of columns equal to the maximum number of accidental scenarios related to each LOC event (j).

$$m_{i,j,k} = \max\ (\ d_{i,j,k}\ ,\ c\ ) \qquad (1)$$

where c is a constant, assumed equal to 5m in the present study. The use of the constant c in equation (1) allows the definition of a "near field" zone, in which the consequences of the event are neglected. This takes into account the unreliability of conventional consequence assessment models in describing correctly the consequences of the events in the "near field", thus avoiding biases in the analysis due to these uncertainties.

A unit hazard vector ($h_{i,k}$) is derived from the matrix, selecting the damage distance of the worst-case scenario of each LOC event considered for the unit:

$$h_{i,k} = \max_{j}(m_{i,j,k}) \tag{2}$$

Physical effect	Threshold Values	
	Effect on humans	Escalation
Flash Fire – transient radiation	½ LFL, %vol	not considered
Fireball – transient radiation	7 kW/m^2	15* - 50 kW/m^2
Jet Fire – stationary radiation	7 kW/m^2	15* - 50 kW/m^2
Pool Fire – stationary radiation	7 kW/m^2	15* - 50 kW/m^2
Vapor cloud explosion – overpressure	0.14 bar	0.16 - 0.22* bar
Physical/mechanical explosion – overpressure	0.14 bar	0.16 - 0.22* bar
Toxic exposure	IDLH, ppm	-

In the domino thresholds, the marked value (*) refers to an atmospheric target equipment; the other value to a pressurized equipment.
IDLH: Immediately dangerous to life and health concentration.
LFL: lower flammability limit.

Table 2: Threshold values assumed for the estimation of the expected damage distances.

Escalation events represent a severe potential hazard for many process installations [12,13]. A specific step of the procedure (step 4 in figure 1) was dedicated to the estimation of an escalation vector ($e_{i,k}$) for each process unit considered. The escalation thresholds reported in table 2, derived from previous studies [13,14], are used to calculate the conventional escalation distances ($v_{i,j,k}$) for each LOC using a procedure similar to step 3.

$$e_{i,k} = \max_{j}\left[\max\left(v_{i,j,k}\ ,\ c\right)\right] \tag{3}$$

The escalation vector will be used in the following step to provide specific domino hazard indexes.

Table 1 summarizes the inherent safety key performance indicators (KPIs) obtained by the procedure (step 5 and 6 in figure 1). The unit potential hazard index (UPI) was defined as the square of the maximum damage distance calculated for the unit. The UPI is thus representative of the maximum impact area that may derive from the worst case scenario considered for the unit.

$$UPI_k = \max_{i}(h_{i,k})^2 \tag{4}$$

A unit inherent hazard index (UHI) was also defined. The UHI was introduced to consider in the analysis the safety scores of the equipment, expressed by the above defined credit factors ($cf_{i,k}$). The UHI is calculated by the following expression:

$$UHI_k = \sum_{i=1}^{n_k} cf_{i,k} \cdot h_{i,k}^2 \quad (5)$$

where n_k is the number of LOC events considered for the unit (k).

The sum of UPI and of UHI values for a group of N units (e.g. the process unit in a plant) is used to calculate respectively the overall potential index (PI) and the overall hazard index (HI) of the process.

$$PI = \sum_{k=1}^{N} UPI_k \quad (6)$$

$$HI = \sum_{k=1}^{N} UHI_k \quad (7)$$

These overall indexes allow the assessment of the expected overall inherent safety performance of the plant (or plant section), based either on a direct assessment of potential worst-case scenarios (PI) or of likely safety performance and release scenarios of the process units (HI).

A similar procedure is applied to calculate the unit and the overall escalation hazard indexes. Similarly to UPI, the unit potential domino index, UPD, was defined as the square of the maximum escalation distance calculated for the unit.

$$UPD_k = \max_i (e_{i,k})^2 \quad (8)$$

The unit domino hazard index, UHD, was defined by eq.(9) in analogy to UHI:

$$UHD_k = \sum_{i=1}^{n_k} cf_{i,k} \cdot e_{i,k}^2 \quad (9)$$

The overall domino potential index, PD, and the overall domino hazard index, HD, were defined summing up respectively the UPD values and the UHD values for all the units.

3. THE CASE-STUDIES

In the following case studies the eight KPIs listed in table 1 were used to provide an overall quantification of the inherent safety of the alternative technologies proposed for hydrogen production and bulk storage. The goal of the analysis is to identify the inherently safer option and to evidence the critical safety issues for each technology.

3.1. Hydrogen Production Processes

This case study considers the comparison of the inherent safety fingerprint of two processes proposed for industrial production of high purity hydrogen (99.5-99.9%) by steam reforming of natural gas. Even if in a long time period renewable resources must be preferred, in a short time perspective the shift toward the widespread utilization of hydrogen requires the optimization and improvement of this kind of technologies, as confirmed by the active research in this field [15-17].

Figure 2 reports the block diagrams of the alternative process options. A hydrogen production potential of $1x10^4$ Nm^3/h was assumed for both processes, in accordance to a survey of plants currently designed. The first analyzed option (*CP - Conventional with PSA*) for hydrogen production is a conventional steam reforming and water gas shift process [18] in which hydrogen separation and purification is realized in batteries of zeolite beds by pressure swing adsorption (PSA). As depicted in Figure 2, natural gas feed is heated and purified by hydrodesulphurization and H_2S adsorption on zinc oxide. Primary catalytic conversion of this stream takes place in the tube side (800°C) of a furnace reactor. Heat for the reaction is provided, on the furnace side, by the combustion with air of natural gas and purged gases from the purification section. The reacted stream is then cooled in a spray quencher with water and fed to the catalytic shift reactor (around 400°C). The syn-gas from this rector is cooled in different stages, producing hot steam, and the condensed water is separated. Hydrogen is purified in the PSA battery and sent to the storage. The exhausted beds of PSA are purged via hydrogen washing [18].

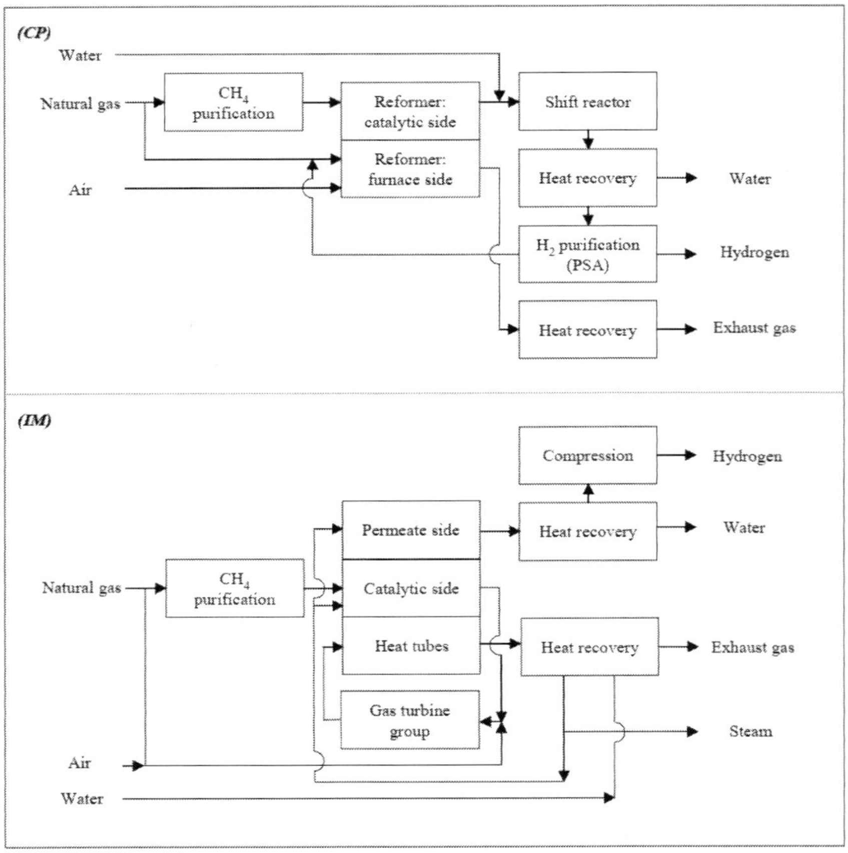

Figure 2: simplified block diagram of the two options for the hydrogen production by steam reforming. CP: conventional steam reforming with PSA; IM: integrated membrane process.

The second analyzed option (*IM - Integrated Membrane,*), is an innovative process based on membrane separation. In the process scheme considered for the option (Figure 2-IM), after methane heat-up and purification, water steam and the methane stream are fed to a single integrated unit where reaction, separation and heat exchange occur. All the reforming reactions take place in the catalytic bed of the reactor and the hydrogen produced continuously leaves the reactor through semi-permeable membranes. The permeated is removed using steam as sweep gas. Furthermore, heat is provided to the reactor by an integrated tubular heat exchanger where hot combustion gases are used as a thermal vector. Heat recovery from material flows leaving the reactor allows hot steam production. Water is separated from the stream of permeated hydrogen by condensation. A final compression and cooling unit is required to reach the same pressure and temperature specifications of the hydrogen output. The mechanical energy for the compression is provided by a gas turbine, that produces also the hot gases required for reactor heating.

A simplified PFD was defined for both options. An example for option *CP* is reported in figure 3. On the basis of the PFD, the heat and mass balances were carried out and the operating conditions and substance inventory of each unit considered in the PFD were preliminary estimated.

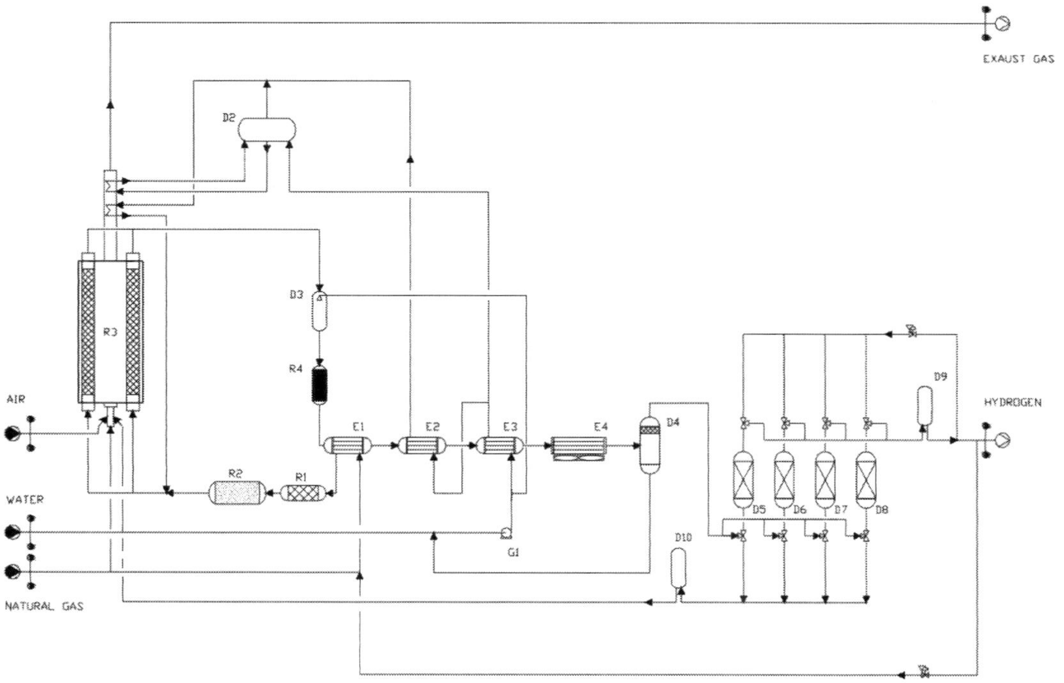

Figure 3: Example of the level of detail required for the simplified PFD used in the assessment of case studies. Hydrogen production, option *CP* (Conventional with PSA). R1: hydrodesulfuration; R2: H_2S adsorption; R3: reforming reactor; R4: water-shift reactor; D3: spray quencher / water mixer; D4: water separator; D5-8: PSA beds battery; D9: hydrogen buffer; D10: off-gas buffer; E1: methane heater; E2-E4: syn-gas cooler

3.2. Hydrogen Storage Options

This case study considers the hydrogen storage usually coupled to a production facility. A bulk storage of 27t of hydrogen was considered in the analysis, on the basis of a survey of existing large-scale storages used in oil refineries.

The first storage option considered is the liquefied cryogenic storage at about 25K (*CS - Cryogenic storage*). The reference scheme for this option (see figure 4) considers two bulk storage tanks (D3 and D4), each containing 13.5t of liquid hydrogen at 1.7 MPa and 20-25K. A dedicated liquefaction section was also included in the reference scheme, since it is necessary for the operation of the bulk storage. The liquefaction process includes two coupled cryogenic cycles, respectively for nitrogen and hydrogen. In each cycle, the gas is compressed (compressors K1 and K2 respectively), cooled (heat exchangers E1-E4) and finally expanded. The liquid fraction is separated from the vapor in a buffer flash chamber (respectively D1 and D2), and the vapor is recycled to the compressor.

The second option considered (*HS – Hydrides Storage*) represents a new technique proposed for hydrogen storage, currently still at pilot stage. The system is based on the storage of hydrogen as a solid inorganic hydride, that may be decomposed yielding gaseous hydrogen by reaction with water [19, 20]. The hydrides are stored in a mineral oil slurry in order to prevent the contact with moisture, thus avoiding unwanted hydrogen release. The proposed route for the production of hydrides [19, 21] is the endothermic reaction, at about 1350K, among the metal hydroxide (i.e. exhausted hydroxide recycled from the hydrolysis at the hydrogen user) and a carbon source (e.g. biomass):

$$MOH^{(l)} + C^{(s)} \rightarrow M^{(l,g)} + CO^{(g)} + \tfrac{1}{2}\,H_2^{(g)} \qquad (10)$$

where M represents the metal (Li, Ca, Na, etc.). The hydride technology is expected to be safer compared to other techniques, due to the inherently less severe operative conditions of the storage [19-21].

The reference scheme considered for the hydride storage (figure 4) is based on two atmospheric tanks (D3 and D4) for the bulk storage of the hydride slurry at room temperature. The tanks were considered to contain 90t of hydride slurry (LiH slurry in mineral oil), equivalent to 27t of hydrogen. In the regeneration of exhaust hydroxide, the carbon based material is contacted with the hydroxide (stored in D1-2) in a decomposition reactor (R1), where reaction (10) takes place. The gases formed in R1 are extracted and compressed (K1). Hydrogen is separated via membrane separation (S1) and reacted in the mixer R2 with the liquid lithium coming from R1 to yield the hydride. The hydride is separated (B1) and mixed with the mineral oil (R3). The heat necessary at R1 is obtained in the burner B2 by the combustion of the carbon monoxide from S1 and auxiliary fuel.

Also in this case a simplified PFD was defined for both options. The heat and mass balances were carried out and preliminary data were obtained for equipment operating conditions and substance inventories.

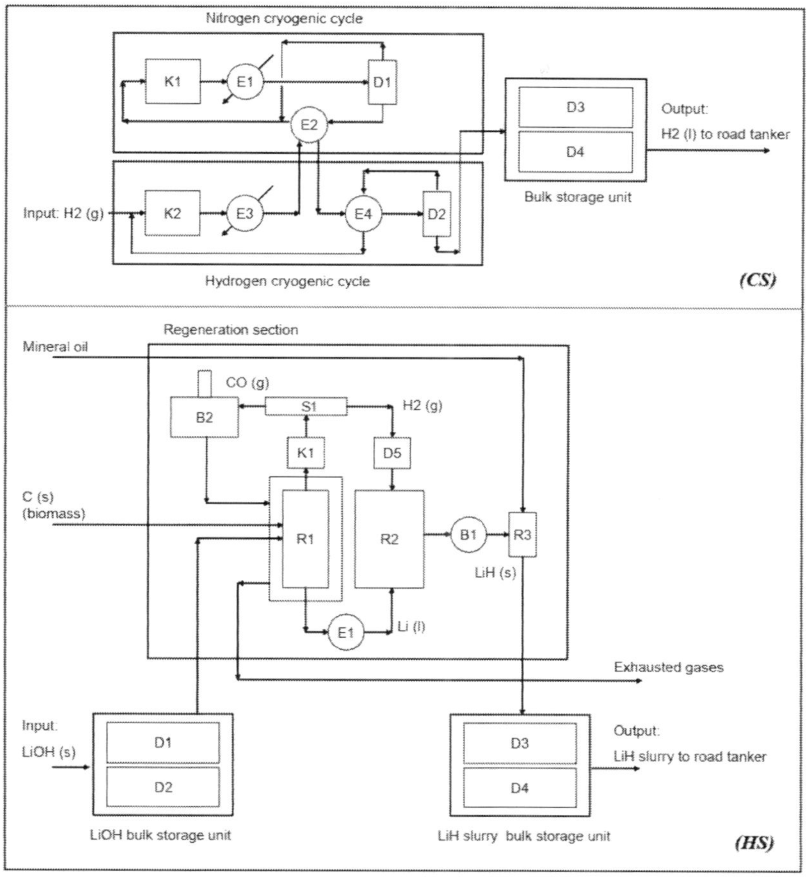

Figure 4: reference schemes for hydrogen storage considered in the case study. (CS) Cryogenic storage and (HS) Hydrides Storage.

4. RESULTS AND DISCUSSION

4.1. Hydrogen Production Processes

The proposed inherent safety assessment method was applied to analyze the alternative options for large scale hydrogen production. According to the procedure in figure 1, the units were identified and are listed in table 3. Table 4 reports some of the standard LOCs that were associated to the process units. The same table shows the reference values used for the credit factors [7,8]. The table evidences that important differences (up to some orders of magnitude) may be present in the credit factors for different LOCs concerning the same equipment item. Clearly enough, higher credit factors are obtained for equipment items that are more prone to cause loss of containment (e.g. heat exchangers, compressors).

The definition of the release mode, of the unit operating conditions and of inventories allowed the calculation of damage distances. Table 3 shows the unit potential and inherent hazard indexes calculated for the alternative options. The table also reports the overall indexes calculated for the two processes. The potential hazard index only gives information on the equipment that may potentially trigger the most severe scenario, while the inherent hazard index also includes information concerning the credibility of the possible scenarios. Thus, the inherent hazard index yields a more realistic description of the credible accidental events that may be associated to plant operation. In particular, this index points out the importance of the safety performance of minor equipment as compressors and heat exchangers. These equipment items may have "per se" relatively small damage distances, but are more prone to start loss of containment events. As shown in the table, these equipment items have inherent hazard indexes that are often higher than those of major process units (e.g. reactors), that may trigger more severe scenarios but with lower credit factors due to the higher safety scores reported.

The results of the assessment evidence that the reforming reactor may be identified as the most hazardous unit in the *CP* reference scheme. On the other hand, the adoption of the membrane technology in the *IM* process proved to be effective in limiting the maximum possible consequences of release scenarios. Due to the unconventional nature of this piece of equipment, a FMEA was performed, indicating that the unit is extremely unlikely to cause LOC events resulting in high damage areas. Focusing on the UHIs indexes, the heat exchangers generally result among the more inherently hazardous units, having severe scenarios associated to rather high credit factors. However, the hydrogen compressor of option *IM* is the unit that has the higher value for UHI, despite of the low UPI value, due to the effect of credit factors.

If the overall indexes are considered, option *CP* results in lower expected inherent safety performances, as calculated by the potential hazard index. If credit factors are considered, the high equipment inherent hazard index of the compressor in option *IM* causes the overall inherent hazard index (PI) to be higher. Thus, the application of the method yields two important pieces of information for the further development and/or the selection among alternatives: on one hand, the potential hazards of alternative *CP* are higher; on the other, a safety critical unit is present in alternative *IM*.

Option CP: Conventional process with PSA				**Option IM: integrated membrane process**			
		UHI_k	**UPI_k**			**UHI_k**	**UPI_k**
R1	hydrodesulfuration	$3.1x10^{-2}$	$4.6x10^{3}$	R1	hydrodesulfuration	$2.3x10^{-2}$	$2.5x10^{3}$
R2	H_2S adsorption	$3.1x10^{-2}$	$4.6x10^{3}$	R2	H_2S adsorption	$2.2x10^{-2}$	$2.5x10^{3}$
R3	reforming reactor	$1.7x10^{-1}$	$9.6x10^{3}$	R3	membrane reactor	$1.0x10^{-2}$	$2.0x10^{3}$
R4	water-shift reactor	$3.7x10^{-2}$	$6.0x10^{3}$	C2	hydrogen compressor	$1.8 x10^{0}$	$7.5x10^{2}$
D3	spray quencher	$3.2x10^{-2}$	$4.9x10^{3}$	D1	water separator	$1.0x10^{-2}$	$1.9x10^{3}$
D4	water separator	$5.3x10^{-2}$	$8.4x10^{3}$	E1	methane heater	$3.4x10^{-1}$	$4.3x10^{3}$
D5-8	PSA beds battery	$1.4x10^{-2}$	$7.0x10^{2}$	E2	hydrogen cooling	$3.9x10^{-1}$	$7.5x10^{2}$
D9	hydrogen buffer	$3.6x10^{-3}$	$4.9x10^{3}$	E3	syn-gas cooler	$9.6x10^{-2}$	$1.9x10^{3}$
D10	off-gas buffer	$2.0x10^{-3}$	$3.8x10^{3}$	E6	syn-gas cooler	$9.4x10^{-2}$	$7.5x10^{2}$
E1	methane heater	$4.7x10^{-1}$	$7.0x10^{3}$	C1	turbogas	$5.0x10^{-2}$	$2.8x10^{3}$
E2	syn-gas cooler	$4.0x10^{-1}$	$6.2x10^{3}$				
E3	syn-gas cooler	$4.4x10^{-1}$	$7.1x10^{3}$				
E4	syn-gas cooler	$1.6x10^{-1}$	$6.2x10^{3}$				
		HI	**PI**			**HI**	**PI**
		$1.9x10^{0}$	$7.4x10^{4}$			$2.8x10^{0}$	$2.0x10^{4}$

Table 3: Case study on hydrogen production: values of unit and overall inherent safety key performance indicators for the considered process options.

LOC	Process pressure vessel (column)	Process gas storage vessel	Shell and tube heat exchanger	Centrifugal compressor
R1: small leak, continuous release from a 10 mm equivalent diameter hole	$1x10^{-4}$	$1x10^{-5}$	$1x10^{-3}$	*n.a.*
R2: catastrophic rupture, release of the entire inventory in 600 s	$5x10^{-6}$	$5x10^{-7}$	$5x10^{-5}$	*n.a.*
R3: catastrophic rupture, instantaneous release of the entire inventory and release from the full-bore feed pipe	$5x10^{-6}$	$5x10^{-7}$	$5x10^{-5}$	*n.a.*
R4: pipe leak, continuous release from a hole having 10% of pipe diameter	*n.c.*	*n.c.*	*n.c.*	$6x10^{-3}$
R5: pipe rupture, continuous release from the full-bore pipe	*n.c.*	*n.c.*	*n.c.*	$6x10^{-4}$

Table 4: LOCs and related "credit factors" for some categories of process equipment of interest in the case study (data for credit factors derived from the literature [7,8]). *n.a.*: not applicable; *n.c.*: not considered.

The same alternative reference schemes were analysed using literature methods for inherent safety index assessment [22-28]. Some results for the reaction unit (reactor R3, in both option) are compared in Figure 5. The application and comparison of alternative methods required further assumptions to integrate the procedures described in the original literature. For instance, an aggregation procedure was introduced for some methods (e.g. ESH, INSET) [4,25]. Several methods do not define clear and homogeneous procedures to calculate the material quantities to be considered in the unit assessment. Significant biases may thus be originated in processes that, as *CP* and *IM*, are based on gas phase streams and have generally low inventories. Since the indexes calculated by the different methods have different scales, the normalized values (obtained dividing by the higher value obtained) are compared in the figure. In the case of I2SI, that has a higher value for an inherently safer option, the inverse of the index was considered.

Figure 5 shows that all the methods qualitatively agree in the identification of the membrane reactor as safer than the conventional shift reactor. However, the agreement is poorer from the quantitative point of view. In particular SWeHI seems to be less sensitive to the differences between the two units. The differences are likely to arise both from the different level of detail required by the application of the method and by the different penalization factors considered in the methods.

Further problems arise in the analysis of the other process units, since not all of them are considered by some of the methods. In particular, some methods do not consider auxiliary units and units having low substance inventories, as compressors and heat exchangers, thus introducing discrepancies in the comparison.

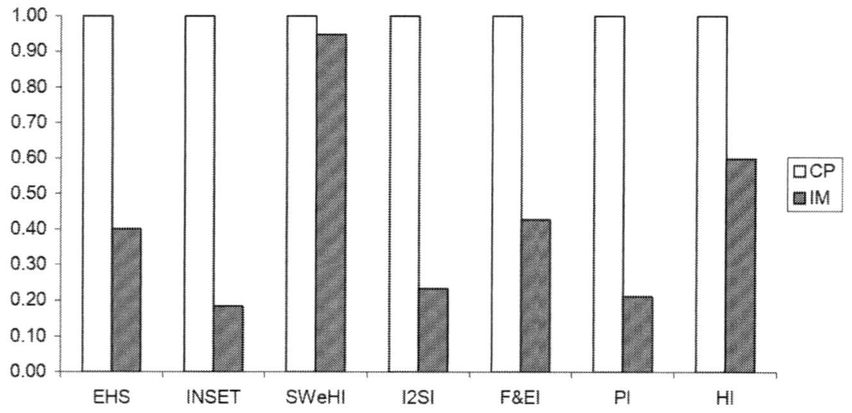

Figure 5: Inherent safety indexes calculated by different methods for the main reactor (R3) in the two analyzed options for hydrogen production.

Figure 6 shows the comparison of the overall inherent safety indexes calculated by the different methods considered. In the assessment, further limitations of the literature methods emerged. The application of PIIS, that requires to consider reforming and shift reactions as two independent steps, resulted in the penalization of the *CP* option, in clear disagreement with the ISI method, even if both methods require similar levels of detail.

The application of the more detailed indexes (EHS, INSET, SWeHI and F&EI) is influenced by the absence of explicit rules for the combination of units indices to yield overall performance indicators. The sum of single scores was adopted as a working hypothesis [6]. Figure 6 confirms the poor quantitative agreement among the methods. Nevertheless, the PI index obtained by the proposed approach and most of the other indexes have a similar behaviour, penalizing the *CP* reference process. This was expected, since most methods consider the potential severity of the scenarios as the prevailing factor in the analysis. Moreover, credit factors derived from equipment safety scores are not considered by the majority of the other tools. As a consequence, only the inherent safety hazard index (HI) introduced in the present study is able to take into account the hazards coming from auxiliary equipment, as compressors, filters, and heat exchangers, that are often overlooked in conventional exclusively severity-based inherent safety assessment methods.

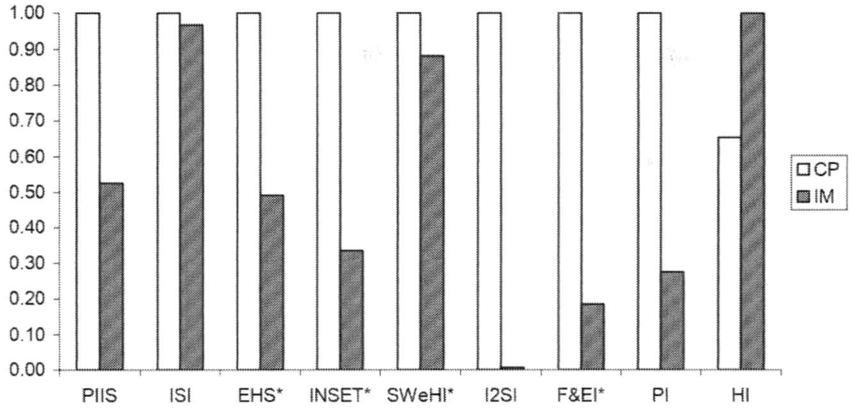

Figure 6: Comparison of the overall inherent safety indexes calculated by different methods for the case study on hydrogen production. (*) Aggregation was obtained by the assumptions discussed in the text.

4.2. Hydrogen Storage Options

The second case study concerned the comparison of the *CS* and *HS* reference schemes for hydrogen storage. The final results obtained for the unit and for the overall hazard indices are reported in Table 5. As expected, both the unit potential and unit hazard indexes are much lower for the hydride bulk storage with respect to the liquefied bulk storage, as expected. Thus, the results obtained for the bulk storage indexes are a concrete example of the effectiveness of the application of the "substitution" inherent safety keyword [29]. Nevertheless, the bulk storage unit does not result the critical item for both the options, due to the more or comparably hazardous units present in the liquefaction or hydride production processes.

The critical units resulted the compression systems for both alternatives. This units are needed in the conventional process for hydrogen liquefaction and in complex hydrides storage for hydrogen gas delivery. As shown in table 5, in both alternatives the compression unit has associated high values for the potential hazard index, comparable or higher than the UPI calculated for the bulk storage tanks. Furthermore, the introduction of credibility factors in the calculation of UHI penalizes equipment items that may more easily cause LOC events, as gas compressors. Hence, the method correctly scores the higher value of UHI for the compression units of both options.

The overall values of the potential and hazard indexes reported in table 5 evidence that the expected safety performance of the hydride storage results higher than that of cryogenic storage, also considering respectively the regeneration and the liquefaction section. Both the potential (PI) and inherent hazard (HI) indexes evaluated for the cryogenic storage result about two orders of magnitude higher than in the innovative *HS* process. One more time, the overall results evidence that in the perspective of an industrial implementation of a new technology, the reliability of the auxiliary equipment will be an important issue to be addressed.

Option CS Cryogenic Storage + Liquefaction Process				Option HS Hydride Storage + Hydride Slurry Production			
		UHI_k	**UPI_k**			**UHI_k**	**UPI_k**
D1	N_2 Buffer storage tank	-	-	D1-2	Hydroxide storage tank	-	-
D2	Buffer storage tank	2.9×10^{-2}	1.6×10^{4}	D3-4	Bulk storage tanks	7.3×10^{-2}	3.6×10^{3}
D3-4	Bulk storage tanks	1.2×10^{1}	1.1×10^{6}	D5	Buffer storage tank	2.7×10^{-4}	3.5×10^{2}
K1	N_2 Compressor	-	-	R1	Reactor for regeneration	1.8×10^{-2}	3.1×10^{3}
K2	Compressor	1.3×10^{2}	5.9×10^{5}	R2	Reactor for hydride production	2.1×10^{-2}	1.7×10^{3}
E1	Heat exchanger	-	-	R3	Mixer	1.4×10^{-2}	1.7×10^{3}
E2	Heat exchanger	1.1×10^{0}	7.8×10^{2}	E1	Lithium cooler	-	-
E3	Heat exchanger	8.8×10^{-3}	6.8×10^{2}	K1	Compressor	6.4×10^{0}	4.2×10^{3}
E4	Heat exchanger	5.6×10^{-2}	7.3×10^{2}	B1	Rotating furnace	8.2×10^{-2}	1.0×10^{3}
				B2	Furnace	-	-
				S1	Membrane separator	2.1×10^{-1}	2.0×10^{3}
		HI	**PI**			**HI**	**PI**
		1.4×10^{2}	1.7×10^{6}			6.8×10^{0}	1.8×10^{4}

Table 5: Case study on hydrogen storage: values of the unit and overall KPIs calculated for the options considered.

4.3 Comparison of Hydrogen Production to Hydrogen Storage hazard

In order to yield a complete overview of the process of production and storage of hydrogen, the PI and HI indices of the *CS* reference process should be combined with the indices for the production of hydrogen by steam reforming. This is allowed in the present method, since the proposed indices are evaluated on an absolute reference. Thus, the method yields the possibility of an absolute ranking of the inherent hazard, and the requirements for active safety devices or other risk control and management measures.

The comparison among the different processes assessed is summarized in figure 7. It is evident from the figure and from tables 3 and 5 that the overall indices for both production processes are negligible compared to liquefied bulk storage hazard, being about two orders of magnitude lower than the corresponding indices for storage process. In addition, these are more than one order of magnitude lower than the sole bulk storage itself. Thus, the applied method outlines as the storage is the critical point for hydrogen production and distribution in conventional processes. On the other hand, the innovative storage reference process based on hydrides, *HS*, appears to reduce significantly the storage hazard, that in this option becomes comparable to that of the production process, as shown in figure 7.

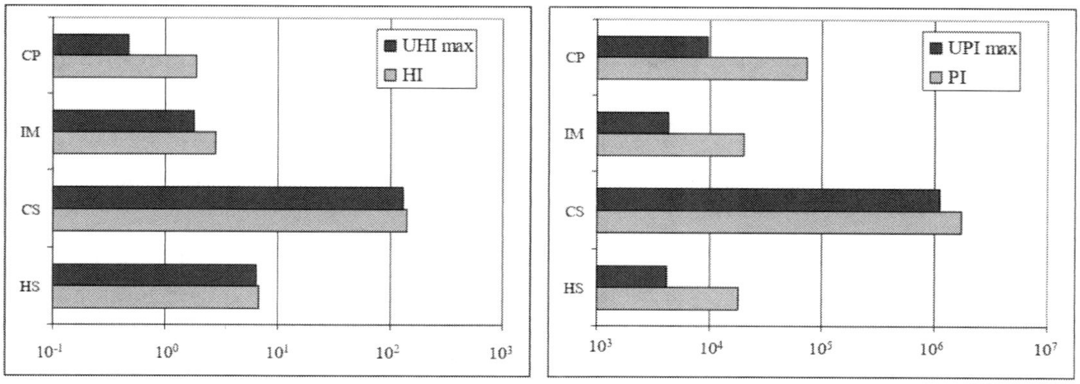

Figure 7: Comparison of the inherent safety fingerprints of all the processes considered in the case studies. UHI max: maximum value of the unit hazard index; UPI max: maximum value of the unit hazard index.

4.4 Domino hazard

Escalation events may give an important contribution to the overall installation hazard [13]. The evaluation of escalation distances and of UPD and UHD indices yields conclusion qualitatively similar to those for hazard indices, as shown in figure 8. The PD index evaluated for the *CS* option is strongly influenced by the bulk storage unit while the HD index, that takes into account the credit factors, is penalized also in this case by the hydrogen compression unit. The escalation hazard KPIs evaluated for the *HS* option result also in this case about two order of magnitude lower than those of the *CS*. It is worth to notice that for escalation hazards, the contribution of the membrane separation unit (labeled as S1 in figure 4) to the overall index is much lower than in the case of hazards to humans (PI and HI), since the toxic dispersion of carbon monoxide is no more taken into as possible escalation scenario.

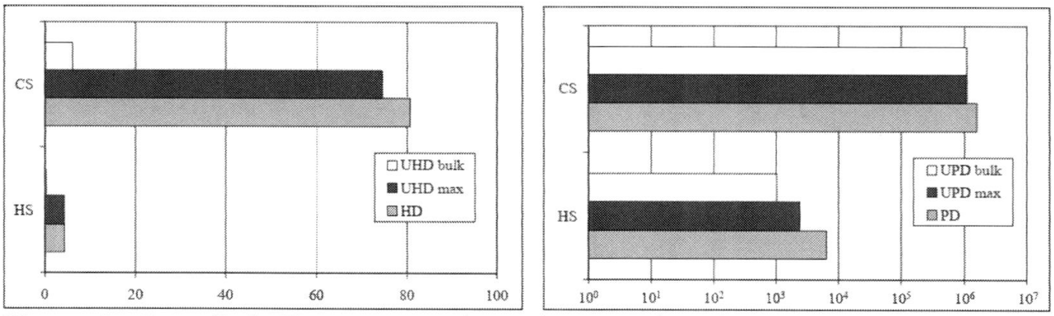

Figure 8: Overall, bulk storage and maximum unit values of KPIs for domino potential and inherent hazard indexes in the case study on hydrogen storage. UHD max: maximum value of the unit hazard index; UPD max: maximum value of the unit hazard index. UHD bulk: value of the unit hazard index for the bulk storage unit; UPD max: value of the unit hazard index for the bulk storage unit.

5. CONCLUSIONS

A consequence-based method for the quantitative assessment of the inherent safety of industrial processes was developed. The method is based on the evaluation of a set of KPIs, related to the severity of potential accidents and to the recorded safety scores of process equipment. A simplified assessment of the hazard with respect to both human targets and domino accident escalation is possible. In the framework of a comparative assessment of the inherent safety of alternative processes, the methodology developed in the present study overrides several problems present in previous methods proposed for quantitative inherent safety assessment. Introducing a direct relation among hazard factors and consequence analysis of potential scenarios, the proposed methodology avoids the use of arbitrary indexes. Moreover it allows a straightforward identification of the critical hazard sources, supporting the design of inherently safety options, as well as a more effective adoption of add-on safety measures. The application of the method to a case study on hydrogen production and storage evidenced that the introduced KPIs allows taking into account the hazards coming from auxiliary equipment, that are often overlooked in conventional inherent safety assessment methods.

5. REFERENCES

1. Kletz, T.A., *Process Plants: A Handbook for Inherent Safer Design* (2nd ed.). Philadelphia, PA: Taylor & Francis, 1998.
2. Khan, F.I., Amyotte, P.R., *How to Make Inherent Safety Practice a Reality*. The Canadian Journal of Chemical Engineering, 81: 2-16, 2003
3. Hendershot, D.C.,. Inherently Safer Chemical Process Design. Journal of Loss Prevention in the Process Industries, 10,3: 151-57, 1997
4. Koller, G., Fischer, U., Hungerbühler, K., *Comparison of methods suitable for assessing the hazard potential of chemical processes during early design phases*, Trans IChemE. 79, Part B: 157-166, 2001.
5. Khan, F.I., Sadiq, R., Amyotte, P.R., *Evaluation of Available Indices for Inherently Safer Design Options*, Process Safety Progress, 22, 2: 83-97, 2003.
6. Tugnoli, A., Cozzani, V., Landucci, G., *A Consequence Based Approach to the Quantitative Assessment of Inherent Safety*, AIChE Journal, 53,12: 3171-3182, 2007.
7. Uijt de Haag, P.A.M., Ale, B.J.M., *Guidelines for Quantitative Risk Assessment (Purple Book)*, The Hague (NL): Committee for the Prevention of Disasters, 1999.
8. American Petroleum Institute, *Risk-bases inspection base resource document, API standard 581*, 1st ed., Washington, D.C: API, 2000.
9. Delvosalle, C., Fievez, C., Pipart, A., Debray, B., *ARAMIS project: A comprehensive methodology for the identification of reference accident scenarios in process industries*, Journal of Hazardous Materials, 130: 200–219, 2006.
10. Van Den Bosh, C.J.H., Weterings, R.A.P.M., *Methods for the calculation of physical effects (Yellow Book)*, 3rd ed., The Hague (NL): Committee for the Prevention of Disasters, 1997.
11. Lees, F.P., *Loss Prevention in the Process Industries*, 2nd ed., Oxford (UK): Butterworth-Heinemann, 1996.
12. Khan, F.I., and Abbasi, S.A., *The World's Worst Industrial Accident of the 1990s*, Process Safety Progress, 18,3: 135-145. 1999.
13. Cozzani, V., Tugnoli, A., and Salzano, E., *Prevention of domino effect: from active and passive strategies to inherently safer design*, Journal of Hazardous Materials, A139: 209-219, 2007.
14. Cozzani, V., Gubinelli, G., Salzano, E., *Escalation thresholds in the assessment of domino accidental events*, Journal of Hazardous Materials, 129: 1-21, 2006.

15. Adhikari, S., Fernand, S., Gwaltney, S. R., To, F. S.D., Bricka, R. M., Steele, P. H., Haryanto, A., *A thermodynamic analysis of hydrogen production by steam reforming of glycerol*, International Journal of Hydrogen Energy, 32: 2875 – 2880, 2007.
16. Reyes, S. C., Sinfelt, J. H., Feeley, J. S., *Evolution of processes for synthesis gas production: Recent developments in an old technology*, Industrial and Engineering Chemistry Research, 42,8: 1588-1597, 2003.
17. Song, X., Guo, Z, *Technologies for direct production of flexible H_2/CO synthesis gas*, Energy Conversion and Management, 47,5: 560-569, 2006.
18. Häussinger, P., Lohmüller, R., Watson, A.M., *Hydrogen*. In: *Ullmann's encyclopedia of industrial chemistry*, 6th ed., Vol. 17, Weinheim (DE): Wiley-VHC, p. 108-123, 2003.
19. Zhou, L., *Progress and problems in hydrogen storage methods*, Renewable and Sustainable Energy Reviews, 9:395–408, 2005.
20. Aiello, R., Matthews, M. A., Reger, D. L., Collins, J. E., *Production of Hydrogen gas form novel chemical hydrides*, International Journal of Hydrogen Energy, 24: 1123-1130, 1999.
21. Conte, M., Prosini, P.P., Passerini, S., *Overview of energy/hydrogen storage: state-of-the-art of the technologies and prospects for nanomaterials*, Material Science and Engineering B, 108: 2–8, 2004.
22. Edwards, D.W., Lawrence, D., *Assessing the inherent safety of chemical process routes: is there a relation between plant cost and inherent safety?*, Trans IchemE 71, Part B, 252-258, 1993.
23. Heikkilä, A., *Inherent Safety in Process Plant Design*, PhD thesis. Espoo (FI): VTT Publications 384, 1999.
24. Koller, G., Fischer, U., Hungerbuhler, K., *Assessing safety, health, and environmental impact early during process development*, Ind. Eng. Chem. Res., 39,4: 960-972, 2000.
25. INSIDE Project. *The INSET Toolkit. Integrated version*, 2001.
26. Khan, F.I., Husain, T., Abbasi, S.A., *Safety Weighted Hazard Index (SWeHI) A New, User-friendly Tool for Swift yet Comprehensive Hazard Identification and Safety Evaluation in Chemical Process Industries*, Trans IChemE, 79, Part B: 65- 80, 2001.
27. Khan, F.I., Amyotte P.R., *I2SI: A comprehensive quantitative tool for inherent safety and cost evaluation*, Journal of Loss Prevention in the Process Industries, 18: 310-326, 2005.
28. Dow Chemical Company. *Fire & explosion index hazard classification guide*, 7th ed., New York, NY: American Institute of Chemical Engineers, 1994.
29. Gupta, J.P., and Edwards, D.W., *Inherently safer design—present and future*, Trans. of the Inst. of Chemical Engineers: Process Safety and Environmental Protection, 80B: 115–125, 2002.

The Evolution and Current Status of Process Safety Management Metrics within DuPont

David E. Cummings, MS, CIH

Global PSM Competency Leader
DuPont SHE and Sustainable Growth Center
1007 Market Street
Wilmington, DE 19898
(302) 774-9558
david.e.cummings-1@usa.dupont.com

ABSTRACT

A broad, consistent set of PSM metrics and leading indicators is important to effectively manage process safety and prevent incidents, help ensure regulatory compliance, meet internal policies, and verify that adequate management attention and resources are allotted to meet program needs. This paper will review the development, current state and future focus areas associated with the use of process safety metrics and leading indicators within The DuPont Company as a global manufacturing organization at a site, business and corporate level. This includes a historical perspective of past metrics and their technical basis with a primary focus on incident classification and audit results, a review of recent improvements involving an expanded group of current metrics, and a general discussion of forwarding looking additional improvements. An overview of lessons learned, guidance, and challenges involving the development and use of process safety metrics and indicators within a global enterprise will also be provided.

1. Background on DuPont

DuPont is a global science based product and services company that operationally is organized around five main business platforms, each with several groupings of businesses, which serve over 200,000 customers worldwide with annual revenues of $30 billion. Founded in 1802, the company operates at more than 175 manufacturing sites in 70 countries involving 64,000 employees.

DuPont operates many different manufacturing technologies involving both higher hazard processes and lower hazard operations (predominantly mechanical systems) which are segmented into five global businesses as illustrated in Figure 1. Manufacturing technologies, products, and associated process safety hazards vary widely including the following examples:

- Chemical Solutions Enterprise (e.g., oleum, aniline, sodium, cyanides, methylamines,)
- Engineering polymers (polyester and acetyl resins, glass laminating products)
- Agricultural products (e.g., herbicides, fungicides, insecticides)
- Fluoroproducts (e.g., refrigerants, electronic gases, fluoropolymer solutions)
- Performance Coatings and Color technology (e.g.; automotive paints and titanium dioxide)
- Advanced Fiber Systems (e.g., Tyvek®, Sontara) and Non-Wovens (e.g., Nomex®, Kevlar®)
- Electronics, Displays, Imaging technologies (electronic polymers, circuit material, LCD's, photo)
- Pioneer Seed (corn, soybeans, alfalfa) and Solae (soy protein technology food products)
- Surfaces (Corian® and solid surface materials)

Operations is also segmented into four worldwide regions including North America, Latin America, Europe, and Asia Pacific. Internal process safety systems and standards are applied uniformly to all of the four regions and manufacturing technologies worldwide through one set of corporate policies and procedures.

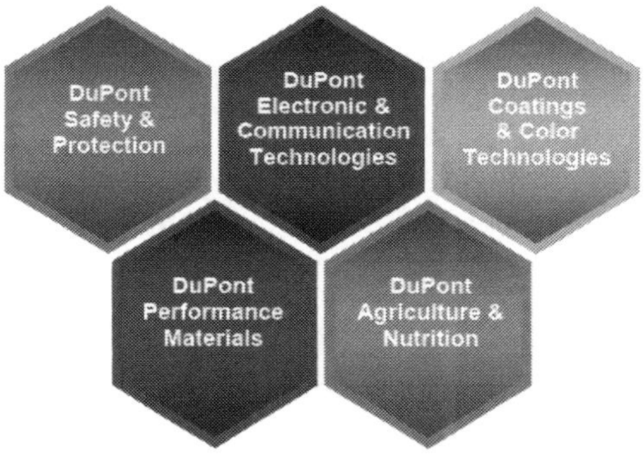

Figure 1: DuPont Global Business Platforms

The corporate structure for PSM relies on a small decentralized and matrixed global PSM competency team which has primary ownership and responsibility for internal standards, technical training, auditing, and internal initiatives. DuPont also maintains a small group of highly skilled PSM consultants within the corporate engineering group who provide complex PSM risk management support to sites and businesses such as QRA's. facility siting , project support, and complex PHA facilitation, however, the company's basic PSM model and philosophy is to develop PSM related knowledge and skills within the manufacturing site line organization and local support functions (operations, maintenance, technical, and R&D).

2. PSM Metrics – The Early Stages (pre- 1990)

Prior to 1990, DuPont's principle data collection and process safety metrics focused on serious process safety incidents, fatalities and lost workday injury cases, and large (financial) property losses due to fires, explosions, and toxic releases. Figure 2 provides a limited example of these summary data. Use of this approach originated in the late 1960's as a result of a major incident which occurred at the DuPont Louisville Works involving twelve fatalities and approximately

$50 million in damages as a result of a series of explosions which started in a compressor handling mono vinyl acetylene.

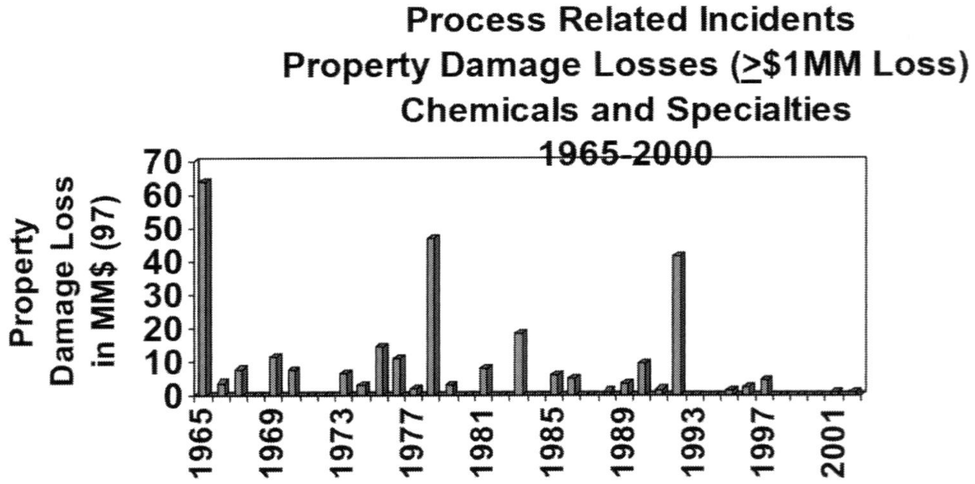

PSM2

Figure 2: Property Damage Losses due to PSM Incidents (1065-2000)
(Source: Arthur F. Burk, DuPont)

3. PSM Metrics – Decade of the 1990's

DuPont's initial set of PSM incident metrics from earlier years expanded significantly in the 1990's as a result of two significant program improvements which included:

1) Development and implementation of a global second (2nd) party process safety auditing program which tracked audit results based on cumulative scores as well as individual scores for each PSM element commencing in 1992.

2) Development and implementation of a global PSM internal incident classification and reporting process for significant and moderate severity events commencing in 1995.

More information on both of these internal practices will be discussed in Section 3 of this paper. These initial data enabled a broad review of annual serious incident performance, as well as a summary and basic trend analysis of audit results. These data were initially collected using paper reports and collated manually. Figure 3 provides an example of a chart tracking the number of significant incidents versus an annual summary of audit score averages.

These summary data were reviewed annually with senior operations leadership and were analyzed in more detail to identify improvement opportunities, need for additional training, and enhancements to existing internal policies and requirements.

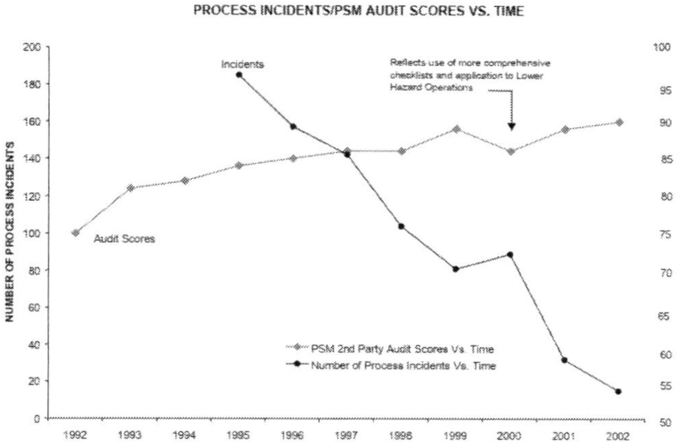

Figure 1: PSM incidents versus 2nd party audit score averages 1992- 2002
(Source: Arthur F. Burk, DuPont)

4. PSM Metrics – 2000 to present

Within the past seven years, DuPont has expanded the internal focus and capabilities to track a more extensive set of process safety metrics and program performance indicators. Each of these indicators will be described in more detail in the following sections.

4.1 Process Safety Incidents

DuPont currently utilizes a detailed PSM incident classification and reporting policy based on incident severity which provides an ongoing lagging metric for year over year PSM incident performance. The incident classification process requires sites to evaluate and score all PSM incidents on a standardized scale of 10- 230 points; the final score determines relative severity and internal reporting and communication requirements.

The current scoring architecture includes an evaluation and summation of ten (10) components for each incident. These components include the following:

- Type of event and material released (flammability, combustibility, toxicity)
- Actual quantity (size) of release
- Potential quantity (size) of release
- Degree of control by site during incident
- Involvement and functionality of lines of defense/layers of protection
- Actual on site impact (injuries, emergency response, evacuation)
- Potential on site impact
- Actual off site impact (injuries, shelter in place, agency reporting, media attention)
- Potential off site impact
- Actual monetary loss associated with the incident

Each incident is evaluated and scored by the site line organization and the final score dictates overall classification as described below. There are also "automatic" classification triggers for specific Category A events (e.g., a fatality, multiple lost work day injury cases, off-site impact, or total costs exceeding $1 million are automatic Category A events regardless of score)

Classification	Description	Internal Reporting
Category A PSM incident (130 points or more)	The most significant class of PSM incident which typically involves major on-site and/or off-site impacts	Corporate reporting to the CEO and corporate SHE organization within 24 hours of classification
Category B PSM incident (75-125 points)	A moderate to high severity PSM incident which typically has a significant on-site impact	Corporate reporting to the SHE organization within 72 hours of classification
Category C PSM incident (70 points or less)	A low to moderate severity PSM incident or near miss which has limited or no impact on site and is used primarily for trend analysis	Reporting into a global electronic incident tracking database after the formal investigation is complete.

Table 1: DuPont process safety incident classification and reporting criteria

All Category A and B incidents are currently included as business and site performance metrics. Higher level Category C incidents (50-70 points) have historically been used primarily for trend analysis in order to identify additional leveraged improvement opportunities across the corporation (for example, revisions to internal standards and guidance, additional technical training, improved or leveraged tools, etc)

Commencing in June 2004, DuPont initiated use of an existing electronic database for recordable injury and illness statistics to report and capture process safety incidents. This database was not specifically designed to manage PSM incident elements, however it did enable expanded trend analysis for a larger data set of events where the type of incident can be sorted based on the following attributes:

- Type of incident (fire, explosion, loss of containment, etc)
- Status of process at time of incident (start-up, normal, shutdown, etc)
- Element(s) of PSM involved which need strengthening
- Element(s) of Operational Discipline involved which need strengthening
- General types of process equipment involved
- Costs associated with the incident

Figure 4 below provides a historical trend on the number of significant process safety incidents within DuPont (Category A and B incidents) for the previous twelve years.

In addition to incidents, DuPont also maintains an annual process safety Total Recordable Rate (TRR) metric which includes a frequency rate for all recordable injuries caused directly by process safety sources such as chemical exposures, burns, and similar events. These data are a small subset of the total recordable injury rates for all types of injuries and illnesses across the company and it includes data for both employees and contractors on a global basis.

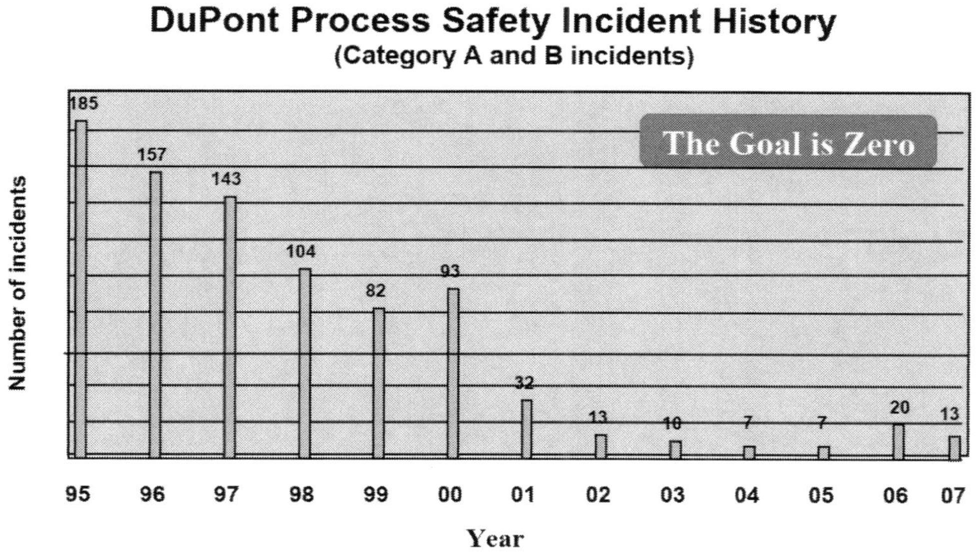

Figure 4: Twelve year global incident trend (Category A and B PSM incidents)

4.2 Process Safety Audits

DuPont has developed and maintains a global second (2nd) party PSM auditing process to periodically evaluate PSM systems and performance at all manufacturing locations. A standard set of audit attributes (protocols and checklists) are utilized based on internal requirements and standards. The audit checklist contains approximately 300 questions across fifteen PSM elements. DuPont conducts approximately seventy-five (75) 2nd party PSM audits each year across all four regions. A 2nd party audit is defined as a formal independent review conducted by trained DuPont personnel who are not assigned to the site subject to the audit using established protocols. 2nd party audits are scheduled by each operating region and tracked using an internal scheduling database. A typical audit involves 4 ½ days with a minimum of two experienced auditors. Completed audit reports and all individual audit recommendations are generated and entered into a central database using commercial software for tracking, closure, semi-annual reporting, and trend analysis (data mining).

Each 2nd party PSM audit is scored by the audit team on a scale of 0-100%. This generates a composite score for the entire audit, as well as individual scores for each of the fifteen PSM elements. PSM audit scores are collected and compiled within each region and summarized for trend analysis to identify further broad based improvement areas. Figure 5 provides an example of PSM audit scores in the North America region for the past three years across all process safety elements. The audit process includes all existing higher hazard and lower hazard manufacturing locations, including recent acquisitions and joint ventures where DuPont has a majority ownership.

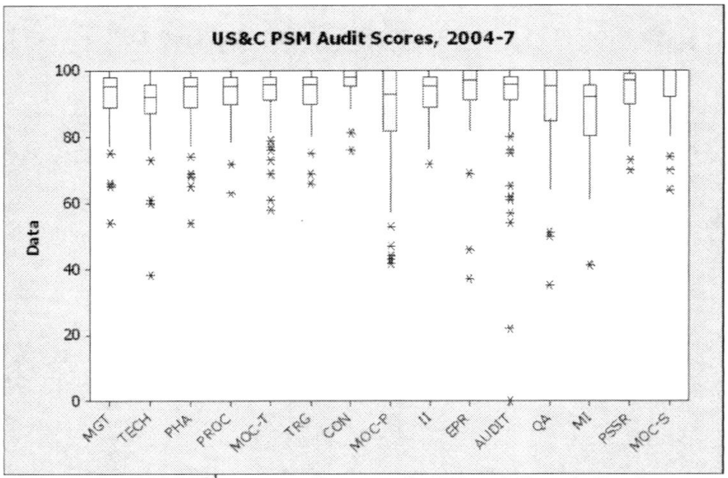

Figure 5 – PSM 2nd party audit scores in US and Canada Region
(Source: James A. Klein, DuPont NA PSM Consultant)

4.3 Site Based and Corporate Quarterly PSM Metrics

In late 2006, DuPont adopted a revised internal policy involving the collection and reporting of a series of standard PSM metrics every quarter for each of the 175 global manufacturing locations. This includes reporting on the following indicators:

- Number of Category A/B/C PSM incidents
- Number of open and overdue first party PSM Audit recommendations
- Number of open and overdue second party PSM Audit recommendations
- Number of scheduled, open, and overdue cyclic PHA's
- Number of open and overdue PHA recommendations
- Number of open and overdue PSM incident recommendations
- Number of overdue Operating Procedure revisions (three year schedule)
- Number of overdue Mechanical Integrity tests and inspections (psm critical equipment)
- Total number of PSM action items with approved extensions by site leadership

These metrics are reported by each site at the end of each quarter and posted in a central database where they are displayed within the DuPont PSM intranet website. Nine data elements are entered into an excel spreadsheet by each operating unit and site, and may be sorted, summed, and displayed at the enterprise, business platform, business unit, regional, and site level. DuPont currently does not maintain a global integrated tracking system for these data, so individual sites are expected to implement local electronic or manual data management systems to track and generate these metrics as part of the site PSM management system.

A key input to these data involves creating a standard set of definitions for each metric which is important to help ensure consistency and provide guidance to all site personnel. These definitions have been developed, communicated to all sites, and are also posted on the internal DuPont PSM website.

Figure 6 illustrates the format and content for the global PSM metrics spreadsheet which is used to generate quarterly status reports. Data may be sorted by site, business, platform and region. Quarterly and annual trend analysis (figure 7) may also be generated to assess aggregate data (open and overdue items) across various entities within the organization.

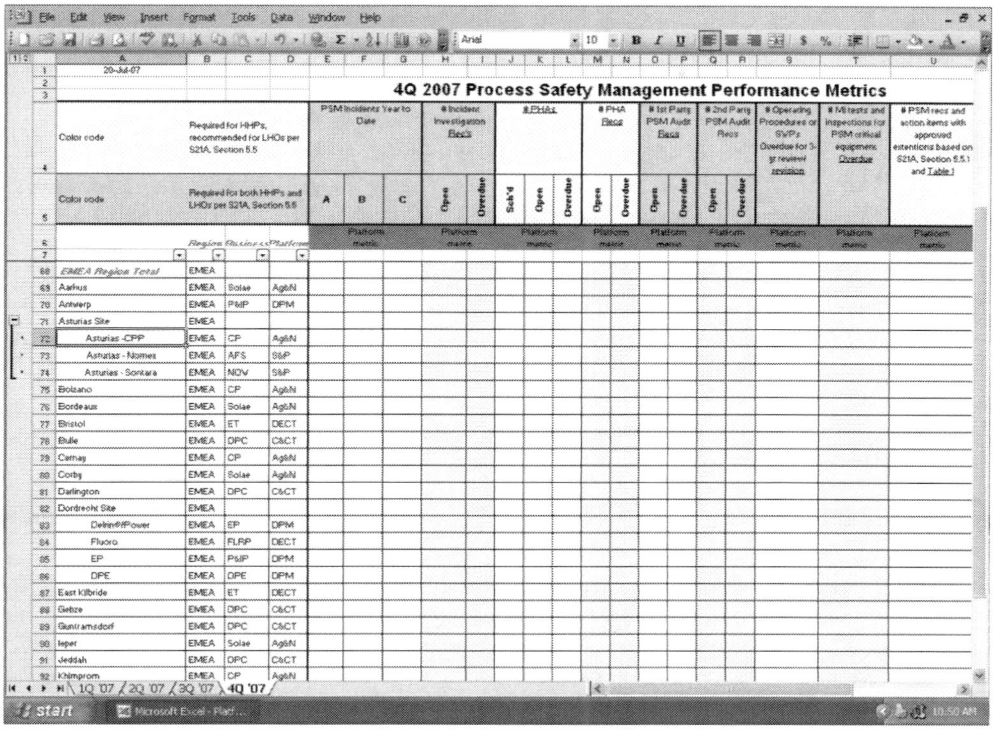

Figure 6: DuPont Global Quarterly PSM Metrics Reporting Spreadsheet

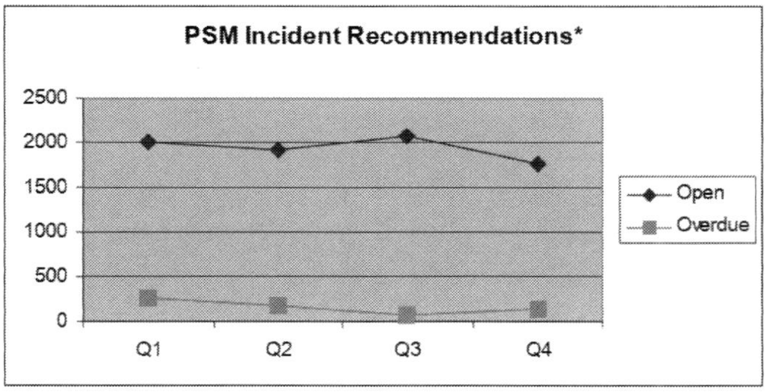

Figure 7: Example of Global Tracking for PSM Incident Recommendations

In addition to these nine quarterly metrics which are reported globally, additional process safety metrics are required within each site for tracking and reporting at the local level only and are reviewed during 2nd party PSM audits. These metrics include the following:

- Personnel (operations and mechanical) skill based refresher training/re-qualifications
- Open and overdue MOC- Technology recommendations and trials
- Open and overdue PSSR recommendations
- Status of MOC – Personnel 90-day competency demonstrations
- Number of open corrective actions or work orders from completed MI tests/inspections
- Open and overdue emergency response drill critique recommendations

4.4 Global PSM Critical Operating Tasks

DuPont maintains a global PSM competency team comprised of key PSM related resources from central functions as well as each of the four operating regions. This team has overall accountability for internal PSM standards, training and development, auditing, and PSM managing systems. The team develops and publishes a set of formal critical operating tasks each year which are periodically reviewed with senior leadership. A final report on the results and accomplishments involving formal goals and objectives is distributed at year end.

Certain critical operating tasks may also lend themselves to a set of one-time indicators or "sub-metrics" which demonstrate completion of a broad based but specific objective. For example, this may include updating of site wide facility siting studies, leadership PSM visits to plant sites, or enhancements involving design reviews and analysis of safety instrumented systems for existing processes as part of cyclic process hazards analyses. These completed goals and objectives activities may also be considered as non routine performance indicators and reported with other traditional metrics for a give time period.

4.5 PSM Related Internal Standards

DuPont has developed and maintains a series of detailed internal standards and policies which provide general and prescriptive requirements for process safety at all sites. Currently, there are twelve (12) specific PSM standards involving technology, facilities or personnel. These PSM standards are periodically reviewed and revised based on internal or external key learnings, or at a minimum every five years. Tracking the progress and updating of these standards is also considered a PSM program performance indicator.

4.6 Changes in Key PSM Personnel

DuPont maintains a decentralized and matrixed approach involving technical personnel who lead and develop PSM related programs and provide risk management subject matter expertise and consultations. These groups include the following:

- A global PSM competency team involving fifteen (15) individuals representing regions, business platforms, and key technical or leveraged functions.

- Eight (8) technology sub-teams who provide guidance on various PSM elements such as PHA's, mechanical integrity, incident investigation, etc which have responsibility for internal technical training, standards, new strategies, and "best practices".

- Twenty-two(22) highly toxic material technology teams that are comprised of a leader and site representatives who handle or manufacture highly toxic materials such as chlorine, phosgene, oleum, anhydrous HF, titanium tetrachloride, etc

- A small group of highly skilled and trained process safety consultants and subject matter experts within DuPont Engineering including the process safety and fire group, explosions hazards laboratory, safety instrumented systems competency, and reliability engineering resources.

In all, the sixty-plus leaders and members of these PSM technical teams comprise the core of DuPont's internal process safety and risk management structure and delivery system, including some of the most experienced resources in the corporation beyond those assigned to individual plant locations. DuPont tracks turnover and transitions across this group of key roles as a higher level metric in order to analyze long term trends in experience and capabilities (e.g., individuals with less than 2 years, 2-5 years, and greater than 5 years of experience in current process safety related roles)

4.7 Third Party PSM Related Assessments

DuPont receives PSM inputs and evaluations from two external third party sources in addition to internal evaluations and metrics. This includes an annual third party audit of our internal second party audit program and an external third party review of the Responsible Care Management System (RCMS) as part or the company's public commitment and membership in the American Chemistry Council. These reviews, while not specifically focused on metrics and performance indicators, do provide an annual independent analysis of corporate and regional performance against internal standards and procedures.

The third party review of Safety Health and Environmental audit programs is conducted each year and includes an analysis of corporate process safety auditing results, a periodic review of each region and key business audit performance and reports, and 3rd party participation on a small number of representative second party audits. The third party auditor evaluates internal audit systems and performance against ten key audit system elements and provides a numerical measure for each item listed below.

- Explicit Top Management Support
- Program and Auditor Independence
- Appropriate audit team staffing
- Defined Audit Program objectives and Scope
- Adequate program resources
- Audit Frequency based on Risk
- Appropriate Audit process
- Documentation of Audit findings
- Resolution of Corrective actions
- Appropriate Quality Assurance Measures

Table 2 provides a sample schedule for 3rd party audit reviews across each operating region and business. The results of 3rd party Audit program reviews are reviewed with DuPont executive leadership and posted on the DuPont internet for public access and review each year.

Region or Business	2004	2005	2006	2007	2008	2009
US/Canada	✓		✓			✓
Europe		✓		✓		
Asia Pacific	✓		✓			✓
Latin America		✓			✓	
Pioneer	✓		✓			✓
Solae		✓			✓	
Oversight Audits	**6**	**6**	**6**	**4**	**6**	**6**

Table 2: Schedule for 3rd Party Reviews - DuPont SHE Audit Programs (Source: Larry Cahill, ERM)

5. Key Points and Limitations with Metrics

A combination of real time, broad based leading and lagging PSM metrics is critical to measure the overall "health" and functionality of the process safety risk management program on an ongoing basis both in a general or macro level within the corporation and with enough specificity within sites and regions to enable targeted improvement actions.

As a first step, senior leadership must establish clear expectations and ground rules for metrics, including timely and accurate reporting, and the development of processes to ensure routine leadership review of metrics and performance indicators, both to recognize and celebrate strong performance and accomplishments as well as institute interventions, where necessary.

In a large enterprise, the approach to measure performance needs to be consistent over time yet flexible enough to address organization changes and updated process safety policies that occur (restructuring, acquisitions and divestitures, new plant site construction, new requirements, etc)

Metrics data collection and reporting systems should be as simple and easy to use as possible so that sites and businesses can access and update the data quickly and focus on the status of PSM improvements rather than expend resources on data entry and statistical compilations; use of an easily accessible and "transparent" information technology data collection and analysis system enhances the use of PSM metrics significantly.

Metrics do have specific limitations which must be recognized across the organization and are better addressed as part of a thorough site based process safety audit. These include:

1) Metrics cannot measure the specific quality of risk management work activities and decision making (for example, one can track timely completion of incident investigation reports and closure of recommendations, but this alone does not help ensure the overall quality and completeness of the investigation recommendations and the supporting root cause failure analysis).

2) Metrics do not measure or guarantee the quality and completeness of actions taken to address recommendations or other improvement actions to achieve closure. Actions to achieve closure based on locally assigned targets must be balanced and not take precedence over achieving the intended risk reduction or mitigation outcomes.

3) Metrics alone cannot provide an absolute assurance of preventing high consequence, low probability catastrophic events (explosions, fires, or uncontrolled chemical reactions).

6. Summary

Dr. Edward Deming emphasized many concepts in his works involving the "seven deadly diseases of management", including the philosophy that one cannot run a company on visible figures (data) alone since some things that are important can't be measured. This has validity to process safety in terms of core values, commitment, and the intrinsic behaviors of individuals, from the operator and mechanic on the shop floor handling hydrogen cyanide, through senior operations and business management. While it is exceeding difficult to measure certain intangibles, DuPont's current set of PSM metrics have been and will continue to evolve over time to serve as a suite of indicators on the internal "state of PSM" associated with all hazardous manufacturing processes.

Current PSM metrics provide a global picture on status, performance, and progress over time for sites, businesses, regions, and the corporation. Table 3 summarizes each of these current metrics.

Metric	Elements	Type
PSM Incidents	• Type (severity) – Category A/B/C incidents and annual trends • Elements of PSM and Operational Discipline involved which need Strengthening • Other incident factors (type of event, equipment, substances, costs) • PSM related injury rates (annual Total Recordable Rate)	Lagging
2nd Party PSM Auditing	• Audits completed versus scheduled (%) • Audit reports finalized and issued per internal timelines • Audit scores for each PSM element • Types of audit findings and trends (data mining) • Status of audit recommendations(open and overdue) • Special focus audits -Highly Toxic Materials handling • 3rd party annual review results	Leading
Site, Business, Regional PSM metrics	• Quarterly reporting of PSM metrics on nine key elements for all sites, businesses and regions (open, overdue and extended action items)	Leading
Annual PSM Critical Operating Tasks	• Review, analysis and reporting results of annual Global and Regional PSM Competency Critical Operating Tasks	Leading
Update/Revisions to global PSM Standards	• Status of updates/revision to global PSM related internal standards and policies	Leading
Changes within Key PSM Personnel	• Changes and experience levels of "top 60" PSM resources (central function, PSM competency team, regions and platforms)	Leading

Table 3: Summary of Current PSM Metrics across DuPont

DuPont's PSM leadership model as described in Figure 8 highlights the foundational role that leadership must provide and maintain to achieve strong and consistent results, including establishing a culture of "the goal is zero", strong operational discipline(OD), adequate resources, and setting accountabilities for performance. This approach, in concert with DuPont's process safety model involving fourteen technical elements, provides a strong framework for managing PSM on a day-to-day basis at all global sites. The performance metrics above complement these technical elements and provides ongoing leading and lagging indicators to demonstrate the overall quality of implementation in a visible and sustainable manner.

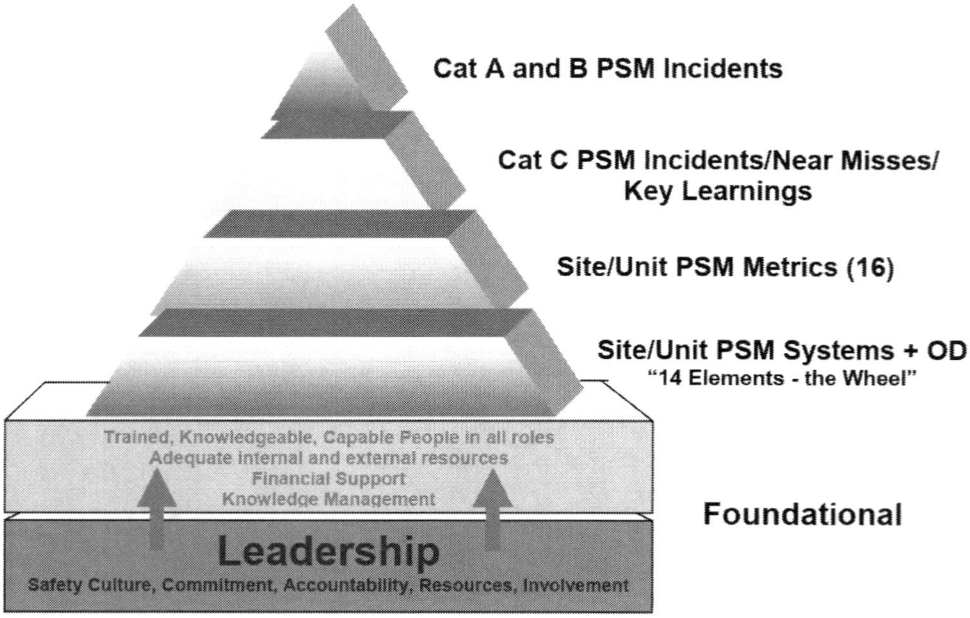

Figure 8: DuPont PSM Leadership and Performance Model

7. Future Focus areas involving process safety metrics and performance

Development, collection, and use of PSM leading and lagging metrics should be an evolving and dynamic process over time as organizations, systems, and stakeholder expectations change. Collection of metrics and review of performance trends must be used to strengthen the existing programs and ultimately ensure that high quality risk management decisions are maintained within each business and across each generation of operations personnel at all levels. Additional areas of consideration involving process safety metrics and performance indicators within DuPont include the following areas:

- Development of broad based metrics with focus on training and development to monitor PSM skills and capabilities across various manufacturing and technical organizations (a measure of "organizational capacity" and resourcing to maintain high quality PSM performance).

- Consideration of improved measures to assess Operational Discipline (OD) and Felt Leadership.

- Full integration of PSM metrics and results (both leading and lagging) into annual HR performance review cycles for line management.

- Development and use of a global integrated web-based central data collection and management system for PSM and SHE activities, including workflow management tools for process safety elements.

- Completion of annual reviews on the "State of PSM" within each of the five major business platforms to analyze performance indicators and results by senior operations and business management.

- Designing and conducting additional voice of the customer and PSM perception surveys to solicit further areas of improvement and focus.

The Use of the Pareto Shape Parameter as a Leading Indicator of Process Safety Performance

Fred Henselwood
NOVA Chemicals
Calgary, Alberta, Canada
henselfw@novachem.com

ABSTRACT

Metrics addressing process safety incident performance typically focus on frequency and severity statistics. Often these lagging metrics are not overly sensitive to actual performance, making trending and forecasting difficult. This paper presents the results from a statistical study of a large incident dataset where changes in the Pareto Shape parameter were observed as a function of time. This approach has been found to give far better insight into process safety performance than traditional incident metrics and readily relates back to concepts such as the "Incident Triangle" and "Layers of Protection". Through the application of this approach, trends within Process Safety Incident performance have been observed earlier, and more accurate forecasting has allowed for the identification of anomalies. In turn, these critical observations have allowed for the better structuring and targeting of process safety programs. Although incident data is generally considered as a lagging indicator this approach has clearly reduced the lag time associated with this type of data and has given valuable insight into the current status of process safety performance.

1. INTRODUCTION

Incidents are typically characterized by two key parameters with these parameters being severity and frequency, (i.e. the incident was this "bad" and occurs this "often") with aggregate data being used to assess process safety performance. For example, there were X incidents last year associated with a total cost of Y dollars. Unfortunately these traditional metrics carry with them significant statistical noise that can make both trending and forecasting of incident performance very difficult to accurately conduct. If one assumes that the number of incidents that occurs in a given time period is a function of random chance (with the expected frequency reflecting actual safety performance) a common modeling approach is to use the Poisson distribution function for estimating the observed number of occurrences.

The Poisson probability distribution function can be written as follows:

$$p(y) = \frac{e^{-\mu}\mu^{y}}{y!}$$

Where *p(y)* is the probability of y occurrences in a given time period and μ is the expected number of occurrences for that time period. When looking at aggregate incident data y is the measured number of incidents while μ is the expected number of incidents. Although the expected value for y is μ the standard deviation associated with y is $\mu^{1/2}$ and as such there can be sufficient statistical noise associated with y so as to blur the estimate of μ. For example, if the safety performance associated with a facility is expected to generate 12 incidents each year, observing 9 incidents one year and 15 incidents the next year would appear to represent a significant shift in performance, yet these values are well within the noise associated with random chance. As such, attempts to gain resolution beyond the square root of the expected sample size are not readily possible. Increasing the sample size by a factor of 4 will reduce the associated noise by a factor of 2, therefore increasing sample size is one strategy that can be used to reduce noise, but there are limits to the improvements in resolution that can be made through this approach.

When viewing incident severity totals such as total incident cost, not only are there limitations associated with the underlying number of event occurrences there are also issues associated with the magnitude of each individual incident. For example, some incidents might only cost a few dollars while other incidents can be on the order of millions of dollars. As such, when this data is aggregated one or two significant events may dominate over the rest of the data pool. Again this situation can be readily described using the Poisson equation in that when one looks at the most extreme incidents where only a few incidents of that magnitude are expected in a given time period the noise associated with random chance is quite large. As an example, if 4 incidents over a million dollars were expected, observations between 2 and 6 occurrences would not be unusual. If the million dollar plus events dominate the observed total incident cost fluctuations between \$2,000,000+ one year and \$6,000,000+ the next year would not be unexpected and as such this statistical noise would blur any true changes in underlying incident performance.

Despite these limitations incident data can provide significant insight into process safety performance. Improvements in resolution can be achieved by moving away from traditional measures such as frequency and severity. The focus of this paper is on an approach that looks at the observed distribution of incidents and how this distribution function relates to process safety performance.

2. THEORY

If one views an incident as a collection of mistakes and/or accidents, incident distributions such as those described by the incident triangle can readily be explained through an assumption that the severity of the incident is a function of the number of mistakes and/or accidents that occurred to produce the incident. Similarly, if one considers the concepts behind the layer of protection analysis (LOPA) model the severity of an incident relates to the number of layers of protection that failed.

Based on these models, the severity of an incident can be mathematically expressed by the following function:

$$Severity = L \times J^{n}$$

Where L is a constant representing a base level of severity, J represents the jump in incident severity associated with each additional failure, and n represents the number of failures that have occurred. For example, if J equals 10 (an order of magnitude) each additional mistake or accident associated with an individual incident would increase the severity of the resulting incident by an order of magnitude. Graphically you can view this as the bottom layer of an incident triangle representing $n = 1$, the next layer of the incident triangle being $n = 2$, and so on and so on, with each layer of the incident triangle representing incidents which are an order of magnitude greater in severity than the previous layer. Within the LOPA model this equation assumes that the magnitude of the incident severity is a function of the number of layers of protection that have failed.

Building from the LOPA model and working with the following assumptions: that each layer of protection is independent; that each layer of protection is responsible for the same jump (J) in severity; and that each layer of protection has the same likelihood of failure, the following series of equations can be developed.

$$P(1) = 1 - P(J)$$

Where $P(1)$ represents the probability that only 1 layer of protection will fail and $P(J)$ represents the probability that a typical layer of protection will fail. If the probability of a layer of protection failing was 1 in 10 then there is a 90% chance that only the original layer of protection will fail. Alternatively, there is a 10% chance that one or more additional layers of protection will fail. If we look at the probability of the second layer of protection failing and not the third layer of protection failing the following equation can be derived:

$$P(2) = P(J) - P(J)^{2}$$

Again if there is a 10% chance of failure for each layer of protection failing the probability that only two layers of protection will fail is 9%. When looking at the probability that the nth layer will fail (and not the nth +1 layer) the following equation can be derived:

$$P(n) = P(J)^{n-1} - P(J)^{n}$$

Which can be expressed as the following cumulative distribution function:

$$F(N) = 1 - P(J)^n$$

Rearranging the severity and the cumulative distribution functions in terms of *n* yields the following two equations:

$$n = \frac{\ln(Severity/L)}{\ln(J)}$$

$$n = \frac{\ln(1 - F(N))}{\ln(P(J))}$$

Equating the two equations to each other gives:

$$(Severity/L)^{1/\ln(J)} = (1 - F(N))^{1/\ln(P(J))}$$

Solving for *F(N)* gives:

$$F(N) = 1 - (L/Severity)^{-\ln(P(J))/\ln(J)}$$

Which has the same form as the Pareto distribution function with the following parameters:

$$Location = L$$
$$Shape = -\ln(P(J))/\ln(J)$$

As such if one assumes that the likelihood of each layer of protection is approximately the same (*P(J)*) and that the increase in the magnitude of the severity (*J*) is the same for each layer of protection then a curve fit of incident data, using a Pareto distribution function, should provide insight into the effectiveness of the related process safety performance.

3. DATA TREATMENT

Eight years of incident data (1999-2006) were utilized in the following study, representing just over 1,000 events each with a severity of \$30,000 (2007 USD) or greater. (Incident costs were corrected for inflation and currency issues.) The dataset primarily represents two olefin manufacturing regions (Joffre Alberta and Sarnia Ontario) the size and operation of which was relatively stable through this time period with some expansion occurring in 2000 within the Joffre area. The selection of \$30,000 as a minimum incident threshold value represented a balance between maximizing the amount of data available for the study while ensuring high data quality and full incident reporting. Lower value incidents are less likely to be fully reported and fully investigated relative to larger incidents and as such data quality typically decreases for lower value incidents. Further, the \$30,000 threshold placed an emphasis on process safety related incidents relative to occupational health related incidents and generally minimized the portion of the reported incident costs associated with the related incident investigation.

The incident data was sorted to form a cumulative distribution plot. This function was then transformed to a linear function allowing for standard curve fitting, which was conducted using Excel.

$$\ln(Cost/Location) = m \times \ln(1 - F(X))$$

Where *Cost* is the observed incident severity measured in 2007 USD, *Location* is the minimum incident value for the study (\$30,000), *F(X)* is the measured percentile divided by 100, and *m* is the slope of the resulting plot. Further, *m* can be related to the Pareto Shape parameter through the following equation:

$$m = -1/Shape$$

As such curve fitting of the observed incident data should readily give information with respect to the quality of the related process safety programs.

Although the Shape parameter and related layer of protection failure probability have been linked to process safety performance the usefulness of these parameters is limited in that the data they express is not readily tangible as their meaning is somewhat abstract. As an example, people cannot readily relate to a change in Shape parameter from 1.0 to 1.2. As such, to improve the insight gained through this analysis an expected cost function has been utilized and represents the summation of a series of uniformly distributed incidents (on a percentile basis) as described by the Shape parameter and an estimated event frequency. The expected cost function is given below:

$$ECF = \sum_{i=1}^{I} Location\left(\frac{1}{1-\left(\left(i-\frac{1}{2}\right)/I\right)}\right)^{1/Shape}$$

Where *I* is the expected number of incidents.

The benefit to the use of this metric is that it approximates the median observed total incident cost resulting in a value that is not overly influenced by the noise associated with extreme incidents within the dataset. This metric also has a benefit relative to the median incident cost for a Pareto distribution function (*Location x 2 ^ (1/Shape)*) in that as the number of expected incidents increases the extreme percentile events become more important to the estimated value and act to skew the expected cost function towards higher values (i.e. two incidents at the 50th percentile do not add to give the same value as one incident at the 25th percentile and one incident at the 75th percentile). As such, as the number of expected incidents rises the average incident cost also rises (to a maximum value as defined by the above Pareto distribution function median value). Although this metric gives a good approximation of the expected cost of a basket of incidents this parameter does suffer from those limitations previously discussed with respect to the noise associated with the expected number of incidents used in the calculations.

4. OBSERVATIONS

Figure 1 illustrates the linear Pareto distribution fit obtained for the entire dataset.

Figure 1.

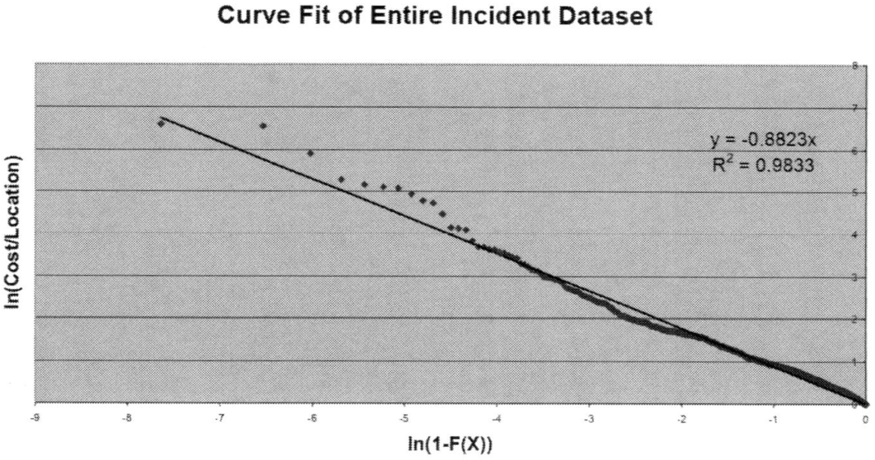

Figure 1 shows the entire incident dataset and the generally good agreement obtained with respect to the alignment between the Pareto distribution function and the observed incident data. Working from the measured slope the Shape parameter can be determined (*Shape* = 1.1334) which then allows for the determination of the probability of a layer of protection failing. ($P(J)$ = 1 in 13.6 +/- 0.1, assumes J = 10 for each layer of protection).

In addition to being able to fit the entire dataset, each month was then analyzed based on the data recorded for the preceding year. In this way it was possible to observe changes in the Shape parameter as a function of time, independent of seasonal factors, so as to assess changes in process safety performance.

Based on observations that the linear fit was at times significantly influenced by the one or two largest incidents within a given dataset, particularly when looking at one-year blocks of data, the linear fits were developed using the 0-98th percentiles of the reviewed datasets. In this way the largest incidents influenced the assignment of the percentiles but did not directly influence the linear fit. The difficulty in fitting these extreme data points is consistent with the limitations already discussed in the introduction with respect to the noise associated with low probability events. As an example, some one-year blocks of data may include a 1 in 10 year event while other may not even include a 1 in 6 month event.

Figure 2 shows the measured Shape parameter as a function of time as measured for the one-year blocks of data by month.

Figure 2.

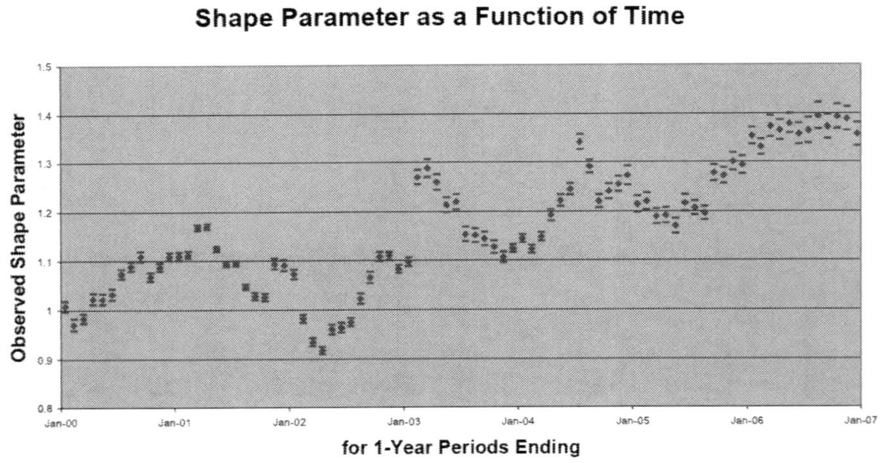

Figure 2 shows the observed trend in the Shape parameter as a function of time. Although associated with some variability the Shape parameter is clearly trending towards higher values. During the study period the likelihood of a layer of protection failing ($1/P(J)$) has gone from approximately 1 in 10 to 1 in 25 (assumes $J = 10$).

As a comparison Figure 3 shows the traditional metrics associated with process safety performance (Frequency and Total Cost).

Figure 3.

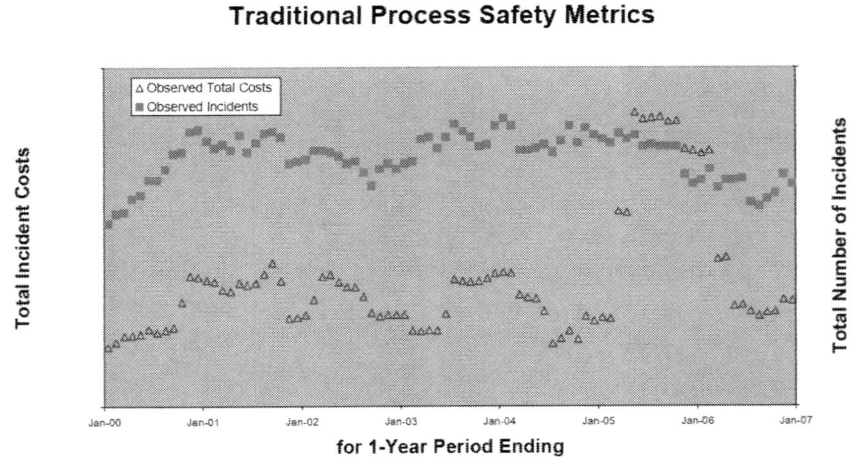

Figure 3 shows the number of incidents observed for the previous year by month (closed boxes) and the total costs associated with those incidents (open triangles).

Clearly the ability to spot trends with these traditional metrics (incident count and total incident cost) is significantly limited relative to the resolution given by the plot of the Shape parameter. Figure 4 illustrates a comparison between the observed total incident costs for one-year periods by month and the expected cost function for these same periods.

Figure 4.

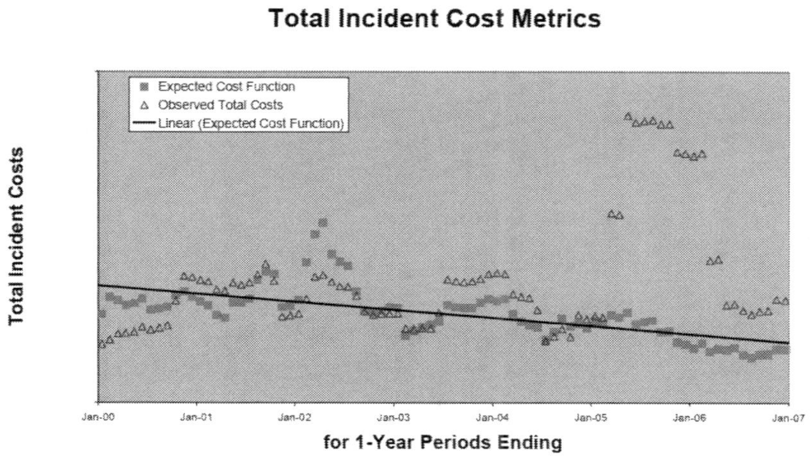

Figure 4 shows the expected cost function calculated for one-year periods by month (closed boxes) and the observed total costs for these same periods (open triangles). In addition a linear trend line has been plotted for the expected cost function.

Clearly much of the noise associated with the observed data has been removed using the expected cost function. The linear fit of the expected cost function indicates that aggregate incident costs have been reduced by over 50% during the course of the study period, representing a significant improvement in process safety performance. More detailed comparison of the incident datasets for 2000 and 2006 shows that approximately 15% of the estimated cost savings is related to the occurrence of fewer incidents (frequency reductions) and the remaining 35% reduction is related to a 40% reduction in the average cost per incident (severity reductions).

A further benefit of the ability to model incident data is that incident forecasting is now possible and the likelihood of incidents of a given magnitude can now be calculated. Figure 5 shows a comparison between the number of expected significant incidents and the observed number of significant incidents as calculated based on the obtained Shape parameters and measured incident frequencies (significant incidents have been defined based on the cost per incident exceeding a set threshold).

Figure 5.

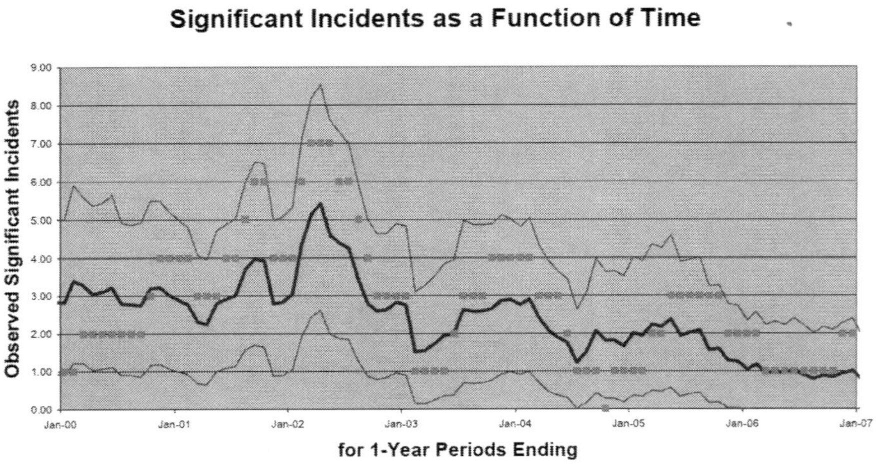

Figure 5 shows the modeled number of significant events for one-year periods by month (heavy line) along with confidence limits representing approximately two standard deviations (light lines). The closed boxes represent the observed number of significant incidents for each period.

As with the expected cost function plot the modeled number of significant events clearly indicates a downward trend. Very notably the number of significant incidents has been reduced from approximately 3 per year to 1 per year over the study period. As significant incidents require multiple layers of protection to fail the frequency reductions observed at this level are greater than those observed for smaller incidents. This is consistent with the originally proposed severity equation where there was an exponent relationship with the number of layers of protection that must fail to produce an incident of a given magnitude.

5. UNEXPECTED BENEFIT

The model approach and assessment of metrics such as the Shape parameter, typical layer of protection failure probabilities, and the expected cost function have given additional insight into the underlying process safety performance. Trends that are detectible with these functions clearly exist over the entire study period and are very clear with respect to the last five years. By comparison, traditional metrics failed to give this type of insight as the associated noise with these metrics overwhelmed the underlying signals they were intended to convey.

When modeling the expected number of significant events (including severities greater than that illustrated in Figure 5) there was a bias towards observing more incidents than what the modeling was predicting. This bias became even greater when larger severities were considered. A review of the incidents potentially making up these anomalies revealed that a few of these incidents were not consistent with the original assumptions. These "freak" incidents typically represented catastrophic events that were initiated outside of the control of the corporation. Examples include extreme weather events and major power outages. In these situations the assumption that a sufficient number of layers of protection existed to prevent the escalation of these incidents does not hold to be true. Figure 6 illustrates a comparison between the expected cost function and the observed total costs after the removal of these freak events from the dataset.

Figure 6.

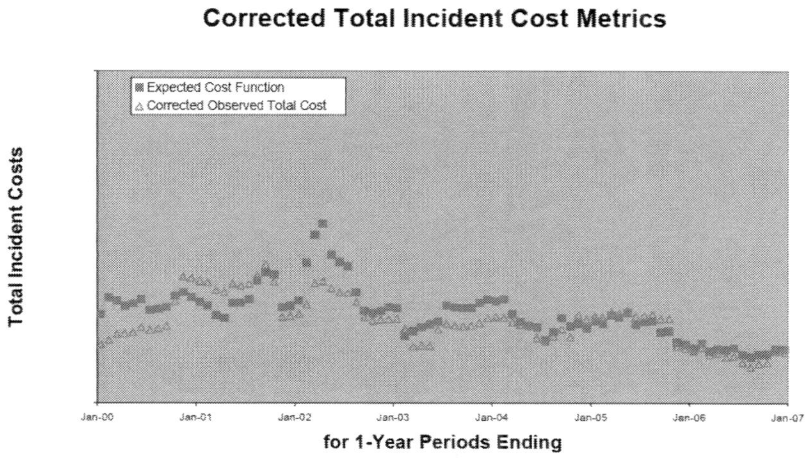

Figure 6 shows the expected cost function calculated for one-year periods by month (closed boxes) and the corrected observed total costs for those same periods (open triangles).

The alignment between the corrected observed total costs and the expected cost function readily shows the strength of this modeling approach. Clearly these freak incidents have always represented a portion of the overall incident costs. However, as the incident costs associated with "normal" incidents involving multiple layers of protection has been significantly reduced the percentage of the total incident costs represented by these freak incidents has increased. Without the use of metrics affording this level of resolution these types of observations are not readily possible and as such the allocation of process safety resources cannot be fully optimized.

6. CONCLUSIONS

The derived relationships and the agreement between theory and the observed data indicates that incidents as a group can be readily modeled (just over a thousand incidents were described by a two parameter curve fit with a R^2 value 0.9833). Further, the derived relationships readily generated far better resolution with respect to the ability to trend process safety performance than that which could be achieved using traditional metrics such as incident counts and related total incident costs. As an example trends showing improvements in performance were not observed based on total incident cost data whereas the derived expected cost functions showed very clear trends. In addition to identifying trends within the data it was possible to identify incidents (less than 1% of all incidents) that were inconsistent with the underlying assumptions through gaining significant insight into the observed distribution of incidents this then allows for the appropriate targeting of process safety related programs.

Secondary to the modeling successes the study also readily demonstrated and quantified the significant improvements that have been made with respect to process safety performance during the study period with the greatest improvements being made with respect to the reduction in the frequency and severity of major incidents.

Applied Risk-Based Process Safety: A consolidated risk register and focus on risk communication

Tracy Whipple [1] and Robin Pitblado [2]
1 BP Exploration Alaska, 900 E Benson Blvd, Anchorage, AK 99508
2 DNV Energy, 16340 Park Ten Place, Houston Texas 77084

1 Introduction

The process industry has been successful in controlling and reducing occupational health hazards, with close to a factor of 10 improvement achieved by several large companies over the past 20 years. The means for control of major hazards however has proven more difficult with a relatively steady number of major accidents, which has not shown a significant decline (e.g. Toulouse, Buncefield, Texas City, P-36) and in general is equally true for the upstream and downstream oil industry, even though these industries have quite different regulatory frameworks.

Given the lack of substantive improvements using traditional methods, there is value in examining alternative approaches where these would add value by complementing, rather than replacing, current process safety initiatives. Current initiatives are primarily driven by OSHA 1910.119 the Process Safety Management regulations, and by similar requirements from the EPA, as well as by a host of standards-based requirements. Some newer ideas from Europe attack the process safety problem in a different manner than PSM and appear to offer the potential for significant improvement.

The EU Directive for onshore process safety is covered in the Safety Case Regulations (e.g. COMAH in the UK). These have been through several revisions since their original issue in the 1980's to take account of actual accidents and lessons learned from other industries (e.g. upstream O&G). The barrier approach to risk management has received extensive treatment in Europe since the early 1990's, with strong reference to Prof Reason's swiss cheese model for accident causation. Shell used the bow-tie method to drive process safety at its Pernis Refinery in the Netherlands (Zuijderduijn, 2000) and at many other facilities globally. The EU sponsored the ARAMIS project that examined the Bow-Tie method in the early 2000's as a means to enhance process safety (Salvi and Debray, 2006). In Norway, a major project is looking at a similar method, titled Barrier and Operational Risk Assessment as a means to enhance process safety (Aven et al, 2006).

BP has had a comprehensive system for process safety at its sites globally since the 1970's and for many years based on a program titled Getting HSE Right. This shared most of the important aspects of PSM but with extra emphasis on risk management. BP updated this standard to align process safety with mechanical integrity and issued a standard titled Integrity Management (BP GS-01) in 2006. This standard implemented best practices in IM in the oil and gas industry based on practices both offshore and onshore, and incorporates several aspects from the barrier approach developed in Europe.

BP operates most of the North Slope facilities in Alaska. The North Slope represents the single largest area of production in the USA. It produces 400,000 bbls/day from Prudhoe Bay and contributes roughly 20% of US domestic supply and about 3% of total US crude requirements.

2 Objective for Hazard and Risk Register

BP's IM standard requires that every BP operation produces a Hazard and Risk Register for its operations to be in conformance with the standard. This paper describes the activities supporting this objective for

BP Exploration Alaska. The goal specified was that the Hazard and Risk Register should provide for transparency where all risks are in the same units for all activities, all areas, all consequences and that decision making would show a logical progression as shown:

Hazard identification → risk assessment → risk ranking → prioritization → decision making

This would be a significant change from the current situation where identified risks are in many different formats and non-comparable units, resulting in sometimes difficult decision making. Since factors other than risk need to be considered in the prioritization, the phrase "risk informed decision making" used in the nuclear industry is appropriate.

The Hazard and Risk Register is seen as a central tool in achieving BP's objective of Continuous Risk Reduction. As well as compiling all important risks, higher priority risks would have additional definition in terms of causes, controls, risk mitigation plans and accountabilities. A flow chart showing the process as would be applied in BPXA is shown in Figure 1.

Figure 1. Hazard and Risk Register role in continuous risk reduction

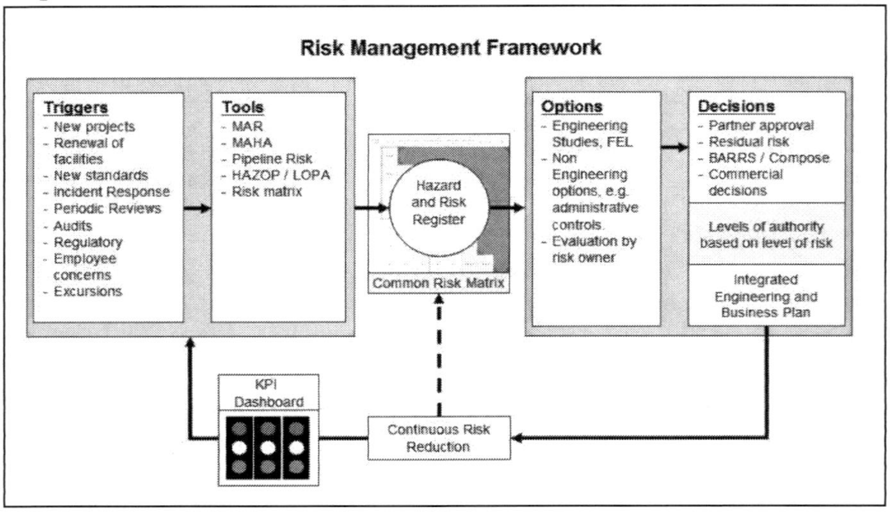

3 Creation of the Hazard and Risk Register

The first step in creation of the HRR is compilation of existing known risks. BP has several processes for generating risk information and these have been developed for different purposes and by different teams. These range from major risk studies with broad coverage and internally consistent methods to single ad hoc assessments using a wide variety of methods. The ad hoc studies were too varied in approach to bring under a common format and the initial HRR was developed using four main risk studies covering all important aspects of North Slope operating facilities. The four risk studies were:

- Export Pipeline risk assessment using the Muhlbauer (2004) method
- Well flowlines likelihood assessment using an inspection methodology
- Major Accident Risk (MAR) assessment using BP GP48-50 method
- Major Accident Hazard Analysis (MAHA) assessment using BP methodology

These methods used three different risk measures and addressed risks in significantly different ways. The desired common risk metric was BP's new 8x8 risk matrix (a simplified form of which is shown in Figure 2). This provides 15 levels of risk, which is considered sufficient to prioritize risks. Smaller matrices, while apparently easier to use, are often deficient for the purpose of prioritization. None of the risk studies used this metric, therefore the first step was to translate the diverse risk results into the common format. Reputation is also considered, but for Alaska the environmental and safety aspects drive reputation issues and this was not therefore separately considered. The BP matrix derives from the ISO17776 standard but has a greater degree of resolution by virtue of the 8x8 categorization.

Figure 2. BP 8x8 Risk Matrix

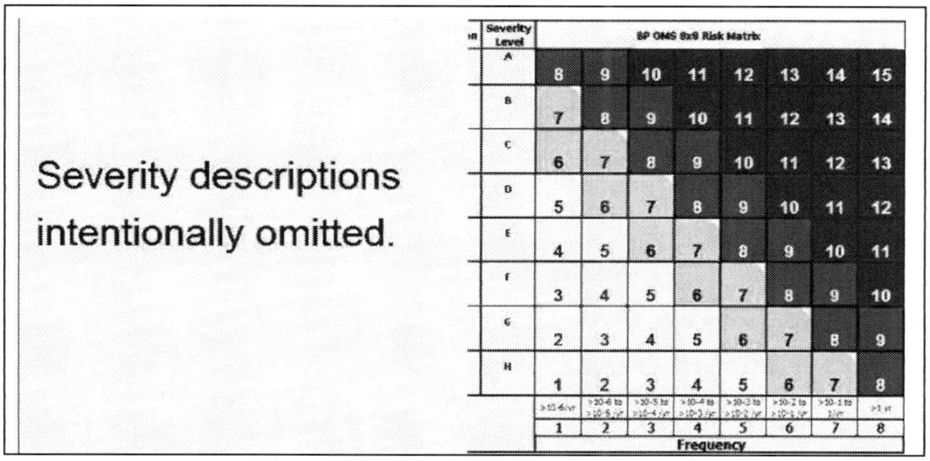

4 Translations

The Major Accident Risk (MAR) and Major Accident Hazard Analysis (MAHA) studies provided fairly direct estimates for all events in terms that could be translated directly into the risk matrix. However the pipeline and flowline studies provided primarily frequency information, and in the form of an index value rather than a quantitative likelihood of failure. These required conversion of an index to a frequency and consequence estimates to be developed for each event.

An objective of the translation was not to introduce any further uncertainty compared to the original risk estimate. Thus judgment was not used for the translation, instead the translation was achieved quantitatively. The likelihood index for the pipeline and flowline assessments was used to modify generic failure rate data which is associated with specific spill sizes and then the frequency and consequence categories could be obtained directly. US OPS and EU EGIG data for pipeline failure statistics were used as the basis for both pipelines and flowlines.

Table 1. Comparison of Pipeline Total Leak Frequencies

Data Source	Years	Total leak frequency All leaks per km - year
USA OPS Hazardous Liquids	1986 – 98	8.1 x 10^{-4} per km-year
USA OPS Gas Pipelines	1986 – 98	1.6 x 10^{-4} per km-year
CONCAWE Oil pipelines	1993 – 97	3.0 x 10^{-4} per km year
EGIG Gas Pipelines	1988 – 92	3.8 x 10^{-4} per km-year

All of these give leak frequencies in the range 1.6 to 8.1 x 10-4 leaks per km-year. The generic average is 4.1 x 10-4 leaks per km-year. Alaska pipelines differ to warmer climate pipes as permafrost issues require pipelines to be suspended aboveground. A detailed analysis for another operator in Alaska suggests a base operation for suspended pipelines might give a leak rate of 5.5 x 10-4 per km-year. For the purposes of this analysis, this base value was used as it gives leak statistics not dissimilar to past operational experience. It is expected that operational enhancements will reduce this value further. An analysis of leak size distributions from OPS statistical data is provided in Table 2.

Table 2 Hazardous Liquid Pipeline Spill Size Probabilities, 1986-98

Spill Size Range (m^3)	Nominal Spill Size (m^3)	Spill Probability
<1	0.3	0.15
1 – 10	3	0.19
11 – 100	30	0.41
101 – 1000	300	0.22
1001 – 10000	3000	0.03
>10000	30000	0.0004
TOTAL		1.00

Given the high sensitivity of the environment to spills in Alaska, a good split for a two leak size distribution would be at $100m^3$ with small leaks being represented by a $10m^3$ spill and large leaks by a $1000m^3$spill. The split on likelihood from Table 2 is:

Small = 0.15 + 0.19 + 0.41 = 0.75 = 75%

Large = 0.22 + 0.03 + 0.0004 = 0.25 = 25%

An example pipeline in BPXA is 27 miles long and using the generic pipeline leak rate, this gives a total leak frequency of 27 miles x 1.61 km/mile x 5.5 x 10-4 per km-year = 2.4e-2 per year and splitting this into small and large = 1.8e-2 and 6.0e-3 per year respectively. In BP Risk matrix terms these are Frequency categories 6 and 5 respectively, and these apply to environment and business interruption. As safety events giving jet fire outcomes can only affect 1% of the length, the safety frequencies are categories 4 and 3 respectively. Safety and environmental impacts can be read directly from the risk matrix. Business interruption was established by the product of expected days of outage and value of crude deliveries deferred.

Based on the analysis for consequence and frequency, categories for each spill size and type were determined and the risk read-off the BP 8x8 risk matrix.

Table 3. Example risk translation for hypothetical pipeline spill event

Leak size	Outcome type	Description	Consequence category	Frequency category	Risk
Small leak (10m3)	Safety	Serious injury	F	4	6
"	Environment	Acute recoverable damage to sensitive environment	E	6	9
"	Business interruption	14 days outage= $1.1m	F	6	9
Large leak (1,000m3)	Safety	3 fatalities	D	3	7
"	Environment	Widespread acute recoverable damage to sensitive environment	C	5	10
"	Business interruption	45 days outage = $3.3m	F	5	8

This translation was extended to all pipelines, to all flowlines, and to the other types of risk study.

The Hazard and Risk Register is laid out in tabular format with one row per hazard and the main columns being:

- Source Risk Tool
- Source Reference (date, revision)

- Event Identifier from Source Tool (Hazard reference)
- Business Unit
- Facility Name
- Hazard description
- Event Type (safety, environment, business interruption)
- Frequency (per year)
- Frequency Ranking (Risk matrix scale: 1 - 8)
- Consequence Ranking (Risk matrix scale: A - H)
- Risk Ranking (Risk matrix level: 1 - 15)

This currently contains 2600 entries and the same again may be expected to be entered in the next few years as additional risk studies are completed (PHA / LOPA 5 year revisions, Facility Siting results, employee concerns, ad hoc risk studies, etc). This becomes a major dataset and the ability to filter the contents for specific facilities, types of risk, etc becomes important for this to be manageable.

However it is relatively easier now to develop comparative priority risks and to communicate these to people who need to know these. This achieves the primary goal of allowing direct comparison of safety, environmental and business interruption risks for all aspects of North Slope operations in the same units of risk.

5 Controls Definitions

The BP IM Standard also calls for build-outs of the top 5 – 40 highest risks to define the causes and controls deployed to manage the risk, and where appropriate improvement measures to obtain continuous risk reduction. Controls include hardware, procedures and administrative controls, and are thus a larger set than safety critical equipment which are more narrowly defined and tend to be generic.

In the case of BP Alaska, the majority of higher risks are associated with environmental and business impacts from pipelines and wells, rather than safety risks from operating facilities for separation, compression, etc. Thus if simply the 20 highest risks were selected from the Register then these would probably only cover 2 or 3 different issues affecting environment and business interruption, but at different geographic locations. There would be many repeated controls. Therefore, judgment was employed to cover a wider variety of risks than simply the numerically highest contributors. The selection aimed to include the highest risks and a selection of higher safety risks from operating facilities where the greatest exposure of people would be expected.

The barrier approach used in the Bow-Tie method was selected and recorded in Excel and the software tool BowTieXP from Governors BV was used for graphics. Once the risk event was selected, the build-out followed a protocol similar to HAZOP. A team-based brainstorm generated the full list of causes and after some review to group those with similar controls, these were consolidated into list of causes. Typically this added to 4 – 8 causes. The team then identified for each cause the set of controls deployed, and typically this generated 3 – 10 controls. A simplified form of the diagram (Zuijderduijn, 2000) is shown in Figure 3 for one cause and one outcome.

Figure 3. Example Bow Tie Diagram

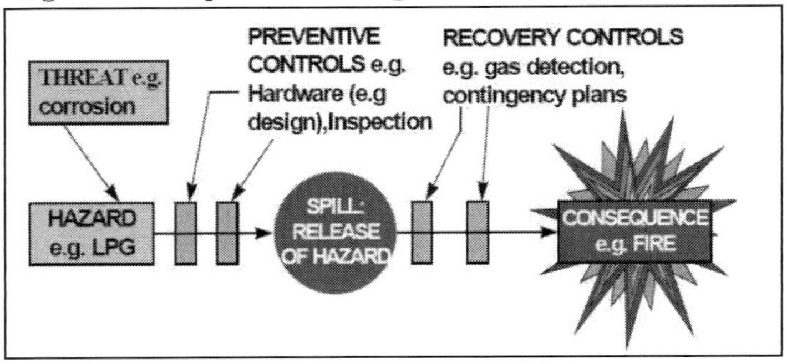

In practice this diagram captures the quantitative risk tools: Fault Tree Analysis (FTA) on the cause side and Event Tree Analysis (ETA) on the outcome side. A barrier on the left side actually represents an AND Gate with inputs being a Demand on the safety system and the Control failing. If desired, the detailed mechanisms of failure of each control can be built-out in the same manner. These are called Escalation Factors, since if these fail the barrier is degraded or fails completely. Development of Escalation Factors is a trade-off as it increases the complexity of the analysis and weakens the communication to the wider audience. The control owner is the person who most needs to understand how their control might be degraded.

Figure 4. Barrier Representations

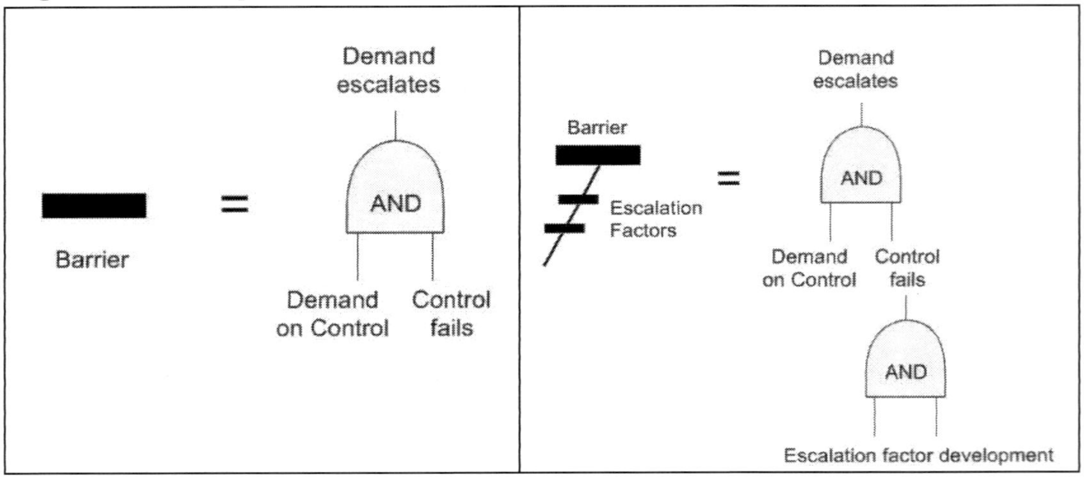

Prof Reason shows that controls are not 100% effective, all controls have holes (like swiss cheese) and it is important to design a system that has a sufficient number of controls and each of these with small enough holes. Since no control is 100% effective, a full control, scoring 1, is defined as being equivalent to a properly operating work permit system. Permits have a defined procedure with a task safety review, identified safety measures, trained issuers and users, and a paper trail to verify compliance.

The next step was to assign control ownership and to value each control. Owners were identified at supervisor to facility management level for the most part. A more difficult task was to determine the value of controls. Shell uses a 5 point system of + and – scores to attribute values.

- ++ : very good to excellent
- + : good to adequate
- ? : unknown
- - : insufficient
- -- : very bad or not existing

DNV developed an alternative scoring method which considers two aspects of control quality and which are relatively easy for the team to determine. The first aspect is how important the control is on that issue, and the second is how effective is the control. Many controls are directly important for the issue (e.g. PSV's for pressure relief or Training for procedures), but others are only indirectly important (e.g. Drug and Alcohol programs for human error). The control value is the product of the two factors. Thus using training as an example, if a specific training course has been developed to address a particular hazard (an importance/focus score would be 100%), but if only 50% of staff have been trained (an effectiveness of 50%), the control score would be 0.5. A simpler chart can also be used, but with less resolution.

Table 4. Control Scoring Chart

		Effectiveness / reliability / coverage of control	
		Weak	Full
Importance / relevance / focus of control	High	0.5	1.0
	Low	0	0.5

The numerical approach allows for some flexibility. There are some controls that should score more than one, since the benchmark is a Work Permit. For example PSV's which are properly designed and subject to a preventive maintenance program are highly reliable and might score 2. If a facility introduces a potential defeat, such as block valves before and after but car-sealed open, then the score might be reduced to 1.5. If the defeats are important then the control can be developed further, and this process is termed an escalation factor – those issues that might degrade the control. This could be examined in more detail using Escalation Factors. To account for dependency a discount factor can be applied. Thus procedures, training and competency are all good controls, with different owners and hence a need to track these separately, but their total control contribution is about 1, not 3, and a discount factor can correct for this.

To date 14 higher risks have had controls built-out using this method. These cover the highest risks (pipeline and flowline spills and well blowouts), as well as a sampling of facility hazards (leaks and explosions in process modules from compressors, vessels and other equipment) and natural hazards (e.g. subsidence). The first 11 have generated over 400 controls, but many of these were duplicates and the number of unique controls was 62, however the owners of the same control may differ as for example Work Permits are deployed and owned by different staff at different facilities. The main controls can be readily polled from the Hazard and Risk Register (Figure 5). A similar figure identifies the main control owners.

BP is using the BowTieXP program for graphical presentation of the risk and this program has additional features. Alternatively VBA tools in the Excel dataset also provide for additional visual treatment and verification of the bow ties.

Figure 5. Main controls identified

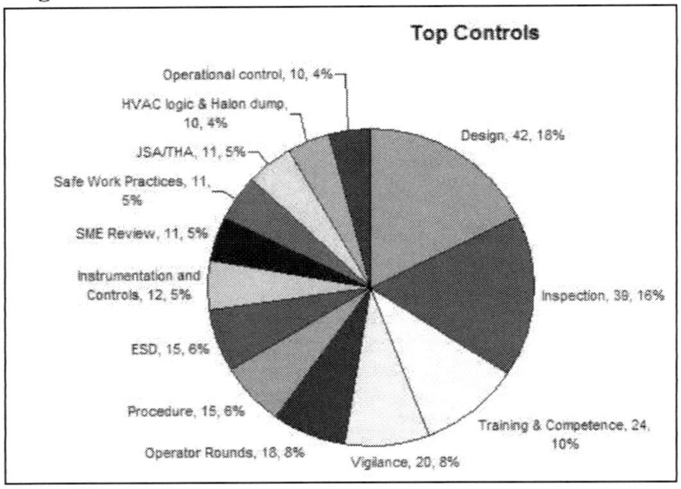

6 Continuous Risk Reduction

Once higher risks are identified and the main controls defined, the next step is Continuous Risk Reduction. Some companies have developed explicit rules for the number of controls required – high risks require 6 controls (3 each on both the cause and outcome side), medium risks – 4 controls, and low risks – 2 controls. Control scores are counted here, not the number of controls, and an excess of controls on one side (e.g. outcome controls) does not compensate for a lack on the cause side.

BP has not formally developed a quantitative target, however as the build-outs apply to the higher risks then a nominal target of 3 controls might be sensible. However some individual causes of an undesired event may be very unlikely compared to the main causes, thus even high risk events can have low risk causes and it is not sensible to apply a single target. For example vehicle impact and dropped object might be important causes of pipeline leak, but meteor impact would be a negligible cause.

The BP objective is Continuous Risk Reduction and this is most effective in reducing total risk when applied to the higher risks. In the past this tended to push risk owners (e.g. facility managers) towards adding extra controls. This may be a good idea, but it also adds burden to operations through extra tasks and complexity for staff. This barrier approach, as it builds-out all existing controls, documents better the degree of control on each separate cause and can identify those causes where there may be an insufficient number of controls and for which additional controls are warranted. However for others, it may be that improving the score of current controls is more effective than adding extra controls. Thus control owners can also focus on low scoring controls, and try to enhance these. An example of this might be a Training control scoring low as not everyone is trained. Rather than adding a new control, it would be better to train everyone and enhance the control count in that manner.

7 Future

The Hazard and Risk Register needs wide dissemination to be effective. Currently the tool resides in Excel and the next step will be to develop a web-based application that can be mounted on the BPXA Intranet.

Another aspect is linking control scores to actual operational experience. In real terms, safety performance tends to be two steps forward and one step backwards, as new controls add to safety but other factors tend to degrade safety.

Currently controls are scored based on team discussions. At best this is assertion and it only reflects the status at the time of the meeting. A future step is to link the ongoing incident records to control scores, as every accident or near miss requires several controls to have failed. Inherently, the swiss cheese model is based on some failures (the size of the holes), but actual failure rates can be derived from an analysis of accident causes. This would be done using Bayesian updates (real failure data to update a prior judgment estimate). This could be very useful as many controls appear in several event sequences. Thus if an accident indicates that for example Preventive Maintenance is deficient at one facility, the same control may be degraded at all facilities. Updating the hazard and risk register with the revised values may highlight that the place of the incident may not be the highest risk due to other controls deployed, but that other facilities using the same control may have no additional controls and are more at risk. The Register can be used to warm operational managers of weaknesses and facilities or activities most at risk and requiring greater attention or special temporary controls. It can also provide senior managers with a near real-time "gauge" of process safety as the total scoring of barriers deployed changes positively due to the CRR process but degrades due to other factors as shown by incidents.

8 Conclusion

A Hazard and Risk Register has been created for BPXA's facilities on the North Slope. It meets the specified objective of collecting together identified risks for all facilities and all activities and expresses these in common risk units covering safety, environmental and business interruption risks. This involved translation of current risk studies into a common set of risk units based on BP's 8x8 risk matrix. This provides for 15 levels of risk, categorized into 4 broad bands. Currently the BPXA register has 2600 entries and future risk studies are likely to double this number.

The Hazard and Risk Register is the most effective way to collect all of these risks and communicate the higher risk contributors to staff, management and state and federal regulators. The higher risks are further expanded to define the causes of the event and all the controls deployed to prevent the event or mitigate its outcome. It becomes feasible now to identify the most important controls deployed against the highest risks and to ensure that these controls get appropriate staff and management attention. Management of Change processes can ensure that no change is introduced that degrades these critical controls.

Identification of risks, knowledge of critical controls and accountability from control owners are key means to manage process safety. The register also provides a clear mean to achieve continuous risk reduction.

9 References

T Aven, S Sklet, J E Vinnem (2006), Barrier and operational risk analysis of hydrocarbon releases (BORA-Release) Part I. Method description, Journal of Hazardous Materials A137, p681–691.

Baker Panel (2007) The BP U.S. Refineries Independent Safety Review Panel. Chairman James A. Baker III. Available from BP.

BP GS-01 (2006), BP Global Mandatory Standard: Integrity Management.

CCPS (2007) Risk Based process Safety, Center for Chemical Process Safety, AIChE, Wiley InterScience.

W K Muhlbauer (2004) Pipeline Risk Management Manual, 3rd Ed, Gulf Professional Publishing, 395pp.

O Salvi and B Debray (2006), A global view on ARAMIS, a risk assessment methodology for industries in the framework of the SEVESO II directive, Journal of Hazardous Materials 130, p187–199.

C Zuijderduijn (2000), Risk Management By Shell Refinery/Chemicals at Pernis the Netherlands, EU Joint Research Center Conference on Seveso II Safety Cases, Athens.

Hydrogen Pipeline Compressor Incident Changes Standard Procedures

A. Daniel Fazekas
Air Products & Chemicals Inc.
Allentown, PA 18051 USA
fazekaad@airproducts.com

For a hydrogen production facility the feed gas and hydrogen product compressors are the heart of the plant, providing power to move gas through the process and the pipeline to the customers. If a mechanical problem occurs within the running gear, machine vibration is the result. In the past our company viewed the vibration switches on a reciprocating compressor as equipment protection. A recent loss of containment incident changed that concept. The incident at Air Products' Port Neches, Texas Hydrogen Pipeline Compressor Station in November 2006 caused a revision to the way we install and test the vibration switches to motor trip interlock and start up flammable gas reciprocating compressors after a vibration trip.

This paper outlines the use of an event tree quantitative risk analysis to determine the reliability required for the vibration switch, relay, and motor disconnect switch interlock for a Safety Instrumented Function (SIF). In addition to the SIF, administrative procedural changes were instituted to achieve a low hazard rate. The demand on the SIF and administrative protection systems was based on one loss of containment incident for our fleet consisting of 10,000 reciprocating cylinder operating years. Quantitative risk analysis required the vibration trip to be redundant per cylinder and proof tested with the motor disconnect switch and interposing relay at least once every 3 years for a 4 cylinder compressor. Existing operating procedures were also changed to include a thorough mechanical inspection with analysis of compressor operating data and start-up of the compressor on nitrogen instead of flammable gas after a vibration trip. Furthermore, new equipment siting procedures excludes sensitive equipment and buildings from direct line up with the cylinder heads.

As a result of one loss of containment incident our company has upgraded the compressor vibration trip to a critical safety interlock requiring increased mechanical integrity testing with more stringent operating procedures and equipment arrangement.

1.0 Introduction

For steam methane reforming hydrogen plants the natural gas and hydrogen product compressors provide the motive power to force the gas through the process equipment and the pipeline to the customer. In the past we classified the vibration switch interlock for motor shutdown of a reciprocating compressor as equipment protection. A loss of containment incident at our pipeline compressor station in Port Neches, Texas changed that strategy. As a result of the incident, the compressor vibration trip interlock was upgraded to a critical Safety Instrumented Function (SIF) requiring more thorough mechanical integrity testing and more stringent operating procedures.

2.0 Incident

On 24 November 2006 the K602 hydrogen compressor at Port Neches, Texas shutdown due to activation of a vibration switch. The mechanics' inspection of the running gear was limited to cylinders #1 and #2 with no indication of the cause for the trip. Thus the inspection did not find the broken crosshead on cylinder #3; when the compressor was restarted, immediate contact between the # 3 piston and cylinder head occurred. The vibration switch was not able to shutdown the motor in time to prevent loss of containment during the restart. One of the two vibration switches was later found not functioning and time to mechanical failure was very fast. To remove any air in the equipment prior to starting the compressor, the operators first purged with nitrogen, then pressure pumped and purged with hydrogen at 600 psig and 70 Degrees F from the pipeline. The compressor recycle valve was wide open, thus the suction and discharge pressure were nearly the same during the startup.

Mechanical failure of the crosshead for reasons still under investigation caused the piston rod and running gear assembly to be elongated and the piston contacted the cylinder head. The 11 bolts holding the cylinder head were stretched and sheared, and then the 250 pound head was dislodged and thrown 89 feet horizontally with a release of 40 pounds of 600 psig hydrogen in 5 seconds. The personnel at the site heard the rush of gas and felt a pressure shock wave in the control building. A limited inventory hydrogen vapor cloud explosion and flash fire occurred creating minor damage to the modular steel control building about 50 feet perpendicular to the gas flow path and on centerline with the motor and drive shaft. Site inspection revealed a smoldering electrical wire located 90 feet from the cylinder and 40 feet elevation and a grassfire along the gas dispersion flow path. Fortunately, the operator and mechanics were inside the control building and not injured. A mechanical diagram and pictures of the incident compressor follow:

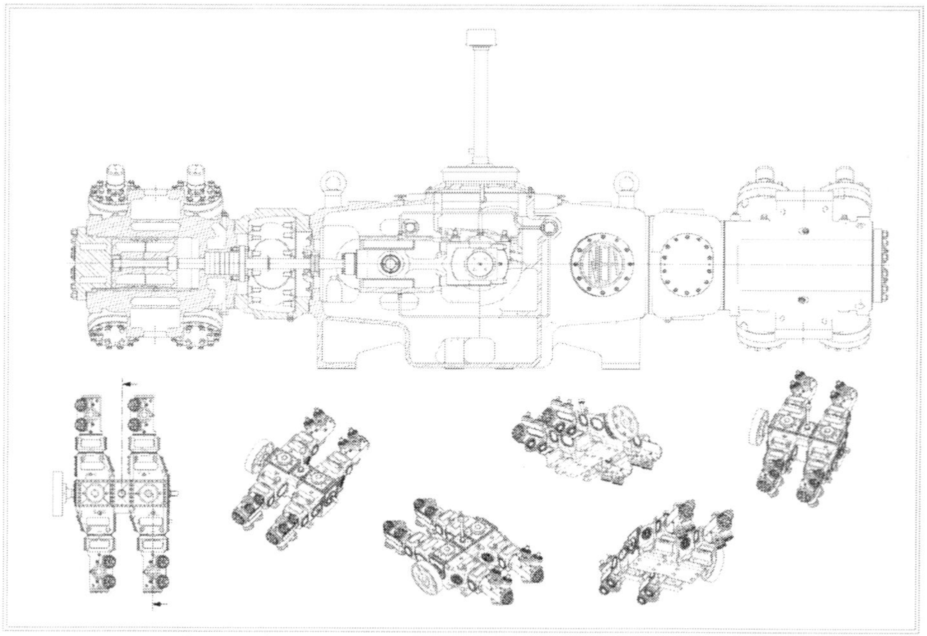

Reciprocating Hydrogen Compressor Mechanical Views

Photograph of the compressor prior to the incident

Photograph of the #3 piston "as-found" outside of the cylinder after the failure

2.1 Frequency of Event

Based on our history for operating reciprocating compressors, the demands on a vibration switch occur once every 10 compressor operating years. Analysis of the operating data for our reciprocating compressor fleet of 10,000 cylinder-years indicated there were three running gear crosshead mechanical failures, but only one of the three progressed to cause a loss of containment. The vibration trip interlock shutdown the compressor for each of the three failures of the running gear. The mechanics found and fixed two of the three running gear failures before restarting. Based on this information the average probability for loss of containment over a 20 year period was estimated to be 0.001 per cylinder based on the following exponential distribution for the single constant failure rate of once per 10,000 cylinder years, thus the loss of containment average probability per compressor was 0.004 for a four cylinder machine, and 0.006 for 6 cylinders, etc.

$$t = 20 years$$

$$\lambda = 1/10,000 cylinder - years$$

$$Pavg = (1/2)(1 - e^{-\lambda t})$$

$$Pavg = 0.001/cylinder$$

$$Pavg = 0.001/cylinder \times 4 cylinders/compressor = 0.004/compressor$$

3.0 Consequences

The consequences from loss of hydrogen containment were evaluated for a generic hydrogen plant compressor where the following events could occur: (1) launching the cylinder head as a missile with release of a physical energy gas pressure shock wave; (2) immediate ignition producing a jet fire or flash fire; and (3) delayed ignition causing an explosion. During normal compressor operation facility personnel can either be outside, randomly located within the 300 by 133 feet (40,000 square feet) compressor area of the facility, or inside the control building located along the motor centerline. The probability of being in the field in the area of the compressor or inside the control building, probability of ignition (immediate ignition producing a jet fire or delayed ignition producing an explosion or flash fire) and the probability of being fatally injured in the locations were analyzed and incorporated into an event tree quantitative risk analysis.

The probability for serious injury from each of the consequences was estimated. However, first a discharge flowrate was determined as follows: For the start up scenario as occurred in the Port Neche's incident when the compressor was shutdown and restarted prior to the loss of containment, the volume of the hydrogen in the compressor was estimated by adding the volume of gas in the equipment (filter, compressor cylinders, and aftercooler) and piping in between the suction block valve and discharge pipeline block valve. Total amount of 6 inch diameter (extra strong carbon steel pipe) was 88 feet on the suction and 131 feet on the discharge side of the compressor. Because of the cylinder internal check valves the hydrogen in the discharge pulsation snubbers and piping flowed through the recycle valve into the suction line and out the opening in cylinder #3 when the cylinder head was pushed off the cylinder by the piston. Density of the hydrogen was 0.224 pounds/cubic feet at 615 psia and 70 Deg. F. prior to release. The total internal compressor mass of hydrogen was estimated to be 43 pounds. Release rate calculations assuming a rupture disc opening of 3.9 inch diameter to simulate two suction check valves restricting the flow resulted in an initial flowrate of 31 pounds per second that reduced to 2.6 pound per second with-in a release duration of 4.8 seconds.

For the scenario where a mechanical failure occurs during normal operation such that the compressor inlet and outlet block valves are open to the pipeline or the equipment supplying the hydrogen and the vibration switch does not shutdown the compressor, the flowrate of hydrogen from the cylinder may last longer than the five seconds to deplete the gas inventory. During continuous operation upon releasing gas from the cylinder the low suction pressure trip may shutdown the compressor. However, an estimated time of 20 seconds was used for the continuous release case consequences to allow for manual interruption of the flow. In a hydrogen plant the inventory of gas at the suction of the compressor is usually thousands of pounds contained in the pressure swing adsorption unit for purification. The gas dispersion for the continuous flow creates a larger cloud of hydrogen downwind from the source than the fixed volume start up case. The time for a fully developed continuous dispersion of the hydrogen was 14 seconds compared to the 5 second puff of flammable gas released during the startup incident.

3.1 Physical Energy Release and Missile

The Port Neches #3 cylinder with a bore of 13.27 inches diameter and suction check valves with an equivalent diameter of 3.9 inches were the basis for the calculations. Total volume of 600 psig hydrogen was 0.79 actual cubic feet (0.18 pounds mass) released to atmosphere when the cylinder head was ejected within the shock wave limiting time of 0.002 to 0.008 seconds. The shock wave had a limited impact area with safe distance of 40 radius feet from the cylinder. However, the cylinder head became a missile that traveled 90 feet, through a chain link fence.

3.2 Jet Fire and Flash Fire

During the Port Neches incident the insulation on the power cable 90 feet away and 40 feet high was smoldering. In addition a grass fire occurred according to the eye witnesses.

For a small cylinder with 13.27 inch diameter bore a jet fire of continuous flow creates a flame length of 140 to 170 feet at 6 foot elevation in the horizontal direction, depending on the wind speed. The thermal radiation lethal area can extend 200 feet in line with the cylinder and 75 feet perpendicular depending on the flowrate and the cylinder diameter.

If the hydrogen should ignite after release and form a flash fire instead of a jet fire, the thermal radiation at a distance from the compressor cylinder of 90 feet was estimated as 7,500 Btu/hr/ft2 which is nearly enough to ignite wood at 8,000 Btu/hr/ft2. The calculated flash fire envelope was the distance the hydrogen travels until it is not flammable (below the lower explosive limit of 4% hydrogen concentration in air). The length of this envelope is 200 feet downwind of the release in the direction of the wind.

3.3 Explosion

A delayed ignition occurred during the Port Neches incident but the hydrogen cloud was not confined by equipment, thus the overpressure shock wave at the control room was estimated at 0.4 psi based on the TNO Multi-Energy Method with a confinement factor of 3 at a distance of 110 feet from the center of the cloud. The building was a portable modular type with steel tubing used in the frame with metal walls. Only minor damage occurred inside the building when shelves fell off the wall. Small windows in the door did not break.

If a delayed ignition of the released hydrogen occurs in congested equipment space with in a hydrogen generating facility an explosion can produce a lethal overpressure wave. In the typical hydrogen facility a pipe rack located 40 feet away and perpendicular to the cylinder centerline can create a congested space filled with hydrogen. The flammable zone of the hydrogen cloud can extend 200 feet from the cylinder thus engulf the piperack. Equipment and piping will cause the cloud to concentrate and not disperse freely. The cooling tower on the other side of the pipe-rack, that runs perpendicular to the released gas flow path, also creates a barrier to gas dispersion from grade to the top pipe row. To calculate the volume of the hydrogen in pipe rack the 30 feet width of the dispersion cloud was used as one dimension of the confined volume with the other two dimensions consisting of the pipe rack 10 feet width and 20 feet height. Total congested volume was 20 x 10 x 30 feet. Thus the confined volume in the pipe rack equaled 6,000 cubic feet and was assumed to contain a stoichiometric concentration of hydrogen (12 vol. % H2 in air), the energy of the explosion using the TNO Multi-Energy Method with a confinement factor of seven generated a shock wave side-on pressure of 4.3 psi at the control building 73 feet away.

The overpressure versus distance was calculated with the Multi-Energy program in the DNV PHAST 6.51 computer software. This pressure is high enough to cause damage to a steel frame, metal siding, pre-engineered control room building per the CCPS vulnerability versus overpressure graphs[1]. The vulnerability of a person in the building was estimated as 85% at 4.3 psi side-on pressure.

4. Event Tree

An event tree developed in Excel worksheets was used to determine the required reliability for the vibration switch Safety Instrumented Function (SIF) to detect a mechanical failure and shutdown the compressor motor. The event tree is provided at the end of this paper, the following is a brief description of the analysis:

4.1 Tree 1 Loss of Containment

The event tree begins with the frequency of running gear problems that place a demand on the vibration switches, estimated by our machinery department as once every 10 years. The frequency is multiplied by the probability of vibration switch failure in the failure branch that leads to an injury, and probability of vibration switch success and compressor shutdown in the success branch that lead to no injuries. The probability for success and failure must add to one in each column of the work sheet.

A fault tree representing the failure rates (L) of the redundant vibration switches with relay and motor controller disconnect interlock at a proof test interval (T) of once every 3 years resulted in an average probability of failing dangerous (PB) on demand of 0.089 as shown in the top "OR" gate labeled SIF in Figure 1. For the success branches in the event tree the probability that the vibration switch was successful in shutting down the compressor was 0.911 (1- 0.089).

[1] Figure 4.6 in CCPS *Guidelines for Evaluating Process Plant Buildings for External Explosions and F ires*

Figure 1
Fault Tree
Vibration Switch Motor Controller Shutdown Interlock

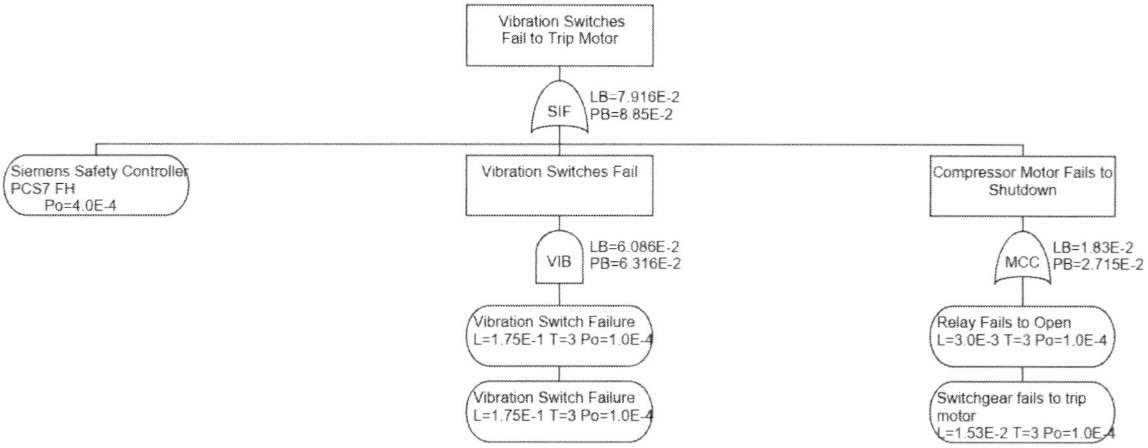

In the vibration switch failure branch of Event Tree 1, the probability of 0.004 that the demand on the vibration switch was caused by crosshead failure leading to loss of containment is multiplied by the demand 0.1/year and 0.089 probability that vibration switch fails. Thus a frequency of 3.56E-05/year leading to loss of hydrogen containment is calculated.

In the success branch for the vibration switch, the crosshead failure probability of 0.004 could still be the cause for the vibration trip. Since the vibration switch works and shuts down the compressors, the mechanics have a 5% probability of not finding and fixing the mechanical failure (95% successful). If the mechanics do not find and fix the failure, the operators have a 10% probability of not following procedures and start up the compressor with hydrogen instead of nitrogen (90% successful). Following the cross head failure branch leading to serious injury, the frequency of failure leading to loss of hydrogen containment 0.1/yr x 0.911 x 0.004 x 0.05 x 0.10 equaled 1.82E-6/year.

The frequency for loss of hydrogen containment is the sum of the two hydrogen release scenarios 1.82E-06 + 3.56E-5 equal to 3.74E-5/year. To determine the frequency of cylinder head ejection causing field operator injury from physical energy this frequency starts the calculations in Tree 2. To determine the frequency of cylinder head ejection causing field operator injury from hydrogen ignition causing either a jet fire or explosion this frequency starts the calculations in Tree 3. To determine the frequency of cylinder head ejection causing control room operator injury from an explosion this frequency starts the calculations in Tree 4.

The frequency for loss of containment for starting up with nitrogen instead of hydrogen was 0.1/yr x 0.911 x 0.004 x 0.05 x 0.90 equal to 1.64E-05/year. Nitrogen start up at a pressure of 150 psig with non flammable gas had less consequence for loss of containment. To complete the analysis for the frequency of impact on the field operator, the calculation continues in Tree 5.

4.2 Tree 2 600 psig H2 Released Physical Energy or Missile impact on Field Operator

Starting with the frequency of 3.74E-5/year for loss of hydrogen containment from Tree 1, the probability for a person being in the field at our occupied facility was estimated at 25% based on a survey at an existing hydrogen plant. The area of impact for the cylinder head was limited to a 2 by 100 foot area out of the 40,000 square foot area that an operator may be when in the field. Probability for serious injury from the missile impact was assumed to be 90%, thus the frequency of injury from the missile was 3.74E-5/yr x 0.25 x (200/40,000) x 0.9 equaled 4.22E-08/year.

The area of impact for the physical energy release from hydrogen is larger than the missile impact area where the safe zone (overpressure of less than 1 psi) is a 30 feet radius from the cylinder. For pressures between 1 and 10 psi the lethal vulnerability for outdoor exposure is assumed to be proportional to the overpressure such that vulnerability equals [0.111 x overpressure (psi)] – 0.111. When integrated over the 30 foot radius the probability of a fatality in the 40,000 square foot zone was 0.016. Thus the frequency of injury from the physical energy release was calculated as 3.74E-5/yr x 0.25 x 0.995 x 0.016 equaled 1.49E-07/year. Adding the missile and physical energy frequency of fatality equaled 1.91E-07/year

4.3 Tree 3 600 psig H2 Released with either immediate or delayed ignition impact on Field Operator

Starting with the frequency of 3.74E-5/year for loss of hydrogen containment from Tree 1, an estimate of 60% of released hydrogen streams will ignite based on our experience with hydrogen leaks was used. If the hydrogen ignites there is a lower probability that it will be a delayed ignition in a confined or congested space. The probability of explosion given ignition for the fixed inventory start up case equaled 8 % based on the estimates in IChem[2] for an average flowrate of 8 pounds per second (3.6 kg/sec). If the probability of explosion given a leak was based on a continuous flow of 31 pounds per second (14 kg/sec), the probability increased to 16% or twice the start up rate, thus the limits of ignition probabilities given a leak was 8% for the start up fixed inventory case and 16% for the continuous release case. Since the continuous release case had a higher frequency of occurrence than the start up case, 16% was used in the quantitative event tree analysis.

The probability for a person being in the field at our occupied facility was estimated at 25% based on a survey at existing hydrogen plants.

[2] Figure 16.1 of IChem *Identification of Hazardous Locations*

Probability for lethal or serious injury from a jet fire was estimated assuming a continuous release and personnel exposure time of 20 seconds. The heat flux generated by a jet fire was calculated with the DNV PHAST 5.1 software using the Shell model. The thermal dosage impact on humans was estimated using the probit equation in PHAST where exposure time is in seconds and heat flux is measured in watts per square meter. The probit value can be converted into a probability of fatality for the ground area where a person is exposed to heat flux from the jet fire. Based on the heat flux converted to a probability of fatality (% lethality) and determining the area for each % Lethality, integrating the fatal area ellipses, then dividing the total area by 40,000 square feet, a probability of fatality from thermal radiation was estimated.

The following equation was used by PHAST for the Probit which is converted into % lethality by an error function as described in CCPS[3]: Probit = $-36.38 + 2.56 \ln (t I^{1.33})$ where t = exposure time in seconds and I = heat flux in watts per square meter. Figure 2 graphs ellipses of constant % lethal probability for the plan view of the compressor operating area at 6 feet elevation. By integrating the concentric ellipses of % lethal probability and dividing by the total area an operator can be located; the average probability for the entire area of 45% was obtained.

Figure 2
Jet Fire Constant %Lethal Probability Plan View
(Smallest radius ellipse is highest % Lethal Probability)

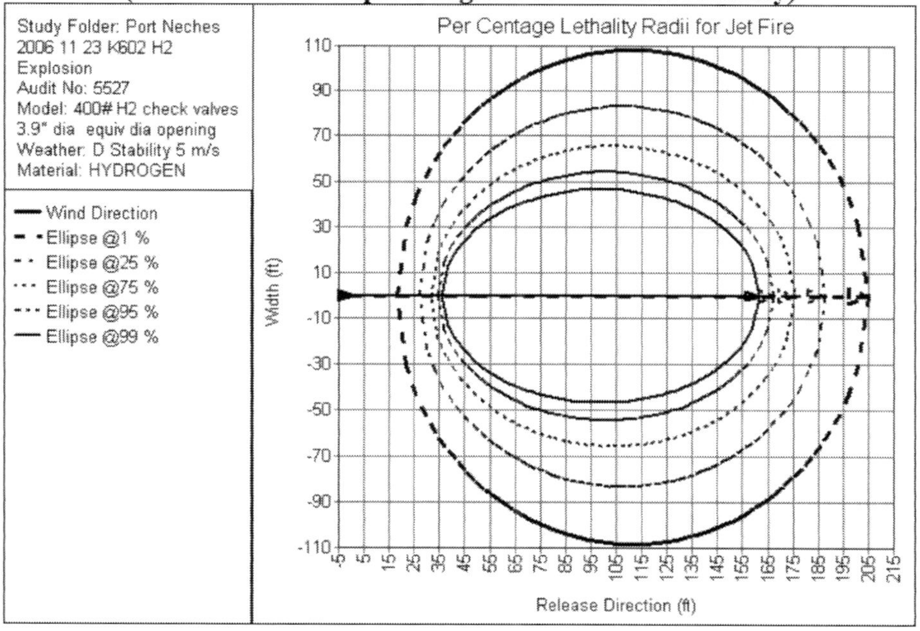

Thus the frequency for fatality from the jet fire 3.74E-5/yr x 0.6 x 0.84 x 0.25 x 0.45 equaled 2.12 E-06/yr.

[3] Table 4.1 of CCPS *Guidelines for Consequence Analysis of Chemical Releases*

The frequency of fatality from an explosion 3.74E-5/yr x 0.6 x 0.16 x 0.25 x 0.63 equaled 5.66 E-07/yr. Where 63% probability of serious injury from an explosion was calculated from concentric circles similar to the heat flux for the jet fire but using the vulnerability for overpressure as a function of distance as described in section 4.2 instead of heat flux % lethal probability. Adding the frequency for thermal radiation and explosion impact on the field operator equaled 2.69E-06/yr.

4.4 Tree 4 600 psig H2 Released with either immediate or delayed ignition impact on Control Room Operator

Starting with the frequency of 3.74E-5/year for loss of hydrogen containment from Tree 1, multiplying by the 98% probability that a person is in the control room and 60% and 16% (probability of delayed ignition given a leak and explosion given ignition for a hydrogen release, respectively) as discussed in section 3.3. The building adds an additional risk since it may collapse or internal walls and furnishings may become missiles to cause serious injury. The vulnerability of people in the building was estimated at 85% based on a 4.3 psi overpressure for a steel frame, metal siding building located 73 feet away per CCPS graph[4]. Thus the frequency for serious injury 3.74E-5/yr x 0.6 x 0.16 x 0.98 x 0.85 equaled 2.99 E-06/yr.

Since the building is metal and insulated it provides some protection from the jet fire thermal radiation for the control room operator who has a much lower probability of exposure indoors than outdoors, so an estimate of 0.5% was used and probability of serious injury if exposed was the same as Tree 3. Thus the frequency for serious injury 3.74E-5/yr x 0.6 x 0.84 x 0.005 x 0.45 equaled 4.24 E-08/yr. Adding the thermal and explosion hazard rates equaled 3.03 E-6/yr.

4.5 Tree 5 125 psig N2 Released Physical Energy or Missile impact on Field Operator

When the operator starts the compressor using nitrogen, discharge pressure is limited to 125 psig, thus the impact on the field operator is considerably less than 600 psig hydrogen if loss of containment occurred. The calculation starts with the frequency of 1.64E-5/year for loss of nitrogen containment from Tree 1, for the scenario where the operator starts up using nitrogen and the crosshead fails. The probability for a person being in the field at our occupied facility was estimated at 25%. The area of impact for the cylinder head was limited to a 2 by 100 foot area out of the 40,000 square foot area that an operator may be located when in the field. Probability for serious injury from the missile was assumed to be 90%, thus the frequency of injury from the missile 1.64E-5/yr x 0.25 x (200/40,000) x 0.9 equaled 1.85E-08/year.

[4] Figure 4.6 in CCPS *Guidelines for Evaluating Process Plant Buildings for External Explosions and Fires*

The area of impact for the physical energy release was less than 13 feet radius from the cylinder for an overpressure greater than 1 psi. For pressures between 1 and 10 psi the vulnerability was proportion to the overpressure such that vulnerability = [0.111 x overpressure (psi)] – 0.111. When integrated over the 13 foot radius the probability of a fatality in the zone was 0.28%. Thus the frequency of injury from the physical energy 1.64E-5/yr x 0.25 x 0.995 x 0.0028 equaled 1.14E-08/year.

Adding the missile and physical energy frequency of fatality equaled 2.99E-08/year

4.6 Summary of Event Tree Calculations

The frequencies of serious injury or fatality from each consequence scenario were added in Table 1 to determine the total hazard frequency rate of 2.9E-6/year for the field operator and 3.0E-6/year for the control room operator.

Table 1
H2 Compressor Crosshead Failure
Hazard Rate Frequency

Person	Event Tree	Scenario	Frequency/ year
Field Operator	2	600 psig H2 Physical Energy & missile	1.9 E-07
	3	H2 Jet Fire or Explosion	2.7E-06
	5	125 psig N2 Physical Energy & missile	3.0E-08
		Total	2.9E-06
Control Room Operator	4	H2 Jet Fire or Explosion	3.0E-06

5.0 Solutions

After completing the root cause analysis and event tree quantification the following key solutions were implemented to prevent recurrence of loss of containment from a compressor crosshead failure:

- Mechanical changes were made to reduce stress on crosshead running gear,
- Began evaluating active transmitters to replace passive switches.
- Instituted a revised start-up procedure after reciprocating compressor vibration trips, where:
  - Machinery Engineer analyzes operating and maintenance data
  - Distance piece and frame extension covers removed
  - Inspection of rods and running gear
  - Manually bar over compressor
  - Inspection and function testing the vibration switch
  - Safe location for people during start up, out of line of site with cylinders and motor terminal box

 - Test run on nitrogen for 10 minutes for mechanical integrity
 - Restart on flammable gas
- Facility equipment siting revised:
 - Hazard zone in line with cylinder travel avoids placing equipment or buildings in direct line with cylinders
 - Control room building location considered relative to flammable gas compressor and congested piping or equipment space
- Vibration switches and motor disconnect switches upgraded to critical safety devices for flammable gas compressors requiring periodic testing:
 - 3 year proof test interval for 4 cylinder compressor driven by one motor
 - 2 year proof test interval for 6 cylinder compressor driven by one motor
 - 1 year proof test interval for 12 cylinder compressor driven by one motor
 - Maximum of 5 year proof test interval for 2 or 1 cylinder compressors
 - Vibration switches are redundant for each cylinder
 - Vibration switch typically located between cylinders so one switch monitors two cylinders arranged side by side
 - Minimum of two vibration switches for single cylinder compressor

6.0 Conclusion

The incident at Air Products' Port Neches H2 Pipeline Compressor Station in November 2006 caused a revision to the way we install and test the vibration trips on compressors, and mechanically inspect and start up a flammable gas compressor after a vibration trip. An event tree quantitative risk analysis determined that the vibration trip should be redundant per cylinder and proof tested with the switchgear. Operating procedures were changed to include mechanical inspection and analysis of compressor operating data prior to starting the compressor on nitrogen after a vibration trip.

7.0 References

1. Cox, A.W., Lee, F.P, Ang, M.L., *Classification of Hazardous Locations,* Rugby, Warwickshire, England: Institution of Chemical Engineers, 1990.
2. *Guidelines for Evaluating Process Plant Buildings for External Explosions and Fires,* New York: American Institute of Chemical Engineers, Center for Chemical Process Safety, 1996
3. *Guidelines for Consequence Analysis of Chemical Releases,* New York: American Institute of Chemical Engineers, Center for Chemical Process Safety, 1999

Tree 1

Event Tree for H2 Compressor Crosshead Failure

Scenario: Major Mechanical Failure Once in 10 years. Average Probability of Crosshead Failure 0.004 (4 cyl per comp x 20 years/10,000 cyl-yrs/2)

Design Basis: Large Hydrogen Plant

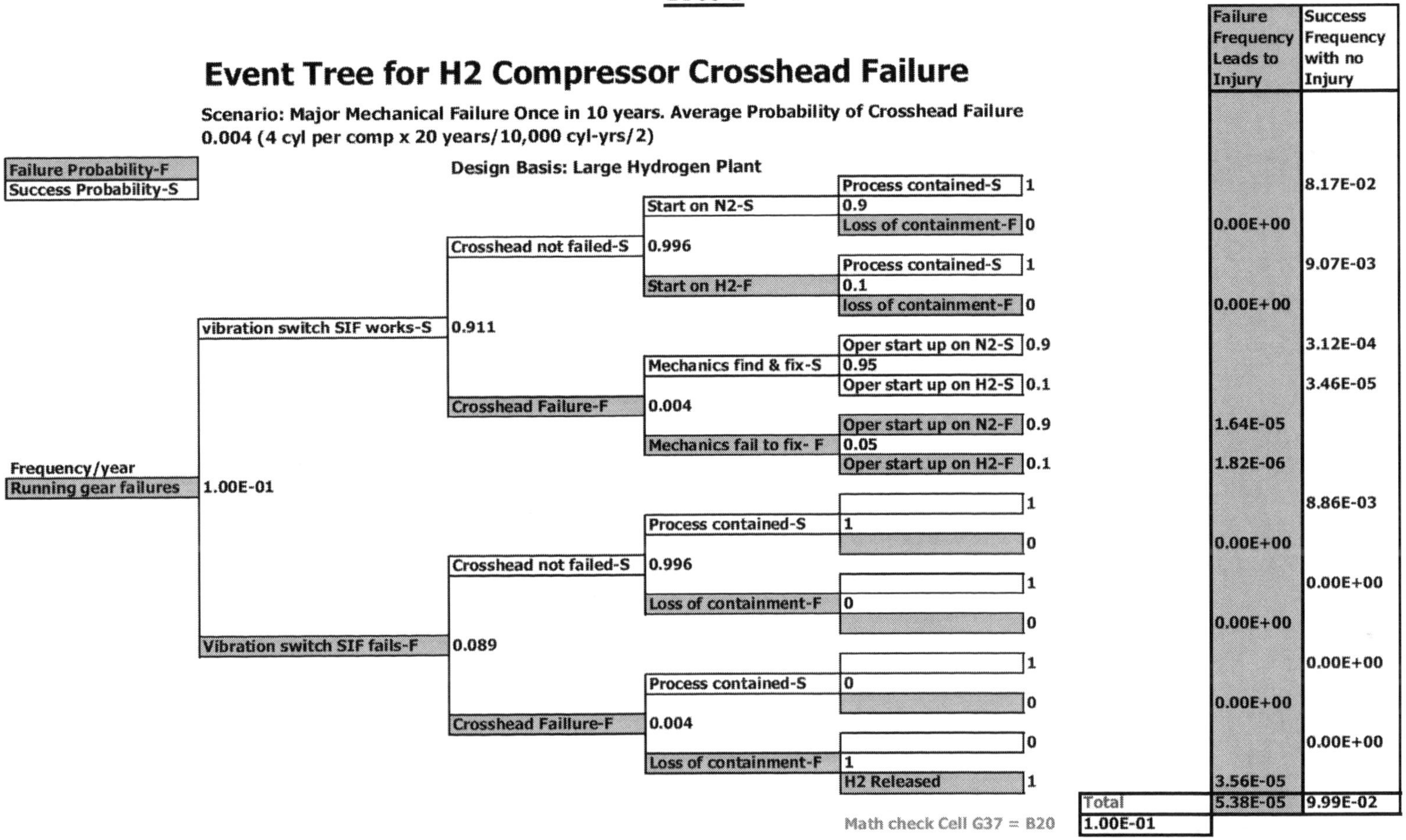

A. Frequency of cylinder head ejection on H2 to determine field operator injury from PV Energy Release and Missile
B. Frequency of H2 Release to determine field operator injury from H2 immediate ignition or delayed explosion
C. Frequency of H2 Release to determine control room operator injury from H2 immediate ignition or delayed explosion
D. Frequency of cylinder head ejection on N2 to determine field operator injury from PV Energy Release and Missile

A go to Tree 2	3.74E-05
B go to Tree 3	3.74E-05
C go to Tree 4	3.74E-05
D go to Tree 5	1.64E-05

Tree 2

Event Tree for H2 Compressor Crosshead Failure

Scenario: 600 psig H2 Released Physical Energy or Missile

Person Exposed: Field Operator

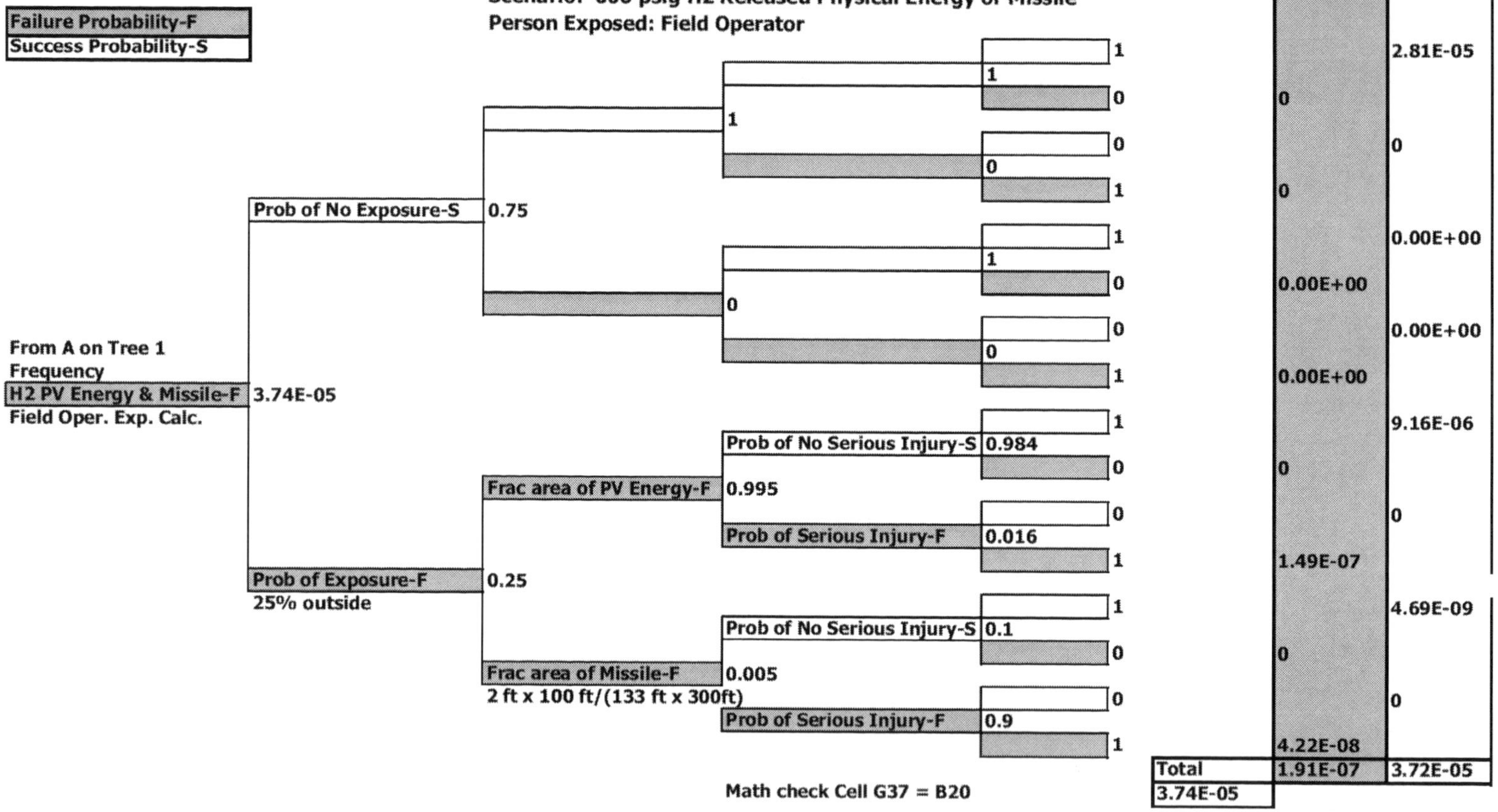

Tree 3

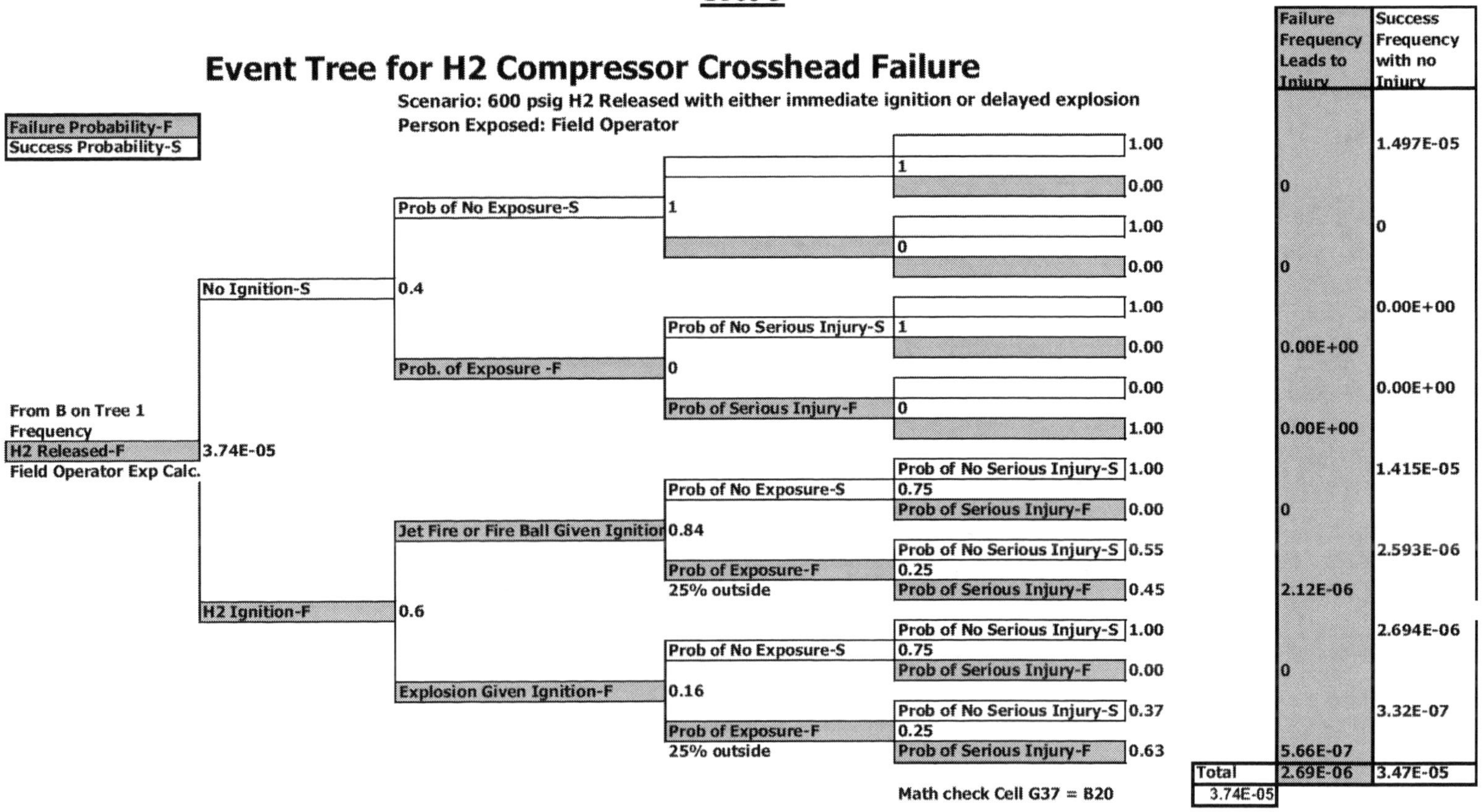
Event Tree for H2 Compressor Crosshead Failure
Scenario: 600 psig H2 Released with either immediate ignition or delayed explosion
Person Exposed: Field Operator
Failure Probability-F
Success Probability-S
From B on Tree 1
Frequency
H2 Released-F
3.74E-05
Field Operator Exp Calc.
No Ignition-S
0.4
H2 Ignition-F
0.6
Prob of No Exposure-S
1
Prob. of Exposure -F
0
Jet Fire or Fire Ball Given Ignitior
0.84
Explosion Given Ignition-F
0.16
Prob of No Serious Injury-S
1
Prob of Serious Injury-F
0
Prob of No Exposure-S
0.75
Prob of Exposure-F
0.25
25% outside
Prob of No Serious Injury-S
Prob of Serious Injury-F
1.00
0.00
0.55
0.45
0.37
0.63
Failure Frequency Leads to Injury
Success Frequency with no Injury
1.497E-05
0
0.00E+00
0.00E+00
1.415E-05
2.593E-06
2.12E-06
2.694E-06
3.32E-07
5.66E-07
Total
2.69E-06
3.47E-05
3.74E-05
Math check Cell G37 = B20

Tree 4

Event Tree for H2 Compressor Crosshead Failure

Scenario: 600 psig H2 Released with either immediate ignition or delayed explosion
Person Exposed: Control Room Operator

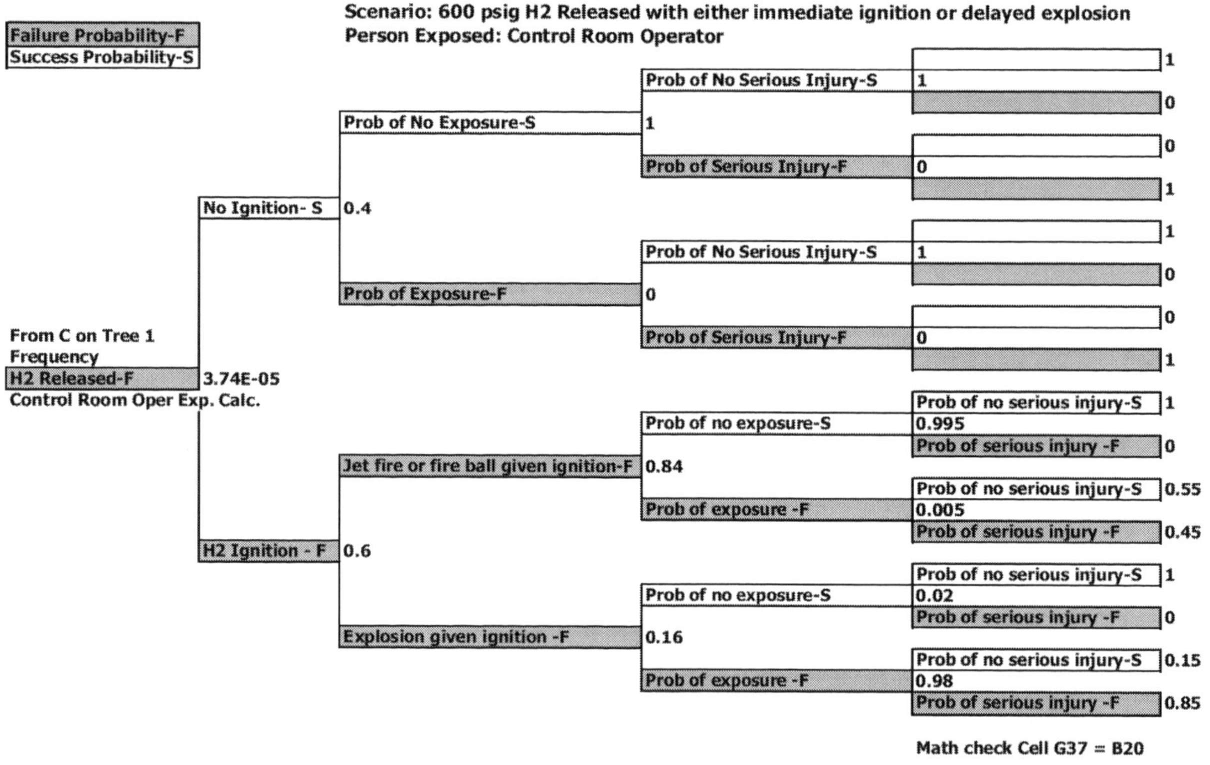

	Failure Frequency Leads to Injury	Success Frequency with no Injury
		1.50E-05
	0.00E+00	
		0.00E+00
	0.00E+00	
		0.00E+00
	0.00E+00	
		0.00E+00
	0.00E+00	
		1.88E-05
	0.00E+00	
		5.19E-08
	4.24E-08	
		7.19E-08
	0.00E+00	
		5.28E-07
	2.99E-06	
Total	3.03E-06	3.44E-05
3.74E-05		

Tree 5

Event Tree for H2 Compressor Crosshead Failure

Scenario: 125 psig N2 Released Physical Energy or Missile
Person Exposed: Field Operator

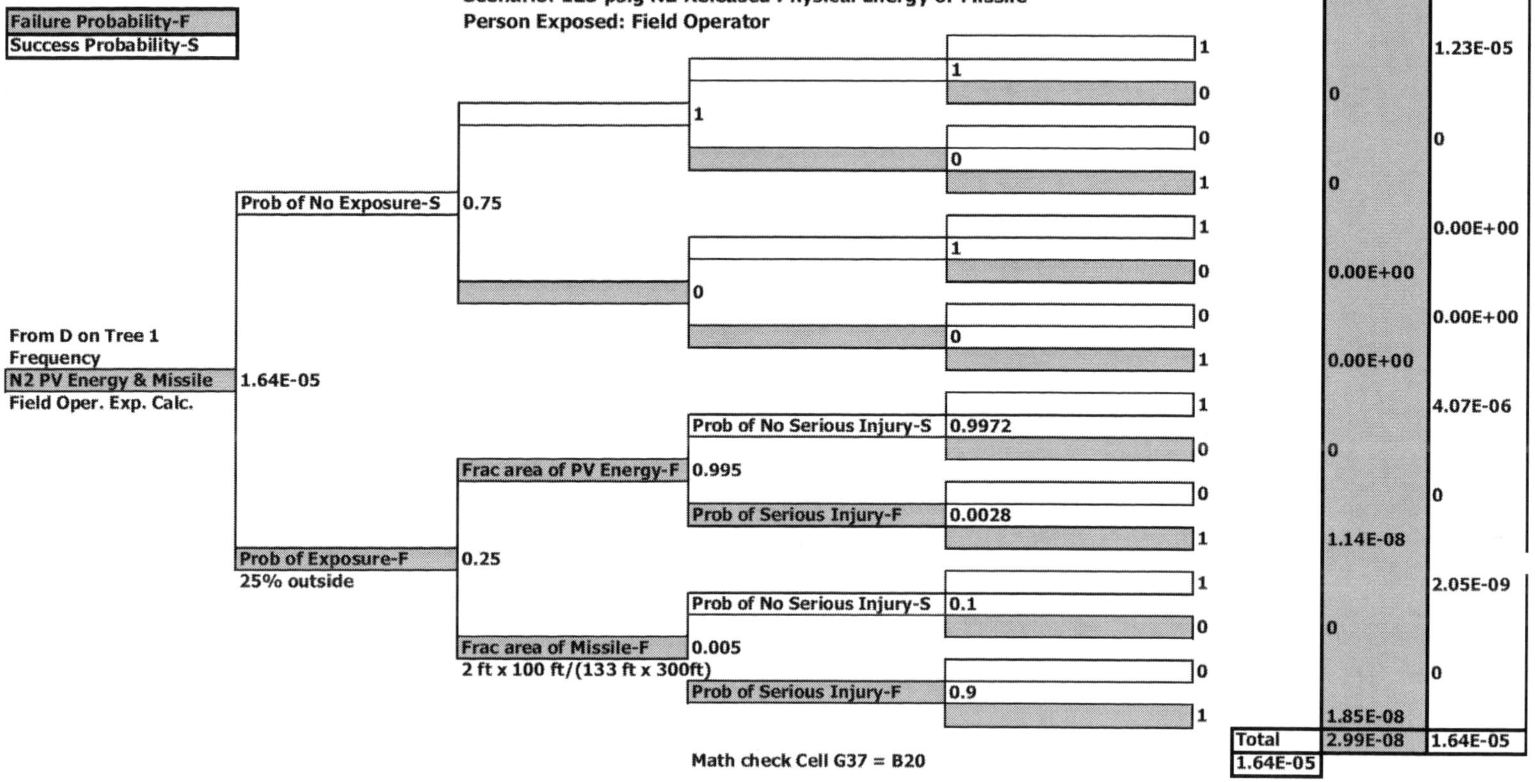

COMMITTING TO PROCESS SAFETY – A RELIANCE EXPERIENCE

Sushil Kumar,
Site President,
Dahej Manufacturing Division,
Reliance Industries Limited

HSE Excellence is adopted as a business strategy in Reliance and the need for more structured & rigorous way is identified to be the Company's need of the hour. Achieving excellence in Process Safety is increasingly important to the organization to cope with rapidly changing environments, more complex and interdependent sets of technologies and integration of diverse manpower to support the growth strategy. This paper brings out the approach adopted at Reliance in realising this objective of reaching world class in process safety i.e. (1) how is this aspect demonstrated at senior management including Board level and, (2) how is this embedded and exhibited at other levels almost as opportunities. We articulate these experiences through this paper.

Reliance Board of Directors have appointed a Board Level Committee on Health, Safety and Environment and have established a Centre for HSE Excellence (CHSEE) as a corporate entity. The objective set by the Directors is to bring quantum jump to reach world class in process safety. This resolve has been communicated by the Chairman of the Board HSE Committee & Executive Director to manufacturing sites and all line managers in his communications and during his site visits.

The Board Committee reviews the Process Safety performance of the company. The Directors ensure that proper resources are committed to demonstrate overall management leadership and commitment, and that adequate attention is paid to PSM during activities such as forming joint ventures, making acquisitions, contract operations.

The CHSEE is responsible for publishing Corporate Standards, Competency building and bringing in best practices. The standards are being prepared to be of international standard. The company is partnering with world best organizations such as Dupont & CCPS. CHSEE is staffed with experts in different fields including PSM competency leaders. As safe behavior and good safety culture are the foundations, behavioral safety and work place safety experts supplement PSM efforts in the CHSEE.

The corporate leadership has the responsibility to lead and manage process safety activities. The corporate leadership also works to integrate other parallel site-based risk management initiatives such as PIAMS, IMS and RCM into PSM systems and processes to the extent appropriate.

The site leadership has the responsibility to assign adequate resources to PSM in line with Corporate vision. Line management in businesses, sites and functions has the responsibility to implement Corporate standards.

In line with the understanding that the management leadership and commitment are basic and essential to successful management of process safety, effective management leadership at Reliance include the following activities:

- Each leader has a personal safety action plan
- Participate personally in activities that visibly demonstrate a commitment to PSM.
- Establishing PSM guidelines to ensure that each site has an effective PSM program.
- Committing resources to implement PSM policies and standards and to sustain continuous improvement of process safety.

This objective and commitment is followed and replicated with required rigour and zeal at manufacturing locations / sites. The responsibility is directly resting with site president with a site PSM committee chaired by a direct report and a full time PSM champion coordinating efforts. Around 10% of the time of line management is typically spent for PSM efforts and resource allocation is prioritized. Senior executives and line managers are extensively visible, demonstrative and committed to participating in steerage, Hazard Awareness & Evaluation and understanding of Process safety, capability building, training and updating themselves and in establishing and administering site systems to achieve and maintain a high degree of operational discipline.

We have embarked on a major exercise on competency building for identified personnel both to have a resource pool to help company achieve quantum jump and to be able to sustain the efforts in long run.

Reliance understands that an effective PSM relies on the individual and collective efforts of everyone involved in managing process safety. Recognizing that employees are uniquely knowledgeable about the process and are "key managers" of process safety (often the only "managers" immediately available), we provide for and encourage a broad spectrum of employee involvement in implementation of the PSM program.

We have accordingly the set up in which ownership of Process safety is with the line organization as a value as important as productivity and quality. First line supervisors and workers are being encouraged both through various structured processes and informally to participate in overall efforts. Some examples involving employees include the following:

- Membership on site PSM committees or subcommittees
- Participation in pre start-up reviews
- Conducting equipment tests and inspections
- Writing and/or reviewing operating and maintenance procedures
- Participation in development of training procedures and programs

As specific immediate measure, operators are being involved in updation of Process Safety Information, Hazard Analyses, Pre - Startup Safety Review, incident reporting, review & updation of Operating Procedures and Management of Change.

Within each element of PSM, site leaderships solicit, on an ongoing basis, participation and input from employees regarding ideas for improving and strengthening site PSM programs and systems.

A LOPA CASE STUDY

Michela Gentile, Ph.D.
BP America
Houston, TX

John Baik, Ph.D.
BP America
Houston, TX

For presentation at:

2008 Spring National Meeting of
American Institute of Chemical Engineers
New Orleans, LA
April 6 – 10, 2008

ABSTRACT

The original concepts that led to today's Layers of Protection Analysis (LOPA) were developed in the US in the late '80s and early '90s. In 2001 CCPS developed the LOPA method as a systematic approach to ensure sufficient protective barriers are available to prevent specific hazardous scenarios or mitigate the consequence.

Different LOPA methodologies have been used: some are based on simple frequency-consequences matrices, others are based on risk targets; some follow the order of magnitude approach, others follow a simplified quantitative risk assessment (QRA) approach.

LOPA is not supposed to be a fancy HAZOP, nor a simplified QRA. This paper presents some often misunderstood LOPA concepts using a case study.

1. Introduction

In the world of process safety, LOPA is a well known acronym that represents a methodology for hazard and risk assessment. The methodology was developed as a simplified form of risk assessment that might be used in cases where a simple HAZOP was deemed insufficient to fully understand the intrinsic hazards and risk associated to a specific scenario. For such scenarios, it was also recognized that a full quantitative risk assessment (QRA) would demand significant expenses in terms of time and resources, with the potential to obtain results that might be approximate because of the lack of accurate data and information.

The Center for Chemical process Safety (CCPS) in 1993 published "Guidelines for Safe Automation of Chemical Processes". This book introduced the concept of Layer of Protection and the approach to analyze if sufficient independent layers were available (1). The LOPA methodology was further defined by the 2001 CCPS book, "Layers of Protection Analysis" (2) and the concepts were retaken in "Guidelines for Safe and Reliable Instrumented Protective Systems" (3), published in 2007 by CCPS. A number of other publications on the topics of layers of protection and LOPAs have also been produced during the last fifteen years.

The literature produced, documents the evolution of the concepts and ideas around the Layers of Protection Analysis. The literature also shows how the application of the methodology has progressed mainly towards a specialized field such as safety instrumented systems.

Today, LOPA is commonly applied to determine the level of integrity required for instrumented systems to close the risk gap in a specific facility or design. In other cases, LOPA is also used as an "advanced" HAZOP with the idea of identifying hazards and risks gaps. A growing interest appears to be developing for "better" or more "accurate" data for protective layers leading the LOPA methodology towards a more quantitative methodology.

In general, the definition of the LOPA acronym is very well understood, but the application of the LOPA concepts and the methodology is challenging and sometimes misused. This paper explores the challenges associated with the use of LOPA in different environments and presents examples of some of the most common "misunderstandings".

2. Back to the basics

The question that a layer of protection analysis seeks to answer is "Are there sufficient protective layers to prevent the scenario or eventually mitigate the potential consequences?" By providing a systematic approach to verify which safeguards can be used as IPLs and how much credit can be assigned, LOPA allows to identify and close risk gaps. LOPA was developed with specific goals, namely to ensure that:

- Appropriate and sufficient risk reduction measures are in place to achieve the risk target

- Existing risk gaps are identified and appropriate measures (i.e., new IPLs) are designed, installed, and maintained.

While performing LOPA, it is frequently assumed that when a risk gap is identified (i.e., the available layers of protections are deemed insufficient to prevent or mitigate the hazardous scenario) it should be closed by designing instrumented functions. In reality, the instrumented functions should be considered as a last resource and only when other type of IPLs cannot be identified or implemented. Inherently safer design should be the first option to be analyzed, followed by non-instrumented options.

Life-Cycle of Protective Systems

Hazard identification and Layer of Protection Analysis are the first steps to identify and design protective system. LOPA is just the beginning of the lifecycle of protective systems. LOPA is the beginning and is not the end. The output from the LOPA, which include the list of IPLs and its requirements, is used to ensure that the existing IPLs are designed, operated, tested, and maintained according to the specifications. The LOPA output also indicates the list of new IPLs that shall be designed and installed in order to close the risk gaps identified by the LOPA.

Hazard Identification

LOPA relies on the identification of hazardous scenarios through the application of other techniques, better suited for the task, such as HAZOP. HAZOP identifies initiating causes and its potential consequences and provides the basis for LOPA. Therefore, if the hazard identification step is poorly performed, LOPA may become quite challenging.

The output from LOPA is a list of existing protective layers that meet the core requirements and are maintained according to the specifications. LOPA provides also a list of risk gaps where additional IPLs are required in order to reduce the likelihood of hazardous scenarios. LOPA is therefore an important intermediate step in the management of process safety.

The study offers the opportunity to ensure that the operational and equipment constraints and needs are fully understood and adequate documentation and management systems are in place.

While all HAZOP scenarios with high consequence rankings should be carefully analyzed, not all of them are candidates for a LOPA. Examples of this are normal corrosion of pipelines or vessels, design errors that lead to release of hazardous chemicals, most weather-related events, acts of God, poor maintenance, etc. These scenarios are better addressed by integrity and quality assurance programs. This issue is addressed in more detail in example 2 of Section 4.

3. Challenges

Since LOPA requires consideration of not only the technical aspects of the process but also human factors and management aspects, it is also subject to the challenges associated with handling of all the information. Therefore, the LOPA facilitator must have good understanding about those aspects and should be able to extract sufficient information from the team and integrate it in the LOPA scenario. In some cases, the daily workload and limited availability of highly trained and experienced individuals in the team must be accounted for as the studies are conducted.

Currently, different LOPA approaches are used depending on the companies involved. Some use a simplified QRA approach that allows the use of modifiers such as probability of ignition, while others use risk matrices, known as order of magnitude approach, that indicate the required number of IPLs for an initiating cause frequency and consequence couple. Theoretically, regardless of the methodology applied, the LOPA study should lead to similar conclusions. However, the outcome of the two approaches can sometimes be quite different.

A challenge in the QRA-like approach is the use of frequency modification factors. People using this approach tend to use the factors by default and may underestimate the risk associated with the scenario. The outcomes of the LOPA can also be quite different depending on the team members involved. Therefore, if this approach is used, it is important that the LOPA team choose the factors based on careful engineering judgments considering the operating environment and conditions.

In the order of magnitude approach, usually no modification factors are used and the results are consistent regardless of the team members involved. However, this approach can sometimes result in an overly conservative result.

Regardless of the approaches used, the outcome of the LOPA should always be checked and confirmed whether it make sense.

4. Case Study

What is a LOPA Scenario?

As mentioned previously, LOPA relies on hazard identification. It is therefore fundamental for LOPA to start from an appropriate HAZOP analysis, where the initiating cause and consequence pair have been clearly identified and the severity of the consequence is ranked without taking into account the safeguards identified for the scenario.

Unless hazardous scenarios and their severity of consequences are identified during the HAZOP, it is not possible for LOPA to assess how many IPLs are required to close the risk gap. The severity of the consequence has to be reasonably conservative. If the severity ranking is estimated too low, then the outcome can be no or less IPLs required. On the contrary, if the severity ranking is estimated in an overly conservative way (e.g., if all flammable releases are assumed to lead to a fatality) it is possible to end up with an excessive number of IPLs required.

For a system shown in Figure 1, a HAZOP scenario for high pressure may look like:

Example 1a:

PARAMETER: PRESSURE

GW	DEVIATION	CAUSE	CONSEQUENCE	SAFEGUARD			
MORE	Excessive flow	Valve failure	No consequence identified				

The scenario presented in Example 1a will not be included in a LOPA since no consequences are identified. In this case, the HAZOP team is assumed to have considered the safeguards in determining the consequence. Since they are assuming that the safeguards would work properly, no consequences were identified.

A better HAZOP entry for the scenario presented in Example 1a, is shown in the following examples:

Example 1b:

PARAMETER: PRESSURE

GW	DEVIATION	CAUSE	CONSEQUENCE	SAFEGUARD			
MORE	High pressure in V-01	Pressure regulator Valve PRV-01fails open	Hazard: vessel/flange failure and release of flammable gas Consequence: Potential for flash fire that will lead to injuries	RV-01 PAH-01			

Based on this HAZOP entry, the scenario will be included in a LOPA study. The team identified the specific deviation (i.e., initiating cause), hazardous events and most credible hazardous consequences based on the type of process fluid and the quantity that could be released. When this HAZOP scenario is analyzed in a LOPA, the identified safeguards will be assessed against the requirements to be credited as IPLs (i.e., independence, specificity, auditability, etc.).

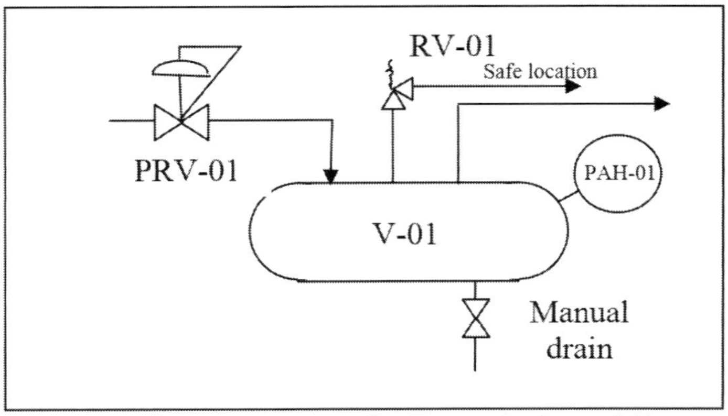

Figure 1: System for Example 1a and 1b.

Validity of Initiating Causes

In cases where corrosion is likely to affect the integrity of a vessel, a HAZOP scenario may be developed as shown in Example 2:

Example 2:

PARAMETER: corrosion/erosion

GW	DEVIATION	CAUSE	CONSEQUENCE	SAFEGUARD			
ISSUES	Excessive corrosion	General corrosion	Possible piping failure, release of hydrocarbons, fire and explosion	Integrity management program Corrosion probes			

From a HAZOP viewpoint, this scenario is valid and the potential consequence may be severe. All HAZOP scenarios with high consequence rankings should be carefully analyzed to ensure that sufficient risk reduction is available to prevent or mitigate the consequences. However, not all HAZOP scenarios that may lead to severe consequences are candidates for a LOPA.

In general, a LOPA is not a useful tool in cases where passive or active protective systems (instrumented or non-instrumented) cannot prevent the scenario or fully mitigate the consequence. When a scenario is better addressed through integrity management and quality assurance programs, a LOPA is not a proper tool to evaluate the scenario. For example, normal corrosion of pipelines or vessels, design errors that lead to release of hazardous chemicals, equipment failure due to lack of maintenance, most weather-related events, acts of God, and poor maintenance cannot be evaluated by LOPA. The concern for vessel failure can be addressed as a High pressure scenario.

IPLs

Let's assume that for a system shown in Figure 2, a HAZOP scenario has been identified for the failure of the control valve, as shown in Example 3. The LOPA team has assessed the LOPA scenario and determined that a risk reduction factor of 1000 is required to close the risk gap.

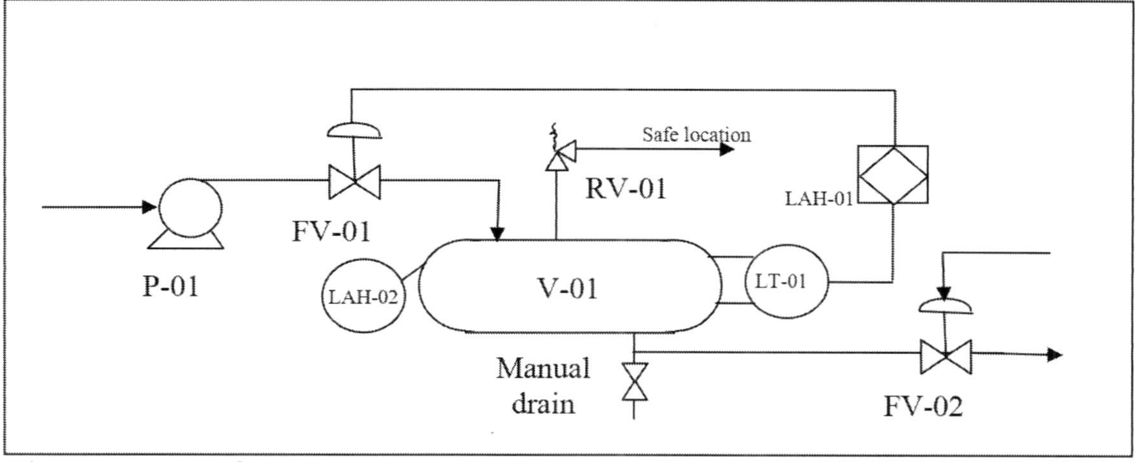

Figure 2: System for Example 3

Example 3:

PARAMETER: FLOW

GW	DEVIATION	CAUSE	CONSEQUENCE	SAFEGUARD			
MORE	Excessive flow	Flow control valve FV-01 fails open	Hazard: vessel overfill and overpressure Consequence: Potential vessel failure, release of flammable liquid, pool fire and potential injuries.	RV-01 (liq.) LAH-01 LAH-02 Manual drain			

Assuming the RV-01 is a valid device (with the required recommendation to ensure the core requirements are met) capable of providing the equivalent of two IPLs (i.e., risk reduction factor of 100), one more IPL is required. The additional IPL may be provided by the supervisory layer. A possibility is given by the high level alarm LAH-2 and operator response. This alarm is independent of the initiating cause. During the discussion of the scenario the LOPA team may discover that given the small size of the vessel it is expected that the level excursion will occur in few minutes. The board operator can detect the alarm but it is not specific to the scenario that is being analyzed.

Time for troubleshooting is required and that may involve sending an operator to the field to investigate the problem, exposing him to potentially dangerous consequences. Additionally, because of operational excursions and spurious trips, the set points for the alarms may be frequently changed. The high level alarm is therefore not specific for the initiating cause and may not provide sufficient time for operator intervention to eliminate the hazardous condition. In this case, the high level alarm should not be credited as an IPL.

To close the risk gap, a SIF may be recommended. However, as mentioned earlier, instrumentation should be the last choice. ISD should be considered as a first option. Brainstorming may lead to a possible solution involving the substitution of a specific length of pipe with a smaller diameter to increase the pressure drop and either prevent the overpressure scenario or increase the process time required to reach the overpressure of the vessel. Another possible ISD option could be to change the impeller of the pump P-01 to reduce the maximum flow and pressure. From the engineering viewpoint, the hazard source has been eliminated (assuming that the hazard has not been migrated to other connected process areas). But the question that now arises is about proper management of those changes. What measures are in place to ensure that the impeller is not changed back to its original size? The answer may be a Management of Change system, recognizing that in the real world things are not perfect. If this ISD is used as an IPL then it would have to be managed according to the lifecycle requirements which imply a more rigorous management process. The following step could be to change the whole pump for a smaller one. In this case it would be more difficult to defeat the ISD measure, and in this case it can be assumed that the hazardous scenario has been eliminated by design.

5. Conclusions

LOPA is used widely in the industry as a risk assessment tool to evaluate the IPLs required to achieve the risk target. This paper presented common problems that are encountered during LOPA. The LOPA concept can be applied to any system but the outcome is not always justified. In general, LOPA is not a useful tool in cases where passive or active protective systems (instrumented or non-instrumented) cannot prevent the scenario or fully mitigate the consequence.

Understanding the life cycle of the protective system is critical for LOPA. If a protective system is credited as an IPL during LOPA, then there should be a management system to ensure that the existing IPLs are designed, operated, tested, and maintained according to the specifications.

In order to close the risk gap, SIFs are usually recommended. However, instrumentation should be the last choice. The LOPA team should always seek for an ISD option first before going to the SIF option.

6. References

1. Center for Chemical Process Safety (CCPS), *Guidelines for Safe Automation of Chemical Processes,* American Institute of Chemical Engineers, New York, NY, 1993.

2. Center for Chemical Process Safety (CCPS), *Layer of Protection Analysis, Simplified Process Risk Assessment*, American Institute of Chemical Engineers, New York, NY, 2001.

3. Center for Chemical Process Safety (CCPS), *Guidelines for Safe and Reliable Instrumented Protective Systems*, American Institute of Chemical Engineers, New York, NY, 2007.

USE OF PROCEDURAL BASED CONTROLS IN LAYER OF PROTECTION ANALYSIS

Raymond Freeman
S&PP Consulting
Houston, TX 77041
email: RAFREE@YAHOO.COM

For presentation at:

2008 Spring National Meeting of
American Institute of Chemical Engineers
New Orleans, LA
April 6 – 10, 2008

Abstract

The CCPS book on the LOPA methodology gives little credit for procedural controls. In batch operations, startups or process shutdowns, the procedural controls are often the only available means of controlling the potential hazards. This paper presents a methodology for the evaluation of procedural based controls as potential independent layers of protection (IPLs) during a layer of protection analysis (LOPA) study. A checklist for the evaluation of procedural controls is presented as well as an example to demonstrate the methodology.

BACKGROUND

The use of the layer of protection analysis (LOPA) technique as a means of evaluating the safety of a process operation has come into wide use in the process industries. Publication of the CCPS LOPA book [Ref. 1], the CCPS controls book [Ref. 2], and the ANSI/ISA S84 standard [Ref. 3] describe in detail the use of LOPA in the evaluation of the number of independent protective layers (IPLs) and the determination of the needed safety integrity levels (SILs) to manage the risk of a process operation. These methods are mainly directed towards the operation of continuous process units. Batch process often rely almost entirely on human based procedures for safety, Startup and shutdown of large continuous plants often done in manual mode with procedural controls for safety. This paper is directed at providing a framework for the evaluation of procedural controls in a LOPA review

WHAT IS A PROTECTIVE LAYER?

The CCPS LOPA book says a procedure is not an IPL. The book also states that an alarm could be an IPL. Both require an operator to take action. Fundamentally an IPL has three requirements:

1. It must be capable of preventing the bad event of concern from occurring
2. It must be independent of other protective layers
3. It must be auditable

Examples of potential IPL include:

- Relief valve
- Basic process control system
- Safety instrumented function (SIF) implemented in a safety instrumented system (SIS)

For a protective system implemented using a computing device (computer, programmable logic solver, etc.) the system consists of three primary sections:

1. Sensor
2. Logic solver
3. Final control element

Figure 1 presents this generic model of a protective layer. Table 1 presents a comparison of the various elements of a protective layer with the protection offered by a procedurally based operator action. Note that there is no difference in the conceptual operation of any of the safeguards in Table 1. The only difference is the use of a human operator as a logic solver in the protective layer.

If there is no conceptual difference between a safeguard based on operator action or a safeguard that uses a computer for the logic solver, how do we evaluate the quality of a specific safeguard that is based on operator action? A number of definitions are needed:

> Procedural control – A layer of protection based on the combination of a field sensor, human operator and final control element (switch, valve, relay) governed by a written procedure.
>
> Interlock – A safety instrumented function (SIF) implemented in a safety instrumented system (SIS) in accordance with the ANSI/ISA S84.01 – 2004 standard.

By itself, a written procedure is not capable of serving as a protective layer. Evaluating a specific procedural control as protective layer requires a review of the operations and tasks that the human operators are conducting. For the purposes of this paper, a procedural control is defined as the entire system that a human operator uses to perform a task and includes:

1. Written procedures that define the task or action to be taken
2. A clear indication that the task must be performed
3. Training on how to perform the task
4. Adequate tools and materials to perform the task
5. Proper and available PPE to perform the task
6. Adequate time to perform the task
7. Time to diagnosis abnormal problems with the task
8. Clear indication of proper performance of the task
9. Ability to verify that the task was performed (auditability)

Table 2 presents a checklist for the evaluation of a procedural based safeguard as a potential protective layer during a LOPA review. Question 1 is to determine if the procedure is capable of preventing the occurrence of the event of concern. The answer to question 1 must be yes for the procedure to be considered as an independent protective layer. Questions 2 through 9 are to determine if the procedure is auditable. Questions 10 through 13 define the time requirements to complete the procedure. Questions 14 through 16 determine if special tools are needed and are available to execute the procedure. Questions 17 through 21 determine how the need to complete the procedure is communicated to the human operator. Finally question 22 verifies that the procedure is independent of other procedures or actions already credited during the LOPA study. Based on the answers to the questions in Table 2, the LOPA analyst can determine the number of protective credits to be given for the procedure.

DEVELOPMENT OF MODEL FOR LOPA CREDITS FOR PROCEDURAL CONTROLS

The importance of the human operator in the safe operation of a process is well understood. Various methods have been proposed for quantitative modeling of the failure of an operator to properly perform their duties. CCPS has published a guideline book on preventing human error in the process industries [Ref. 4]. The CCPS Quantitative Risk Analysis [Ref. 5] book devotes a section to the prediction of of human failure probabilities.

Suttinger and Sossman [Ref. 6] have published a methodology for evaluating the quality of human response in safety instrumented systems. The integration of human based procedural controls into LOPA is thus a natural extension of existing risk analysis techniques.

Consider the simplified response model presented in Figure 2. The time available to prevent an accident must be longer than the sum of time to detect/diagnosis the initiation of an accident sequence and the time to respond. For control console operator response to a clearly annunciated abnormal event, the response time is generally short. The critical time is the time to diagnosis the problem. Swain [Ref. 7 and 8] has developed a diagnostic response model for the console operator which is presented in Figure 3. The three curves plotted are for the simultaneous occurrence of one, two or three abnormal events. Note that the three curves are the upper 95% confidence response models and actual operator response would be expected to be much better (faster and more reliable) than that presented in Figure 3. In a LOPA analysis, we are looking at a single cause-consequence pair and the appropriate diagnostic response model is the curve labeled "one event." For short times available to diagnosis a problem (less than 10 minutes), the probability of the operator failing to diagnosis an abnormal event is 1. For the case where the available diagnostic time is between 10 minutes and 40 minutes the probability of failure to diagnosis is below 1×10^{-1}. When greater than 40 minutes are available, the failure to properly diagnosis probability drops rapidly below 1×10^{-2}.

The situation of human operators or maintenance workers failing to follow a written work practice is similar to the operator response to an alarm. The basic human error rate is presented by Swain as 3 errors per 1000 attempts [Ref. 8]. This is the basic frequency of failures of people to properly perform a single task (procedural step, mechanical step, etc.). For an n step procedure, the overall probability of completing the entire procedure [Ref. 9] without error is:

$$P_s = (P_i)^n \qquad \text{(Eqn. 1)}$$

Where:

P_s = overall probability of success

P_i = probability of success at each step of the procedure

n = number of steps in the procedure

The probability of success of each step is related to the basic human error probability as:

$$P_i = 1 - HEP \qquad \text{(Eqn. 2)}$$

Where:

HEP = basic human error probability = 0.003

The probability of failure of completing all n steps of the procedure is:

$$P_f = 1 - P_s \quad \text{(Eqn. 3)}$$

$$P_i = 1 - HEP \quad \text{(Eqn. 4)}$$

For a 10 step procedure, the overall probability of failure to complete the procedure properly is:

$$P_f = 1 - P_s \quad \text{(Eqn. 5)}$$

$$P_f = 1 - (1 - HEP)^n \quad \text{(Eqn. 6)}$$

$$P_f = 1 - (1 - 0.003)^{10} = 0.03 \quad \text{(Eqn. 7)}$$

Rounding up to the nearest order-of-magnitude yields 1×10^{-1}. This assumes that there is no chance of the operator detecting an error or in recovering from a mistake in following the procedure.

If two independent groups are cross checking each other using a 10 step procedure, the expected error probability would be:

$$\text{Human Error Probability} = (0.03)(0.03) = 9 \times 10^{-4} \quad \text{(Eqn. 8)}$$

Rounding up to the nearest order-of-magnitude yields 1×10^{-3}.
However, the potential for common cause failure (non independence) of the two groups exists. Providing a factor of 10 increase in the human error probability yields 1×10^{-2}.

The results from the above calculations can be used as a first estimate upper probability for the failure of a procedural control. Table 3 presents a suggested set of LOPA credits for procedural controls. The frequencies in Table 3 are over a factor of 10 higher than those proposed in the Human Factors Handbook (Ref. 8).

METHODOLOGY FOR INCLUDING PROCEDURAL CONTROLS IN LOPA

The analysis process for including human procedural controls in a LOPA review is straightforward.

1. Define the Event of Concern
2. Define the sequence of events or procedures that must be violated or compromised for the Event of Concern to occur. These are the procedural controls that may become independent protection layers (IPLs).

3. Evaluate each of the step of the sequence of events or procedures using the questions in Table 2. Those steps that are found to be adequate for inclusion as a protective layer should be evaluated for LOPA credit in step 4 below.
4. Evaluate the procedures defined in step 3 above for LOPA credit using Table 3.

If the same human operator is the initiator of the event of concern, a protective layer would not be claimed for that operator to respond to the problem (not independent). In general, no more than three LOPA credits (a reduction in the frequency of 1×10^{-3}) should be given during a LOPA review.

EXAMPLES

The easiest way to understand the application of the LOPA methodology to procedural controls is to look at a set of examples. The following three examples follow a LOPA team as they review the raw material handling for a hazardous batch process.

Example 1

During the review of a batch process system, the question of emergency shutdown of raw material addition was discussed. The high level alarm is activated using the BPCS. The operators are trained in the use of the emergency shutdown button on the control panel which will stop all flows to the reactor. The alarm will be activated 15 minutes prior to overflowing the reactor. A high level shutoff will be automatically tripped by the BPCS prior to the high level alarm.

How much LOPA credit should be given for the procedural control to shutoff the flow to the reactor? Based on the facts described above, we may conclude that the same BPCS is being used to activate both the alarm and the high level automatic shutoff. These are not independent and therefore no additional LOPA credit should be given for the procedural control.

Example 2

As the discussion of high level in the reactor of example 1 continues, the LOPA team proposes to change the wiring of the high level alarm and operator response. The LOPA team proposes that the high level alarm will be routed from a sensor into the SIS logic solver associated with the batch process. An alarm will be activated on the SIS control panel and an alarm signal will be sent to the batch process alarm panel. A second valve in raw material "A" supply line to the reactor will be tied into the emergency shutdown button on the control panel.

How much LOPA credit should be given for the procedural control to shutoff the flow to the reactor? This revision to the system will eliminate the common element of the BPCS for activation of both the high level automatic shutdown and the high level procedural control by the operator. A review of the questions in Table 2 finds that the proposed system is auditable and properly documented with a written procedure that the operators have been trained upon. No special tools are needed to perform the required action. The time required to complete the action is very short once the decision to take action is made. The alarm is clearly annunciated using two different paths (BPCS alarm remains with the new SIS alarm added). Therefore, based on a time available to take action of 15 minutes, one LOPA credit (a reduction in the frequency of 1 x 10^{-1}) is taken for this procedural control (see Table 3).

Example 3

During the review of the clean out step for the batch reactor, the potential to double charge the reactor was found to be possible. To double charge the reactor the following sequence of events must occur:

1. The operations group on Shift A must fail to empty the reactor
2. The maintenance group on Shift B must fail to clean out the reactor
3. The operations group returning on Shift A must fail to recognize that the controls indicate that the reactor is still full and attempt process a new batch.
4. The operations group on Shift A must bypass the high level BPCS permissive set to prevent addition of raw materials on high level in the reactor. This would require the intervention of control engineering group to reprogram the BPCS logic solver.
5. The operations group on Shift A must then charge raw material into the reactor and ignore the alarms installed in Examples 1 and 2 above.

How much LOPA credit should be given in the above sequence of events resulting in an overcharge of the reactor? Discussion with the LOPA team finds that the operations group has failed to empty the reactor once in 10 years of operations. This single failure to empty the reactor was immediately detected when the Shift B maintenance group started their reactor clean out task. No overflow of the reactor occurred. During the ten years of operations over 2000 batches of material have been processed. Since operations Shift "A" is the cause of the problem (failure to empty the reactor), they will be assumed to violate all of the procedures that they are involved with. The LOPA credit analysis follows the steps of the above sequence.

- Step 1 - The initiating event frequency of 1 event in 10 years was chosen to represent the start of this sequence of events.
- Step 2 - Since a second independent group (maintenance Shift B) must fail to detect that the reactor is still full. The maintenance group would attempt to clean out a full reactor and would sign off that the reactor is clean and ready for use in the next batch. This requires at least two failures to follow existing written procedures by a group independent of the operations group Shift "A" who initiated the sequence of events. Only one LOPA credit (a reduction in the frequency of 1 x 10^{-1}) is taken for this procedural control (see Table 3) as the operations group Shift "A" has already failed to follow procedures.
- Step 3 - Since the operations group Shift "A" has already violated the procedures, no LOPA credit is taken for the continued violation of procedures and the high level alarms.
- Step 4 - A third group (control engineering group) would have to violate their procedures and bypass a reactor high level shutoff system without verification that the system was empty, clean and in a safe mode. Only one LOPA credit (a

reduction in the frequency of 1×10^{-1}) is taken for this procedural control (see Table 3) is taken,

- Step 5 - Since the operations group Shift "A" has already violated the procedures, no LOPA credit is taken for the continued violation of procedures and the alarms.

In summary, the LOPA analyst would provide credits for this example 3 as:

- The initiating event frequency of 1 event in 10 years is to represent the start of this sequence of events.
- Two LOPA credits (for steps 2 and 4) or a frequency reduction of 1×10^{-2} is taken for the two procedural controls.

CONCLUSIONS

This paper has presented a methodology for estimating the frequency of failure of human procedural controls during a LOPA review. The methodology is conservative and may be expected to overestimate the frequency of human error. If the results of the LOPA analysis are found to be tolerable, the analyst would expect the actual frequency of human error to be lower and the resulting risk to be lower than calculated. If the results of the LOPA analysis are found to be intolerable, the analyst may need to use more refined risk evaluation tools (such as CPQRA, Ref. 5) or to recommend changes in the system to reduce the risk.

REFERENCES

1. Center for Chemical Process Safety, *Layer of Protection Analysis – Simplified Process Risk Analysis*, American Institute of Chemical Engineers, New York, NY 2001
2. Center for Chemical Process Safety, *Guidelines for Safe and Reliable Instrumented Protective Systems*, Wiley, New York, NY 2007
3. ANSI/ISA-S84.00.01-2004 Part 3 (IEC 61511-1 Mod) "Functional Safety: Safety Instrumented Systems for the Process Industry Sector – Part 3: Guidance for the Determination of the Required Safety Integrity Levels – Informative
4. Center for Chemical Process Safety, *Guidelines for Preventing Human Error in Process Safety*, American Institute of Chemical Engineers, New York, NY 1994
5. Center for Chemical Process Safety, *Guidelines for Chemical Process Quantitative Risk Analysis – 2nd Ed*, American Institute of Chemical Engineers, New York, NY 2001
6. Lawrence Suttinger and Carl Sossman, "Operator Action Within a Safety Instrumented Function," Westinghouse Savannah River Company Report WSRC-MS-2002-00091, 2002, Available at: http://sti.srs.gov/fulltext/ms2002091/ms2002091.html
7. NUREG/CR-4772, Accident Sequence Evaluation Program Human Reliability Analysis Procedure, U.S. Nuclear Regulatory Commission, February 1987
8. NUREG/CR-1278, Handbook of Human Reliability Analysis with Emphasis on Nuclear Power Plant Applications, U.S. Nuclear Regulatory Commission, August 1983
9. Paul Meyer, *Introductory Probability and Statistics*, pg 236, Addison-Wesley Publishing Co., 1972

Table 1. Comparison of Different Types of Protection Layers

Type of Safeguard	Sensor	Logic Solver	Final Control Element
Interlock in Basic Process Control System	Field sensor – temperature, pressure, etc.	Basic Process Control System computer	Remotely operated control valve, block valve, motor starter switch
Interlock in Safety Instrumented System	Field sensor – temperature, pressure, etc.	Safety Instrumented System Interlock	Remotely operated control valve, block valve, motor starter switch
Operator action procedural control – control room	Field sensor – temperature, pressure, etc.	Human action based on response to annunciated alarm from BPCS or SIS	Remotely operated control valve, block valve, motor starter switch
Operator action procedural control – Field Operator	Field sensor – temperature, pressure, etc.	Human action based on response to local alarm, sensor reading or observation. May involve communication with console operator via radio	Remotely operated control valve, block valve, motor starter switch. OR manipulation of manual valve by field operator.

Table 2. Checklist for the Evaluation of the Quality of a Procedural Based Safeguard

No.	Question	Yes/ No	Comments
1	Is the procedure capable to stopping the bad event of concern from occurring?		Answer must be yes to be considered as an IPL
2	Is the procedure auditable? Is the trail of written procedure, operator training, re-training and certification of operator understanding complete?		Answer must be yes to be considered as an IPL
3	Is there a written procedure? Does the procedure clearly define the steps that need to be completed in the proper sequence to successfully prevent the undesired event of concern from occurring?		Answer must be yes to be considered as an IPL
4	Have the operators been trained in the use of the procedure? Re-training as needed?		Answer must be yes to be considered as an IPL
5	Is a hard copy (paper) of the procedure available for the immediate use by the operator? Is this procedure clearly identifiable as an emergency procedure (for example, printed on a different color of paper)?		Answer must be yes to be considered as an IPL
6	Is the procedure based on a written check list?		Optional, checklists are a way to reduce frequency of operator forgetting to do something
7	Is the procedure a two person (done-by checked-by) written checklist?		More LOPA credit is given for a two person checklist procedure
8	Is the failure to complete the procedure clearly and immediately presented to the person doing the procedure? Can they tell that they have made a mistake in doing the procedure?		
9	Is a record completed that indicates the procedure was completed?		There needs to be a record that the procedure was completed. Without a record, another group may attempt to complete the procedure again.

Table 2. Checklist for the Evaluation of the Quality of a Procedural Based Safeguard (cont.)

No.	Question	Yes/ No	Comments
10	How much time will it take for the operator to take action to stop the bad event of concern, once the need for action is determined?		Must be long enough for an operator to take action. Normally a minimum of 10 minutes
11	How long does it take to complete the procedure once started?		Normally same as item 10
12	Is there any travel time to the site where action is required?		Must be included in total time to complete the procedure
13	How much time is available from the initiating event until the undesired event occurs?		Must be long enough to allow for 10 minutes to diagnosis the problem and the time to take action (items 10 and 11 above)
14	Are any special tools required to perform the procedure?		If special tools are needed, they must be readily available for the procedure to be counted as an IPL.
15	Are any special tools available for immediate use?		
16	Are special protective equipment (gloves, respirators, air packs, etc.) needed at the work site? Are these items immediately available and how long will it take to gather them?		
17	Is the need for the completion of the procedure clearly annunciated?		Must be clear and not confusing
18	Is there an alarm on a control panel to indicate the initiating event has occurred? How is the alarm implemented? Using the BPCS or the SIS?		See questions 20 and 21
19	Is the control panel normally manned?		Alarms do no good if the operator is absent when the alarm is activated
20	If there is an alarm on a control panel, is the alarm activated using the basic process control system (BPCS)?		If the alarm is activated via BPSC, a maximum of one LOPA credit should be given.
21	Is the alarm activated using a SIS? If yes, what is the SIL rating of the alarm?		If the alarm is activated via SIS, the maximum LOPA credit is equal to the SIL level of the system

Table 2. Checklist for the Evaluation of the Quality of a Procedural Based Safeguard (cont.)

No.	Question	Yes/ No	Comments
22	Is this procedure independent from other procedures or actions taken and claimed as an IPL during the LOPA analysis?		Answer must be yes for the procedure to be considered an IPL

Table 3. Suggested LOPA credits for Procedural Controls

Description of Procedural Control	LOPA credits	PFOD
Single human operator with less than 10 minutes to diagnosis and take action on a clearly annunciated alarm event.	0	1E0
Single human operator with 10 minutes to 40 minutes to diagnosis and take action on a clearly annunciated alarm event	1	1E-1
Single human operator with 40 minutes to 8 hours to diagnosis and take action on a clearly annunciated alarm event activated by a SIS.	2	1E-2
Single human completing a simple routine task with between 10 minutes and 40 minutes to complete the task	1	1E-1
Two operators acting independently completing a short (10 items or less) written checklist procedure (done-by/check-by) with between 10 minutes and 40 minutes to complete the task	2	1E-2
Two independent operations groups completing a task with one group independently checking the work of the other group (operations checking maintenance) with adequate time (greater than 10 minutes) to detect any issues	2	1E-2

Figure 1. Generic Model of a Protection Layer

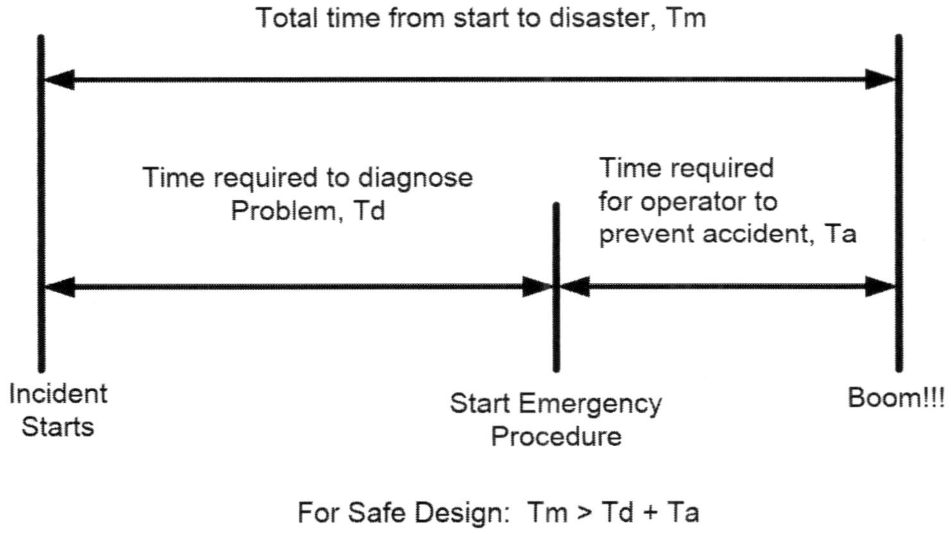

Figure 2. Time Available to Prevent an Accident

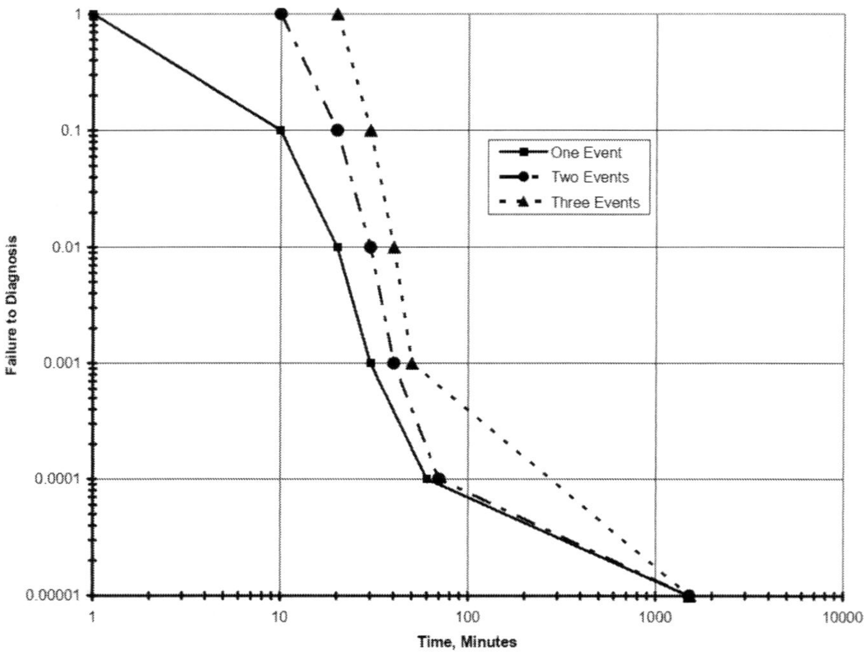

Figure 3. Probability of Failure by Control Room Personnel to Correctly Diagnosis an Abnormal Event or Events

Improvements in the safety screening of resin manufacturing processes

George Kalfas
DuPont Marshall Research Laboratory
Philadelphia, PA 19146
George.A.Kalfas@usa.dupont.com

Todd Krieger and Ross Wilcox
DuPont Engineering Research and Technology
Wilmington, DE 19898

ABSTRACT

Consequences of acrylic resin reactor runaways can be devastating, as it is evident in investigation reports of industrial accidents. Critical in preventing major accidents is the safety screening of any acrylic resin formula to be manufactured in large scale. Computer model simulations facilitate the evaluation of "layers of protection" against acrylic polymerization runaways.

This work describes the adaptation of a chemical process dynamic simulator (DuPont™ TMODS™) for use in simulating acrylic polymerization runaways. "Loss of cooling" and "monomer pooling (accumulation)" scenarios are the causes of these runaways. Simulations show that scenarios leading to "monomer pooling" result to more energetic runaways and to larger emergency relief capacity requirements than "loss of cooling" scenarios.

1. Introduction

Acrylic polymerizations are highly exothermic free radical polymerization processes with heat of reaction ranging between 50 and 80 kJ/mole depending on the type of monomer [1]. A typical semi-batch acrylic polymerization process flowsheet is shown in Figure 1. Two feed tanks are shown, one holding the monomer mixture and one the initiator solution. The "kill" (short stop) tank contains a concentrated inhibitor solution designed to terminate the free radical polymerization by quickly consuming the free radicals present in the reacting mixture. In a reflux process, heat generated by the reaction is removed mainly by the cooling water flowing through the condenser, whereas in a sub-reflux process, heat is removed by the heat transfer fluid flowing through the reactor jacket. Usually, the rate of heat generation is controlled by the monomer feed rate to the reactor.

"Loss of cooling" is one root cause that can lead to a runaway reaction. However, equipment malfunction or human error can lead to accumulation of unreacted monomer(s) in the reactor; a situation termed "monomer pooling". If polymerization were to occur in a sizeable monomer pool, then a large amount of heat would be released in a short period of time. Such a situation has the potential to overwhelm the heat removal capability of the system and to result in a pressure build-up followed by bursting of the rupture disc and release of the reactor contents to the containment system (*catch tank*) through the emergency vent line. During a runaway event, there is an increased probability of a vapor leak leading to the formation of a flammable vapor cloud with subsequent explosion and fire.

At least two layers of independent non-procedural safeguards above normal controls are recommended for hazards with catastrophic consequences, such as those arising from acrylic polymerization runaways [2]. Layers of protection in an acrylic reactor system may include the following:

- Automatic feed interruption
- Automatic emergency cooling
- "Kill" (short stop) injection system
- Emergency relief and containment system
- Pressure rated vessels and piping
- Vapor Cloud detection and mitigation system

2. Acrylic polymerization runaways

An analysis of chemical runaway incidents in industrial processes conducted by Barton and Nolan in 1989 showed that polymerization runaways occur by far more often than any other type of reaction runaway [3]. Out of 167 US incidents recorded during the period from 1980 to 2001 in the CSB Reactive Incident Database (http://www.csb.gov/completed_investigations/docs/Reactive%20Incident%20Data.xls), 130 incidents are classified under known reaction types. Polymerization is the third most frequent reaction type accounting for 21 of those 130 incidents. Seven of those are attributed to polymerizations involving acrylic monomers (*acrylate esters* and *methacrylate esters*) or styrene. Acrylic polymerizations are particularly hazardous processes due to the high heat of polymerization and the rapid reaction rate. Descriptions of some acrylic polymerization runaways can be found in published sources. For illustration purposes, four industrial accidents are described here.

On October 30, 1991 at a paint factory in Kawagoe, Saitama, Japan [4], incorrect amounts of monomers (ethyl- and butyl- acrylate) and solvent were charged into a 10 m^3 acrylic resin reactor due to computer "misoperation". The reaction ran out of control. A 90 kPa_g rupture disc installed in the emergency vent line ruptured and the reactor contents were discharged into a catch tank. However, flammable vapors leaked in the operating area from a condenser flange leading to an explosion. The extent of the damage was estimated at 16 million yen.

On October 4, 1997 at a manufacturing facility of the Cook Composites and Polymer Co. in Houston, an acrylic polymerization runaway led to a vapor cloud explosion and to serious injuries of an operator [5]. Approximately 7000 lbs of monomer (n- butyl acrylate) accumulated unreacted in the reactor (along with 4000 lbs of xylene), because the initiator (di t-butyl peroxide) solution feed was not mixed and no initiator was fed into the reactor along with the monomer. The runaway was extremely fast, overwhelmed the condenser and over-pressurized the reactor. The rupture disk burst, but the emergency relief system was incapable of quickly relieving the pressure. A welded joint between the reactor head and side wall failed and the head was launched upward. The subsequent vapor cloud explosion resulted to a pressure wave that damaged surrounding structures.

On May 18, 2001, a runaway reaction in a 6-ton acrylic reactor in a resin manufacturing facility in the northern part of Taiwan led to a vapor cloud explosion and subsequent fire that injured over 100 people and caused damages to many nearby facilities and residences within half-a-kilometer range [6]. The power of the blast was estimated be equivalent to 1000 kg of TNT. The reactor contained methyl acrylate, acrylonitrile, acrylic acid, methacrylic acid, methyl alcohol, isopropyl alcohol, and benzoyl peroxide.

On January 31, 2006, an acrylic polymerization runaway in a 1500 gallon reactor resulted to the formation of a vapor cloud that exploded and destroyed the entire facility of Synthron LLC in Morganton, NC. The acrylic reactor produced powder coating and paint additives. The recipe being manufactured was 6080 pounds, about 12% more than a typical batch in this reactor, according to the Chemical Safety Board (CSB) investigation [7]. The reactor contained a monomer with 147°C normal boiling point (NBPT) and two solvents, one with 111°C NBPT and one with 81°C NBPT. The components are not named in the report. However, they are likely *n-butyl acrylate*, *toluene*, and *cyclohexane*, respectively. Undisclosed in the report was also the chemical initiator of the polymerization.

The day before the accident, a number of adjustments were made in the original recipe followed by a brief and inaccurate review process that did not recognize the hazards. The changes had the following effect to the initial reactor charge:

- increased monomer amount by 45%
- increased monomer concentration by 27%
- increased reflux temperature by 5°C

According to CSB estimates, the combined effect of these changes was to increase the maximum heat output 2.3 times. In addition, the cooling water side of the condenser was fouled reducing its ability to remove heat by 25%. The reactor manway lid was secured with only four of the eighteen clamps specified by the vessel manufacturer. As the pressure increased during the runaway, flammable solvent vapors began to leak from the reactor manway gasket. A few minutes later the flammable vapor cloud exploded fatally burning one employee, injuring fourteen and destroying the entire facility.

While multiple failures are usually at play during major accidents, a common theme in the account of these four accidents is the presence of a large "pool" of unreacted monomer. It should also be noted that in all of these cases, the monomers involved are fast reacting *acrylate ester* monomers. Finally, safeguards were either lacking or they were overwhelmed by the force of the runaway. In the Synthron runaway case study, CSB recommended Layers of Protection Analysis (LOPA) [8] as a good practice for the review of process safeguards.

3. Process modeling with DuPont™ TMODS™

TMODS™ is a dynamic process modeling software package. It contains ready-to-configure models for a variety of process equipment including reactors, tanks, condensers, pumps, flow control valves, process charts, PID controllers and other process units or control functions. A free radical solution polymerization model based on a comprehensive modeling approach for addition polymerization processes [9, 10] is also included. TMODS™ models can be interfaced to control systems for control system checkout and for operator training applications. Although, originally developed for continuous processes, TMODS™ has been augmented with features to readily and robustly simulate batch processes [11], as well.

Typically, acrylic resin manufacturing consists of recipe-driven semi-batch processes. The batch recipe format in TMODS™ employs the BatchML dialect of XML developed by the World Batch Forum [12]. Coatings resin processes employ a great variety of raw ingredients (monomers, solvents, initiators, chain transfer agents, and inhibitors). Upon loading a batch recipe into TMODS™, physical and chemical kinetic parameter values for the recipe ingredients are automatically extracted from a proprietary component database. Pre-configured failure scenarios leading to runaways can be selected and sequentially simulated by means of a failure scenario "executive" shell.

4. Emergency relief calculations

Using a dynamic process model of the reactor and the surrounding equipment, the time evolution of reactor pressure can be simulated by modeling the emergency relief pipe as a large capacity vapor valve. However, in these simulations single phase vapor venting is assumed, significantly underestimating the emergency relief rate. Taking into account the effect of two-phase flow on relief system capacity according to the methodology developed by the Design Institute for Emergency Relief Systems (DIERS), one can make a more realistic estimate of the relief rate. An application of the DIERS technology in estimating emergency relief requirements for acrylic resin manufacturing through process simulation has been introduced [13], previously. "Loss of cooling" was assumed as the "worst case" system failure scenario in this earlier work. Simulations of several examples of acrylic reactor runaways were presented and discussed.

We used a similar approach to calculate the required relief rate and evaluate the emergency relief system capacity assuming two-phase homogeneous venting. Methods detailed in [14] were adapted using the thermodynamic output available from the simulator. These methods are essentially steady-state calculations. Therefore, they do not account for changes in the reactor as it relieves. The required relief rate is estimated using the following equation:

$$W_{req} = \frac{M * Q_a}{\left[\left(\frac{V}{M} * \frac{\lambda_v}{\Delta v_v}\right)^{0.5} + \left(C_p * \Delta T\right)^{0.5}\right]^2}$$

where:

W_{req}	required vent flow
M	mass of reactor contents
Q_a	heat generation rate
V	reactor volume
λ_v	heat of vaporization of the mixture
Δv_v	specific volume difference between liquid and vapor
C_p	heat capacity of the mixture
ΔT	temperature rise corresponding to the "overpressure" (during the relief)

During a relief event, the reactor is assumed to discharge its contents via the relief vent piping system to the *catch tank* as a homogeneous two-phase flashing stream. In the catch tank, the two phases separate and the vapor is vented to the atmosphere via the catch tank vent stack. The actual vent rate through the specified pipe is obtained by integration of the following equation [15]:

$$\frac{dP}{dx} = \frac{\rho \cdot g \cdot {}^{dz}\!/\!_{dx} + 4 \cdot f \cdot {}^{G^2}\!/\!_{(D \cdot \rho)}}{1 - \left({}^{G}\!/\!_{\rho}\right)^2 \cdot {}^{d\rho}\!/\!_{dP}}$$

where:

P	pressure
x	distance along the pipe
ρ	density
g	acceleration of gravity
z	elevation
G	mass flux
f	friction factor
D	pipe diameter

Pipe diameter, length, fittings, and elevation change associated with both the reactor emergency vent pipe and the catch tank vent stack are defined in the model. Drops in elevation are ignored to simplify the calculations. Thermodynamic calculations generate vent stream properties, including liquid and vapor density and vapor quality at ten intervals from the "overpressure" down to atmospheric pressure. A polynomial curve fit for each property's profile is used for the numerical integration. A fixed Darcy friction factor of 0.02 is used for the two-phase mixture.

Vapor vent calculations are estimated using compressible flow models. Flow rate and pressure in the reactor vent pipe are modeled using the following equations [16]:

$$\frac{dMa^2}{Ma^2} = k \cdot Ma^4 \cdot \frac{\left[1+(k-1)\cdot\frac{Ma^2}{2}\right]\cdot 4 \cdot f}{\left(1-Ma^2\right)\cdot D}dx$$

$$\frac{dP}{P} = -\frac{1+(k-1)\cdot Ma^2}{2\cdot Ma^2\cdot\left[1+(k-1)\cdot\frac{Ma^2}{2}\right]}dMa^2$$

where:

Ma	Mach number, and
k	heat capacity ratio.

An iterative procedure is used to solve simultaneously the flow equations in the emergency vent line of the reactor and the vent stack of the catch tank. In addition to calculating the flow through the actual vent piping, the required diameter of a single pipe segment with the same elevation change and fittings as the specified emergency vent line is computed. This calculation determines whether the existing emergency vent line is capable of accommodating the required vent flow.

5. Simulations of failure scenarios

For polymerization processes run at reflux temperature (boiling point), the concept of "condenser capability rating" (CCR) has been introduced as a measure of the condenser adequacy under "normal" process conditions [1]. Subsequently, adiabatic simulations were employed to estimate the required emergency relief capacity during "loss of cooling" failures. However, failure scenarios leading to "monomer pooling" give rise to much more energetic runaways and even larger emergency relief capacity requirements.

In an example process, consider the polymerization of a mixture of methyl methacrylate (MMA) and ethyl acrylate (EA) in toluene (TOL) with benzoyl peroxide (BPO) as the initiator. In a 5000 gallon reactor, 8000 lbs of TOL is charged and heated to reflux. The monomer mixture (18000 lbs) is fed from the monomer feed tank over a period of two (2) hours simultaneously with the initiator solution fed from the initiator feed tank over the same period.

Consider a failure scenario that delays the initiator feed thirty (30) minutes after the monomer feed begins. During these thirty minutes, unreacted monomer accumulates forming a monomer "pool". Upon commencement of the initiator feed, the monomer in the reactor begins to polymerize releasing heat. Polymerization rate is a strong function of the monomer composition. The pressure profile during the runaway of a 100% ethyl acrylate monomer pool is shown in Figure 2, along with the set-points for the layers of protection. In this runaway, pressure reaches its peak value only three minutes after the runaway begins. It takes only a minute for pressure to reach the peak value after it crosses the "kill" system trigger point. Therefore, the "kill" system is essentially rendered ineffective in this particular case, because of the fast polymerization rate of a purely acrylate monomer "pool". The emergency relief system diameter proves to be too small to relieve the force of this runaway. It cannot stop the pressure rise. The reactor pressure peaks above the maximum allowable working pressure (MAWP) for this reactor set. Hence, the probability of developing a vapor leak that can form a flammable vapor cloud is high.

The simulated pressure profile during the runaway of a 75:25 EA/MMA monomer "pool" is shown in Figure 3. The presence of the methacrylate monomer makes this runaway much slower than the pure acrylate monomer pool runaway in Figure 2. In this case, it takes about twenty minutes for the reactor pressure to reach its peak value. Simulated temperature and pressure profiles during runaways of monomer pools with different compositions are shown in Figure 4 and Figure 5. Obviously, the higher the percentage of the acrylate monomer in the acrylate/methacrylate mixture, the faster and stronger is the runaway.

In the simulated runaway of a 75% EA monomer pool formed after feeding monomer for 45 minutes without initiator, the effectiveness of the emergency relief system is demonstrated, when the "kill" system fails to "fire" (Figure 6). The reactor pressure rise is stopped by the emergency relief at the rupture disc burst pressure. If the "kill" system fires as designed, then the rupture disc burst pressure will not be reached.

The summary output from simulating a series of runaways of 90:10 EA/MMA monomer "pools" caused by increasing delays in the initiator feed is shown in Figure 7. As the monomer "pools" become progressively larger, the relief rate and the emergency vent diameter requirements become larger, as well. Emergency relief capacity requirements are clearly lower, when the cause of the runaway is "loss of cooling", as shown in Figure 8. It should be noted that these results are not from merely adiabatic simulations, but from dynamic simulations that take into account the thermal inertia of the reactor system including that of the condenser. The effect of the "kill" system injection has not been taken into account in this set of simulations and the feeds were allowed to continue.

6. Conclusions

In an earlier work [13], acrylic polymerization runaways due to "loss of cooling" scenarios were examined. However, a common theme in major acrylic polymerization accidents is the presence of a large *pool* of unreacted monomer. In this work, scenarios leading to runaway because of either "loss of cooling" or "monomer pooling" were simulated. Runaways of acrylic monomer pools were found to be much more energetic than "loss of cooling" runaways. These runaways can overwhelm certain protection safeguards of polymerization reactors. Therefore, "monomer pooling" scenarios need to be taken into account in the design of these safeguards.

Monomer mixture composition has a strong effect on the maximum (*peak*) heat generation. Computer simulations showed that in "monomer pooling" situations, *acrylate-rich* monomer mixtures present the highest hazard, because of their fast polymerization rate. Injection of inhibitor solution can become ineffective in these fast runaways and the emergency relief system can be incapable of relieving the reactor pressure. Measures that prevent the formation of large acrylate-rich monomer pools are necessary to reduce the probability of a catastrophic event below acceptable limits. Such measures include process interlocks, on-line or at-line analytical testing, or even, major redesign guided by concepts of inherently safer processes.

7. References

1. J. A. Klein and A. S. Balchan, *Safe Formulation and Manufacture of Acrylic Resins*, International Conference and Workshop on Process Safety Management and Inherently Safer Processes, Orlando, FL, AIChE, pp.329-342, 1996.
2. CCPS, *Guidelines for design solutions for process equipment failures*, AIChE, New York, NY, 1998.
3. J.A. Barton and P.F. Nolan, *Incidents in the chemical industry due to thermal runaway chemical reactions*, Hazards X: Process safety in fine and speciality chemical plants, Symposium series No. 115, pp. 3-18 (IChemE, Rugby, UK), 1989.
4. Japan Science and Technology Agency (JST) knowledge database of accidents and failures in the fields of science and technology (http://shippai.jst.go.jp/en/Detail?fn=0&id=CC1200077)
5. M. Gromacki, *Acrylic polymer reactor accident investigation: Lessons learned and three years later*, CCPS International Conference and Workshop on Process Industry Incidents. Orlando, FL, pp. 411-433, 2000.
6. C. S. Kao and K. H. Hu, *Acrylic reactor runaway and explosion accident analysis*, Journal of Loss Prevention in the Process Industries, 15, pp. 213–222, 2002.
7. United States Chemical Safety and Hazard Investigations Board, *Runaway Chemical Reaction and Vapor Cloud Explosion*, Case Study No. 2006-04-I-NC, 2007.
8. CCPS, *Layer of Protection Analysis - Simplified Process Risk Assessment*, AIChE, New York, NY, 2001
9. D. J. Arriola, *Modeling of Addition Polymerization Systems*, Ph.D. thesis, Univ. of Wisconsin-Madison, 1989.
10. Congalidis, J.P., J. R. Richards and W.H. Ray, *Feedforward and Feedback Control of a Solution Copolymerization Reactor*, AIChE J., **35**, pp. 891-907, 1989.
11. R. K. Cox, J. F. Smith and Y. Dimitratos, *Can simulation technology enable a paradigm shift in process control?*, Computers and Chemical Engineering, **30**, pp. 1542-1552, 2006.
12. World Batch Forum, *Batch Markup Language*, http://www.wbf.org/
13. A. S. Balchan, D. A. Paquet and J. A. Klein, *Emergency Relief Adequacy for Acrylic Polymerization Processes*, Process Safety Progress, **18**, 2, pp.71-77, 1999.
14. J.C. Leung, *Simplified Vent Sizing Calculations for Emergency Relief Requirements in Reactors and Storage Vessels*, AIChE J., **32**, 10, pp. 1622-1634, 1986.
15. CCPS–DIERS, *Guidelines for Pressure Relief and Effluent Handling Systems*, AIChE, New York, NY, 1998.
16. A. H. Shapiro, *The Dynamics and Thermodynamics of Compressible Fluid Flow*, Vol. I, Ronald Press, New York, pp. 165-167, 1953.

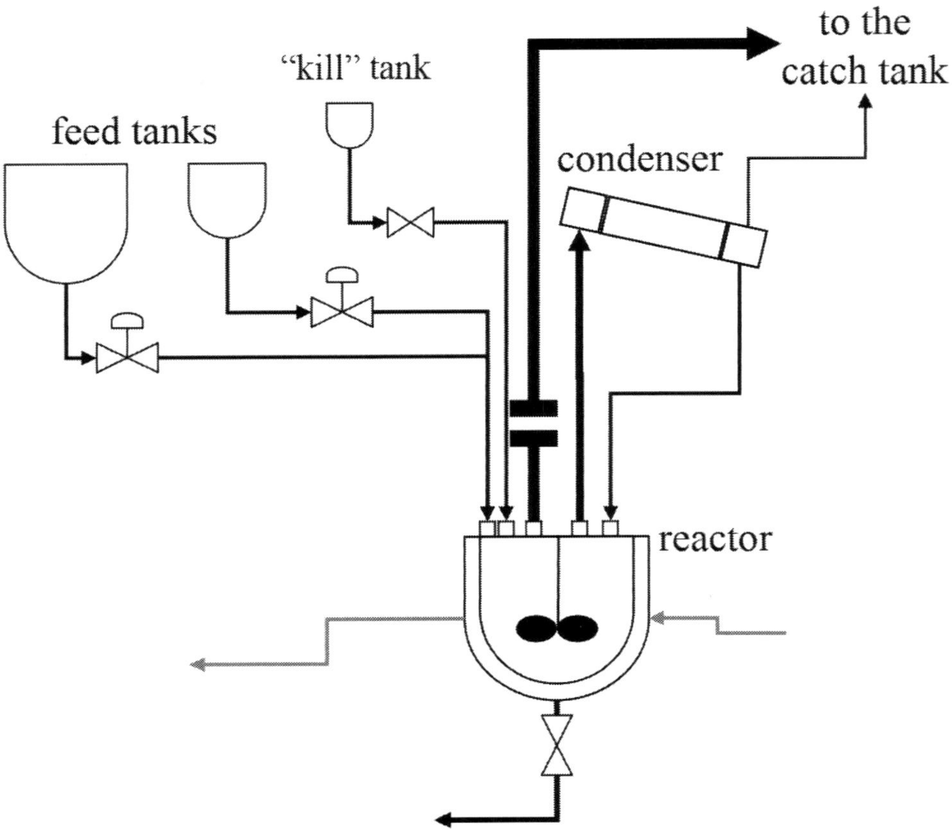

Figure 1 Sketch of a typical semi-batch acrylic polymerization flowsheet

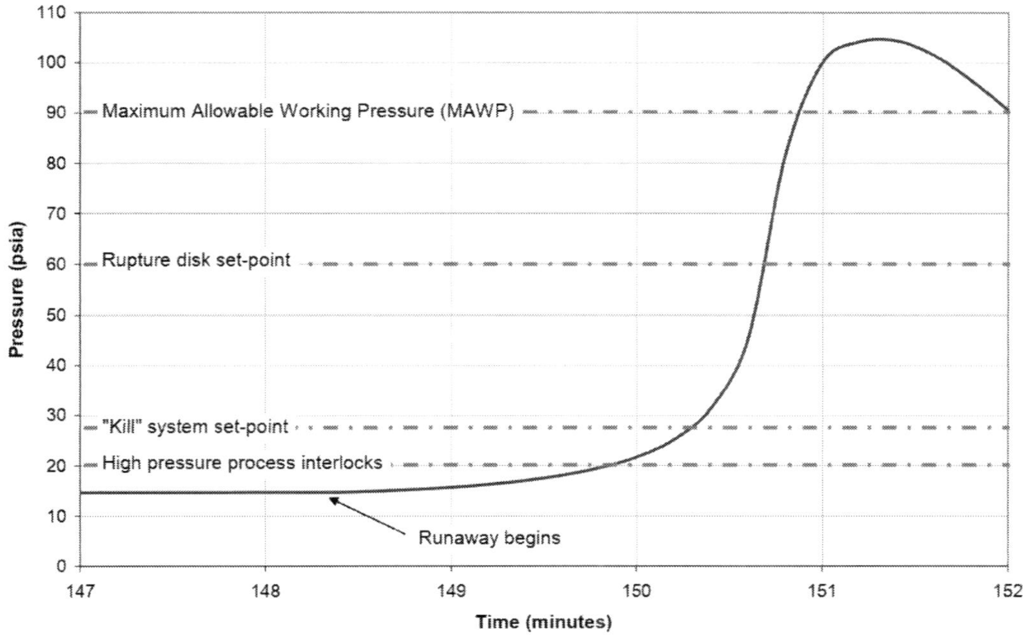

Figure 2 Simulated pressure profile during the runaway of a 100% ethyl acrylate monomer "pool" along with the set-points for the layers of protection

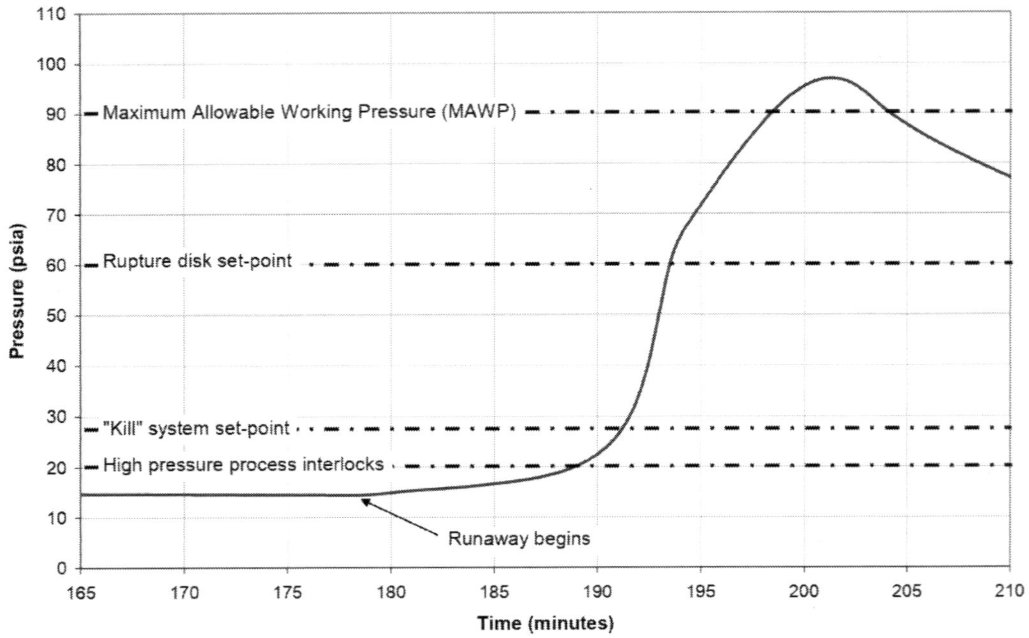

Figure 3 Simulated pressure profile during the runaway of a 75:25 ethyl acrylate / methyl methacrylate monomer "pool", assuming that all of the "layers of protection" fail to activate.

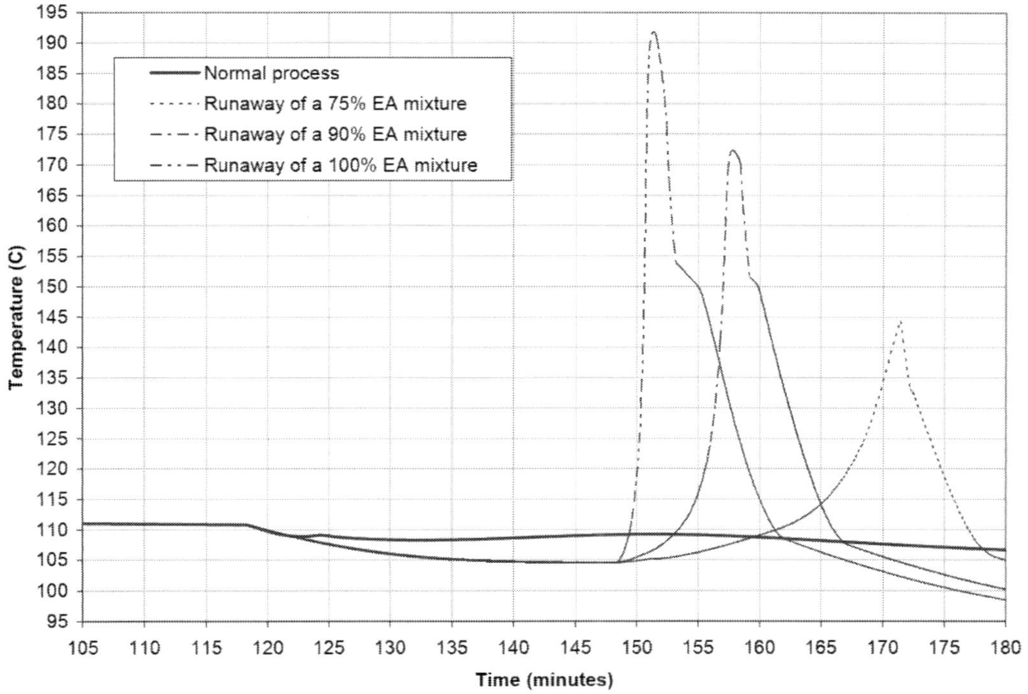

Figure 4 Simulated temperature profiles during runaways of acrylic monomer mixtures with different composition

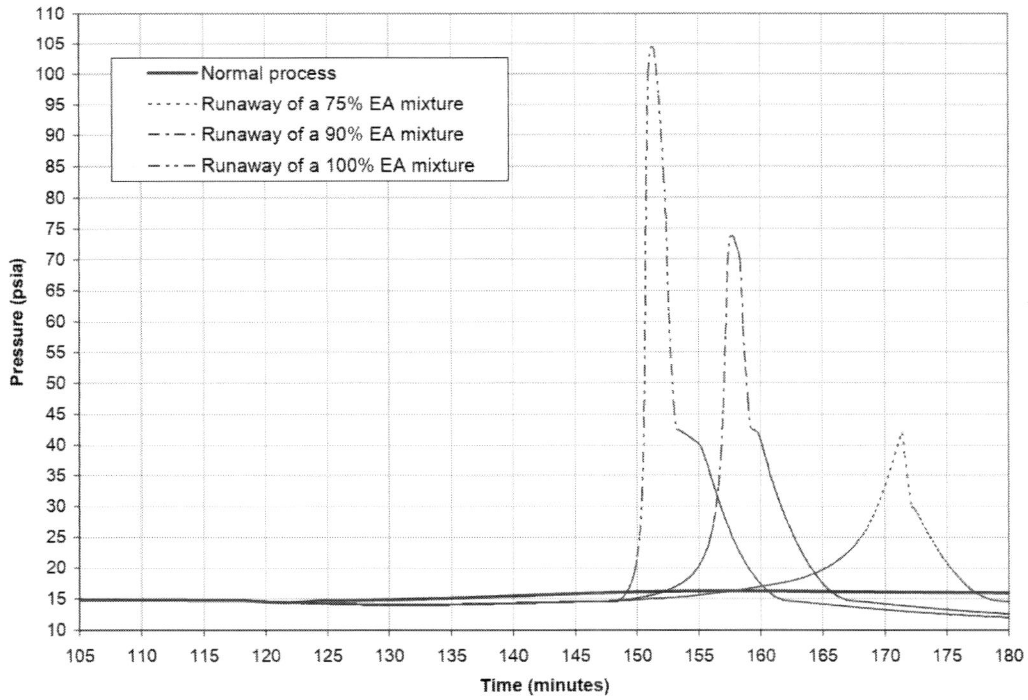

Figure 5 Simulated pressure profiles during runaways of acrylic monomer mixtures with different composition

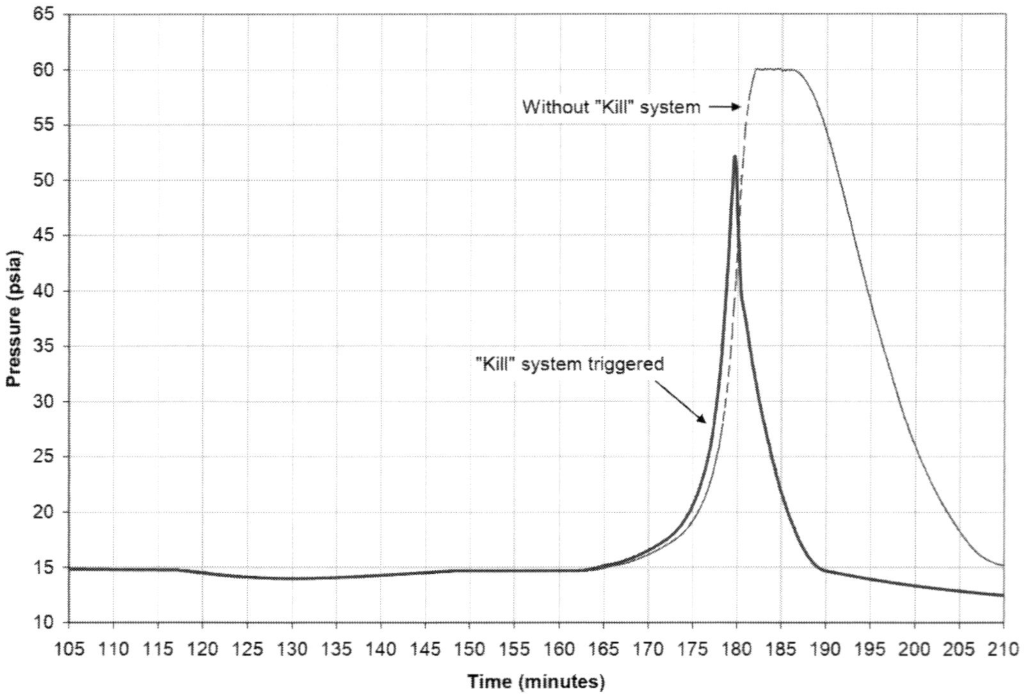

Figure 6 Simulated pressure profiles from a 75:25 EA / MMA monomer "pool" runaway with and without the effect of the "kill" system

```
Safety Analysis for rxr5000
      Completed: Sat Jan 12 09:52:52 EST 2008
      Recipe Run: X123

      Batch Size:          4125 Gal
      Peak Exotherm:       1538.1 BTU/MIN/100 GAL
      Reactor Temp:        113.92 Deg C
      CC Rating:           16.037 BTU/HR/sq.ft./F
____________________________________________________
Failure Case: Initiator Feed Delay 10.0 mins

Did Not Reach Set Pressure
Peak Pressure was 37.904 psia
____________________________________________________
Failure Case: Initiator Feed Delay 20.0 mins

   Max Pressure:           59.797       psia
   Temperature:            163.41       Deg C
   Req Relief Capacity:    31146.       LBm/min
   Req Diameter:           15.360       inches
...
____________________________________________________
Failure Case: Initiator Feed Delay 30.0 mins

   Max Pressure:           59.814       psia
   Temperature:            161.39       Deg C
   Req Relief Capacity:    48435.       LBm/min
   Req Diameter:           18.832       inches
...
____________________________________________________
Failure Case: Initiator Feed Delay 45.0 mins

   Max Pressure:           60.134       psia
   Temperature:            159.37       Deg C
   Req Relief Capacity:    67151.       LBm/min
   Req Diameter:           21.877       inches
...
____________________________________________________
Failure Case: Initiator Feed Delay 60.0 mins

   Max Pressure:           59.995       psia
   Temperature:            157.6        Deg C
   Req Relief Capacity:    85789.       LBm/min
   Req Diameter:           24.338       inches
...
```

Figure 7 Summary output from simulating 90:10 EA / MMA monomer "pool" runaways caused by different delays in the initiator feed

```
Safety Analysis for rxr5000
      Completed: Sat Jan 12 09:52:52 EST 2008
      Recipe Run: X123

      Batch Size:         4125 Gal
      Peak Exotherm:      1538.1 BTU/MIN/100 GAL
      Reactor Temp:       113.92 Deg C
      CC Rating:          16.037 BTU/HR/sq.ft./F
```

```
Failure Case: Cooling Water Failure at 105.0 mins

   Max Pressure:          59.701      psia
   Temperature:           164.2       Deg C
   Req Relief Capacity:   3774.       LBm/min
   Req Diameter:          5.650       inches
...
```

```
Failure Case: Cooling Water Failure at 120.0 mins

   Max Pressure:          59.702      psia
   Temperature:           163.43      Deg C
   Req Relief Capacity:   3940.       LBm/min
   Req Diameter:          5.706       inches
...
```

```
Failure Case: Cooling Water Failure at 150.0 mins

   Max Pressure:          59.703      psia
   Temperature:           160.18      Deg C
   Req Relief Capacity:   4785.       LBm/min
   Req Diameter:          5.983       inches
...
```

```
Failure Case: Cooling Water Failure at 180.0 mins

   Max Pressure:          59.703      psia
   Temperature:           158.05      Deg C
   Req Relief Capacity:   5464.       LBm/min
   Req Diameter:          6.164202    inches
...
```

```
Failure Case: Cooling Water Failure at 210.0 mins

Did Not Reach Set Pressure
Peak Pressure was 39.316 psia
...
```

Figure 8 Summary output from simulating 90:10 EA / MMA runaways caused by "loss of cooling" at different times during the batch process

Inherent Risk Assessment (IRA)

Azmi Mohd Shariff* and Chan T. Leong

Process Safety Research Group, Department of Chemical Engineering, Universiti Teknologi PETRONAS, 31750 Tronoh, Perak, Malaysia
**Tel : 05-3687570; Fax : 05-3656176; e-mail: azmish@petronas.com.my*

ABSTRACT

Quantitative Risk Assessment (QRA) has been a very popular and useful methodology which is widely acceptable by the industry over the past few decades. QRA is typically carried out at a stage where complete plant has been designed and sited. At that time, the opportunity to include inherent safety design features is limited and may incur higher cost. This paper proposes a new concept to evaluate risk inherent to a process owing to the chemical it uses and the process conditions. The risk assessment tool is integrated with process design simulator (HYSYS) to provide necessary process data as early as the initial design stages, where modifications based on inherent safety principles can still be incorporated to enhance the process safety of the plant. The risk assessment tool consists of two components which calculate the probability and the consequences relating to possible risk due to major accidents. A case study on the potential explosion due to the released of flammable material demonstrates that the tool is capable to identify potential high risk of process streams. Further improvement of the process design is possible by applying inherent safety principles to make the process under consideration inherently safer. Since this tool is fully integrated with HYSYS, re-evaluation of the inherent risk takes very little time and effort. The new tool addresses the lack of systematic methodology and technology, which is one of the barriers to designing inherently safer plants.

Keywords: *quantitative risk assessment, inherent risk, inherent safety, consequence estimation, inherently safer design*

1.0 QUANTITATIVE RISK ASSESSMENT (QRA)

Quantitative Risk Assessment or commonly known as QRA has gained a wide acceptance as powerful tool to identify and assess the significant sources of risk and evaluate alternative risk control measures in chemical process industries. QRA is part of Process Safety Management System (CCPS, 1999) and is considered a valuable tool in decision-making processes, to communicate among the experts involved, to quantify opinions and to combine these effectively with available statistical data. Shell International Exploration and Production B.V (1995) views this technique as systematic approach to identifying hazards, potentially hazardous events and estimating likelihood and consequences to people, environment and assets, of incidents developing from these events. Lees (1996) in his review of various studies, concluded that QRA is an element that cannot be ignored in decision making about risk as it is the only discipline capable of enabling a number to be applied and comparisons to be made in quantitative manner.

Many governments and local authorities require industries which are hazardous in nature to produce a document to demonstrate that risks are as low as reasonably practicable (ALARP). In the UK, the QRA found its principal application in process industries safety cases (Lees, 1996). In Malaysia, the Control of Industrial Major Accident and Hazards (CIMAH) Regulations, 1996, require industries classified as Major Hazard Installation to submit a written Safety Report to the Director General at least three months before commencing the industrial activity (or before introducing hazardous substances into the plant). QRA is a component of the Safety Report and it can be prepared using any suitable method (International Law Book Services, 2000).

Lees (1996) noted that there are considerable variation between one QRA to another mainly due to the specific problems they try to address and the boundaries of the problems. Arising from such factors, Shell International Exploration and Production (1995) noted QRA studies can take between 40 to 1500 man-hours depending on the level of details. Despite the manhours spent, it is important to understand that the methods like QRA cover only specific elements of the aspects involved in the safety of a plant. This method also requires subjective judgment and can only provide a partial idea about the safety present in a facility.

CCPS (1999) established a general guideline for QRA. Individual organizations and companies may have procedures that are specific to their respective needs. Generally for a QRA to be meaningful, it is carried out after significant process design and main equipment layout tasks have been completed.

2.0 INHERENT RISK ASSESSMENT (IRA)

From the above deliberation, it has been observed that by the time QRA is carried out, much of the design work has been completed. Even equipment layout has been determined and to large extent, some operating philosophy has been set. At this juncture, one of the most common methods to mitigate risk and its consequences is by means of adding protection devices such as instrumented protective functions or mechanical protection devices such as relief valves, etc. However, this approach alone is unable to avoid or reduce the risk of serious chemical accidents (Zwetsloot and Askounes-Ashford, 1999). These protective measures added late in the design, often require regular preventive maintenance to detect revealed failures in order to prevent catastrophic events. The preventive maintenance throughout the life of the plant, adds to the operating cost as well as necessitating repetitive training and documentation upkeep.

In order to detect hazards proactively to allow for the opportunity to proactively reduce their magnitude or likelihood of occurrence, this paper proposes a method to integrate risk quantification with process design simulator. With the integration, process design engineers can assess the risk which is inherent to their design from the beginning of the design stages. With the early detection, proactive measures based on Inherent Safety principles can be implemented. This new methodology is named as Inherent Risk Assessment (IRA) and will adopt very similar approach as the conventional QRA to encourage rapid adoption by the industries.

Inherent Safety is a proactive approach for hazard/risk management during process plant design and operation. Inherent Safety aims to reduce or eliminate the root causes of the hazards by modifying the design (hardware, controls, and operating conditions) of the plant itself instead of relying on additional engineered safety systems and features, and procedural controls which can and do fail. Kletz (1991) formalized several principles defining Inherent Safety (IS) which includes intensification, substitution, attenuation, simplification, limitation of effects and error tolerance.

Inherent safety has become an important aspect and is deemed as the best method to design a plant safe for operation, with no harm to the environment and health.

2.1 Similarity and Comparison with QRA

Even though IRA adopts the structured approach from QRA, the two methodologies have fundamental differences. The key different is the stage when the methods are applied. Traditional QRA is applied after detail engineering design has been completed when Process & Instrumentation Diagram (P&ID) is fully available. QRA also takes into consideration the historical site-specified weather conditions and plant layout. In contrast to this, the IRA technique that developed in this research can be used as early as process simulation begins during preliminary design stages in parallel with selection of process route and development of heat and material balances. Unlike QRA, the IRA does not account for safety control measures such as procedures and instrumented protective functions. It merely reflects the inherent risk due to the inherent properties of the chemicals involved and process conditions of the design.

Owing to the different timing that the two assessments are carried out, the results are used for different purposes. More often than not, QRA results are used for reasons presented in Section 1.0 above, while results for IRA can be used to provoke inherent safety modifications during process simulation stages. Process design engineers have considerable flexibility to improve the design in simulation stage as the manhours and cost to do so is relatively small. IRA results are suitable to be used as a quantitative method to screen processes and improvements.

Since QRA exercise is very extensive in nature and requires significant amount of time to produce; only few selected credible cases are studied and documented. Due to the lack of integration between risk assessment software and process design simulator, all the information for process conditions need to be manually transferred. The IRA can be effectively implemented using risk models if it is fully integrated with process design simulator such as HYSYS. Data is automatically transferred from the process design simulator to the risk models hence reducing the chances of error. With the integrated model, inherently safer design options can be quickly evaluated in a short time. For instance, inherent risk before and after pressure modification of a unit operation can be promptly assessed. Due to the different requirements and purpose, IRA cannot be used to replace QRA but they are meant to complement each other and are used in different timing along the process design stages. Table 2.1 below summarizes key distinction between QRA and IRA.

Table 2.1: Comparison between QRA and IRA

Criteria	QRA	IRA
Stage to be applied	After completion of detail engineering design	During preliminary design / simulation stage
Purpose	To demonstrate or prove "safety case" as required by regulatory agencies	To proactively identify risk inherent to the design and guide its reduction by adopting inherent safety principles
Regulatory requirements	Required by regulatory agencies, for example Department of Occupational Safety & Health in Malaysia and the Health and Safety Executive in the UK	Presently, there is no regulatory requirement
Information required	Process & instrumentation diagrams (P&ID), detail historical weather data	Simulation data and predicted piping and equipment sizing
Scenario	Only few credible scenario to be studied in detail	Basic scenario such as pipe or equipment leak
Duration of analysis	Relatively long. Ranging from 40 to 1500 manhours.	Relatively quick as it is carried out in parallel with simulation work
Result representation	3-region Frequency-Number (FN) curve	2-region Frequency-Number (FN) curve

The inherent risk level calculated by IRA needs to be presented in a comprehensive and easily understood manner to promote its adoption by the industries. Since the 3-region FN curve has been commonly accepted as graphical representation of risk result for QRA, the same concept is used in IRA to represent the inherent risk. A three region FN Curve based on

Malaysian limits is shown in Figure 2.1. However, for IRA, slight modification is proposed in which the FN curve uses only two regions instead of three regions. It can be noted that the region below the diagonal line in Figure 2.2 covers the two regions on "Tolerable if ALARP" and "Tolerable" of Figure 2.1. These two regions are not separately represented in an IRA result because during this stage, safety measures and control mechanisms are not yet in place to reduce risk to ALARP.

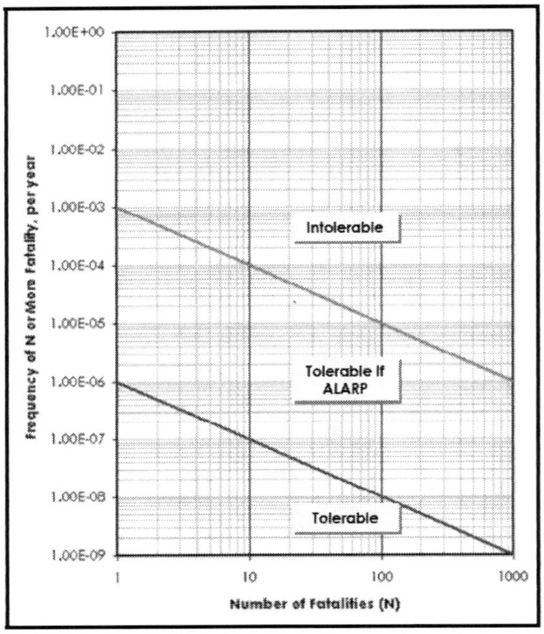

Figure 2.1: Malaysian FN curve (DNV, 1993)

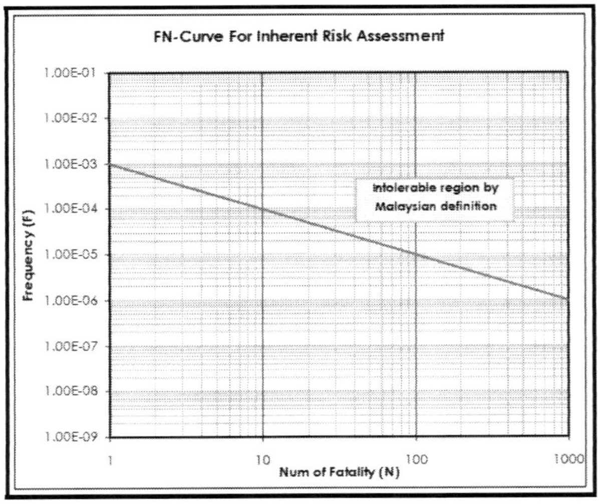

Figure 2.2: FN curve to represent inherent risk

Figure 2.2 is developed based on the numerical limits dividing intolerable and tolerable regions set by the Malaysia authorities. Should the inherent risk of a process route fall above the top diagonal line, then it is clear that the route is not acceptable unless subsequent modifications can bring it below the line. For designs that have risk level below the line initially, there is still room for further improvement, again by adopting the principles of inherent safety.

2.2 Determination of Inherent Risk during Design Stage

CCPS (1999) defines risk as a measure of human injury, environmental damage or economic loss in terms of both the incident likelihood and the magnitude of the loss or injury. The Health and Safety Executive UK (2001) refers risk as the chance that someone or something that is valued will be adversely affected in a stipulated way by the hazard. Wentz (1990) once proposed risk as a mathematical functions as follow:

$$risk = probability \times consequences \qquad \text{Equation 2.1}$$

In all the definitions above, it can be concluded that in one way or another, risk has components of probability (uncertainty) of the event and consequences (effects) resulting from the event. It may be difficult to precisely quantify the risk parameters e.g. consequences of an explosion, so they must be estimated by adopting a systematic approach as described in detail by Crowl and Louvar (1990) in their work.

In order to estimate consequences and probability and hence estimating risk, a spreadsheet based tool is developed which is fully integrated with process design simulator. The integration allows seamless transfer of process data from simulator to the spreadsheet based tool for a quick and efficient estimation of risk. The technique behind the integration has been deliberated by Mohd. Shariff et. al. (2007). The same paper also provided examples of estimation of consequences due to explosion from a process being simulated. The early quantification of explosion consequences was used as a guide to determine location of control building.

The same concept is expanded in this paper to include the probability component thus allowing risk estimation as per Equation 2-1. The integration with a process design simulator, in this case HYSYS, allows for the inherent risk assessment to be determined early in the design based on the availability of process design simulation for a particular project.

2.3 The Integrated Explosion Event Tree

This paper has been customized to demonstrate the risk due to explosion hence the spreadsheets have been developed to determined probability of an explosion event and consequences should such an event occur. This section describes the probability model used in the IRA for the proposed case study in this paper.

The two most popular methodologies to determine probability of accident are Event Tree Analysis (ETA) and Fault Tree Analysis (FTA). In this work, the ETA is chosen over FTA due to its advantages in modeling events like explosion, which are sequential in nature. FTA implies sequence of event as irrelevant and has less obvious logic (Smith, 2003) compared to ETA. Furthermore, ETA has been used by many authoritative publications in this area of research including Lees (1996), Hendershot (2000) and CCPS (2000) to describe fire and explosion events.

In order to ensure consistency in terms of factors being considered in one case to another, this paper adopted an Event Tree Analysis (ETA) from Hendershot (2000), as shown in Figure 2.3. It presents a simplified basis to describe the probability of an explosion in the event of loss of containment. However, it is important to note the possibility for mechanisms in any other explosion case to deviate from the one presented here. Consistency is important so that various process options and modifications can be compared on equal basis from the event sequence aspect leaving only the chemical and process condition aspects as variables in the IRA. The sequence used in the ETA blocks in Figure 2.3 is unsuitable for modeling other forms of hazards.

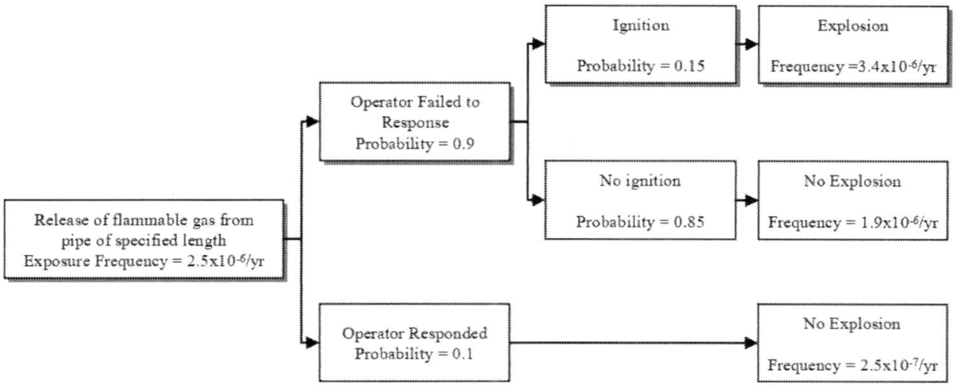

Figure 2.3: ETA for potential explosion due to the release of flammable gas from pipe rupture.

The frequency and probability data shown in figure 2.3 is only for illustrative purposes and are based on information in example described here. It is important to note that during the simulation stage, the actual sizing of pipe diameter and length is no yet available. Therefore, a good engineering judgment and estimation is required to determine the length and diameter to be used in the simulation.

A case of a rupture in a 25 meter long 300 mm diameter pipe is estimated. Base failure frequency is referred from database provided by Allen (1998) and CCPS (1989). From calculation below, the exposure frequency is expected to be $2.5\text{x}10^{-6}$ per year.

exposure frequency = base failure frequency x length x duration

= $1\text{x}10^{-7}$ failure year^{-1} m^{-1} x 25 m x 1 year

= $2.5\text{x}10^{-6}$ per year

Cox (1990) concluded that there is a 0.9 chance for operator failure to respond within 5 minutes. The probability of explosion in Figure 2.3 is calculated from Equation 2-2, developed based on data provided by Withers (1988) on ignition and explosion of leaked hydrocarbon.

$$P_{ex} = 0.00175 \times (0.9999)^{m_f} \times (m_f)^{0.4582} \qquad \text{Equation 2-2}$$

where
P_{ex} = probability of explosion
m_f = mass of combustible material

For the purpose of illustration, it is assumed that the flammable material determined from HYSYS is 100 tons. This gives the probability of an explosion of 0.15. The frequency of an explosion resulting from an event due to operator failure to response and probability of explosion is 3.4×10^{-6} per year as given in Figure 2.3.

2.4 The Integrated Consequence Estimation Tool

Following Equation 2-1, the other component to determine risk is the consequence aspect. In order to quantify inherent risk, it is therefore crucial to estimate the consequence of an incident. An explosion consequence model is developed in spreadsheet which is integrated with HYSYS for ease of data transfer and to reap benefits of the built-in database within HYSYS. The steps taken to estimate the consequences of an explosion adopted in the tool is provided in Figure 2.4.

3.0 CASE STUDY

This case study aims to evaluate Inherent Risk from all streams within a simulation of a hydrocarbon fractionation plant. The simulation is at the initial stage of a design where a series of distillation columns are used to separate different components of hydrocarbons as given in Figure 3.1. This simulation has 27 process streams that undergo a screening process to relatively rank streams according to their potential of causing damages in the cases of explosion. Detail of the screening process has been separately discussed in another paper (Chan and Mohd Shariff, 2008). The top five streams from that rank are further analyzed for their respective inherent risk based on the inherent properties of the chemicals used and process conditions of the design.

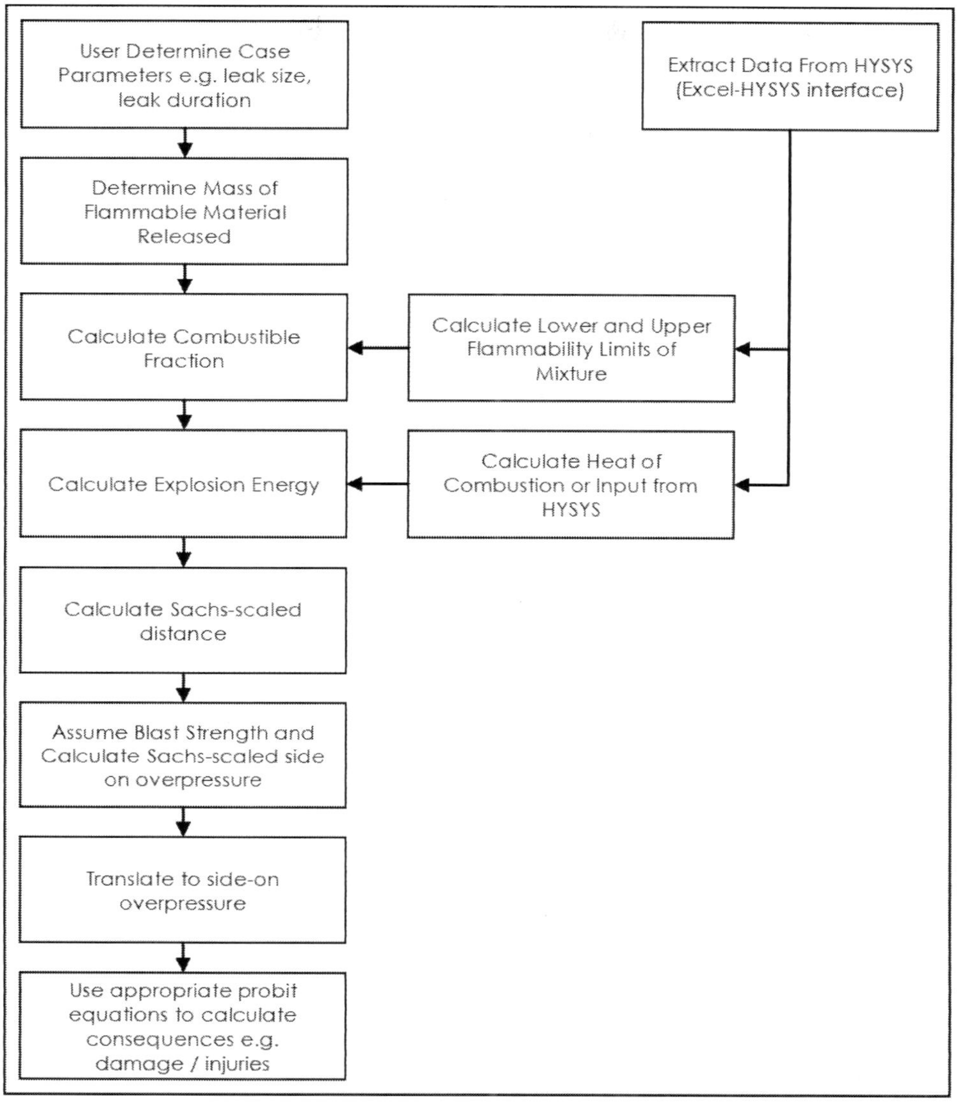

Figure 2.4: Integrated consequences estimation tool for explosion study.

The same potential accident scenario is selected as a basis to evaluate the inherent risk, which in this case is a catastrophic rupture of a 300 mm diameter pipe resulting in hydrocarbon leakage for 1 minute. Owing to the different process parameters such as pressure and composition of each stream, the amount of hydrocarbon leaked differs. This in turn influences the probability, the consequence (overpressure) of the explosion incident and the overall event frequency (F) calculations. The number of fatality (N) can be obtained by further assumed that 100 workers are exposed to the hazard due to the explosion.

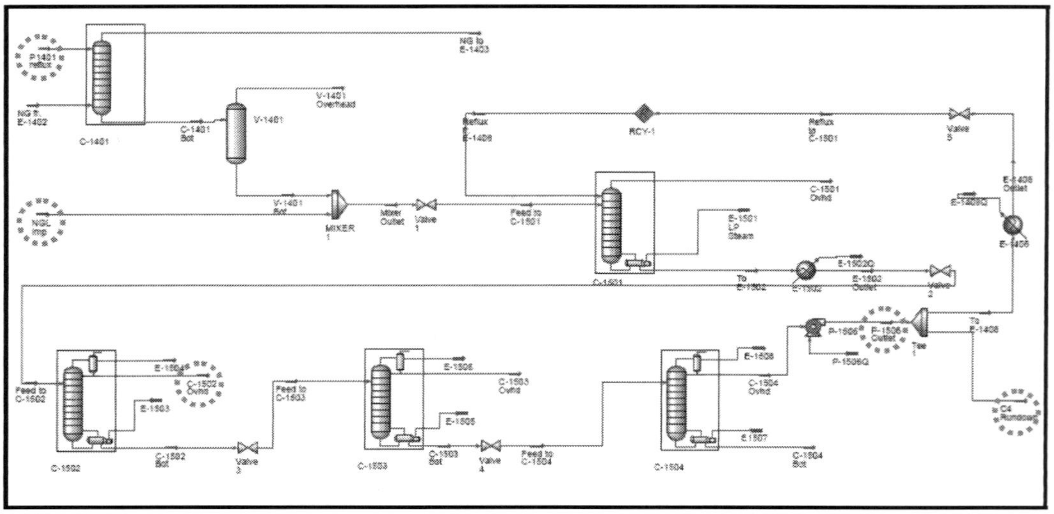

Figure 3.1: Hydrocarbon fractionation plant simulation

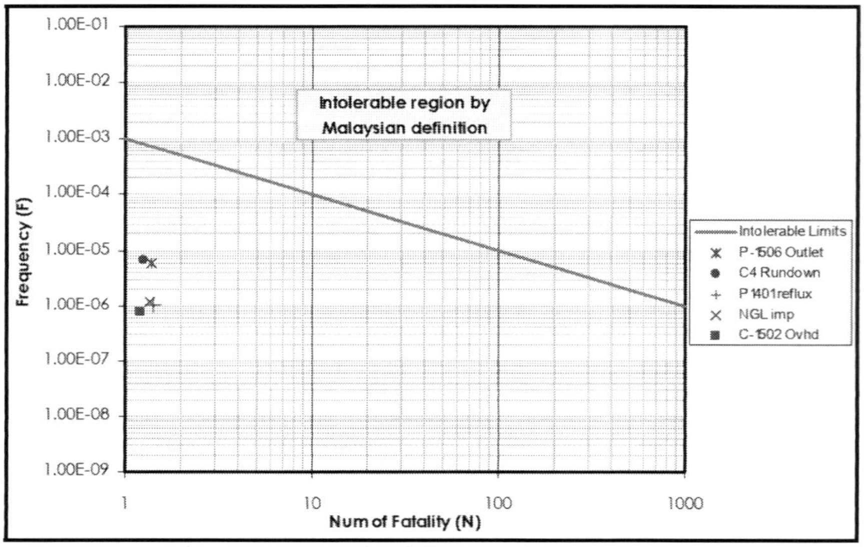

Figure 3.2 : Example of FN Curve for Given Simulation Case

The FN values representing the inherent risk due to explosion for all the top five streams are given in Figure 3.2. It is found that all the calculated risks are below the intolerable region set by Malaysian definition. It can be interpreted that the design does not have any streams that breaches the Malaysian FN limits. Therefore, further improvement of the design is not really required at this juncture. However, further consideration should be given to assess inherent risk due to other form of accidents such as fire, toxic release, etc. if necessary. If the evaluated inherent risk is above the intolerable region, therefore it is necessary to improve the design by applying the inherent safety principle. By implementing this new concept in the design stage, perhaps potential major accidents can be eliminated or reduced.

As mentioned earlier, the IRA only assesses initial acceptability of the design based on safety aspects. It is expected that as the design progresses with more detail design information available, more safety features, process control measures and work procedures will be put in place to ensure risk is as-low-as-reasonably-practicable (ALARP).

4.0 CONCLUSION

This paper proposed and demonstrated a technique to quantify risk which is inherent to the process condition and stream composition early in process design stage by means of integration with process design simulator for seamless data transfer thus allowing early determination of the associated risks. The inherent risk assessment will allow process designers to proactively consider and implement modification to improve the safety of the process plant.

5.0 ACKNOWLEDGEMENTS

The authors would like to thank Universiti Teknologi PETRONAS for providing the financial support, which has made the presentation at this conference possible.

6.0 REFERENCES

1. Allen, D. (1998), ***Web-Based Process Engineering Risk Calculator***, http://www.cheque.uq.edu.au/ugrad/theses/1998/DaveA/index.html (accessed 20 Feb 2007).
2. Chan T. Leong and **Azmi Mohd Shariff** (2008), *"Process Route Index (PRI) to Assess Level of Explosiveness for Inherent Safety Quantification"*, **Journal of Loss Prevention in Process Industries**, (Submitted for publication on 15th Jan 2007)
3. Center for Chemical Process Safety (CCPS) (1989), ***Guidelines Process Equipment Reliability Data with Data Tables,*** American Institute of Chemical Engineers, New York.
4. Center for Chemical Process Safety (CCPS) (1999). ***Guidelines Consequence Analysis of Chemical Releases***, American Institute of Chemical Engineers, New York.

5. Center for Chemical Process Safety (CCPS) (2000). ***Guidelines for Chemical Process Quantitative Risk Analysis***, 2nd Ed., American Institute of Chemical Engineers, New York.
6. Cox A.W., Lees F.P. and Ang, M.L. (1990), ***Classification of Hazardous Locations***, Institute of Chemical Engineers, UK.
7. Crowl D.A., Louvar J.F. (1990), ***Chemical Process Safety : Fundamentals with Applications***, Prentice Hall, USA.
8. DNV (1993), ***MLNG Tiga QRA Report***, Malaysia.
9. Hendershot, D. C. (2000). ***Process minimization: Making plants safer***. Chemical Engineering Progress, 96, 35–40.
10. HSE (Health and Safety Executive UK) (2001), ***Reducing Risk, Protecting People, HSE's Decision Making Process***, Crown, United Kingdom.
11. International Law Book Services (2000), ***Occupational Safety and Health Act 1994 and Regulations and Orders***, Kuala Lumpur, Malaysia.
12. Kletz, T.A., ***Plant Design for Safety – A User Friendly Approach***, 2nd edition; Taylor and Francis, 1991.
13. Lees, F.P. (1996), ***Loss Prevention in Process Industries,*** Volume 1, 2nd Ed, Butterworth Heinemann.
14. Mohd Shariff, A., Rusli, R., Chan, T.L., Radhakrishnan, V.R., Buang, A., (2006), ***Inherent Safety Tool For Explosion Consequences Study***, Journal of Loss Prevention in the Process Industries 19 p 409–418.
15. Shell International Exploration and Production B.V., ***Quantitative Risk Assessment, Volume 3***, The Hague, Netherlands, 1995.
16. Smith, D.J., (2003), ***Reliability, Maintainability and Risk***, 6th Ed., Butterworth Heinemann, Great Britain.
17. Wentz, C.A., ***Safety, Health and Environmental Protection***, McGraw Hill, New York, 1999.
18. Withers, J., ***Major Industrial Hazards : Their Appraisal and Control***, Gower Technical Press, England, 1988.
19. Zwetsloot, G.I.J.M. and Askounes Ashford, N., ***Encouraging Inherently Safer Production in European Firms: A Report from the Field,*** Journal of Hazardous Materials, Special Issue on Risk Assessment and Environmental Decision Making, A. Amendola and D. Wilkinson (eds.), 1999, pp 123-144.

Topical 4 - Paper No. 147c

APPLICATION OF AN EMERGENCY EGRESS RISK INDEX SYSTEM AND PROTECTION LAYER EVALUATION IN THE DETERMINATION OF LIFE SAFETY EGRESS REQUIREMENTS FOR OPEN STRUCTURE INDUSTRIAL OCCUPANCIES

Prepared by:
Thomas V. Rodante, P.E.
&
Kelly G. Mansfield

Baker Engineering and Risk Consultants, Inc.
3330 Oakwell Court, Suite 100
San Antonio, TX 78218-3024
www.BakerRisk.com

Prepared for presentation at:
2008 Global Congress on Process Safety,
CCPS Plenary Session - Inherent Risk

Copyright © (*Baker Engineering and Risk Consultants, Inc.*)
Unpublished

Abstract

NFPA Life Safety Code® requirements are for most occupancy classes, specific and detailed. However, for industrial facilities extensive exemptions are granted from prescriptive requirements. Such exemptions are not directly related to the industrial hazard exposure beyond a gross distinction between high and low/ordinary hazard contents. Additionally, the use of hazard mitigation features is not considered as protection layers. An egress design based solely on the NFPA Life Safety Code® may not be commensurate with the inherent risk. While NFPA Life Safety guides do exist for performance-based analysis, none were written for an industrial application. As an alternative approach, BakerRisk devised an Emergency Egress Risk Index system. An analysis of protection layers is used to evaluate alternatives for obtaining an acceptable resultant risk level.

1.0 Introduction

NFPA Life Safety Code® (NFPA 101®) requirements are specific and detailed for most occupancy classes. However, for industrial occupancy classes[1] particularly in open structures, extensive exemptions are granted from prescriptive requirements. Such exemptions are not related to different levels of exposure within the general "high hazard industrial occupancy" designation for say a petroleum or chemical facility, or to existing mitigation features provided. Rather, exemptions and liberal occupancy load factors[2], are granted solely on the basis of the general occupancy classification.

An egress design based solely on the NFPA 101® may not be commensurate with the inherent risk of say a petroleum refinery. While NFPA 101® does recognize an "equivalency"[3] concept, limited guides exist for performance-based analyses. NFPA 101A "Guide on Alternative Approaches to Life Safety" does not contain direction for implementation in an industrial occupancy. As an alternative approach, an Emergency Egress Risk Index system is offered for consideration. In addition to establishing a base risk, an analysis of protection layers is used to evaluate alternatives for obtaining an acceptable resultant risk level.

1.1 NFPA 101® Industrial Occupancy Requirements

There are a few chapters in NFPA 101® that apply to Industrial Occupancies. Petrochemical facilities would be classified as a High Hazard Industrial Occupancy[4]. In Chapter 11, "Special Structures and High-Rise Buildings", most areas of petroleum and chemical refineries would fall into the category of "Open Structures"[5], for which the requirements are minimal. For instance, while liberal occupancy load factors are granted for industrial occupancies, open structured industrial occupancies are totally exempt from such considerations. NFPA 101® Chapter 11 Section 11.2 goes on to exempt Means of Egress from requirements such as capacity, number (at grade level), travel distance limitations, marking, illumination, etc. Despite the fact that petrochemical facilities are high hazard and may be multi-level, there are no requirements for the protection of vertical openings or vertical means of egress. Additionally, features not otherwise normally allowed as primary components are permitted (e.g. ladders, fire escape ladders, and elevators, each when serving three or fewer occupants). Areas of refuge are allowed in any industrial occupancy but their application is not practical for open structures of petrochemical facilities.

Chapter 40, "Industrial Occupancies", would also be applicable to petrochemical facilities and provides prescriptive requirements for egress components such as doors, passageways, stairs, ramps, and ladders. Compared to other occupancy classes, industrial facilities are allowed more flexibility. As an example, allowable components include spiral stairs, industrial equipment access stairs (including ramps, platforms, walkways),

[1] NFPA 101® 2006 Edition Section 3.3.168.8
[2] NFPA 101® 2006 Edition Section 7.3.1.2 (e.g. special purpose industrial, storage,)
[3] NFPA 101® 2006 Edition Section 1.4
[4] NFPA 101® 2006 Edition Appendix A3.3.168.8.2 and 40.1.4.1.3
[5] NFPA 101® 2006 Edition Appendix A.3.3.240.7

alternating tread devices, slides, etc.) There are no construction requirements for any industrial occupancy. In addition to the liberal occupancy load factors, floor space occupied by machinery, equipment, or process vessels, need not be included in the computation.

1.2 Potential Shortcomings of NFPA 101® to Petrochemical Facilities

Given the exemptions in Chapter 11 and flexibility provided by Chapter 40 of NFPA 101®, it is possible to provide less than optimally protected means of egress in a petrochemical facility. For example, a tower with a single ladder meets the intent of NFPA 101®. However, in an example of two towers side by side, one of which has a pump at its base containing a hydrocarbon above its auto-ignition temperature, the risk imposed by each tower to an operator working at an elevated platform is not the same. The increased risk resulting from self-igniting materials might warrant additional layers of protection for safe egress, or protective features to decrease the likelihood of a fire.

Another example in which the requirements of NFPA 101® may not be stringent enough would be an elevated platform where an operator takes samples. NFPA 101® would allow a single ladder or stairway with unlimited travel distance. If the sample were a highly toxic stream, or product above its auto-ignition temperature, or perhaps at very high pressures, it may be justifiable to provide an additional means of egress, rescue provisions, or hazard reduction to the sampling technique.

1.3 Alternative Approach

As an alternative approach, an Emergency Egress Risk Index system can be applied as described below. Risk rankings can be assigned to the means of egress and an analysis of protection layers carried out. The suggested process identifies existing risk and additional protective layers needed to obtain an acceptable resultant risk level. The key to this study is to reduce the overall risk by manipulating, through the application of additional Layers of Protection (LOP), the likelihood that an event could result in harm to an individual.

Most traditional risk matrices consist of something similar to Figure 1. In a traditional risk based approach, both the likelihood and consequence of an event are manipulated to achieve the desired level. The difference in applying the Emergency Risk Index is that only the likelihood scale is evaluated and manipulated. Regardless of the number of rows and columns used (4 by 4, 5 by 5, etc.) in the case of emergency egress, the consequence is fixed and assumed to involve a permanent disability or fatality. Given the limited employee density of most petrochemical operations, a consequence of multiple fatalities from a non blast event generally would not seem probable. However, if justifiable by specific hazards present and design release case, other consequences should be considered. Essentially, the remainder of this paper will consider a 3 by 1 (likelihood by consequence) matrix model.

The cited objective of the NFPA Life Safety Code® is to "... minimize danger to life ... " (Section 1.1.1), to provide egress facilities that "... allow prompt escape ..." (Section 1.1.3), and to "... provide adequate egress time or protection for people ..." (Section

1.1.4). Given these stated conditions, it would appear that an alternative approach based upon manipulating the likelihood of that consequence would be considered to meet the "equivalency" provisions.

Risk (L x C)					
LIKELIHOOD	4	4	8	12	16
	3	3	6	9	12
	2	2	4	6	8
	1	1	2	3	4
		1	2	3	4
		CONSEQUENCE			

Figure 1 Risk Matrix

2.0 Methodology

BakerRisk, in agreement with its industrial clients, proposed an approach to emergency egress evaluations based upon a risk indexing system combined with application of an evaluation of protection layers to reduce the risk where it was deemed necessary. This type of analysis evaluates the existing condition in the area of concern and compares the current risk exposure to a more acceptable level of risk. Additional Layers of Protection are added if needed to achieve a lower acceptable risk.

Once the inherent risk index is determined for each area, credit can then be given for the existing protective features and for adding additional Layers of Protection.

The Emergency Egress Risk Index is defined as: **R = L x C and L = E x F x H** where:

R = Risk Index
E = Exposure time
F = Frequency of a process fluid release (equipment related)
H = Hazard inherent to the fluid release
C = Consequence of exposure to the release

A spreadsheet is used (see Appendix 1) to document each potential egress area. Field data is collected to document existing conditions against local, federal, company, and NFPA Life Safety Code® requirements. Deficiencies are noted for potential abatement during the Risk Indexing and Layer of Protection evaluation phase of the study.

Conditions influencing the likelihood include occupancy time (E), type of equipment in the affected area (F), and the physical and chemical composition of the process fluid involved (H). Consequence is always considered to be the same (permanent disability / fatality), and the three likelihood conditions are evaluated for each identified location. Sample likelihood values are provided below.

A team of plant personnel should be assembled to conduct the egress analysis. More importantly, this team must first define and agree upon a set of likelihood values, allowable protective features, and the credits to be given for each protective feature. Consideration for team composition should include plant fire & safety, operations, process specialists, etc.

2.1 Exposure Time

Table 1 – Exposure Likelihood Values

Exposure Time		E Value
< 1 hour per shift	An area where operations or maintenance visit infrequently, such as areas where field checks are made daily or instruments are calibrated or repaired	1
1 to 4 hours per shift	Areas where operations or maintenance personnel do routine duties within this time frame	2
Continuous – More than 4 hours per shift	Areas where operations or maintenance personnel conduct routine duties for more than 4 hours per day	3
Manned enclosed space / room	Areas where operations or maintenance personnel conduct routine duties of an operation within an enclosed structure	4

2.2 Release Potential

Table 2 – Release Likelihood Values

Type of Equipment		F Value
Piping/Equipment in non-erosive or non-corrosive service	Piping/Equipment in low sulfur hydro treated process service	1
Piping/Equipment in corrosive or erosive service	Piping/Equipment in sour service or piping in abrasive service	2
Rotating equipment	Any pump or compressor in the affected area	3
Frequent flange opening (Coker)	Coker drill deck and switching deck where flanges are separated daily	4

2.3 Hazard

Table 3 – Hazard Values

Process Fluid Condition		H Value
Combustible	Process fluids being processed at a temperature below their flammable point	1
Flammable, hot coke, high pressure releases (e.g. cutting operations, 600# steam, etc.)	Process fluids being processed at a temperature at or above their flammable point, but less than their auto-ignition point. Also includes hot coke.	2
Auto-ignition	Process fluids being processed at a temperature above their auto-ignition point	3
Toxic	Process streams, when released would cause an atmosphere at or above Immediately Dangerous to Life or Health (IDLH) concentrations (e.g. H_2S >100 ppm)	4

2.4 Consequence

Table 4 – Consequence Values

Consequence	C Value
Permanent disability / Single fatality	1
Multiple fatalities	2

2.5 Layers of Protection (LOP) Credits

Values for protection layers are determined on the relative level of protection provided based on interpretation of local and federal regulations, NFPA Life Safety Code® requirements, and industry practices. No credit is given for a primary means of egress, as this is considered to be a base level safeguard that must be present at all locations. Credits for the level of protection provided or recommended are as follows:

Table 5 – Protection Credits

Protection Layer	LOP Credit Value
Provide two means of egress	-3
Provide a motor operated valve (MOV) as an emergency isolation valve (EIV)	-2
All other levels of protection (LOP), E.G. water deluge spray systems, barrier walls, automatic shutdown, etc.	-1

2.6 Study Assumptions

Vapor Cloud Explosions (VCE) are not considered. It is assumed that egress is not a viable solution for a VCE initiating at the point of, or within, the means of egress under consideration. Similarly, evacuation away from blast waves traversing an area from an initiation point elsewhere is also not considered to be viable.

Inspection and maintenance data are not considered; hence, no credit is given for equipment history. The assumption is that all equipment has the same mechanical integrity throughout a given plant. This assumption is partially justified by the condition that emergency egress would be necessary if no failure occurred. However, if plant history dictates, (e.g. high mineral content in the feed streams, chemical injection, etc.) a category can be made for non-erosive/non-corrosive service vs. corrosive or erosive service.

Considerations are based on normal operation and not for turn-around situations. It is assumed that during turn-around periods, while the occupancy loads would increase, temporary scaffolding, process fluid removal, and other safeguards would be in place to reduce the emergency egress risk.

2.7 Analysis, Credits and Abatement

Initially, the existing "as built" conditions are calculated and compared to the expectations of the team based upon their combined experience and knowledge of egress standards. An adjustment is made to the initial assumed values and then the process re-applied to test for validity. The values presented above are the tested and adjusted values as used in a particular facility. Each facility intending to apply this methodology should begin by establishing and validating site specific values.

An acceptable residual risk index must also be agreed upon before beginning the analysis. For the study example presented herein, an index of 6 or less is acceptable from the standpoint of meeting minimum regulatory compliance and pacesetter industry practice. An Emergency Egress Risk Index value of 6 represents an area where personnel exposure time is less than 1 hour per shift, exposed to piping with an erosive process fluid operating above the auto-ignition temperature. This level is also representative of an area where personnel operate for up to 4 hours per shift and are exposed to non-erosive / non-corrosive piping with a process fluid operating above the auto-ignition temperature. Again, both of these situations would be required to only have a single means of egress per NFPA Life Safety Code® requirements for open structure industrial facilities.

The initial Emergency Egress Risk Index is represented by:

R(i) = (E x F x H x C) + Σ(LOP Credits) where:

R(i) = Resultant Emergency Egress Risk Index
E = Exposure time
F = Frequency potential of a process fluid release
H = Hazard inherent to the fluid release
C = Consequence of exposure to the release
LOP = Credits for protection layers above the minimum required single egress

The abated Emergency Egress Risk Index is denoted as **R(a)** after credits are applied for the installation of recommended additional LOPs.

R(a) = (E x F x H x C) + Σ(LOP Credits initial) + Σ(LOP Credits additional)

3.0 Examples

3.1 Furnace Example

In this example, consider a furnace having a single stairway leading to a multi-level platform with dual ladders. Operators on an elevated aerator deck are required to perform spot checks on heater conditions and instruments. As shown in Table 1, the exposure (E) is 1. Personnel are anticipated to spend less than one hour per shift on the platform. The burner fuel is of low-quality refinery off-gas, therefore, it is considered that the piping can be in corrosive or erosive service. These conditions define a release potential (F) of 2, as shown in Table 2. Since this is a process heater, the hydrocarbon is heated above its auto-ignition temperature. Thus the hazard value (H) in Table 3 is 3. The consequence as shown in Table 4 is 1.

The initial risk index (R(i)) is a 3 and since it is within the acceptable risk index of 6, no further recommendations are required. The analysis is presented below. Refer to datasheet in Appendix 1.

LOP: Secondary egress is provided on aerator deck. LOPA Credit = -3
Risk Index (initial)= **(E x F x H x C) + Σ(LOP Credits)** = (1x2x3x1)-3 = 3

3.2 Coker Switching Deck Example

A coker switching deck is enclosed and unsprinklered. All other decks are open with a stairway on one side and a ladder on the other, separated by a distance of more than 50 feet. There is an adjacent elevator.

3.2.1 Scenario 1: Fire while de-heading coke drum

The switching deck is continuously manned, giving it an exposure value of 3. As part of the normal operations on the switching deck, flanges must be opened and closed, resulting in a release potential (F) of 4. The hazard value (H) is 2, since the process fluid is hot coke. The consequence is 1.

LOP: Secondary means of egress is provided. = -3
Risk Index (initial): **(E x F x H x C) + ∑(LOP Credits)** = (3x4x2x1)-3 = 21
Recommendations: A full evaluation of this area against the criteria stipulated in NFPA's Life Safety Code® enclosed structures is recommended. The following, while not comprehensive, is offered for consideration:

- "Area of Refuge" and "Horizontal Exit" concepts can only be applied if these areas have integral protected means of egress components.
- The platform area and means of egress area should be fully protected by an automatic sprinkler or deluge system.
- Fire barrier requirements for egress components should be rated against a petroleum type "High Rise" fire test.

3.2.2 Scenario 2: Flange leak on live coke drum. Possible fire.

The area around the coke drum is continuously manned giving it an exposure value (E) of 3. A flange would involve mechanical failure and given the erosive nature of coke gives a release potential (F) of 2. The hazard value (H) is 2, since the process fluid is hot coke. The consequence is 1.

LOP: Secondary means of egress is provided. = -3
Risk Index (initial) = **(E x F x H x C) + ∑(LOP Credits)** = (3x2x2x1)-3 = 9
Recommendations: Same as above for Coker Switching Deck Scenario 1.

3.2.3 Coke Drum - Nuclear Gauge Platform

The nuclear gauge is located on a platform accessible by a single ladder connected to the main access platform having a stairway on one end and a ladder on the other. Since the scenario is an emergency while performing maintenance on the nuclear gauge, it has an exposure value (E) of 1. The erosive nature of coke provides a release potential (F) of 2. The hazard value (H) was increased to 3, because in addition to the process fluid being hot coke, the gauge is a nuclear device. The consequence is 1.

Scenario: Emergency while maintenance is working on the nuclear gauges.
LOP: Secondary means of egress is provided. = -3
Risk Index (initial) = **(E x F x H x C) + Σ(LOP Credits)** = (1x2x3x1)-3 = 3
Recommendations: None, meets requirement of two means of egress.

3.3 Diesel Stripper

The diesel stripper column has access from a single ladder. At the base of the column, there are two dual seal sour naphtha reflux pumps. There is fire detection and deluge over the reflux pumps. The potential scenario is a pump seal or pipe flange failure. Since the diesel stripper is manned less than one hour per shift the exposure value (E) is 1. The rotating equipment warrants a release potential (F) of 3. The hazard value (H) is 4, because the H_2S is above IDLH levels (Immediately Dangerous to Life & Health). The consequence is 1.

LOP: Ladder on column, Fire detection & deluge (-1), Seals (-1) = -2
Risk Index (initial) = **(E x F x H x C) + Σ(LOP Credits)** = (1x3x4x1)-2 = 10.
Recommendations: Provide H_2S detection around reflux pumps (-1). Provide secondary egress via ladder on reflux drum deck (-3), add a MOV EIV to the suction of pumps (-2). Note: It is an industry international standard to provide secondary egress in areas where high concentrations of H_2S are present.
Risk Index (abated): 10-1-3-2 = 4

4.0 Conclusion

Application of a risk indexing system approach to emergency egress considerations in an open industrial structure can provide more suitable safeguards than that contained in prescriptive NFPA 101 requirements. Through consideration of the inherent process risk and potential release scenarios, along with allowing credit for layers of protective features, appropriate credit can be given for standard industry loss prevention methods. The petrochemical industry routinely employs remotely operated emergency isolation valves, gas detection systems, deluge systems, dual pump seals, Safety Instrumented Systems, etc. to reduce the likelihood and severity of potential fire incidents. These same features should also be considered to supplement the minimum egress requirements for industrial occupancies.

Appendix 1. - Sample Data Collection

Issue	Reference	Requirement	Reference	Reference	Requirement	99F901 Coker Diluent Tower Heater	99C901-4 Coker Drill Deck	99C901-4 Coker Nuclear Gauge Platform
	Local Code Part 8 Entrances, Walkways, Stairways and Ladders		**NFPA 101 Life Safety Code 2003 ed. (Enclosed Structure, Special Purpose, High Hazard Occupancy)**	**NFPA 101 Life Safety Code 2003 ed. (Special Structures) Open Structure**		Single stairway and dual ladders	Enclosed structure, deck has stairway and elevator on one end and ladder on other end	Ladder and platform to secondary egress
Occupant Load			40-1.7		In high hazard 100sqft per person Maximum number anticipated under probable conditions	NA / C	C	NA / C
				11-1.7	Maximum number anticipated under probable conditions	C	NA / C	C
Plans	**Section 115 Plans**	employers to establish an emergency response plan for responding to an emergency that may require rescue or evacuation.				Note 1	Note 1	Note 1

Issue	Reference	Requirement	Reference	Reference	Requirement	99F901 Coker Diluent Tower Heater	99C901-4 Coker Drill Deck	99C901-4 Coker Nuclear Gauge Platform
	Local Code Part 8 Entrances, Walkways, Stairways and Ladders		**NFPA 101 Life Safety Code 2003 ed. (Enclosed Structure, Special Purpose, High Hazard Occupancy)**	**NFPA 101 Life Safety Code 2003 ed. (Special Structures) Open Structure**		Single stairway and dual ladders	Enclosed structure, deck has stairway and elevator on one end and ladder on other end	Ladder and platform to secondary egress
Components			40-2.2.1 through 40-2.2.13		Doors per 7-2.1 Delayed egress locks per 7-2.1.6.1 Access-controlled doors per 7-2.1.6.2 Horizontal sliding doors Stairs per 7-2.2 can have open grating Spiral stairs per 7-2.2.2.3 if <= 5 occupants Winders per 7-2.2.2.4 Smoke proof Enclosures per 7-2.3 Horizontal exits per 7-2.4 Ramps per 7-2.5 Passageways per 7-2.6 Escalators and moving walks per 7-2.7 Fire Escape Stairs per 7-2.8 Fire Escape Ladders per 7-2.9 Slide Escapes per 7-2.10 (up to 100% of capacity) Area of refuge per 7-2.12	C	NC	NC

Issue	Reference	Requirement	Reference	Reference	Requirement	99F901 Coker Diluent Tower Heater	99C901-4 Coker Drill Deck	99C901-4 Coker Nuclear Gauge Platform
	Local Code Part 8 Entrances, Walkways, Stairways and Ladders		**NFPA 101 Life Safety Code 2003 ed. (Enclosed Structure, Special Purpose, High Hazard Occupancy)**	**NFPA 101 Life Safety Code 2003 ed. (Special Structures) Open Structure**		Single stairway and dual ladders	Enclosed structure, deck has stairway and elevator on one end and ladder on other end	Ladder and platform to secondary egress
Means of Egress	Subsection 119(4)	Multiple entry and exit points are required in situations where a worker could become isolated from a primary escape route and unable to return to it.				C	C	C
			4-5.3.1 40-2.4.1		Minimum of 2; at least 1 w/o traversing another story except - low & ordinary Indus occ wherein an exit can be reached w/in 50' (unsprinklered or 100' (sprinklered)	NA / C	NC	NA / C
				11-2.2.4.1	Above, except grade level of open structures considered to have infinite number	C	NA	C
				11-2.2.4.2	Above, except open structures w/ occupancy <= 3 and means of egress <= 200'	C	NA	C
				11-3.2.4.1.	Towers can have single if conditions in reference are met	NA	NA	NA

Issue	Reference	Requirement	Reference	Reference	Requirement	99F901 Coker Diluent Tower Heater	99C901-4 Coker Drill Deck	99C901-4 Coker Nuclear Gauge Platform
	Local Code Part 8 Entrances, Walkways, Stairways and Ladders		**NFPA 101 Life Safety Code 2003 ed. (Enclosed Structure, Special Purpose, High Hazard Occupancy)**	**NFPA 101 Life Safety Code 2003 ed. (Special Structures) Open Structure**		Single stairway and dual ladders	Enclosed structure, deck has stairway and elevator on one end and ladder on other end	Ladder and platform to secondary egress
				11-3.2.4.2	Towers w/ 360 degrees of visibility can have a single for a max 75' travel distance (100' if sprinklered)	NA	NA	NA
			40-2.4.2		occupancy >500 requires multiple as per 7-4.1.2	NA	NA	NA
			40-2.4.3		High haz requires 2 separate and EQUAL as per 7-11	C	NC	C

C	In compliance with code requirements	NA / C	Code not applicable, but if applied situation would be compliant
NC	Not in compliance with code requirements	NA / NC	Code not applicable but if applied, situation is not compliant
NA	Code or situation is not applicable		
Note 1	Management systems are not included in project scope	Note 2	Company standards do not make exemptions for platforms attached to columns

The Benefits of Comparing Similar Hazards across 'Sister' Plants

A.M. (Tony) Downes, Andrew Twarowski
FMC Corporation
Princeton, New Jersey
Anthony.Downes@FMC.com
Andy.Twarowski@FMC.com

ABSTRACT

Companies often have multiple 'sister' plants in different locations, sometimes scattered around the globe. Generally, these companies are striving for similarly low risk among all these sister plants, even though they may have been built at different times, using evolving technology, and often have different capacities. In the pursuit of achieving acceptably low risk at all the chemical plants within a company, Process Safety specialists tend to put each under the PHA 'microscope' separately. FMC has several businesses that operate multiple "sister" plants, and has found that there are substantial benefits to be gained by also <u>*comparing*</u> *the hazards at similar plants.*

We first applied the comparison technique that is the subject of this paper to one of our world-wide chemical operations that had several sister plants. Our assumption was that any given high consequence scenario should appear as a common element at all the plants unless there was a fundamental difference between plants. Comparing each of these common scenarios one by one across all similar plants allowed us to ensure each scenario had been recognized, evaluated and safeguarded in a consistent manner. Some surprising inconsistencies were discovered.

This paper briefly describes the general organization of FMC's Process Safety Risk Profile program, including some recent improvements, then focuses on our scenario comparison methods in sufficient depth for others to apply them.

1. INTRODUCTION

About ten years ago FMC developed and implemented a chemical process hazard analysis application called Process Safety Risk Profile (PSRP), and presented it as a topic to this forum. We will now share our experiences gleaned from the implementation and growth of this tool. This paper will provide a little refresher on what a PSRP is, how it can enable comparisons between similar plants, and the experiences and lessons learned, complete with some examples[1] and anecdotes to reinforce key points. Lastly, we will identify the critical success factors of this endeavor, should you be interested in applying this tool and concept to your business.

2. Identifying the High Consequence scenarios (Scenarios vs Deviations)

FMC developed Process Safety Risk Profiles (PSRPs) at all its chemical manufacturing locations world-wide using the PSRP methodology presented by FMC to the CCPS, several years ago. However, even for sites that aren't required by regulations to perform Process Hazards Analyses (PHAs), the PSRP is a useful tool for performing an economical hazard review. The PSRP is simple yet comprehensive. It takes less time to conduct, and allows for a compact report to be generated. The PSRP at a plant can be developed using one of two different methods.

2.1 Team-based PSRP Development (Method 1)

FMC typically develops the scenarios, at sites not governed by PSM, using PSRP methodology, by using a team that functions much like a "What-if" PHA. This group is facilitated by a corporate Process Safety specialist familiar with the process and, in particular, with estimating risk.

PSRP's use techniques similar to Process Hazards Analyses (PHA's), but do not focus so closely on the details of the process as PHA's have to. PSRP's are 'coarser', tend to be done at the Flowsheet level, and are looking only for scenarios that have the potential for very significant consequences. Similar techniques have been used for years, and have been referred to using several different terms, including "Coarse Scale HazOps"[2]. The significant differences between these 'predecessors' and our PSRP's is that we have estimated the magnitude of the potential consequence (both on- and off-site) and the likelihood of the scenario, before and after considering active safeguards. See Table 1 for highlights of the differences between HAZOP and PSRP approaches to a hazard analysis

1 Examples used in this paper are real but are not drawn from FMC operations.

2 Kletz, "Hazop & Hazan"

Scenarios vs Deviations

• HAZOP	• PSRP
1. Causes/Events	1. Hazard Matrix
2. Deviations	2. Consequences
3. Scenarios	3. Scenarios
4. Consequences	4. Recommendations
5. Recommendations	
Node level	Flowsheet or unit level

Table 1

Before we can compare scenarios at various sites, we have to bring them to some sort of common basis. It's difficult to compare HazOp or What-If results directly, because issues can surface at different points in the discussion. For example, the hazard of over-pressuring receiver vessel V-2 in Fig 1 may show up under several different Hazop Deviations. However, every high-pressure polyethylene plant in the world has plenty of safeguards against the hazard of rupturing its low pressure separator in case they blow through from the high pressure separator.

Low pressure separator High Consequence Scenario

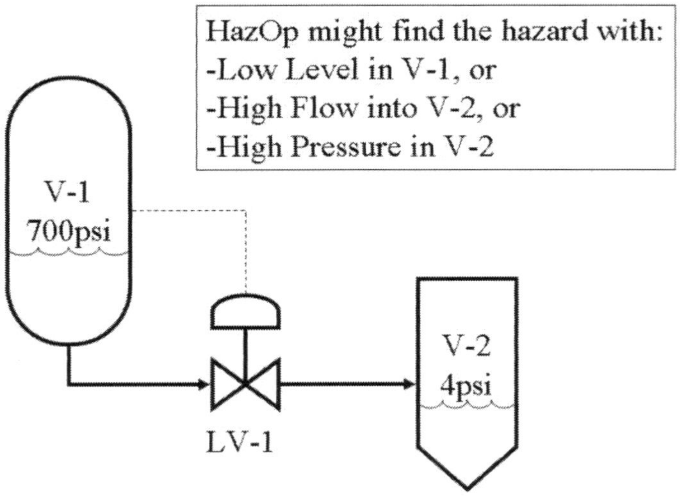

Figure 1

2.2 Extracting High Consequence Scenarios from existing PHA's (Method 2)

At PSM/RMP sites in the US, or in any other jurisdiction where required, site staff (sometimes with the help of a consultant) conduct the mandatory PHA's. When this is the case, we extract the PSRP scenarios from the PHA using a small team composed of a corporate Process Safety specialist, a senior process engineer and/or the site's process safety coordinator. This is not an onerous task, requiring only 2-4% of the time needed for the PHA itself.

Most PHAs document all the deviations they consider, regardless of the consequence/risk level. They may choose to flag any recommendations against lower consequence deviations as optional "Operations only" recommendations and manage these outside the PHA's "safety recommendations" system. This dramatically increases the effort required by the corporate safety staff to manage the corporation's High Consequence Scenarios (Fig 2). We needed a way to "separate the wheat from the chaff".

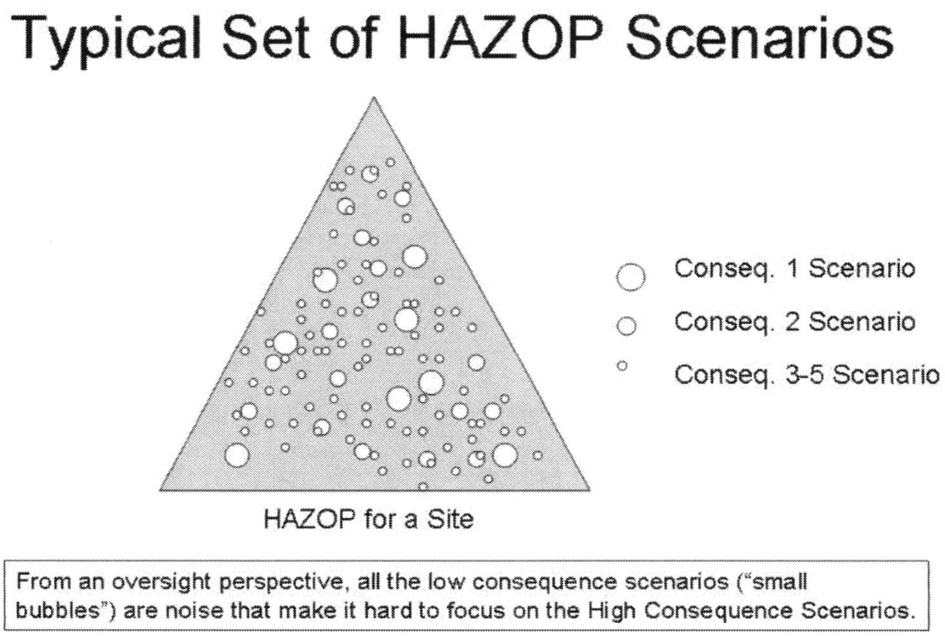

Figure 2

Providing your company estimates the risk levels (at least for deviations, or What-if cases, with potentially high consequences), you can easily skim through hours of PHA teamwork to identify those "important" scenarios that you may wish to compare from site to site (procedure depicted in Figure 3). Some software packages for recording PHAs allow you to apply filters so only the high consequence deviations are displayed.

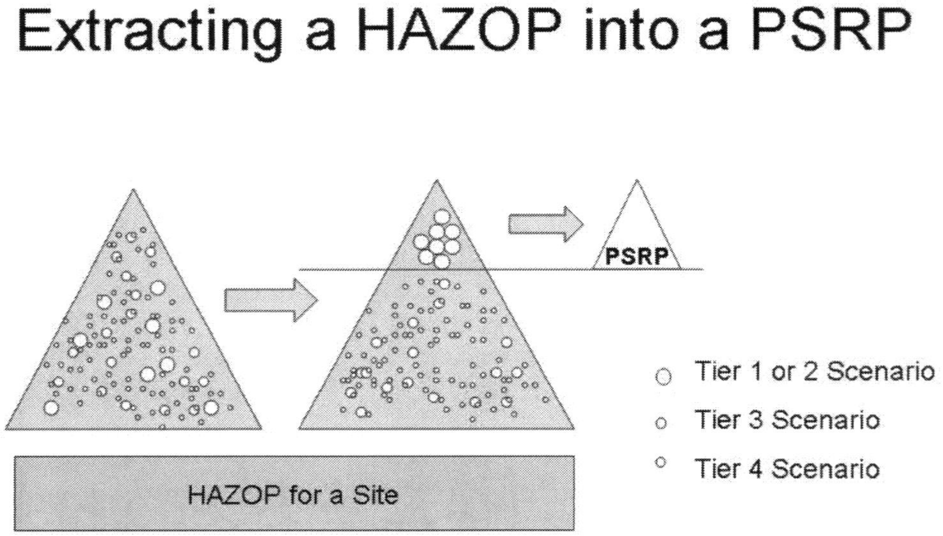

Figure 3

2.3 FMC Verified the Validity of PHA → PSRP Conversion

To make sure the process to extract important scenarios worked, we put it to a test. Several FMC employees, with varying experience in Process Safety (from 5 to 20 years), each independently performed the procedure. It only took about 2 hours to identify the high consequence scenarios from a week-long HazOp and extract the Description, Cause, Safeguard & Risk information, as well as the important Recommendations to reduce that Risk. Everyone identified the same High Consequence Scenarios.

3 CONDUCTING THE COMPARISON OF SISTER PLANTS

Once the information about each plant's High Consequence Scenarios has been analyzed, we can compare similar scenarios at similar plants. This should be done by a group of experienced process engineers, each representing their own plant. In our comparison we used a mix of technical department managers, senior process engineers, and site process safety specialists, all coordinated by a corporate process safety specialist. To provide focus for their discussions, we compiled all the High Consequence Scenarios at each of the plants into a spreadsheet table.

3.1 Making the comparison table

Prior to the comparison meeting, we developed a spreadsheet (as shown in Figure 4) to collect and display the information. The first four columns (A-D) were:

A) The Row Number - For easy reference when using paper copies of the comparison. We left this blank until we had added all the scenarios.
B) The System name - e.g. Feed Prep, Catalyst Injection, Reaction, etc.
C) The Sub-system name - depending on the complexity of the Plants, this may or may not be necessary
D) The Brief Description - We filled this in as we were populating the comparison spreadsheet with the scenarios from the plants.

We placed the scenarios from the Plant A into the table using 5 columns (E-I):

E) The Scenario number - used to refer back to each plant's PSRP information if/when further details were needed.
F) The Scenarios column - containing the text of the Scenario Description as copied from each plant's PSRP.
G) The Consequence category - From 1-5 where 1 is the highest Consequence
H) The Likelihood category - from A – E where A is the highest Likelihood
I) The Risk tier - where 1 is the highest Risk (Fig 6)

A	B	C	D	E	F	G	H	I
				Plant A PSRP Results				
#	System	Sub-system	Brief Description (by Corporate Safety)	Scen#	Scenarios	C	L	R
14	Separation	Low Pressure Separator	Rupture LPS due to blow-throughof gas from HPS	PlantA-LDP-04.06	Lose liquid level in HiPress Separator. Safeguards are LSLL to LV and RD to atmosphere.	1	D	2
15	Separation	Recycle compr'r	Draw air into the low pressure compressor leading to Decomp in the reactor if [O2] exceeds ~10ppm	PlantA-LDP-06.02	Draw air into suction of 1st stage Recycle compressor for any reason, leading to Decomp in Reactor. Safeguards are lab sampling, RD's on reactor and reactor blast walls	3	B	2

Figure 4

Next, we looked for similar scenarios at Plant B and placed them into the table on the corresponding rows so that the same scenario was on each row for all plants. If Plant B had identified a scenario that wasn't identified by Plant A, then we inserted a blank row in the appropriate System/Subsystem (Fig 5) and placed the new scenario in the blank row, in the Plant B columns. We repeated this process with each of the other sister plants. We added two additional columns to the right edge of the spreadsheet to record Comments and Recommendations for use during the comparison meeting.

			Plant A PSRP Results					Plant B PSRP Results				
System	Sub-system	Brief Description (by Corporate Safety)	Scen#	Scenarios	C	L	R	Scen#	Scenarios	C	L	R
Separation	Recycle compr'r	Draw air into the low pressure compressor leading to Decomp in the reactor if [O2] exceeds ~10ppm	PlantA-LDP-06.02	Draw air into suction of 1st stage Recycle compressor for any reason, leading to Decomp in Reactor. Safeguards are lab sampling, RD's on reactor and reactor blast walls	3	B	2	PlantB-LDP-02.04	Oxygen level in reactor exceeds 10ppm due to air ingress at Recycle compressor suction eg. Draw vacuum & leak into system from extruder	3	C	3
Separation	Recycle compr'r	Draw air into the low pressure compressor due to Unloader air leak -> Decomp (Plt B experience)		*No equivalent scenario identified*				PlantB-LDP-02.04	Oxygen level in reactor exceeds 10ppm due to air ingress (eg. Due to air leak through recycle complessor 1st or second stage unloaders)	3	C	3

Figure 5

3.3 The Comparison Meeting

The division involved has periodic technical exchange meetings. We added a day and a half to one of the participants to do the comparison using a team as described in section 2. There were 47 high consequence scenarios in the table, so we averaged about 15 minutes per scenario. As you might expect, some scenarios were essentially identical, uncontroversial, and only required a minute of discussion. The most controversial scenario took about 45 minutes and we eventually agreed to launch a study to resolve some thorny questions outside of the meeting.

The general flow of the discussion for each row was to:

A) Describe the scenario and ensure everyone understood what we were talking about,

B) Look for any differences in risk rating (C, L or R) and ask the relevant site representative to describe their situation to the group.

C) Discuss how the scenario was handled at each of the sites and look for common patterns and/or best practices.

D) Record any notes in the Comments column and any required actions in the Recommendations column.

Recommendations to an individual plant might be to consider revising their PSRP and/or add a Safeguard that they didn't have, or to look for a better technical solution for a problem that was common to all the plants, or even to begin a test program to develop reactivity data.

4. FINDINGS

The following are examples of issues that surfaced during the comparison meeting.

4.1. Missed scenarios

The first problem we noticed was that a number of significant scenarios had been identified at some plants, but not at others. Some teams had just missed them - not just once, but in several revalidations.

Example: Potential for a pool fire in the Purification Unit

Each of the plants in our comparison operates a Purification Unit. One hazard common to all of the units is that they contain substantial volumes of organic solution above the flash point. In cases like this, FMC always assumes that a pool fire is possible. Since there is little containment or confinement for vapors, and the vapor generation rate is small anyway, we do not consider a vapor-cloud explosion to be a realistic scenario in this area.

Using industry data, we've estimated the likelihood of a pool fire to be about 1 per 100 unit-years (0.01 per year at each unit). In FMC's risk profile system, this is in the "C" region. At most of the sister plants, we estimated that A) there would not be off-site consequences, B) a fatal injury to an employee was unlikely and C) that the combined Property Damage and Business Interruption from such a fire might be in the region of $500,000. This combination of consequence and likelihood (risk) is considered by FMC to be generally acceptable. Note that the risk analyst is estimating a point along a continuum of likelihood-consequence pairs for this unit. In this case, the area is properly classified electrically, there are no unusual hazards and there is an automatic deluge system. We assume the deluge will limit the damage in most fires. A deluge failure might result in approximately 10 times more damage, but is considered to be at least 10 times less likely. For our purposes this estimate is close enough.

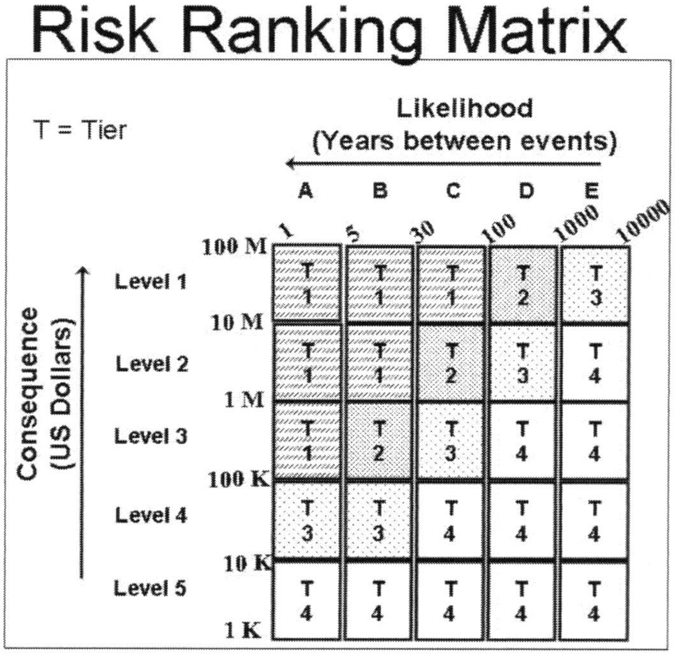

Figure 6

The PHA team at one of the facilities had not considered a pool fire at all. They just missed this Loss of Containment scenario. The worst scenario they identified in the Purification Unit was a small flash fire inside one of the purifier vessels. They rated the consequences as well as the likelihood of such a flash fire to be low. Therefore their PSRP did not contain this scenario. This oversight was not identified until we did the comparison of all the facilities. They added the scenario to their analysis. At this point, you may be asking yourself, "So what? The risk of this scenario was acceptable anyway." As a general principle, emergency response plans should be based on broad versions of the scenarios identified in the hazard analysis, so this difference is significant to Emergency Response planning, even though it is considered an "acceptable risk".

At another of the plants, the physical arrangement of the Purification Unit is different, it's significantly closer to the rest of the process areas, and there is not good drainage to prevent the possible spread of a pool fire to adjacent units. In this case, the likelihood was left at "C", but the consequences of a pool fire were estimated to be significantly higher. The risk ratings at the two facilities were different, but that difference was documented and was explainable. It may be necessary to explain such differences to division-level management who wonder why they need to spend money to address a risk at Plant A but Plant B doesn't appear to have a problem. A table showing that the hazard exists at both places but is less at B can be very helpful.

4.2 Taking credit for Active Safeguards before estimating Consequences

A typical error that inexperienced PHA teams make is to estimate the consequences of a scenario after considering the safeguards that are in place. For example, in most multi-stage gas compression systems, there is a possibility of liquids accumulating between each stage. For this reason, there is usually a suction drum ahead of the next stage, complete with a liquids drain and a high level interlock to shutdown the compressor. If the liquid is not all drained out, then it may carry over to the next stage and can cause a catastrophic failure of the compressor. One PHA team missed this and rated the worst consequences of high level as a shutdown of the compressor due to the active safeguard, which they rated as a low consequence. Unfortunately, the consequence level was also used to determine (at a first pass) those devices that must be scheduled for periodic inspection in the Mechanical Integrity program. Since the high level interlock wasn't deemed to be protecting against a high consequence safety event, it wasn't initially included.

5. LEARNINGS

It's obviously a good idea for sister plants to cooperate on process safety issues. Beyond "two-heads-are-better-than-one", there are the opportunities for multiple plants to learn lessons from unusual incidents that happen at one of them.

The following are some generalized learnings about the comparison process, its requirements and benefits.

5.1 Critical Success Factors

If you've read this far you are probably interested in applying the PSRP concepts in some form at your company. Keep these critical success factors in mind:

- FOCUS - Use a compact, efficient hazard scenario identification process that focuses attention on high consequence scenarios. Consider using a coarse-scale what-if method or an extraction process from existing PHA work.
- UNIFORMITY - A uniform process facilitates cross-site comparisons
- EXPERIENCE – The person driving this process must have technical and operational experience combined with strong process safety knowledge. The team members must be very knowledgeable about their own processes and plant systems including key safeguards.
- TEAMWORK – The corporate oversight person needs to work effectively with site process safety resources to conduct the PSRP process and the cross-site comparisons depicted in this paper.

5.2 One-time-only (probably).

Once we've completed the comparison study and established the standard safeguard requirements for the process, there seems to be no need to revisit it. The ongoing maintenance of the study results should be done via each plant's process safety programs. When a new plant is built, we may choose to add its list of scenarios to the comparison table, though in practice, we have used the scenario list from one of the existing plants as an input to the design and hazard assessment process for the new plant.

6. CONCLUSION

It's possible to compare high consequence process safety scenarios across sister plants to ensure they have treated these situations uniformly. Using the methods described in this paper, the required effort is small in comparison with the benefits. If your company does PHAs using completely independent teams, this comparison may identify high consequence scenarios that have been missed, or mis-rated, through several revalidations.

7. REFERENCES

1. Fryman & Roopchand "Process Safety Scenario Evaluations", CCPS Conference, 1999 NB. PSSE has since been renamed PSRP within FMC, but the methodology is largely unchanged.
2. Kletz, T.A. "Hazop & Hazan"(4th ed) IChemE 1999,

Conventions on multi-operator sites : an efficient risk management tool

R. Gaucher, O. Dolladille
INERIS
Verneuil-en-Halatte, France
rodolphe.gaucher@ineris.fr, olivier.dolladille@ineris.fr,

ABSTRACT

Until the end of the 20th century, single user sites were the norm in Europe, but in the last 10 years, the situation has changed. Today, especially in the chemical industry, sites are operated by several companies with facilities close to one another.

Beyond business interests, multi-operator sites can offer great advantages to deal with environment, safety, security and emergency management. But, this change generates new risks, because of the interdependence between different companies, and calls for a new organization of risk management at site scale.

For managing overall risk, it is necessary to define a minimum of common rules. In particular, companies should be encouraged to conclude clear and efficient contracts between themselves. These contracts under private law are very flexible and can be easily updated to accompany the development of sites.

The aim of this paper is to highlight the main issues brought by this new kind of organization and to present INERIS' conclusions on appropriate contracts between operators (conventions) and efficient risk management on each site.

1. INTRODUCTION

Over the last decade, many industrial sites in Europe, which had been owned and operated by a single company, became multi-operator sites : several companies operate on a same site with a certain degree of interactions between their activities. This trend, which first concerned huge chemical parks, can now be seen in all industries. In the context of globalization, such changes, which are driven by economic strategies, can be explained by successive sales and purchases, focus on core businesses, outsourcing of utilities or services, integration on site of contractors' activities, and so on.

In many cases, multi-operator sites were the result of the voluntary implantation of some companies close to other ones to benefit from the links they could establish between their activities. This could lead, when framework conditions were favorable, to the formation of clusters of industrial sites more or less closely integrated. In a few cases, the development of a multi-operator site can be decided, planned and coordinated. It is for instance the case of shared warehouses or industrial harbors or sometimes chemical parks.

The economic interests of multi-operator sites are obvious : shared infrastructure and facilities, easiness for an operator to set up, change or transfer its activities… But these sites raise different questions about environment, occupational health and safety (EHS) issues. Installations operated by different companies close to one another and/or interconnected by pipe networks generate new risks on interfaces.

Moreover, multi-operator sites constitute a new phenomenon for the implementation of European and French EHS regulation, in particular the Seveso II Directive. These regulations do not answer all the questions raised by these new forms of industrial organization. On the contrary, their implementation to the letter on such sites could purely and simply lead to the closure of some of them.

In 2005, the French Environment Ministry asked INERIS to look into the issue. A first set of observations was made and INERIS identified key in-depth studies to be carried out. The main ones are presented in the first part of this paper. Among them, the ministry asked INERIS to study more specifically private agreements between operators in the field of EHS risk control.

2. KEY ISSUES AT STAKE ON MULTI-OPERATOR SITES

To identify the questions raised by this new situation, INERIS decided on the one hand to examine how French and European law takes into account multi-operator sites, and, on the other hand, to interview a large panel of those involved in several French regions and different industrial sectors. The graph below represents the various stakeholders involved.

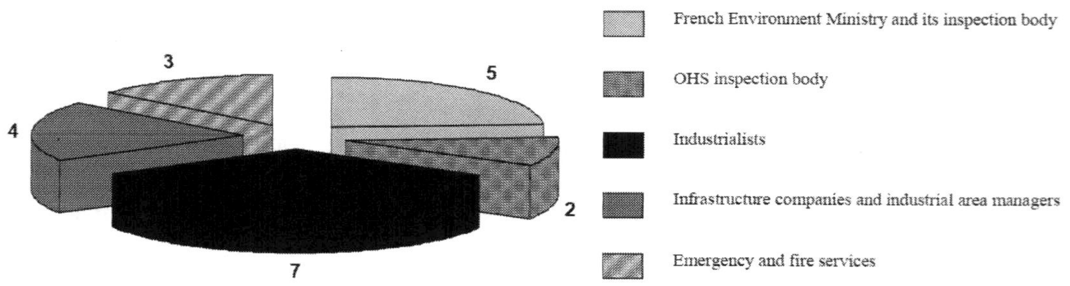

Figure 1: Distribution of interviewees

2.1 French and European Legislation

First of all, both legislation only talk about **individual** operator, not multi-operator. Also, in the French legislation, there is no definition of what a "site" is. The environmental legislation only talks about "installation", and "establishment" for operators concerned by the Seveso II Directive.

There is no clear legal definition of a multi-operator site. It seems that, in view of the great variety of such sites, it would be difficult to find an unambiguous definition. In some cases (huge chemical sites for instance), operators share infrastructure, fence and access control. Their activities are also closely linked as far as their use of materials and energy is concerned. The complex overlapping of their activities makes it difficult to identify the boundaries of each company.

Consequently, the difficulty lies in identifying individual responsibilities at interface level, from the new risks generated by the interdependence between companies to overall risk at site scale.

The second main issue is the problem of industrial neighbors. In the eye of the law, there is no difference between neighbors, whether they are industrial or not, or whether they are inside or outside a fence. This means that an operator must respect environmental regulation on the protection of its industrial neighbors ; in terms of emission limits for hazardous chemicals and noise, in terms of exposure to accidental risks of fire, explosion and toxic dispersion, etc … These requirements are much more restricting than those of OHS regulation which the operator must abide regarding its employees. But, on sites where activities are closely linked, employees of different operators may be exposed to the same emissions or risks. Some should be protected by OHS regulation, others by environmental one (more restricting). As said in the introduction, a strict implementation of the law could lead to the closure of some sites because of the prohibitive costs to ensure that employees on a site get the same level of protection as the one required for neighbors.

2.2 Conclusions for further in-depth analysis

The interviews confirmed the observations above : there are as many definitions of what a multi-operator site is as they are interviewees, ranging from closed site by common fence and access (called "closed site" in the paper) to side by side companies without linked activities. Another interesting observation is that the interviewees were not often aware of all the issues raised by multi-operator sites.

On the basis of these interviews, INERIS formulated the following recommendations :

- To provide a definition of multi-operator sites and typical multi-operator site models ;
- To find the origins of multi-operator sites and trends ;
- To define a methodology to guarantee a systematic identification and control of boundaries of each company on site ;

- To address administrative issues : for instance, several operators without addition rules of chemical quantities to compare with "Seveso II thresholds" ;
- To identify solutions to the problem of industrial neighbors from both a regulation and a risk prevention point of view : see § 2.1 ;
- To look into the consequences of a change of operator in terms of EHS risk management ;
- To consider waste water and waste gas emission issues: interest of shared facilities, risk assessment of cumulative emissions, identification of responsibilities… ;
- To identify specific needs for public information ;
- To study voluntary and compulsory management systems : interest of coherence, links or common part between management systems of each operator, especially Seveso II establishments concerned by a compulsory safety management system ;
- To provide a methodology to identify, prevent and control interference EHS risks[1] ;
- To investigate land use planning issues deriving from the French regulation on Seveso II establishments which introduces "technological risk prevention plans" ;
- To consider emergency management at site scale : common organization, shared services and resources ;
- To examine conventions/contracts between operators.

In October 2007, INERIS met the chairman of the European Process Safety Center who worked on the issue focusing on German chemical parks (which are much more developed than in France). It appeared that our respective conclusions were very similar.

Among the recommendations detailed previously, the Ministry asked INERIS to study two issues in priority : emergency management and conventions between operators. Only the second one is detailed in the next part of this paper.

3. CONVENTIONS

3.1 Scope of the study

For the next stage of this study, the French Environment Ministry asked INERIS to focus on :

- multi-operator sites where at least one Seveso II establishment operates ;
- major accident hazard control first, without forgetting other hazards and impacts ;

The question asked by the ministry were :

- Why do operators establish such conventions ?
- What is the scope of such conventions ?
- How could they contribute to a better management of interference and overall risks on multi-operator sites ?
- How the environment inspection body could take into account these conventions when it assesses how operator's activities comply with legal requirements ?

[1] When two or more companies operate close or related installations, new risks can appear due to simultaneous activities (co-activity). These new risks are called interference risks.

A working group was created by INERIS with industrialists, environment inspectors and fire brigades, selected on account of their knowledge and experience of conventions. About twenty multi-operator sites and associated conventions were considered. Several meetings took place to discuss with each member of the working group the reasons behind conventions, their content, their implementation and the way they work.

3.2 Definition and scope of conventions

"Convention" means every contractual document dealing directly or not with EHS risk issues and established between two or more companies operating on the same site. Contracts between a company and its subcontractors are not considered here. In France, these relationships are well covered by OHS regulation in terms of risk prevention.

Conventions may have the following objectives :

- To clarify responsibilities of each operator who owns a permit and each owner of a given installation, especially at interfaces ;
- To specifying financial and operational rules on shared infrastructures, facilities and services ;
- To define common EHS rules applicable at site scale ;
- To defining organizational governance to deal with rules on site and disagreements between operators.

Moreover, conventions can be seen as a guarantee for an operator of :

- continuity of shared infrastructures, facilities and services (called "shared resources" in the paper). ;
- risk prevention and control of other operators' activities which could have an impact on its employees and installations.

3.3 Definition of responsibilities

The first goal of a convention is to clarify the role, responsibilities and commitments of each operator and owner regarding shared (or linked) infrastructures, installations, activities and associated resources. This could concern also land owners and the ones who are responsible for past pollution as far as they are involved in site issues.

Responsibilities to be defined concern the operation, maintenance and financing of shared and linked resources (for instance : pipe racks, effluent networks, fire protection devices, exchange of materials or energy, waste water treatment plant, retention pond, utilities, etc.). In particular, conventions must be clear about the financial contribution of each operator, depending on their respective needs, and the possibility of reviewing their commitments subject to potential compensation.

To ensure the continuity and efficiency of shared resources vital to EHS prevention, the convention can oblige operators to use these ones rather than to operate their own resources.

This clarification aims at preventing operators from a weakening of their responsibilities and at allowing them to act efficiently on interfaces between their activities and those of other operators.

3.4 Common EHS rules

In order to guarantee risk control at interfaces, operators should define and implement common EHS rules that apply to everyone. For instance, on a closed site, common rules will first concern access and circulation requirements (people, goods, vehicles).

Specific rules should also be added, only for certain operators, as regards interference risks. For instance, Seveso II establishments on site could define rules concerning shared internal emergency plan and services to deal with major accidents and domino effects.

Since conventions are private agreements, it has to be noticed that all operators who sign them will have to fulfill common requirements, even those who have few EHS obligations. The result is that most of operators benefit from shared resources and additional competencies to prevent risks they wouldn't have otherwise.

A long term benefit could be the development of shared practices, vocabulary and culture of risk.

INERIS identifies several good practices that could make conventions more efficient. First, the convention which defines the EHS rules should be distinct from the one which deals with shared resources. The reason is that all operators are not necessary concerned by shared resources, depending on their needs, whereas they are all involved in the compliance with EHS rules.

Conventions should also state the obligation for an operator who wants to sale its company to inform the purchaser of the obligation to sign the convention and comply with common EHS rules. Thus, rules are linked to installations and not to operators. It is a guarantee of continuity for operators on the site.

3.5 Interest of conventions from authorities' point of view

Many benefits can be derived from the definition of conventions. As said previously, the authorities can only address individual operators. The clarification of responsibilities at interfaces, which allows to know well boundaries of each company, fits in with authorities' goals.

The French law gives few opportunities to the environment inspection body to control changes of operators and their potential consequences in term of risks. As a result, conventions also appear as an interesting tool to manage changes of operator both for industrialists and the inspection body.

When it deliver the permit required to operate installations, the environment inspection body may impose the establishment of a convention in certain conditions. In that case, it is preferable to impose only the objectives of the convention. The objectives should primarily concern, risk control on interfaces and the definition of responsibilities.

On the contrary, imposing content could restrict the flexibility of conventions which is necessary in a fast moving industrial environment and could weaken the responsibilities of operators.

Moreover, if the permit specifies that operators must inform the inspection body before any change in the convention, the inspection body will be assured to have its word on private contracts when conventions are drawn up and updated.

Finally, from an operator point of view, a convention can be a way of showing to the inspection body that some legal requirements are met with the assistance of a third party (e.g., shared fire-fighting capacities).

3.6 INERIS' observations

During this study, INERIS highlighted that most conventions are limited to commercial issues (contracts on services between operators). Risk prevention and control on interfaces are often forgotten.

This situation led INERIS to make another observation. When a company gives birth to a multi-operator site by selling some business units, there are only few changes on the site : people, culture of risk, habits, relationships remain the same. The need to define contractual rules to manage EHS risks on interfaces doesn't appear as a necessity : "we come from the same family !". But, within a few years, companies follow different paths, people move and relationships are more and more business oriented. It is then much more difficult to elaborate EHS rules than to change rules defined at the creation of the multi-operator site.

The last observation is that, among multi-operator sites, chemical parks are the most advanced in the implementation of conventions taking into account EHS risk management.

4. DEFINING EFFECTIVE CONVENTIONS

4.1 Infrastructure company

Among operators, an infrastructure company can operate and maintain shared resources : pipe networks, roads, sewers, power supplies… It can also provide services to other operators such as authorizations and general training of subcontractors before they enter the site (chemical hazards on the site, traffic rules, emergency procedures…).

Most of the activities of an infrastructure company are concerned by interfaces with operators' activities. This implies that the infrastructure company is especially involved in interference risk control. Consequently, the more an infrastructure company has activities, the easier it is for it to act as a site manager and to organize cooperation between operators in the field of EHS management.

The experience of chemical parks shows us that an infrastructure company is in a stronger position to manage overall risk as long as :

- it shares the same industrial risk culture, because operators often originate from a single historical company ;
- its status specify that it is a non-profit company (less sensitive to free market forces) ;
- it imposes some compulsory services and shared resources to all operators, preventing them from setting up some less effective installations than a centralized one for instance ;
- it has the power to check how operators comply with common EHS rules.

It can be assumed that when all these conditions are met, the infrastructure company can really manage the global interest of the site. The global interest of the site may go against individual interest. For instance, it may be relevant, when an accident occurs, to protect sensitive installations of a particular operator rather than those more threatened of another operator with a view to limiting consequences outside.

4.1 Governance

An efficient convention needs an efficient governance at site scale.

The convention must refer to specific decision-making bodies (committees) dealing with :

- shared resources : short-/long-term financing, implementation, operation and maintenance ;
- common EHS rules : definition, implementation, control and update.

These committees can eventually merge into a single one. But it should be taken into account that issues at stake are not the same for each operator.

On the one hand, for instance, only the main users of shared resources (often shareholders of the infrastructure company when it exists) have voting rights, which are proportional to their needs. In some cases, the convention dealing with shared resources can be included into the status of the infrastructure company.

On the other hand, the weight of each operator in decision-making regarding common EHS rules should not only be based on financial aspects and payroll. Seveso II establishments, as regard to the potential hazards of their activities, should see their position strengthened in decision-making, according to the consequences on risk control.

It must be kept in mind that a weakening of responsibilities must be prevented. The balance between all these constraints should be found after several amendments to of the convention, on condition that the convention and the governance bodies are defined at the creation of the multi-operator site.

Finally, the rules concerning committees which are defined in the convention should provide for the resolution of disagreements or conflicts. In particular, role of corporate level and how operators waive any claims between themselves in specific cases should be included.

It has been observed that strong infrastructure companies (when they exist) have often a driving role in committees because of their knowledge of activities and interfaces on sites.

The figure 2 shows how the governance could be define at site scale and beyond.

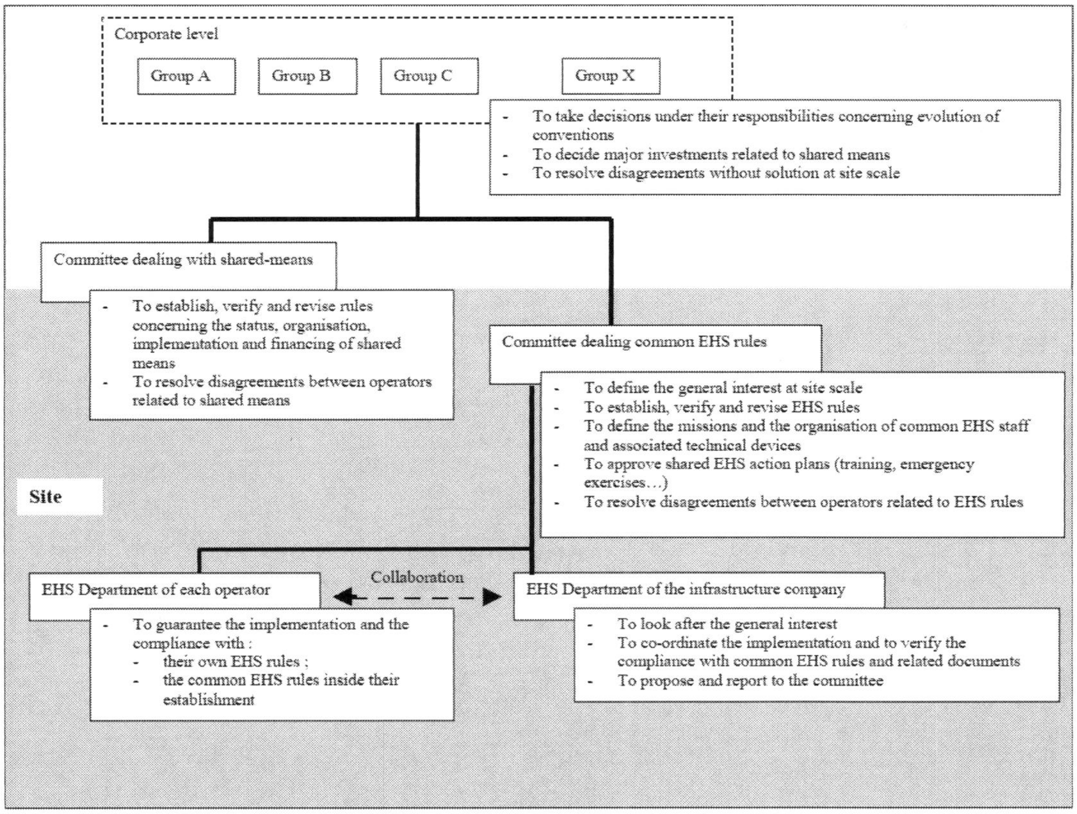

Figure 2 : Committees and responsibilities to be defined in conventions

INERIS believes conventions should concentrate on the definition and rules of the exercise of individual and collective responsibilities (committees described above). It should refer to operational documents defining common EHS rules rather than trying to include them in the convention. Otherwise, operational rules cannot be updated as fast as changes on site require.

Indeed, it is obvious that conventions must define rules on the way they should be updated. It is also clear that regarding what is at stake for operators, the update process of a convention takes a long time. This is why the update frequency of conventions is 5 years on average.

4.3 Checking process

The checking process aims at assessing how operators met their commitments as defined in convention(s). Each step of the checking process, including how to deal with nonconformities, must be described in the convention.

The assessment can be made at different level :

- Direct checks : this concerns one-off verifications of compliance with some common EHS rules (speed limit, loading/unloading procedures…) ;
- Internal audits : their scope is the organization at operator and site levels to implement and comply with the convention.

These two types of checking are complementary. Direct checks make it possible to verify the day-to-day observance of common rules by people on site. Internal audits are essential to highlight progress of convention' requirements necessary to improve EHS performance and the convention itself.

On a multi-operator site, direct checks can be carried out by the infrastructure company. Internal audits can be conducted by the infrastructure company, accompanied or not by some operators, an external expert or by operators from other multi-operator sites to encourage exchanges of views. In any case, the authority of the infrastructure company to conduct such checks must be stated in the convention. This is necessary to ensure that results of the checks will be followed by corrective actions.

On the basis of the conventions it studied, INERIS observed that the checking process is the most critical step regarding the methodology known as Plan-Do-Check-Act (PDCA) applied to the management of the convention. The interviews showed that, if some direct checks are made when an infrastructure company operates on a site, there is no audit decided and planned in common by operators to check the application of conventions. Moreover, audits conducted by operators are rarely carried out at interfaces.

In addition to internal checking process, it can be noted that, in France, the inspection body can verify some or all the requirements of the convention when it is quoted in the permit of an operator.

5. Conclusion

Changes in the European economic environment have accelerated over the last 10 years the development of multi-operator sites. If many environmental and economic benefits are already clearly identified, this new situation raises questions about EHS and overall risks at site scale that are not handled by European and especially French regulations.

Following an identification of interfaces between the different companies and an in-depth analysis of interference risks, the establishment of convention(s) between operators can provide the conditions of the continuity of the site. Conventions must clarify the individual responsibilities of each operator and intend to create the necessary governance to discuss and improve risk management at site scale and to guarantee the protection of industrial neighbors equal to non industrial ones.

On an economic point of view, conventions under private law are flexible enough to react to fast changes in company's strategies. On an EHS risk control point of view, they can contribute to increase the effectiveness of shared resources and rules and to develop or maintain a common culture of risk, aiming at reaching a high level of performance.

INERIS has shown through its observations on field that an important key to success is the checking step which is essential to allow a real continuous improvement of the convention. The following figure proposes the application of the PDCA methodology to convention :

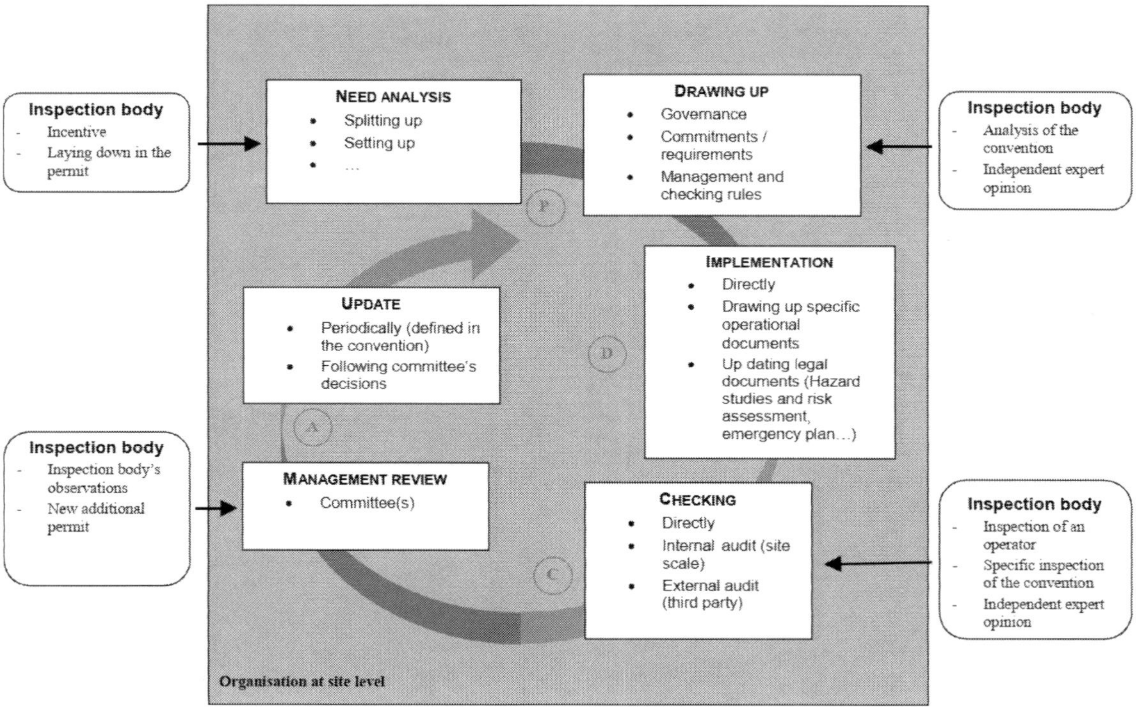

Figure 3 : Plan-Do-Check-Act (PDCA) methodology applied to the management of the convention

On multi-operator sites, conventions can be seen as a useful tool in the service of operators but also authorities. Their development should be promoted, insisting on the quality of their content and on their real implementation and check on field.

An important issue remains : if conventions allow to handle responsibilities between companies, they do not solve the acceptance by individuals of interference risks and the consequences in term of employer's responsibility.

6. REFERENCES

1. Dolladille O., Moreau E., Etat des lieux relatif à la gestion des risques au sein des sites multiexploitants / Risk control on multi-operator sites : state of the art, July 2006
2. Dolladille O., R. Gaucher, Sites multiexploitants : état d'avancement et perspectives, May 2007
3. Dolladille O., R. Gaucher, Sites multiexploitants : synthèse et premiers résultats des travaux en cours, July 2007
4. Dolladille O., R. Gaucher, Gestion des risques au sein des sites industriels multiexploitants : conventions entre exploitants / Risk control on multi-operator sites : conventions between operators, January 2008
5. Dolladille O., R. Gaucher, Prévention des accidents majeurs : mesures de maîtrise des risques sous influence d'un tiers sur un site multiexploitant, February 2008
6. Law 76-663 19th of July 1976 and its Decree 77-1133 of 21st September 1977, today in the Code de l'Environnement
7. SEVESO II Directive (1996) – Council Directive 96/82/EC
8. Law 2003-699 30th of July 2003, Loi 2003/699 du 30 juillet 2003 relative à la prévention des risques technologiques et naturels et à la réparation des dommages
9. Jochum C., Process Safety / Risk Management of Chemical Parks in Europe, Report n°29, EPSC, 2004
10. Jochum C., Friedenstab T., Spindler G., Peter J., Industriepark und Störfallrecht / Industrial Parks and the Law of Hazardous Incidents : criteria for regulation of safety responsibilities with establishments governed by the law of hazardous incidents in industrial parks , August 2002
11. Störfall Kommission, Arbeitshilfe für die Anwendung der Störfall-Verordnung bei Industrieparks, 2005
12. Seminar on Industrial Park and Multi-Operator Sites in Cagliari (Italy), Proceedings, 2003

Recent Developments in the Analysis of Fires, Explosions and Production Disruption Incidents in Chemical Plants and Oil Refineries

By

Dale R. McIntyre[1] and
Emory Ford[2]

ABSTRACT

Significant changes in computing power, regulatory climate, safety culture and globalization of business ventures have occurred since the original publication of MTI Publication No. 30, Guidelines for Assessing Fire and Explosion Damage. The implications of these changes are reviewed with regard to the work of investigators charged with the analysis of fires, explosions and major production disruption incidents in chemical processing plants and refineries. A review of advances in computer modeling of large-scale energy release events is presented. Storm damage assessment and the analysis of incidents involving sabotage are discussed.

[1] ConocoPhillips, Bartlesville, OK. dale.r.mcintyre@conocophillips.com
[2] Materials Technology Institute, St. Louis, MO. eaford@comcast.net

INTRODUCTION

The Materials Technology Institute (MTI) of the Chemical Process Industries first published its Publication No. 30, "Guidelines for Assessing Fire and Explosion Damage", in 1990. Later, portions of the publication were incorporated verbatim into the API 579 "Fitness for Purpose" guidelines. A second printing was prepared when the first sold out. In 2006, with demand for Publication No. 30 continuing to be strong and the number of copies from the second printing dwindling, MTI decided to prepare an updated second edition rather than simply reprinting the original yet again.

The selection of new material for inclusion in the second edition forced the authors to consider the nature of the changes that have taken place in the chemical and oil refining industries in the years since 1990. The impacts of these changes on the work of investigators handling fires, explosions and major production disruption incidents are the subject of this paper. They include effects relating to

1) Changes in computing power with implications for event modeling
2) Globalization of the industries
3) Changes in regulatory climate
4) Changes in safety culture
5) Dealing with sabotage and storm damage assessment

Some of these changes have been highly beneficial to the work of safety incident investigators. Others have been very much of a mixed blessing. Reflections on these changes and their implications are presented below.

Improved Computational Methods and Concepts for Modeling Explosions

The exponential improvements in computing power which have transformed the business world in the years 1990-2008 have also transformed safety investigators' ability to model fires, explosions, pressure vessel ruptures and other major production disruption events. Reconstructions and scenario testing are now possible at a level of detail which would have been impossible two decades ago. This has been almost uniformly beneficial for the work of safety investigators.

Of course, the fundamentals of fire, combustion and energy release events have not changed during the time period in question. The older methods of calculation and estimation are still as valid as they ever were. What has changed is the recognition that the limits of these methods and their uncertainties need not be the last word on any particular event.

Traditional methods of calculation and estimation of event parameters should still be learned because they ground the investigator in the physical principles of the fire and explosion processes. These principles and basic equations will form the background of the analysis whether done by hand or on a supercomputer. Various hypotheses on the nature and mass of the reactants will need to be tested against the observed damage. If the potential energy available from release of a hypothetical reactant of a certain assumed mass is insufficient to cause the damage observed, then obviously additional sources of energy must be postulated to explain the event.

Mass of Fuel in a Spill

The mass of material in the system may be estimated from simple physical principles, based on the volume and process conditions of the train of vessels and piping connected together at the point of the incident. [1] For gaseous reactants,

$$W_g = nM \tag{1}$$

Where

W_g = the weight of the gas discharged
n = number of moles of gas = PV/RT for ideal Gay-Lussac gases
M = average molecular weight of the gas

For liquid spills the weight is simply

$$W_L = \rho V_L \tag{2}$$

Where

W_L = weight of liquid spilled
ρ = liquid density at process temperature
V_L = container volume

For liquefied gases having an atmospheric boiling point less than the ambient temperature, assume that 100% will be vaporized and use equation 2. For liquid fuels with an atmospheric boiling point greater than the ambient temperature, the mass available to feed a vapor cloud can be calculated as

$$W_v = \frac{W_L C_{pm}}{\Delta H_v}(T_1 - T_2) \qquad 3)$$

Where

W_v = weight of vapor from spilled liquid
W_L = weight of liquid spilled from eq. 2
C_{pm} = mean specific heat over T_1 to T_2, = $(C_{p1} \cdot C_{p2} \cdot \ldots C_{pn})^{1/n}$
T_1 = lesser of liquid process temperature or liquid boiling point at pressure
T_2 = liquid's atmospheric boiling point
ΔH_v = liquid's heat of vaporization at T_2

Calculation of Vapor Cloud Size

The volume of a vapor cloud from a gas or volatile liquid release may be estimated using the equation

$$V_c = \frac{WRT_{atm}}{vMP_{atm}} \qquad 4)$$

Where

V_c = Vapor cloud volume
W = weight of vapor in cloud from eq.1 or 3
P_{atm} = ambient atmospheric pressure
T_{atm} = ambient atmospheric temperature
M = molecular weight of the gas
v = (LFL + UFL)/2
LFL = lower flammable limit per NFPA 325M, as vol. fraction
UFL = upper flammable limit per NFPA 325M, as vol. fraction

The size of the vapor cloud can be estimated by making some assumption as to the cloud shape, based on observation if possible. A cylindrical cloud would have size

$$D_c^2 h = \frac{V_c}{\pi} 4 \qquad 5)$$

Where

D_c = diameter of vapor cloud
h = height of vapor cloud
V_c = vapor cloud volume from eq.4

By contrast, a hemispherical cloud would have

$$r = \sqrt[3]{\frac{V_c}{2.0945}} \qquad 6)$$

Where

r = vapor cloud radius
V_c= vapor cloud volume from eq. 4

The basic computational method described above makes no assumptions about the actual size of the leak or rupture which caused of the spill; it is simply assumed that the contents of the system have spilled somehow. This limits its utility for the testing of hypotheses on potential leak sources such as leaking flanges of different sorts, ruptures of pipe of different sizes, etc. which must often be done for real accident situations. For these, more detailed studies of potential discharge rates, aerosol formation, rainout, cloud dilution and concentration profiles are in order.

For single-phase spills the discharge rate equations are regarded as well-defined and analytical solutions are available for release rates. Two-phase systems are significantly more complex [2]. In general terms,

$$\frac{dM}{dt} = C_D\left[\frac{2\rho_o(P_o - P_{atm})}{1-\beta^4}\right]^{1/2} A \qquad 7)$$

Where dM/dt = general discharge rate

C_D= nozzle discharge coefficient
ρ_o = initial density
P_o = process pressure

P_{atm} = discharge or atmospheric pressure
B = ratio, orifice diameter to vessel or pipe diameter
A = effective cross-sectional area of the leak or rupture

The nozzle discharge coefficient C_D is a function of the fluid, approaching 0.975 for gases but as low as 0.61 for the discharge of a liquid through a smooth orifice. For two-phase flow, C_D will increase between these two limits depending on the gas-to-liquid ratio. An additional complication comes when the leak is not a smooth circular hole [2, p.33]. Variations of C_D with leak shape and orientation have been postulated [3] as follows:

Table 1 – Nozzle Discharge Coefficient for Liquid Flow, Variations with Leak Shape

Leak Shape	Edge Condition	Petal orientation	Leak Orientation	C_D
Circular	smooth	none	any	0.61
Circular	Uniform petals	Jagged in	any	0.622
Circular	Random petals	Jagged in	any	0.577
Rectangular	smooth	none	horizontal	0.619
Rectangular	smooth	none	vertical	0.625
Rectangular	Uniform petals	Jagged in	horizontal	0.629
Rectangular	Uniform petals	Jagged in	vertical	0.626
Rectangular	Random petals	Jagged in	horizontal	0.609
Rectangular	Random petals	Jagged in	vertical	0.613
Rectangular	Uniform petals	Jagged out	horizontal	0.826
Rectangular	Uniform petals	Jagged out	Vertical	0.828

Dispersion is a mixing process in which turbulent eddies of long wavelength decay into eddies of progressively smaller wavelength. The basic source terms for dispersion modeling are release rate, duration and temperature [4]. Dispersion is influenced by the same factors which promote flame speed acceleration [2, p.60], to wit:

1) Jet mixing
2) Ambient temperature and pressure, wind speed & direction
3) Terrain and local plant congestion
4) Time
5) Impingement
6) Wake effects.

Dispersion models predict different effects for the various release classes [4] .Chemical releases are variously classed as

a) Two-phase gas releases
    1. choked
    2. unchoked
b) Two-phase pressurized liquid releases
c) Two-phase refrigerated liquid releases
d) Single-phase gas releases
    1. choked
    2. unchoked
e) Single-phase liquid releases
    1. high volatility
    2. low volatility

There are additional complications due to the release orientation (vertical or horizontal, ground level versus elevated in a structure), the plume characteristics (dense-gas versus neutrally buoyant). For all these reasons, choosing a dispersion model is not a simple or straightforward process.

In general, however, computer software packages available in the public domain for dispersion modeling or of two types:

1) Dense-gas models, and
2) Neutrally-buoyant models

Early dense gas models included DEGADIS [5] and SLAB [6]. More recently, an improved universal dispersion model has been developed for coupling with DNV's PHAST consequence-analysis software package. Neutrally buoyant models readily available include INPUFF [4] and AFTOX [7].

Dense gas clouds are those heavier than air. Such clouds behave distinctly differently from lighter-than-air gas releases. Dense gas clouds tend to slump down toward the ground, after which gravity-driven non-Gaussian dispersion and their negative buoyancy makes the center of the cloud dip and the edges to bulge, forming a "pancake" shaped cloud, a stratified flattened cylinder (height-to-diameter ratios of typically 0.02 to 0.05) which may very well include a secondary pool of condensed liquid at its base.

Lighter-than-air and neutrally buoyant gas clouds can be modeled using Gaussian dispersion methods typically applied to air pollution problems such as the dispersion of a smoke plume from a stack [8].

Table 2, from Dai [4], gives a brief review of appropriate dispersion models for various types of gas release.

Table 2 – Use of Dispersion Models for Different Release Types

Source Type	Continuous Release	Finite Release	Transient Release	Instantaneous Release
Ground level	DEGADIS SLAB AFTOX	DEGADIS SLAB AFTOX	DEGADIS	AFTOX*
Evaporating Liquid spill	DEGADIS SLAB AFTOX	DEGADIS SLAB AFTOX	-	DEGADIS SLAB AFTOX
Vertical Jet	DEGADIS SLAB INPUFF	DEGADIS SLAB INPUFF	-	-
Horizontal Jet	SLAB	SLAB	-	-
Instantaneous	-	-	-	SLAB

*Also models gas releases from elevated heights

Calculation of Energy Release- The Classical TNT Equivalent Method

The classical method for estimating the possible energy release from a vessel burst or vapor cloud explosion was the TNT equivalent method. As will be detailed below, this method has significant problems when applied to common industrial vapor clouds rather than ideal condensed-phase explosives. However, the TNT equivalent method is computationally simple, was formerly widely used by insurance companies for probable maximum loss (PML) calculations [1] and thus in the past was a common point of departure for hypothesis testing on explosion investigations. Moving beyond the limits of the TNT equivalent method is one of the great benefits of the computational dynamics models developed in recent years.

The TNT equivalent method involves calculation of the weight of TNT which would produce an explosive force equivalent to that produced by a vapor cloud explosion of a known size and type. To calculate the TNT equivalent for a vapor cloud,

$$W_{tnt} = \frac{W \Delta H_c}{\Delta H_{ctnt}} f \qquad 7)$$

Where

W_{tnt} = equivalent weight of TNT

W = weight of vapor in cloud

ΔH_{ctnt} = TNT's heat of combustion, = 4648 kJ/kg

ΔH_c = heat of combustion of vapor to CO_2 and H_2O

f = Efficiency, or explosive yield factor

The factor "f" in equation 7, for efficiency or explosive yield, is one of the problematic points in applying the TNT equivalent method to common industrial gas explosions. Investigations of actual chemical plant incidents gave "f" valves ranging from 0.01 to 0.05 with a suggested average value of 0.02 [1]. Spillage of rocket fuels in the early days of the US space program suggested that f~0.1. Consequently it would appear that "f" may vary widely depending on the contents of the spill.

For a pressure vessel burst, the stored energy in the contents may be calculated from the initial pressure. The TNT equivalent may then be calculated using the formula

$$W_{tnt} = \frac{P_1}{\Delta H_{tnt}} V \ln \frac{P_1}{P_{atm}} \qquad 8)$$

Where

W_{tnt} = equivalent weight of TNT

ΔH_{ctnt} = TNT's heat of combustion, = 4648 kJ/kg

V = vessel volume

P_1 = vessel internal pressure at failure

P_{atm} = atmospheric ambient pressure

Once a TNT equivalent for an assumed mass of a certain reactant has been calculated, circles of equal overpressure can be drawn using the cube root scaling law [Figure 1, from reference 1].

Using the observations of damage to the site (i.e. broken windows, structural damage, injuries, fatalities etc) the incident investigation team can develop empirical plots of apparent overpressure (see tables in MTI Publication 30) versus distance from the explosion center. If the empirical plots of overpressure versus distance correspond reasonably well to the circles of overpressure from the TNT equivalent calculations, then the assumed mass of the hypothetical reactant is a plausible source of the damage observed.

Beyond the TNT Equivalent Method; Computational Fluid Dynamics Modeling of Explosions

The TNT equivalent method was once widely used for probable maximum loss (PML) estimates because the calculations are relatively simple. It has a broad and deep base of research from military applications. The TNT equivalent method may have some remaining utility as a rough estimate for dealing with chemical plant explosions of energetic condensed phases such as liquid peroxides, or of sabotage by military explosives. But for many other fuels and situations, the TNT equivalent method has been widely criticized as simplistic and possibly misleading.

A study using the FLACS computer program [9] for comparing the TNT equivalent method to three-dimensional computerized numerical simulations for the modeling of hypothetical plant explosions of propane-air mixtures found [10] that the TNT equivalent method was overly simplistic because it does not readily account for areas of particularly high overpressure generated by the turbulence produced by regions of high equipment density. Thus the TNT equivalent method overestimated the peak overpressure in some cases and underestimated it in others. The primary problems with applying the TNT equivalent method to incident investigation were judged to be 1) the earlier noted uncertainties in choosing a yield, or efficiency, factor when the reactant is not a true high explosive, and 2) regions of high overpressure, where damage will be greatest, do not relate simply to the volume of gas contained in plant equipment but rather also are strong functions of local equipment density. Therefore the TNT equivalent method may not be adequate for predicting explosion overpressures from industrial releases of flammable gas clouds.

Another study [11] researching the Flixborough incident in the UK found that computerized fluid dynamics simulations of the event (using the EXSIM software simulation tool [12]) gave significantly higher peak overpressures than estimates derived from visual observations of the damage. The authors concluded that assumptions of overpressure estimations based on visual inspection alone may lead to large assessment errors. The reason for this is that tables of damage overpressures to various items of equipment typically give threshold pressures, and cannot give peak pressures. As the authors commented, "There is little evidence of the possible maximum explosion pressure when the structure is demolished."

Because the cost of computing power has fallen so dramatically in recent years, the use of computerized fluid dynamics for the numerical simulation of chemical plant explosions is becoming more and more practical. Specialized computer software packages dedicated to the simulation of explosions, or readily adaptable to such simulation, are now widely available commercially. Examples include the quantitative risk assessment tools from TNO [13], the widely recognized Baker-Strehlow –Tang explosion prediction method [14] and the Shell SCOPE (Shell Code for Overpressure Prediction in gas Explosions) and CAM (congestion assessment method) software [15]

The computational fluid dynamics method to explosion simulation uses the same basic procedure as that for fluid flow and combustion problems, in that

1) The physical boundaries of the problem must be defined
2) The space under study is divided into discrete cells, known as a "mesh"
3) The basic equations that define the physical modeling are identified, for the conservation of
    a. Mass,
    b. Momentum and
    c. Total energy [16].

    These are time-variant non-linear differential equations
4) The terms of the equations are divided into four categories
    a. The transient term giving the time rate of change of the conserved quantity
    b. Convective terms for flow due to pressure gradients and velocity
    c. Diffusive terms for heat conduction, viscosity and molecular diffusion
    d. Chemical source terms and gravity
5) Boundary and initial conditions are defined.
6) The equations are solved iteratively as steady-state or transient
7) Analysis and visualization of the resulting solution.

Criteria for a computer program designed to simulate explosions have been defined as follows [17]:

1) It must be able to describe the extremely non-linear response of gas and heat generation from energetic materials as a function of temperature,
2) It must be able to accurately describe the large deformations that occur in structures at the high strain rates associated with explosions, in a way that maintains separate flow fields for different materials.

3) If the reactants are condensed-phase materials, the program must have a method for treating the close mechanical coupling of gases and solids so that ignition and explosion which occurs at interfaces where there is initially no gas can be treated.
4) The program must be able to describe vessel ruptures which release compressed gases from one part of the domain into another with the processes of mixing, reaction and transfer of momentum and energy.

On that last point, the ability to simulate vessel ruptures, some practitioners caution that traditional Lagrangian formulations such as the finite-element method may exhibit difficulties with large vessel deformations at high strain rates, which may lead to mesh lockup and entanglement [18]. These workers recommend a "material point method", a type of particle-in-cell method they judge more suitable for problems involving large deformations of history-dependent materials such as pressure vessels and storage tanks. The advantage is that there is a computational basis for predicting dynamic fracture based on the energy release rate available for crack propagation. Crack growth is predicted when

$$G_c < G_I + G_{II} \qquad 9)$$

Where

G_c = material toughness determined experimentally
G_I = energy release rate for tensile loading, eq 10
G_{II} = energy release rate for shear loading, eq. 11

The terms G_I and G_{II} are calculated from equations 10 and 11, respectively, where

$\delta\alpha$ = crack opening length
σ = stress
μ = displacement

$$G_I + \lim_{\delta\alpha \to 0} \frac{1}{2\delta\alpha} \int_0^{\delta\alpha} \sigma_{\theta\theta}(\delta\alpha - r, 0)\mu_\theta(r, \pi)dr \qquad 10)$$

$$G_{II} + \lim_{\delta\alpha \to 0} \frac{1}{2\delta\alpha} \int_0^{\delta\alpha} \sigma_{r\theta}(\delta\alpha - r, 0)\mu_r(r, \pi)dr \qquad 11)$$

Industry Globalization

The globalization of the world's chemical industry and the oil refining industry has been truly dramatic in the last 20 years. No new oil refineries have been built in the United States since 1976, and only one large new chemical plant is presently under construction in the US. Yet the world's demand for industrial chemicals and refined oil products has not dwindled but continued to grow strongly. That growth has been filled by a number of world-class enterprises built in resource-rich countries where raw materials and labor are competitive. Many of these plants are joint ventures between mature North American or European companies and the host countries. Some are largely or wholly owned by the host countries with technical services bought from all over the world.

Thus the days when an investigator of fires and explosions could expect to work almost wholly along the US Gulf Coast are gone. The work may take one to any continent in the world except Antarctica. In a globalized world full of conflict, the threats to the physical assets have changed as well. A recent survey of ignition sources in chemical plant fires [19] found that arson or sabotage ranked ahead of such well-known hazards as electrostatic discharge, welding and torch cutting as a cause of fires or explosions.

Safety incident investigators should never raise the possibility of sabotage lightly, or on scant evidence. Sabotage is an emotive word, reeking of treason and betrayal. It is a crime in all nations, and a capital offense in some. The merest suspicion of sabotage may have fatal consequences for a suspect in some countries where summary justice is still imposed.

Since sabotage is a crime, a sabotaged chemical plant or oil refinery is not only an incident site but a crime scene. Local or national law enforcement agents may insist on control of the site if sabotage is a serious possibility.

When the word "sabotage" is in the air, and when law enforcement becomes involved, operators and craftsmen may cease all cooperation with the company investigation for fear of criminal prosecution. Without the cooperation of the operators on duty at the time of an incident, it is highly unlikely that the company's incident investigation will uncover the truth in a timely fashion.

Consequently the subject of sabotage must be approached, and even mentioned, with utmost caution. Yet sabotage is a factor which cannot be ignored by incident investigators. Sabotage may impinge of the work of fire and explosion investigators in one of four possible contexts:

1) Sabotage by external agents
2) Sabotage by insiders
3) Sabotage of investigations and
4) Perception of the investigator as a saboteur

The last context, a perception of the investigator as a saboteur, can lead incident investigators into adventures. The following case history gives an example.

Case History

An investigator from the UK was dispatched to analyze a piping rupture in a Middle Eastern natural gas processing plant partly owned by his employer. The investigator found that the cause of the rupture was corrosion of the carbon steel piping from H_2S-bearing ("sour") water condensing out of the gas. The investigator believed that the corrosion was not confined to the failed piping spool alone, and launched an extensive inspection effort to locate other areas in the piping array which might also be at risk for future failure. This inspection effort was projected to keep the plant idled for several weeks.

However, the plant in question was a vital source of foreign exchange for the host country, and every day of lost production deepened the economic anxiety of the government. This nation had an active, powerful and paranoid internal security agency deeply suspicious of foreigners. The agency began to question the plant workers about the activities of the British investigator. The tone of the questions made it clear that the security agents viewed the inspection program as a Western Imperialist plot to sabotage the economy of the host country. The British investigator fled the country in the dead of night shortly before he was due to be arrested, and his parent company divested itself of their equity in the plant in question [20].

In dealing with such situations, it is invaluable to have trustworthy local contacts that speak the local language and regularly deal with the host country government agencies. In the absence of such an early warning system, the personal cost to the investigator may include an uncomfortable interrogation, an uncertain term of imprisonment or worse.

Changes in Regulatory Climate and Safety Culture

A side-effect of globalization is that many chemical companies and oil refiners now deal with many different national regulatory systems. The regulatory climate in most developed nations has become steadily more stringent during the last two decades. Consequently incidents of fire or sudden energy release attract much more government scrutiny than in the past. The result is that incident investigators from the owners of the asset can expect to spend significantly more time in liaison with the various inspectors and investigators from the sometimes overlapping government agencies now involved.

The pace of the investigation will inevitably be affected by the need to obtain consensus on equipment condition assessments and any required destructive testing. This will tend to delay repair and restart decisions compared to a more streamlined process driven solely by the judgement of asset owners' internal engineers and safety specialists. On the other hand, some of the government safety entities have significant resources which can aid in the analytical reconstruction of complex incidents. The US Chemical Safety Board, formed during the mid-1990's, is one good example of a repository of analytical expertise which can be very beneficial especially to small or mid-sized corporations challenged for experienced incident investigators.

Changes in safety culture are noticeable over the last 18-20 years in many chemical plants and refineries in mature, developed nations. There is a strong trend towards formalization, with carefully documented root cause failure analysis committees, more health-safety-and-environment professionals, and more risk management studies. There is, in general, less tolerance for risk and more tools made available to operating organizations to reduce it. The bonuses of upper managers are often closely tied to specific safety metrics, so serious action to meet them is made. If these metrics are meaningful, then real improvements in safety performance, with consequent reductions in fires, explosions and production disruption events, are observed.

Safety culture in the many world-class chemical plants and oil refineries being built around the world in newly industrialized nations is highly variable. In some plants, safety culture is very much a work in progress. Risk tolerance is much higher in some instances. For example, a mega-refinery now a-building in Asia has reportedly suffered many dozens of fatalities in the course of construction. Work proceeds apace despite a safety situation which would attract crushing government intervention in many developed countries.

Storm Damage Assessment

The world-famous hurricanes that came ashore on the US Gulf Coast in 2005 led to a widespread recognition in the industry that methods and techniques for storm damage assessment are equal in importance with methods for analyzing fires and explosions. MTI hosted a series of lectures and round table to share experiences on this topic, and the National Association of Corrosion Engineers (NACE) and the American Petroleum Institute (API) held committee sessions to capture lessons learned. Many of these lessons are presented in a new chapter on storm damage assessment in the upcoming 2nd edition of MTI Publication No. 30.

Figure 1 – Hurricane flooding in a chemical plant on the US Gulf Coast, August 2005

Assessing plant equipment for storm damage has similarities to fire or explosion damage assessment. There is the same economic need to return the plant to service as soon as safely possible. There is also the need for prudence in performing necessary repair and refurbishment so that excessive downtime in later service does not negate the advantages of an early re-start.

Hurricanes affect chemical processing plants in the same ways that they affect residential and office structures, by

1) Flooding from storm surge,
2) Wind pressure damage
3) Wind-driven rain damage

Plant equipment must be assessed for evidence of damage from each of these mechanisms before return to service.

Storm damage assessment differs from other types of production disruption incidents (fires, explosions, etc.) in that virtually all the services and infrastructure on which normal industrial activity depends will be erratic, damaged or destroyed. Consequently, an on-site assessment team for storm damage must be self-reliant to a degree far greater than normal for a team assessing the aftermath of a fire or explosion. In the very early days after a storm, the damage assessment team must be prepared to be largely or completely self-sufficient with regard to

1) Communications
2) Accommodations
3) Food and water
4) Electric power
5) Transportation and fuel

After the 2005 hurricanes Katrina and Rita hit Louisiana and Mississippi, chemical plants and oil refineries in the area reported loss of electrical power, loss of communications, lack of access by road or water, lack of security and widespread organizational disruption to plant employees [21].

In the storm-damaged zone, land telephone and electrical power lines were down, the poles carried away by wind or water. Cellular telephone service ceased a day after the storm as the cell tower battery back-ups depleted their charges. Satellite telephones were the only reliable means of communication. Text messaging was found to be a surprisingly useful tool.

No hotel accommodation was available within 50 miles of the affected plants. Food was unobtainable since all restaurants and grocery stores were closed or swept away, and all catering contractors fled the area. All drinkable water had to be carried in. One major oil and chemical processing company hired a passenger ship and moored it in the Mississippi River next to their damaged plant to serve as accommodation for the assessment and clean-up crew.

Transportation into the storm-damaged zone was problematic. Many of the bridges collapsed due to storm surge. Roads were intact to the west of the storm's track, but blocked by downed trees and debris in the earliest days. All rental cars within a 100 mile radius of the storm's track were either submerged or in use by evacuees fleeing the area. Truck transportation was commandeered by military and government agencies to aid in rescue efforts. The only access to some of the affected plants was by helicopter, and helicopters were available only to those firms with the foresight to lease them well in advance of the storm. .

Loss of electrical power after the 2005 US Gulf Coast hurricanes made all operations more difficult. Plants reported power outages ranging from 4 days to three months. Those plants with back-up diesel generators had an obvious advantage, but high fuel consumption was a problem. There are no gas stations or truck stops open after a hurricane.

Contract security guards failed to report to work after the storm, and local police were unable to answer calls for assistance since many of their officers were themselves trapped by flood water. Looting was rife. At one chemical plant in the New Orleans area, company employees fetched personally-owned hunting rifles from their homes and mounted security patrols themselves around their plant for five days after Hurricane Katrina, to deter looters, until the arrival of Military Police squads from the Louisiana National Guard six days after the storm. Another chemical plant organized armed convoys under military escort to convey its recovery workers from a dry-land base camp to the damaged plant each working day. These convoys were considered necessary for three weeks after the storm struck land.

When physical entry to the site becomes possible, downed power lines are deadly risks to storm-damage assessment teams. Identifying and neutralizing downed power lines must precede all other activities.

After the site is safe to work in, investigative tasks for the storm damage assessment team include:

1) Mapping the height of storm surge floodwater
2) Checking mechanical equipment for distortion from rushing floodwater
3) Documenting wind damage to buildings, tanks, piping and stacks.
4) Identifying business records at risk for loss or corruption
5) Investigating rain-driven salt intrusion on steel surfaces and under insulation. [22]

A problem peculiar to storm damage assessment is the need to assess the extent of damage to the plant's business records. Unlike fire or explosion damage, flooding from storm surge is not likely to confine itself to process units. Control rooms and administration buildings may be inundated as well. Thus one vital task for the storm damage assessment team will be a judgement on the extent of damage to vital business records such as start-up and operating procedures, operating logs, maintenance records, contracts for supplies and raw materials and so forth. A records management specialist should join the storm assessment team to rate the various records into three categories:

1) Destroyed
2) Salvageable
3) Endangered

Destroyed records are those which are

a) No longer physically intact – blown away by wind or swept away by flooding

b) Records below the water line soaked in oil or chemicals

c) Records on the hard drives of computers exposed to floodwater

Salvageable records or those which were

a) in the high-humidity zone just above the flood line
b) submerged or wet but with rainwater or clean water
c) records showing minimal mold (these may be freeze-dried to salvage them)

Endangered records are those which are

a) located on the upper floors of buildings above the flooding
b) free of mold but subject to high humidity due to power loss

If they are to be preserved for future use, endangered records must be removed from the plant site as soon as possible and stored under temperature and humidity control. With the power loss so common after severe hurricanes, the air conditioning systems in plant buildings will often be inoperable, and rampant mold may soon destroy records which survived the storm itself.

The urgency for removing and saving the salvageable records and replacing destroyed records will, of course, depend on the back-up documents available. Most companies make considerable efforts to duplicate vital business records and store the duplicates in some safe location. However, such duplicates are not always completely current, and their safety must be carefully assured. During the 2005 US Gulf Coast hurricane season, some companies stored their back-up records in buildings located in downtown New Orleans. Many of these duplicate records were themselves destroyed, damaged or made unavailable when broken levees flooded the city.

CONCLUSIONS

Over the last 18-20 years, dramatic changes in the chemical industry and oil refining industry have lead to equally significant changes in the work of safety investigators studying fires, explosions and production disruption incidents. New challenges have arisen from the globalization of the industry with its multitude of regulatory regimes and a wide variation in risk tolerance. At the same time, incident investigators are now better equipped than ever to do their work, with new analytical tools and a supportive safety culture from most government agencies and corporate managers.

ACKNOWLEDGEMENT

Gregg Sapp of ConocoPhillips reviewed a draft of this paper and made some very helpful comments on it. His contribution is most gratefully acknowledged.

REFERENCES

1) Oil and Chemical Properties Loss Prevention Estimation Guide, IM. 8.0.1.1, March 2, 1992, Industrial Risk Insurers, 85 Woodland St., Hartford, CT 06105-1226
2) Woodward, J.L., Estimating the Flammable Mass of a Vapor Cloud, Am Inst. Chem. Engr., New York, NY, 1998,p.32
3) Dodge, F., Bowels, E., and White, R., "Release Rate of Hazardous Chemicals from Damaged Cargo Vessels" *Proc. 1980 Natl. Conf. on Control of Hazardous Materials Spills,* 1980
4) Dai, W., "Modeling Accidental Releases" *Chemical Engineering Progress*, Nov. 2005, pp 33-39
5) Spicer, T. et al., "DEGADIS: Heavier-Than-Air Gas Atmospheric Dispersion Model" presented at the 79th Meeting of the Air Pollution Control Assoc., Minneapolis, MN (June 1986)
6) Ermack, D.L., "User's Manual for SLAB: An Atmospheric Dispersion Model for Denser-than-Air Releases: DE91-008443 (1990), Lawrence Livermore National Laboratory, US Dept. of Energy, Livermore, CA (1990)
7) Kunkel, B.A., "User's Guide for the Air Force Toxic Chemical Dispersion Model (AFTOX)", PL-TR-91-2119, Environmental Research Paper No. 1083, Phillips laboratory, Directorate of Geophysics, Air Force Systems Command, Hanscom AFB, MA (1991)
8) Stern, A.C., Wohlers, H.C., Boubel, R.W., and Lowry, W.P., Fundamentals of Air Pollution, p.276-281, "Models of Diffusion and Dispersion from Point Sources", Academic Press, New York, 1973
9) FLACS – Flame Acceleration Simulator – Christian Michelson Institute in Norway
10) Savvides, C.N., and Tam, V.H.Y, "The Modeling of Gas Explosions: A Comparison of a 3-D Numerical Model with Simpler Models"
11) Høiset, S., Hjertager, B.H., Solberg, T., and Malo, K.A., "Flixborough Revisited – As Explosion Simulation Approach" *J. Hazardous Materials*, A77 (2000) pp1-9
12) EXSIM Website: http://www.exsim.com
13) Van Der Voort, M.M., Klein, A.J.J., dMaaijer, M., Van den Berg, A.C., Van Deursen, J.R., and Versloot, N.H.A., "A Quantitative Risk Assessment Tool for the External Safety of Industrial Plants with a Dust Explosion Hazard", *J. Loss Prevention in the Process Industries*, Vol. 20, Issues 4-6, July-Nov. 2007, pp 375-386
14) Perorazio, A.J., Kelly-Thomas, J., Baker, Q.A., Ketchum, D.E., "An Update to the Baker-Strehlow-Tang Vapor Cloud Explosion Prediction Methodology Flame Speed Table", *Process Safety Progress*, Vol. 24, Issue 1, pp 59-65, 4 Jan 2005

15) Puttock, J.S., Yardley, M.R., and Cresswell, T.M., "Prediction of Vapor Cloud Explosions using the SCOPE Model", *J. Loss Prevention in the Process Industries*, Vol. 13, Issues 3-5, May 2000, pp 419-431
16) Huld, T., Peter, G., and Städtke, H., "Numerical Simulation of Explosion Phenomena in Industrial Environments", *J. Hazardous Materials,* **46**, (1966) pp185-195
17) Wright, C. A., "High Performance Simulations of Fires and Explosions", Dept. of Chemistry, Univ. of Utah, Salt Lake City, Utah
18) Henderson, T.C., McMurtry, P.A., Smith, P.J., Voth, G.A., Wight, C.A., and Pershing, D.W., "Simulating Accidental Fires and Explosions", *Computing in Science and Engineering*, March/April 2000, pp 64-76
19) Woodward, J.L., Estimating the Flammable Mass of a Vapor Cloud, American Institute of Chemical Engineers, 3 Park Ave., New York, NY 10016-5901, p. 76, after, "Large Property Damage Losses in the Hydrocarbon-Chemical Industries: A 30-Year Review", 14th Ed., p. 12, Marsh & McLennan Protection Consultants, New York, 1992
20) Swidzinski, M.W., ConocoPhillips, personal communication, May 2004
21) Hurricane Damage Panel Discussion, Materials Technology Institute Technical Activities Forum, June 7th, 2006, St. Louis, Mo.
22) STG 34 Petroleum Refining Information Exchange, 2005 Corrosion Technology Week, National Association of Corrosion Engineers

Controlling Risk during Major Capital Projects

William Bridges
President; Process Improvement Institute, Inc. (PII)
Knoxville, Tennessee
wbridges@p-i-i-i.com
www.p-i-i-i.com

Revonda Tew
Senior Engineer/Instructor; Process Improvement Institute, Inc. (PII)
rtew@p-i-i-i.com

Abstract

This paper describes the best practices for scheduling and performing PHAs/HAZOPs, etc., during various key phases of major projects. The paper outlines the scope and content of each project phase hazard review and what the outcomes should be. The concepts are not new, but many contractors and vendors around the world do not seem to understand the concepts well. The paper presents the basics, and then shows best practices and examples from various companies. The paper also outlines how information related to process safety should be developed during and then delivered from a major project.

Background

The construction of a new plant/facility or a large capital expansion to an existing production facility represents significant risk. That risk increases when construction takes place in an operating environment and when critical tie-ins are made to "live" equipment. There have been many excellent papers over the past two decades on how to manage major capital projects to achieve a process design and the supporting operating procedures and process safety information that will control process safety. These papers include:

- The paper presented by Olin Chemicals in 1989, which provided a basic outline of risk control steps implemented during 5-6 stages of project (many companies followed and continue to follow such a process for controlling risk)[1]
- The paper presented by BP Oil in 1992, which provided a basic outline of risk control steps implemented during 5-6 stages of project[2]
- The comprehensive paper provided by Syncrude in 2001[3]

CCPS has provided excellent guidance in their original PSM resources[4] and in many subsequent textbooks and other publications.

All of these resources refer to the industry standard practice of conducting multiple risk reviews (also called hazard reviews, preliminary hazard analyses, process hazard analyses [PHAs], project risk reviews, design safety reviews, and many other titles). Typically there are one to two risk reviews within a project related to a small project and up to 6 risk reviews building upon each other to yield the "initial PHA" for a large project (such as building a new plant or new unit/facility). Conventional risk analysis tools must be applied at strategic project milestones, but these alone will not ensure a safe and seamless startup and a positive return for investors. A holistic approach to managing risks on a large project requires some innovative measures that must be supported by the entire project team. Fortunately, there is a wealth of experience to draw on. The paper discusses lessons learned in the past two decades on the keys for controlling risk during projects.

Types of Projects

There are various sizes and structures of projects, depending on the scope of the endeavor, the urgency, the nature of the business, the company culture, the company sophistication, and many other factors. The two most important project types (factors) for purposes of this paper are (1) project size (expressed usually in expenditure expected or size and number of pieces of equipment to be installed) and (2) type of financial control for the project.

Typical Project Sizes:

This paper focuses on major projects, though most of the best practices translate into moderate and small projects as well. **Table 1** below provides a basic definition of projects sizes and the typical number of risk reviews conducted during the project:

Table 1: Project Size, Example Scopes, Relative Project Length, and Risk Review Effort

Project Size	Example Project Scope	Example Project Length/Duration (concept→ commissioning)	Number of Risk Reviews
Major	Major projects handled **external** to an affiliate/plant, such as expansions and new facilities	12-36 months	5-7
Moderate	Works **engineered by an affiliate/ plant** (installing a new design of knockout pot for a feed to a unit)	6-9 months	2-3
Small	**Minor** affiliate/plant works (installing piping to bypass a control valve)	1-2 months	1

Financial Control of Projects:

Financial control of a project can also influence the way risk is managed. There are various styles of financial control, but the most common are **(1) Sequentially Loaded,** where the estimate of the project and the next phase is adjusted after each successive phase is complete and **(2) Front-end Loaded,** where the project is essentially run as a "fixed cost, turn-key" contract where any change to the scope affects the contract cost and schedule, so each change requires a scope change approval. Many financial managers believe the "Front-end Loaded" approach is best, but typically the inertia against making scope changes is so strong, that the unit delivered typically has smaller long-term benefit to the owners. Below is a brief summary of the two types of projects:

- **Sequentially loaded**
  - Requires more calendar time
  - Allows for risk approvals (gates) at major project steps
  - Allows more flexibility of design during project phases
  - Usually more cost effective in the short-term
  - Almost always more cost effective and safer in the long-term, reaping much more benefits over the life of the project

- **Front-end loaded**
  - Supported by financial officers
  - Shorter calendar time
  - Firm budgets; easier to sell to stake-holders
  - Scope changes are difficult, even for safety
  - Not always cost effective due to after-project changes and losses resulting from compromises in process safety and process reliability
  - Performance standards/measures are critical to success
  - Requires full-time process-safety expert from start

Project managers or team members likely do not have any say in the style of financial control used by the company. For a front-end loaded project, the best alternative is to make sure a full time process safety expert (preferably one from the owner/ultimate plant location) is available to help ensure that, by the final project phase, the needs of the facility are met. The project manager or process safety expert should have the authority or have ready access to the proper authority, to approve scope changes related to minimizing the inherent and residual risk of the process.

General Concept of PSM Development during Major Capital Projects

All of the PSM elements apply to, or are affected by, capital projects to varying degrees, especially:

- Process Safety Information – developed throughout the project phases
- Operating Procedures – developed in the last phases of a project

- Management of Change – heart of project risk control
- Process Hazards Analysis – heart of MOC
- Pre-Startup Safety Review – before highly hazardous chemicals enter and before production starts
- Emergency planning and response – cannot happen properly if the process layout, design, and operation do not consider it ahead of time

Most companies learned many years ago that the PSM elements are the same management practices required for operational and reliability excellence – so building for these during each project phase makes excellent business sense. Also, by addressing PSM early in the project development, it is easier to expand the goal of risk management to "sustainable control of human error," the key to controlling risks. This not only provides excellent process safety control, but excellent reliability, quality, and efficiency control.

Scope of Risk Reviews and PSM Development For Each Project Phase

As mentioned earlier, major projects can have 5 to 7 or more phases and these can be spread over 12 months to 36 months or more depending on the project size. However, decisions for controlling risk made during the project phases echo through the next 20 to 50 years of operation, because design features, automated control features, and human interactions must be controlled continuously to control the inherent hazards of chemical processes.

Figure 1 presents an example of a major capital project's phases for a large new chemical process unit or plant, with six "in-project" phases and one "post-project" phase. Though not universal, this approach appears to be a widely accepted view of major project phases. For smaller projects, condense this approach to 5,4, or a minimum of two phases. **Table 2** on the following pages provides much more detail on the risk review for each project phase.

Figure 1: Example Project Phases and Related Scope of Risk Reviews (RR)

	Phase 1	Phase 2	Phase 3	Phase 4	Phase 5	Phase 6	Phase 7
	Project Responsibility						Plant/Unit Responsibility
	Phase 1 Conceptual Design	**Phase 2** Feasibility & Detailed Specification	**Phase 3** Preliminary Design	**Phase 4** Detailed Design	**Phase 5** Construction	**Phase 6** Pre-Commissioning	**Phase 7** Post-Commissioning
Project Concerns and Deliverables	Fit to business strategic plan Fit to existing operations Review of available technology Inherent safety options Site planning Raw material resourcing options	Detailed feasibility study (availability of technical staff; marketing plan) Detailed technology review and specification development Preliminary plot plan and tie-in plan +/- 40% cost estimate Preliminary schedule & milestones	Preliminary construction & operation plans Initial process flow diagrams (PFDs) Initial material and energy balances Raw material planning Utility planning Candidate vendors for major components Fire protection plan	Process design and generation of Rev 0 P&IDs; continuing to Rev 5 (nominal) in this phase Revise material & energy balance Technical specs for all components Basics design of process controller Fabrication started and major components ordered	Execute fabrication and installation plan Develop detail plans for tie-ins to OSBL Continue to develop specs for OEM manuals and operating procedures Set up CMMS/ database for ITPM Draft PSM management systems/procedures	Complete fabrication/ installation Complete development and closure of PSSR and other punchlists Commission/ validate equipment (dry, wet, with HHC) Populate CMMS and other PSM/ reliability databases Complete operating and MI procedures (by SMEs); validate	Assist commissioning team Ensure training by vendors/OEMs are completed in the field and proficiency of plant staff is validated Complete performance measure of initial operation (to ensure contract commitments are met) Manage changes Closeout project
Risk Reviews	**Conceptual RR** Strategic plans Inherent safety Plot plan review for Facility Siting; consequence modeling for major releases Begin Human Factor consideration		**Preliminary Design RR** What-if analysis of each major unit operation HAZOP/FMEA of selected scenarios LOPA of selected scenarios & review options for inherent safety	**Detailed Design RR** HAZOP/FMEA of most nodes (focusing on continuous mode of operation) LOPA of 1-5% of scenarios Final SIL (if needed) determination	**Final Detailed Design RR** HAZOP/FMEA of changes since previous RR, including rec. resolutions; place special attention to changes in field Begin human factor and facility siting (HF&FS) checklists	**Commissioning RR ("Initial PHA for new unit")** HAZOP/FMEA of changes HAZOP and/or What-if of non-routine operating modes startup, emergency shutdown, etc.) Complete HF&FS checklists	**Post-Startup RR (3 to 6 months after startup)** Close any recommendations that were rated as post-startup issues Review each MOC for its impact on the "Initial Unit PHA" Perform critique of risk review efforts during project.

Table 2: Details of Risk Review Phases of a Major Capital Project

Project Phase #	RISK REVIEW Phase Name	Goals of RISK REVIEW	RISK REVIEW Methodology	RISK REVIEW Team Membership (in addition to leader & scribe)	Key PSM Element Development for each Phase (using US OSHA PSM as template/example, OSHA PSM regulation in parentheses ( ))
1&2	Conceptual	Choose inherently safer option, ensure overall feasibility, estimate impact on neighbors	• Consequence modeling (to help on next project phase) • What-If (no guidewords) • Selected checklist for judging inherent safety	• Senior operator for unit or from similar unit • Senior process engineer for unit or from similar unit • Process/design engineer from project • Process Safety specialist (if not already listed above)	• Process safety information, including chemical hazards, chemical reactivity, hazards of inadvertent mixing, inventories, applicable codes and standards (d) • Baseline info for future PHA (e) • Baseline info for future MI (j) • Begin inherently safer concept • Begins leadership • Begins employee participation (c)
3	Preliminary Design	Identify and resolve most expensive design alternatives, including layout of plant, facility siting concerns, environmental protection issues, and major tie-ins	• What-If (no guidewords) • HAZOP/FMEA of selected scenarios • LOPA of selected scenarios	• Senior operator for unit or from similar unit • Senior process engineer for unit or from similar unit • Process/design engineer from project • Process Safety specialist (if not already listed above)	• Process safety information, including applicable codes and standards, process flow diagram, thermal/kinetic chemistry information, material and energy balances, and materials of construction (d) • Facility siting basis set • Beginning emergency response plans and procedures (n) • Baseline info for future PHA (e) • Baseline info for future MI (j) • Continue employee participation (c)
4	Detailed Design	Begin detailed identification of potential accident scenarios, primarily focused on normal (usually continuous) mode of operation. Begin risk assessment for scenarios with large residual risk	• HAZOP/FMEA of equipment nodes, focusing on normal (usually continuous) mode of operation • What-If of lower consequence & lower complexity systems • LOPA of 1-5% of the scenarios; determine SIL, as necessary	• Senior operator for unit or from similar unit • Senior process engineer for unit or from similar unit • Process/design engineer from project • Process Safety specialist (if not already listed above)	• Process safety information, including applicable codes and standards, P&IDs, revised materials of construction, safety interlocks and controls, equipment design basis and some final equipment details (d) • Multiple layers of protection • Detailed info for future PHA (e) • Detailed info for future MI (j) • Begin detailed emergency planning and response (n) • Continue employee participation (c)
5	Final Design	Update results of previous RISK REVIEW for new details,	• Complete HAZOP, FMEA, or What-If for nodes started in previous RISK REVIEW	• Senior operator for unit or from similar unit	• Process safety information (revisions) (d)

Project Phase #	RISK REVIEW Phase Name	Goals of RISK REVIEW	RISK REVIEW Methodology	RISK REVIEW Team Membership (in addition to leader & scribe)	Key PSM Element Development for each Phase (using US OSHA PSM as template/example, OSHA PSM regulation in parentheses ( ))
		identify potential accident scenarios for nodes not previously reviewed, primarily focused on normal (usually continuous) mode of operation. Resolve most previous recommendations Complete risk assessment for scenarios with large residual risk	• Perform HAZOP, FMEA, What-If for nodes not covered in previous RISK REVIEW (due to previously missing information) • Begin Human Factors and Facility Siting checklist • Perform general Utility Failure checklist • LOPA of 1-5% of the scenarios; determine SIL, as necessary	• Senior process engineer for unit or from similar unit • Process/design engineer from project • Process Safety specialist (if not already listed above)	• Multiple layers of protection (revisions) • Revised details building toward initial PHA (e) • Revised data for MI, including development of inspection, test, PM plans and populating databases (j) • Begin detailed consideration of human factors • Data and recommendations for developing operating and maintenance procedures (f) & (j) • Continue employee participation (c)
6	Commis-sioning	Conduct full hazard/risk review of operating procedures to control risk of errors during startup, shutdown, emergency shutdown, and other non-routine modes of operation Close out previous RISK REVIEW issues (from earlier phases of project) and complete the human factors & facility siting checklist This RISK REVIEW creates the "Initial PHA" of the process	• HAZOP (2 guideword or 8 guideword) or What-If (no guideword) of operating procedures (choose method based on hazard and complexity of each task) • Complete HAZOP, FMEA, or What-If for nodes started in previous risk reviews • Perform HAZOP, FMEA, What-If for nodes not covered in previous risk reviews (due to previously missing information) • LOPA of 1-5% of the scenarios; determine SIL, as necessary • Complete Human Factors and Facility Siting checklist	• Senior operator for unit or from similar unit • New/junior operator for unit • Senior process engineer for unit or from similar unit • Process/design engineer from project • Process Safety specialist (if not already listed above)	• Process safety information (revisions) (d) • Multiple layers of protection (revisions) • Complete the initial PHA (e) • Revise MI plans, procedures, and database (j) • Complete detailed consideration of human factors • Continue employee participation (c) • Complete development of emergency planning and response procedures (n) • Complete operating procedures (f) • Complete initial training (g) & (j) & (h) • Develop remaining elements of PSM, including MOC procedures and incident reporting and investigation system (l) & (m) • Conduct initial PSSR (i)
7	Post-Startup	Conducted 3-6 months after startup similar to the future Revalidations, but with the goal of compensating for weaknesses in MOC process at the initial startup of the new unit/process	• Audit of MOCs (and P&IDs and SOPs) since "Initial PHA" (since commissioning RISK REVIEW) to ensure nothing has been missed by MOC • HAZOP or What-If of missed or poorly reviewed changes • Update PHA for the entire set of	• Senior operator for unit or from similar unit • New/junior operator for unit • Senior process engineer for unit or from similar unit • Possibly project/design engineer (for QA of project)	• Initial PHA Revalidation (completed much earlier than the required 5 years cycle) (e) • MOC (l) • Incident investigation (for lessons learned in first 3-6 months) (m)

Project Phase #	**RISK REVIEW Phase Name**	**Goals of RISK REVIEW**	**RISK REVIEW Methodology**	**RISK REVIEW Team Membership** (in addition to leader & scribe)	**Key PSM Element Development for each Phase** (using US OSHA PSM as template/example, OSHA PSM regulation in parentheses (_))
			changes (looking at whole picture for effect of all changes) • Close any pending recommendations (if possible)		

Risk Control – Initial Phases of a Major Capital Project

As described in **Figure 1** and **Table 2**, the first two project phases are critical for establishing the inherent safety of the process, and therefore an opportunity for company leadership to show their true colors. These first two project phase design reviews are typically called "Conceptual Design Reviews" and "Preliminary Risk Reviews," and are frequently referred to as "Preliminary HAZOPs" outside of the USA. The figure and tables list some of the activities during Conceptual Design and follows into the Preliminary Design phase. A partial list of these and other activities include:

- Select process (inherently safe or at least inherently safest design):
  - Identify low waster chemistries
  - Low inventory
  - Select non-hazardous solvents and reagents
  - Low energy reactor configurations[5]
- Decide on the maximum level of inherent risk (not residual risk) that will be accepted
- Establish siting and layout (with control of inherent risk to neighbors considered)
- Establish grading and drainage
- Establish process design (preliminary)
- Establish circuit isolation requirements
- Establish process control philosophy
- Establish strategy for control of ignition sources (electrical area classification)
- Establish relief and blowdown requirements
- Establish fire protection strategy

In 1996, Trevor Kletz noted that little literature exists for describing these front-end studies.[6] Perhaps this is true for the "public domain literature," such as textbooks and papers/articles. However, most companies have internal documents that explain how to grow a PHA throughout the project from concept to startup. These internal procedures have been summarized in various papers and textbooks listed earlier[1,2,3,4] and in many other forums over the years. The point Trevor Kletz was making appears to be that despite the past 35 plus years of adoption of HAZOP (and a longer period for adoption of What-If and FMEA), the implementation of risk review methods during the early stages of a project has been lacking in a great many companies. This is especially true for the owners who allow the "prime contractor" to have main oversight of the risk review process, a conflict of interest. No one will care about controlling the risk of the delivered process/unit **more that the owner/operator**.

Project level risk reviews have been performed to increasing levels of care for the past 40+ years. Over that time, methods have been improved or customized for theses early- phase Risk Reviews. In the past 15 years, considerable focus has been placed on controlling inherent safety at the initial concept phase of a new unit/process. Now there is a major focus not only on inherent safety, but also inherent reliability, and inherent environmental protection during the **Conceptual Design phase and Preliminary Design phase.**

Table 3 below (adapted from Syncrude, 2001)[3] presents a more detailed list of considerations during the Risk Reviews for the earliest phases of a project.

Table 3: Conceptual and Continuing into Preliminary Design Risk Reviews – Key Concerns and General Methods to Guide these Reviews

Key Concern	Ways to Address/Evaluate Key Concern during Risk Review
Inherently safe/reliability process selection	▪ Evaluate low waster chemistries ▪ Evaluate low inventory ▪ Evaluate use of non-hazardous solvents and reagents ▪ Evaluate low energy reactor configurations[5]
Plot Location Relative to Other Units	▪ Process opportunities – reduced pipe runs ▪ Operating logistics – number of operators ▪ Distance from control room and tie-in to control room ▪ Backup contingency considerations – begin Utility Failure risk review ▪ Knock-on effects from major incidents in nearby units; including consequence modeling is necessary ▪ Begin Facility Siting risk review (checklist-based) here, if possible
Size of Plot Areas – based on preliminary process designs	▪ Access to utilities and other support systems, sewers, etc. ▪ Cost of real estate, site preparation ▪ Elevation considerations
Precise Spacing Between Unit Boundaries – standoff distances between battery limits or between major equipment in nearby units	▪ Insurance and Industry spacing guidelines ▪ Knock-on effect to nearby units ▪ Business interruption considerations
Inner Unit Layout	▪ Ease of Access (begin Human Factors considerations) ▪ Operability and maintainability ▪ Constructability and Repair considerations (considering future lifting activities for construction and repairs) ▪ Equipment decking and layering (don't expect operators to climb through caged ladders to take routine process reading)
Fire Protection Review	▪ Emergency access routes ▪ Emergency escape ▪ Fire fighting or toxic rescue capability ▪ Placement of detectors ▪ Strategy to protect neighbors from your releases ▪ Construction of control to protect "stay-behinds" (See consequence modeling mentioned in "Ploy Location, earlier)

However, weaknesses still persist among these improved early-phase Risk Review practices. Below is a listing of keys to optimize the Phase 1 and 2 Risk Reviews for a project.

Keys to Performing Excellent Risk Reviews in the Conceptual and Preliminary Phases of a Major Capital Project:

- **On the risk review team, include senior operator, not just supervisory personnel from an existing or very similar unit, even if the person must be contracted or must travel a great distance to attend the Risk Review.** There have been many projects where this did NOT occur at the Conceptual or Preliminary Design stages and the missed considerations led to either a flawed project (i.e., the scope of the project is limited to 3 year vision instead of 5-8 year future vision) or an inherently unsafe and/or unstable process due to range of typical operating parameters for pressure, temperature, or level; or a poorly planned pathway for foot traffic due to plot plan errors of omission, etc.). This should be a company requirement for all risk reviews (all project phases, all PHA revalidations, all MOC risk reviews). As an example, this was the SINGLE best rule for project risk management developed within Amoco Oil and Phillips following accidents in their facilities in the late 1980s.

- **On the risk review team, include senior process engineer from an existing or very similar unit, even if the person must be contracted or must travel a great distance to attend the Risk Review.** This is for the same reasons listed for the senior operator. However, these two senior staff members will see the process quite differently, due to their typically very different experiences in the same process units. This should also be a rule at your company for all risk reviews.

- **Allow contingency in budget and schedule for possible changes.** If the project managers are put in an "overly" uncomfortable position with respect to changing scope, adding or changing features, etc., then the resulting project could be an "on-time/on-schedule" new unit, but may be unsafe or run poorly long-term. Recall the comparison of front-end loaded versus sequentially loaded projects described earlier.

- **NEVER let the contractor/vendor manage the risk review or provide the risk review leader.** Typically, the contractor does not have the owner's/operator's interests in mind since their chief concern is to deliver a project on schedule, for a stated price, and meet an initial performance warranty. The owner/operator should always chose the Risk Review leader and should ensure the leader is:
  - Independent of the project management team
  - Independent of contractors/vendors
  - Independent of the Unit/Process/Plant that the major project is related to
  - Fully capable PHA team leaders/facilitators (well trained and practiced in the HAZOP, FMEA, and What-If methods) – a dedicated scribe is also necessary to help the leader, if the risk review meeting is expected to last more than 8 hours total.
- **Force the consideration of inherently safer/more reliability alternative.** For all of the reasons mentioned earlier.

There are many other factors that lead to successful risk reviews in the early project phases, but the considerations listed above are key.

Risk Control – Detailed Design Phases of a Major Capital Project

The risk reviews are a major risk control feature of the design phases of a project as well. These risk reviews can be one to three progressive efforts over one to three project phases, depending on the size of the "major capital project," with the Risk Review report building toward the "initial" official hazard review report for the process unit (discussed in the next section). The Risk Reviews during the detailed design phases can typically include:

- Using HAZOP, FMEA, and/or What-if (brainstorming methods) in progressively more detail
- Initiating and then progressively improving (from phase to phase) the risk review record (HAZOP tables, What-If tables, checklist tables)
- As mentioned in **Figure 1** and **Table 2**, the risk reviews during detailed engineering will evaluate the risk of any design modifications and/or newly identified hazardous scenarios, which have been added since the previous reviews. This includes making sure the final detailed design resolves the occupational safety, process safety, reliability, and operational risk issues identified in previous reviews
- Maximize inherently safer design in the selected process:
  - Design tanks, and piping to make overflow hydraulically impossible
  - Optimize reactor conditions and recycle loops to minimize waste
  - Design manual charge stations at ergonomic work height
  - Site hazardous operations away from zone impacted by possible explosion
  - Design tanks to withstand maximum possible pressure[5]
- Performing a final review of equipment, ventilation, containment, and environmental safeguards, including instrumentation, interlocks, fail-safe decisions, detailed layouts, and fire protection provisions
- Begin the Human Factors risk review (checklist-based)
- Continue the Facility Siting risk review initiated earlier (checklist-based and modeling-based risk reviews)
- Apply Layers of Protection Analysis (LOPA) to complex risk scenarios and use this to define Safety Instrumented Systems' needs

As before the Risk Reviews during the detailed engineering phase require intensive participation by operations' senior staff, including operators, supervisors, and process engineers.

Keys to Performing Excellent Risk Reviews in the Detailed Engineering Phases of a Major Capital Project:

- **As for Risk Reviews in the earlier project phases, continue to have your most senior operator and process engineers from the unit or similar unit on the Risk Review team. Maintain the rules mentioned earlier for the Risk Review leader/facilitator.**

- **Catch design problems before they create operational traps.** This is the goal through all of the Risk Reviews for a new process unit, but given that the inherently safest and more reliable option has most likely been chosen, the detailed design phase becomes the time when to most effectively detect and eliminate such traps. **To accomplish this in the detailed design/engineering phases, equal focus must be given to hazards and operability/quality issues.** Look for ways to keep the final process easy to operate, easy to maintain "online" and ways to avoid process upsets." Two-thirds of accidents occur during non-routine modes of operation, so by avoiding deviations from the norm, the plant's exposure to these higher risk modes of operation are reduced.

- **Write Operating Procedures Designed to Control Risks.** After the Risk Reviews are complete, the next important step in controlling risk before start up is developing the written procedures for controlling operations, controlling troubleshooting, controlling emergencies, and maintaining equipment. The data needed for writing procedures is mostly developed by the end of the **Final Design Phase** and the finalization continues into the **Commissioning Phase** (other papers address best practices for drafting, formatting, and validating procedures using subject matter experts (SMEs)[7]). In summary, operations must write procedures containing the right content (right instruction for each step, in the right sequence) and format the instructions (steps and pages) properly to lower the chances of someone making errors when following the procedures. There are about 25 rules for formatting procedures and these are readily available in the published literature. If done right, these written procedures will become the basis for performing the initial training of the new unit's operators and serve as a refresher guide for these experienced staff over the long-term. This in turn will minimize (but never eliminate) human errors of skipping steps and doing steps wrong, and is important in controlling the human error portion of process safety risk.

Risk Control – Pre-Commissioning / Commissioning (Initial Startup) Phase of a Major Capital Project

The pre-commissioning Risk Review builds upon the previous Risk Reviews in the project. As the equipment design is completed the fabrication and construction begins. During this same period, initial training of the new or transferred staff occurs, using the procedures mentioned in the previous section. The pre-commissioning Risk Review can begin just prior (4-6 weeks prior) to start-up of a new facility, or a little earlier if possible.

The key consideration for this project phase is to complete the risk review of non-routine modes of operations. The project Risk Reviews to this point will not have covered these modes of operation very well. (Note that in perfect world, the risk review of the non-routine modes of operations, which uses the operating procedures as a basis, would be completed before training begins. However, in most cases, the training begins as the procedures are being completed and as the risk review is done.) The risk review of non-routine operating modes can be performed using a full 8 guideword HAZOP, a streamline

2 Guideword approach (which is what was used before HAZOP was invented in the 1960s), or a No Guideword What-if. All of these approaches are described elsewhere and will be explained in some detail in the 3rd Edition of the Guidelines for Hazard Evaluation Procedures, CCPS, due late 2008. References are available to explain these methods (Bridges[8]). This procedural analysis is to ensure that hazards due to human error in association with the process design have been identified and analyzed.

During this final risk review before start up, the project team must also ensure that all the PSM requirements for initial PHAs have been met. If all preceding Risk Reviews have followed the guidelines discussed in this paper for controlling risk at each project phase, meeting the "initial" PHA requirements will not entail a tremendous amount of effort other than the risk review of non-routine modes of operation. PHAs must address the hazards of the process; therefore hazards during all modes of operation must be analyzed. The resulting report will be the "initial PHA" of the process unit, which is required to meet PSM standards.

This pre-commissioning risk review should not be confused with the pre-startup safety review (PSSR), which is also necessary but the purpose of the PSSR is to validate that the process design and specifications have been met.

This final Risk Review session before startup consists of:

- Reviewing and evaluating changes made during construction, ensuring that no new hazards have been added since the last hazard review. High priority is given to detecting details which may have been overlooked, and to concentrating on the adequacies of plans to cope with operating emergencies that might arise
- Maximizing inherently safer design in the selected process, such as planning for rework of initial product[5]
- Completing reviews for Facility Siting and access issues
- Completing the review for Human Factors issues
- **Reviewing (HAZOP/What-If) of start-up, shutdown, emergency shutdown, and online maintenance procedures.** Some believe this risk review to be another "validation" review of procedures, to ensure they are correct. But that is not the purpose of this risk review of procedures. Our aim in the HAZOP/What-If of the procedures is to ensure we have adequate safeguards (hardware, interlocks, SIL, and/or independent administrative safeguards) to offset the errors of skipping steps and performing steps wrong – such human errors WILL occur, it is just a matter of when.[8]

The lack of risk review of non-routine mode of operations, such as by HAZOP or What-if of procedural steps, is the most frequently observed weakness in the project risk review cycle.

After this risk review, the project team can proceed to close recommendations, decide which (if any) of the recommendations can be deferred until after initial startup, close the PSSR (not part of the risk review, but part of PSM in general), and finalize the initial PHA report for the new process unit. Typically, the plant MOC system begins to take over control of new risks after the pre-commissioning Risk Review meeting is closed.

There are of course many deliverables from the project team, including the finished equipment, ready to commission and then smoothly commissioned, operating and maintenance procedures, populated databases for mechanical integrity (MI), Process Safety Information - files of all necessary design bases for relief valves, completed drawings, complete equipment files of all types, etc.

Risk Control – Post-Commissioning Phase of a Major Capital Project

Many companies require a final risk review related to a new process unit about three to 6 months after startup. This was implemented when managers realized that the MOC system (which starts at the close of the pre-commissioning RISK REVIEW meeting) typically becomes overwhelmed by the magnitude of changes necessary in the final rush to startup a new unit. To avoid missing a key risk factor for very long, it is wise to schedule a post-startup Risk Review. This can also serve as the first PHA Revalidation for the new process unit. The Post-Commissioning PHA is conducted the same as a PHA Revalidation.[9]

Factors That Determine the Number of Risk Reviews for a Major Project

This paper has dealt with Major capital projects (12 months length or large). However, projects come in all scopes and sizes and there are varying scopes within the project categories. As mentioned earlier, the number of risk reviews for a major capital project varies by the combination of project size, process complexity, and process risk/hazard.

Example 1 shows that for a **large process** (more than $100 million USD of installed cost and two or more years of project time) that is **complex** (high pressure system, high temperatures, complex control) and that **handles hazardous chemicals**, about 5 Risk Reviews are needed of varying scope before the plan is fully commissioned. The table shows actual expenditures of Risk Review team labor expended for each review. This expenditure includes the time to document the Risk Reviews.

Example 1: New Ethylene Plant

- Size: 250 nodes of equipment (nodes are vessels, columns, fired heaters, lines/exchanger circuits, etc.) plus the typical number of operating and maintenance procedures

HAZARD REVIEW ESTIMATES

	Conceptual	Preliminary	Detailed	Construction	Pre-Startup	Total
Meeting Time	1 week	3 weeks	5 weeks	3 weeks	3 weeks	15 wks
Team Size	8	8	8	6	6	
Staff Hours Total	164	1152	2120	864	984	5284
Cumulative Schedule	2 months	6 months	14 months	18 months	22 months	24-30 mon
Focus	Plot plans, process options, tie-in options (WI and some modeling)	Cursory HAZOP, FMEA, and WI	Rigorous HAZOP, FMEA, and WI; start checklists	Changes since previous; finish checklists	SOP Hazard Review & Changes since previous	

Example 2 shows that for a **small process** (much less than $100 million USD of installed cost and 1 year of project time) that is **relative simple** (conveying of iron ore from a port to a steel plant) and that **handles low hazard material** (i.e., rock/ore), only two Risk Reviews are needed, each low intensity efforts. **Example 2** shows actual expenditures of Risk Review team labor expended for each Risk Review. This expenditure includes the time to document the Risk Review.

Example 2: New Ore Conveying

- Size: 10 nodes of equipment (nodes are conveyors, screen, crushing, storage, dedusting, etc.) plus the typical number of operating and maintenance procedures

HAZARD REVIEW ESTIMATES

	Detailed	Pre-Started	Total
Meeting Time	2 days	2 days	4 days
Team Size	10	10	
Staff Hours Total	218	202	419
Cumulative Schedule	3 months	9 months	10-11 mon
Focus	Rigorous WI, some FMEA; start checklists	SOP Hazard Review & Changes since previous	

Conclusions

For new process units, only one opportunity may exist to adequately assess and manage risk, and that is during the project phases before start up. Sound industry practices have existed for some time pertaining to managing the risk of the final process operation by risk management efforts during a major capital project. Many have learned these lessons, but many have not. The renewed focus on designing inherently safer and more reliable processes, and renewed focus on controlling risk during non-routine modes of operation may not have migrated to the project control system of some companies. There are many key points related to controlling risk during the design and startup of a new process unit, but the most important appear to be:

- Effective company leadership and effective project management, where the primary goal is an efficient operating unit (long-term process reliability and safety) – where this primary goal far exceeds the secondary goals of bringing in a project on schedule and on budget.
- Including experienced operators and experienced process engineers on the risk review teams, from the very start of the project.
- Develop and apply PSM elements at each project phase to better design the project for process safety.
- Having subject matter experts develop procedures (operating, maintenance, and lab procedures) and having the risk review of each of these procedures to ensure there are sufficient safeguards to protect the new process when the imperfect humans have to follow the procedures perfectly.

One "unwritten" deliverable from a project team (including the risk review teams) is a process and related procedures that will sustainably control human error during all modes of operation. The project team cannot ensure these safeguards will stay in place long-term, but if human error is predicted and controlled at each phase of a project, by following the steps outlined here and elsewhere, the control of human error, and therefore the control of accidental losses, will at least have a good start.

References:

1. P. F. McGrath, Olin Corporation, *"Using Qualitative Methods to Manage Risk,"* CCPS/AIChE, 1989.
2. *"Management of Change as Applied to Capital Projects,"* API PSM Implementation Workshop, 1992.
3. *"Risk Management on a Major Facility Expansion,"* Syncrude Canada Ltd., CCPS/AIChE, 2001.
4. *"Plant Guidelines for Technical Management of Chemical Process Safety,"* Chapter 5, CCPS/AIChE, 1992.
5. S, Berger, et al., Rohm and Hass Company, *"A Cost and Time-Efficient Framework for Inherent Safety and Pollution Prevention During Process Development and Engineering,"* CCPS/AICHE, 1996.

6. *Trevor A. Kletz, "Inherently Safer Design Achievements and Prospects," CCPS/AICjE, 1996.*

7. W. Bridges, et al, *"Create Effective Procedures and Manual,"* Chemical Engineering Progress, 1997.

8. W. Bridges, et al, *"Addressing Human Errors in Process Hazard Analyses,"* Chemical Engineering Progress, 1994.

9. "Revalidating Process Hazard Analyses," CCPS/AIChE, 2001.

Case Study of a Facility Siting Analysis carried out for a FSRU

Anish Ravi

Consultant, Scandpower Risk Management Inc.
Houston, TX
ara@scandpower.com

Joar Dalheim

Manager, Scandpower AS
Kjeller, Norway
jda@scandpower.com

ABSTRACT

Consequence models have been applied to facility siting studies only in the recent past .Traditionally more simplified methods such as TNT equivalency methods have been used which have their own limitations and thus cannot be consistently applied to all types of processing facilities. This paper discusses a case where computational fluid dynamics has been used in conjunction with the more traditional Multi-Energy method to give realistic and credible blast loads for a LNG processing vessel.

1. INTRODUCTION

The study intends to propose design loads/relative safe distances for occupied buildings or regions where presence of personnel is inevitable on a FSRU. FSRU is a converted LNG carrier imbibing advantages of both onshore based LNG receiving terminals and offshore based FPSO technologies [1]. A schematic of the vessel in consideration is shown in Figure 1.1. The schematic also shows the placement of regasification unit on the vessel which for this study is the primary source for vapor cloud explosion.

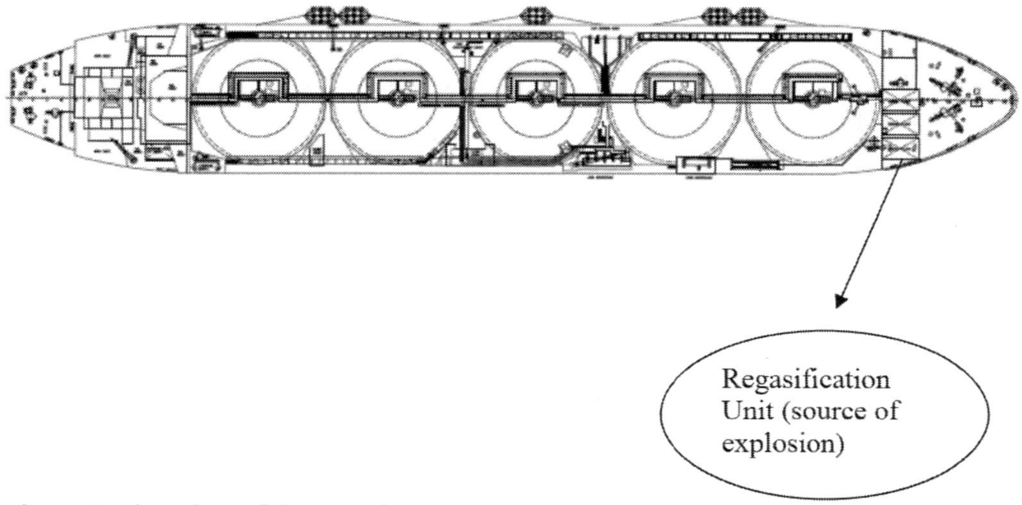

Figure 1: Plan view of the vessel

2. METHODOLOGY

The study was carried out in two stages: (a) FLACS simulations [2] to determine the source blast strength from VCE and (b) Multi Energy Method [3] uses the source blast strength obtained from FLACS simulations (maximum overpressure from all scenarios evaluated) to give a plot of decay of blast loads with distance from the center of explosion (in this case the regasification unit).

3. GENERAL LINES ON MULTI-ENERGY METHOD and FLACS

The ME model was developed by van den Berg in 1980's. The method is based on numerical simulation of a blast wave from a centrally ignited spherical cloud with constant velocity flames. The model can be used to estimate maximum static and dynamic pressures, duration of positive pressure phase and overpressure/time profile of explosion. Unfortunately it cannot be used as a stand-alone tool because both the source strength and charge (amount of energy) participating in the blast wave are unknowns and have to be estimated using other tools and that is where a CFD code like FLACS (abbreviated for Flame Acceleration Simulator) comes in handy. FLACS is a 3-D computational fluid dynamics code based on fundamental principles of physics i.e.

conservation of mass, momentum, angular momentum and energy. It uses the so called k-epsilon method to estimate turbulent quantities. A system of equations is generated which is solved by dividing the entire geometry/computational domain into small volumes called control volumes. Fundamental principles of physics are checked for each of these cells. The cells are delimited by lines which in CFD terms are referred to as grids (Figure).

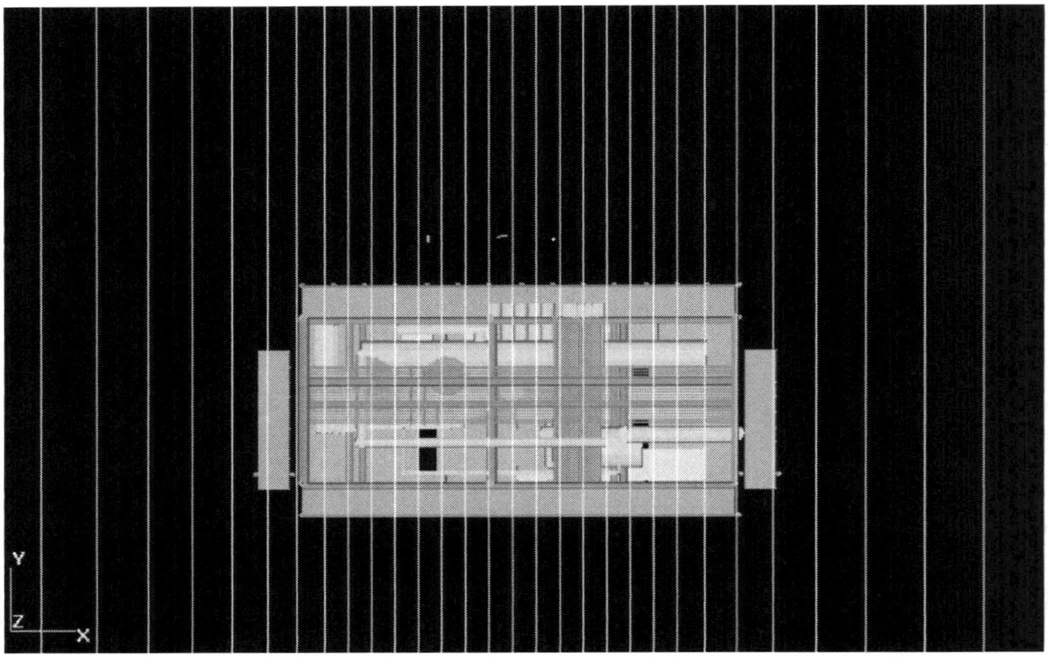

Figure 3.1: Example of a grid imposed on the module in FLACS

4. GENERAL LINES ON PHSYICS OF EXPLOSIONS

When a leak/blow out occurs a certain quantity of inflammable gas is released into the atmosphere. This gas in combination with air and in presence of ignition source can result in fires/explosions depending on concentration of flammable gas and strength of ignition source [4].

Inherently safe designed topside structures should be able to withstand credible explosion pressures [5]. A lack of inconsistent/ incomplete CAD database can highly under estimate explosion pressures especially during early phases of the design. The design of topside structures/FSRU is indeed a problem especially when blast loads are required for design but at the same time proper design needs to be defined for explosion pressure analysis. A sensitivity study performed by GexCon (developers of FLACS) revealed the fact that explosion pressures can grow by a factor of 30 from conceptual design stage to the final construction phase. A major contributor to the rise in pressure was found to be 2" piping.

5. DETAILS OF SIMULATION PERFORMED IN FLACS

5.1 **Geometry and Meshing**

CAD models of the regasification unit were imported into FLACS.

Dimensions of one regasification Unit: 10.5 x 6.1 x 8.0 m.

Figure 5.1: Regasification Unit as viewed in Microstation

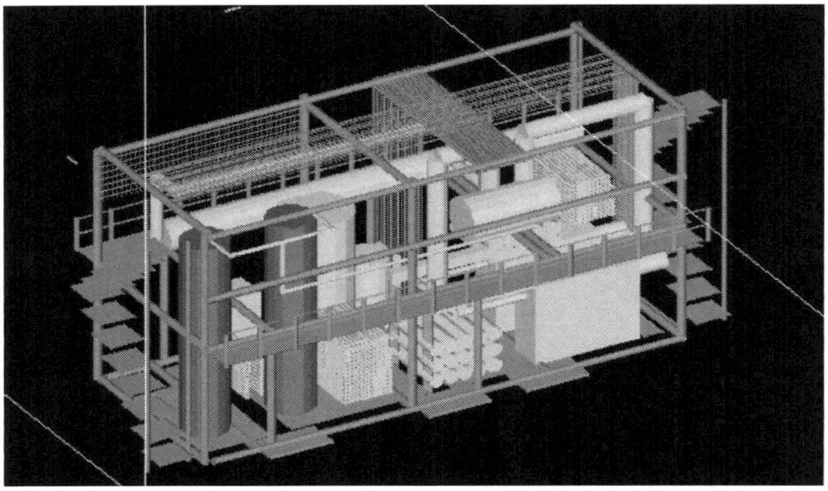

Figure 5.2: Regasification Unit modeled in FLACS

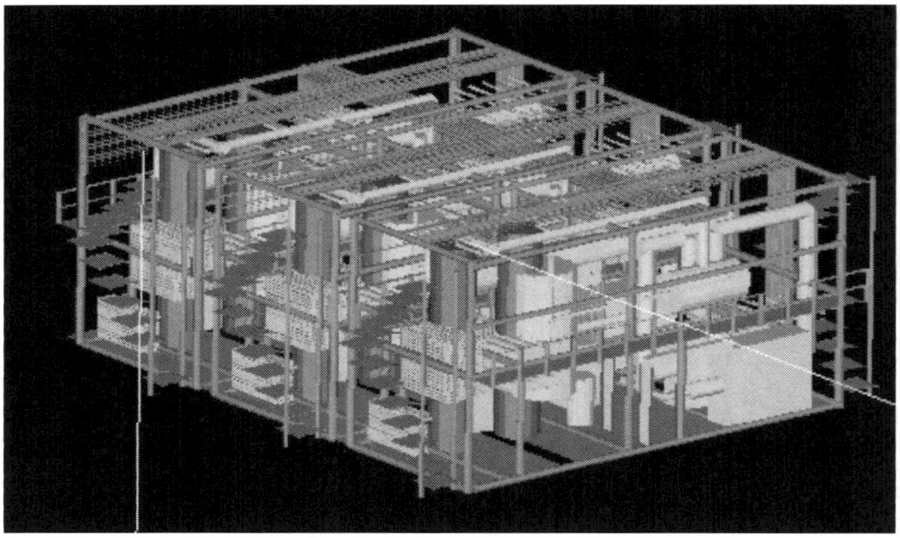

Figure 5.3: 3-D View of all three Regasification Units

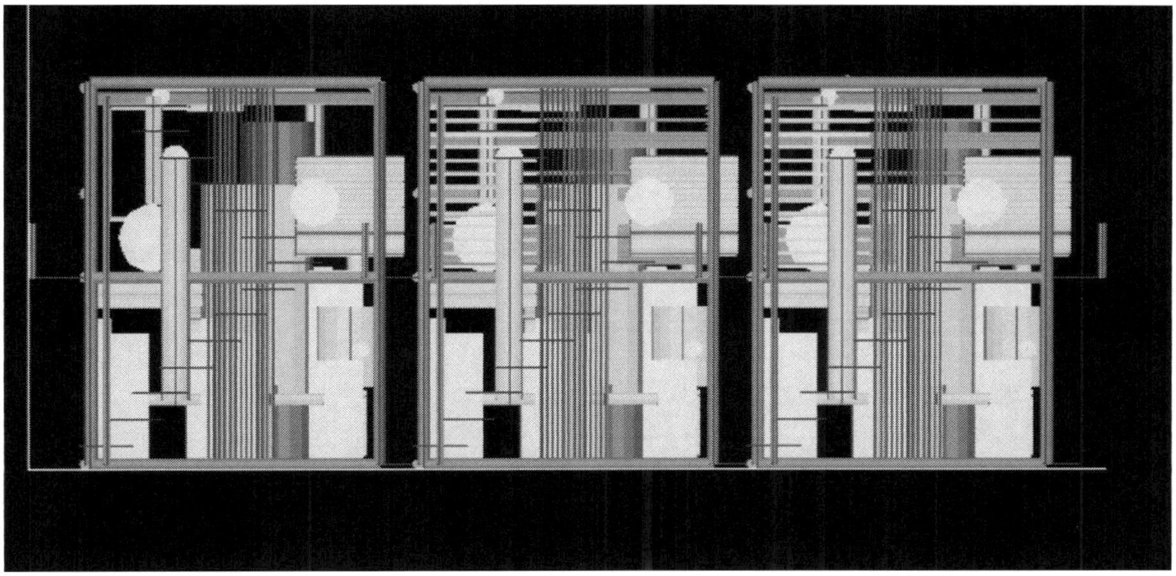

Figure 5.4: Side View of all three Regasification Units

Also the model was updated to include congestion in the form of piping, equipment and valves according to engineering judgment and estimates received from Hamworthy as shown in Table 5.1

Table 5.1: Estimate of Model completeness provided by Hamworthy (manufacturers of regasification systems for this project)

Piping class/equipment type	Model completeness in %,weight and length
Steelwork	40 %
Instruments	1,000 kg
Ladders & Platforms	2,000 kg
Valves	5,000 kg
Small piping (< 2")	20 meter
Piping (8" sch 10s)	10 meter
Piping (8" sch 160)	15 meter

As discussed in Section 3 grids is an integral part of the code. For this study a grid of 0.75 meter is used throughout. Grid sensitivity tests with finer grids did not reveal any great deviations.

5.2 **Gas Composition of the gas leaks**

All explosion simulations were performed with the gas composition as shown in Table 5.2 ..

As is evident from the table the mixture is predominantly methane.

Also a couple of simulations were performed with 100 % propane to visualize the deviations in blast loads if any

Table 5.2: Gas Composition of Components used for Explosion Simulations

Component	Mole fractions
Methane	0.8685
Ethane	0.085
Propane	0.03
Butane	0.0122
Pentane	0.0023
Nitrogen	0.002

6. SCENARIOS EVALUATED

More than 200 different scenarios were simulated in FLACS in order to determine the magnitude of dimensioning blast load in the Regasification Unit. The cloud size, cloud location and ignition location were varied. The Table 6.1 below depicts the gas cloud size, gas cloud coordinates, ignition location for just the dimensioning scenario (scenario that resulted in maximum overpressure).

Table 6.1 : FLACS Scenario Description

Simulation # (in FLACS)	Cloud coordinates (m)			Cloud size (m^3)	Ignition location (m)			Orientation of regasification unit (m)
	X	Y	Z		X	Y	Z	
060129	15.6	-7.8	-3.9	1573	20.4	1.75	-2.9	11.75, -9.75, -3.92

7. RESULTS

The results outlined in the following sections are given in accordance with the methodology explained in Section 2. To summarize, three steps form the core of the results:

Step 1: Congest the regasification unit on engineering judgment and estimates provided.

Step 2: Perform explosion simulations on the "Fully Congested Model" using FLACS.

Step 3: The source strength or maximum blast pressure obtained from the Step 2 forms the basis of Multi-Energy Method.

8. CONGESTION MEHODOLOGY

For a reasonable explosion pressure analysis, the geometry model needs to be complete and accurate. The design loads extracted from the analysis need to be conservative enough to allow future growth of design pressures without compromising the safety of the structures. Thus, the fact that the regasification unit being still in design phase; it becomes imperative to add piping and other secondary structures like valves, equipment etc. according to experience gained from previous projects. Table 8.3 and Figure 8.3 show the cylinder/pipe length varied through different levels of congestion in a regasification unit. Figure 8.4 shows the variation of overpressure with cylinder length during various stages of congestion for a regasification unit.

Table 8.3:Cylinder /Pipe Congestion at various Levels of Model Development

Congestion level	Diameter range of the piping (inch)	Cylinder/pipe length (meters)	Cylinder count	Maximum overpressure (barg)	Congestion parameter = total length/volume of the module
Original model & Hamworthy estimates			153	0.01	0.2
	0-2.5"	91			
	2.5-5"	116			
	5-9"	36			
	9-15"	10			
	15-31"	15			
	31"->	2			
Congestion one level up			547	0.015	0.62
	0-2.5"	636			
	2.5-5"	116			
	5-9"	36			
	9-15"	10			
	15-31"	15			
	31"->	2			
Fully congested model			607	0.0251	1.47
	0-2.5"	642			
	2.5-5"	933			
	5-9"	232			
	9-15"	25			
	15-31"	70			
	31"->	33			

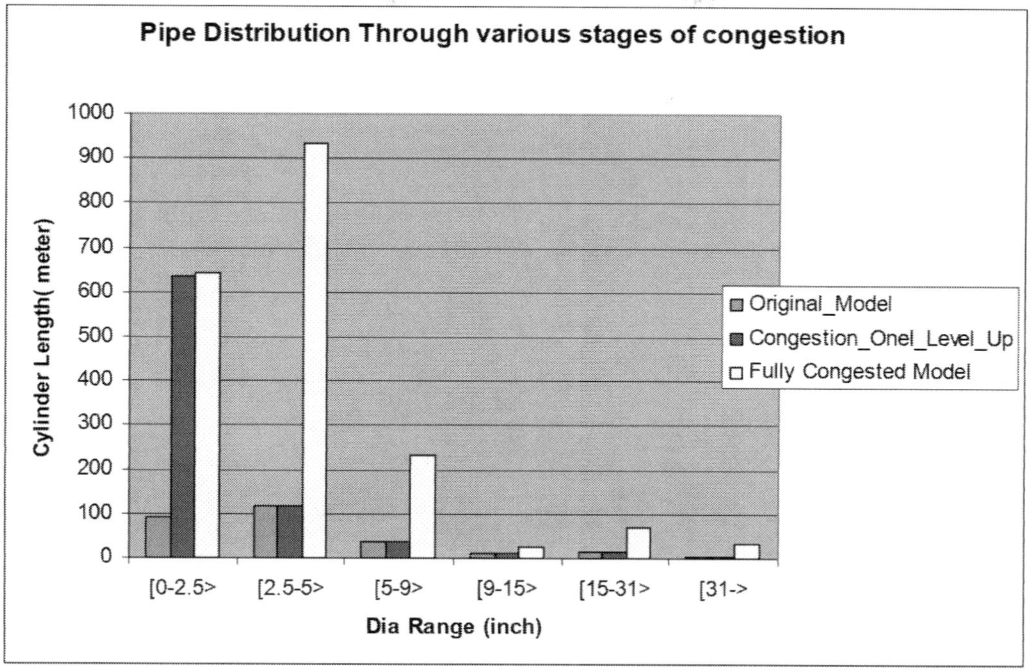

Figure 8.3: Graphical Representation of Pipe Congestion

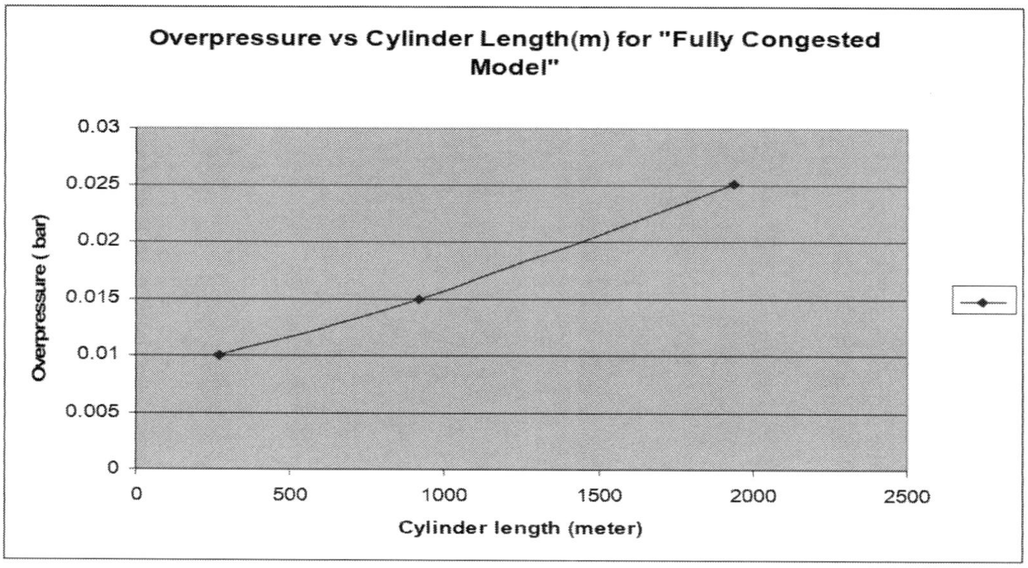

Figure 8.4: Graphical Representation of Overpressure Variation with Cylinder/Pipe Length for a single regasification unit.

Figure 8.3 shows the increase in overpressure as the number of regasification modules increases for a 50 % gas cloud cover from one to three.

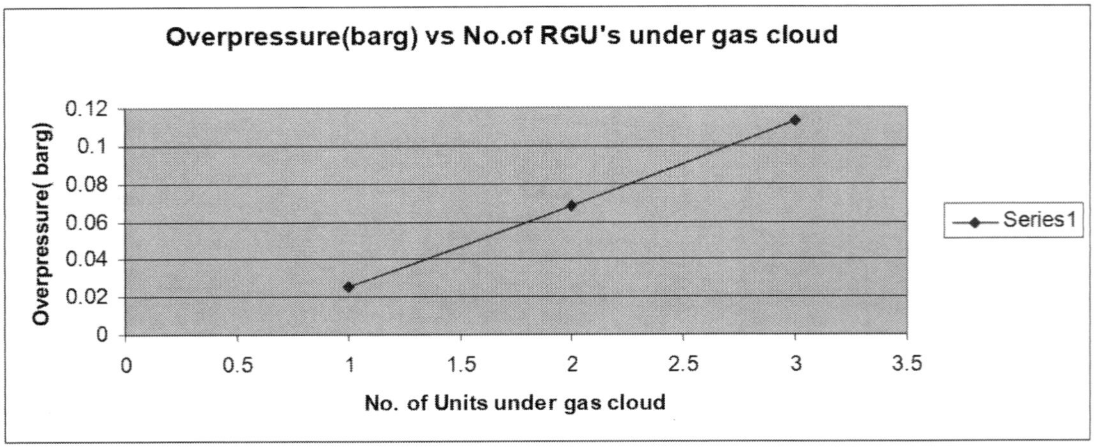

Figure 8.3: Variation of Overpressure as the Number of Regasification Modules increases from one to three.

9. SOURCE BLAST STRENGTH RESULTING FROM AN LNG LEAK INSIDE THE REGASIFICATION MODULE

Scenario 500210 is the dimensioning scenario that yields the maximum overpressure, where LNG is used as the fuel gas.

Table 9.1:Overpressure Results from FLACS

Scenario # (in FLACS)	Cloud size (m^3)	Percent (%) filling the module	Maximum overpressure (barg)
500210	1573	50%	0.113

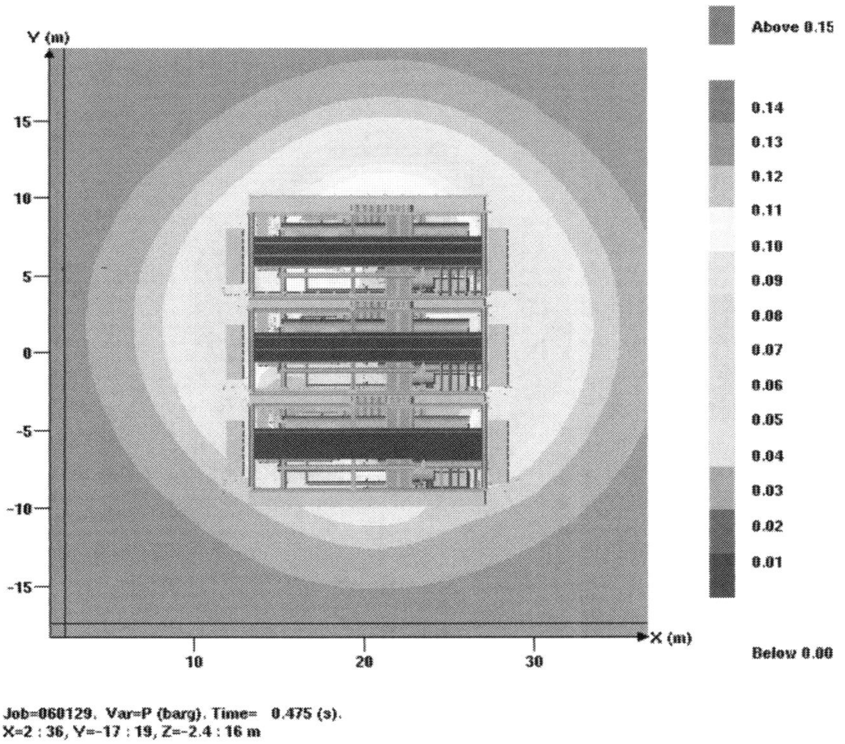

Figure 9.1 : Plan view of pressure contours developed around regasification module at a certain time step of 0.475 seconds. This necessarily is not the time at which the maximum pressure is observed. The legend on the right of the figure shows colors ranging from red representing maximum overpressure to white denoting least pressure

10. SOURCE BLAST STRENGTH RESULTING FROM A PROPANE LEAK INSIDE THE REGASIFCATION MODULE

The above scenario when simulated with 100 % propane in the fuel yielded overpressures in the order of magnitude 0.1965 barg. This source strength is used as input to the Multi Energy Method described in Section 11 below where it forms the basis for design load calculations.

Table 10.1: Overpressure Results from FLACS

Simulation # (in FLACS)	Cloud size (m^3)	Percent (%) filling the module	Maximum overpressure (barg)
060129	1573	50%	0.1965

Therefore, there is a clear increase of 74 % in blast loads when there is a propane leak in the regasification units.

11. MULTI-ENERGY METHOD

The Multi-Energy Method describes the theoretical decay of energy in a perfect blast wave only when the source strength and total energy of combustion is known. And these variables can easily be identified using FLACS simulations as illustrated in Sections 9 and 10. Using the determining overpressure value (0.1965 barg) as described in Section10, design loads can be predicted for buildings/equipments located at varying distances from the regasification unit. Table 11.1 and Figure show the decay of blast wave radially from the center of explosion. In other words, Table gives an indication of safe location/design loads of buildings/equipments if the regasification unit did indeed become a source of explosion.

Table 11.1: Design Loads as a Function of Distance from the Center of Explosion (Regasification Unit)

Distance (m)	Side on pressure (barg)	Duration (ms)
2.22	0.1965	110.2
4.43	0.1965	110.2
6.65	0.1965	99.1
8.86	0.1965	88.1
11.08	0.1941	77.1
13.30	0.1828	77.1
15.51	0.1678	77.1
17.73	0.1489	77.1
19.95	0.1325	77.1
22.16	0.1191	77.1
24.38	0.1083	77.1
26.59	0.0993	77.1
28.81	0.0916	77.1
31.03	0.0851	77.1
33.24	0.0794	77.1
35.46	0.0744	77.1
37.68	0.0701	77.1
39.89	0.0661	77.1
42.11	0.0627	77.1
44.32	0.0595	77.1
46.54	0.0567	77.1
48.76	0.0541	77.1
50.97	0.0518	77.1
53.19	0.0496	77.1
55.41	0.0476	77.1
57.62	0.0459	77.1
59.84	0.0441	77.1
62.05	0.0425	77.1
64.27	0.0410	77.1
66.49	0.0398	77.1
68.70	0.0384	77.1

70.92	0.0372	77.1
73.14	0.0361	77.1
75.35	0.0350	77.1
77.57	0.0341	77.1
79.78	0.0331	77.1
82.00	0.0322	77.1
84.22	0.0314	77.1
86.43	0.0305	77.1
88.65	0.0298	77.1
90.86	0.0290	77.1
93.08	0.0283	77.1
95.30	0.0278	77.1
97.51	0.0271	77.1
99.73	0.0265	77.1
101.95	0.0259	77.1

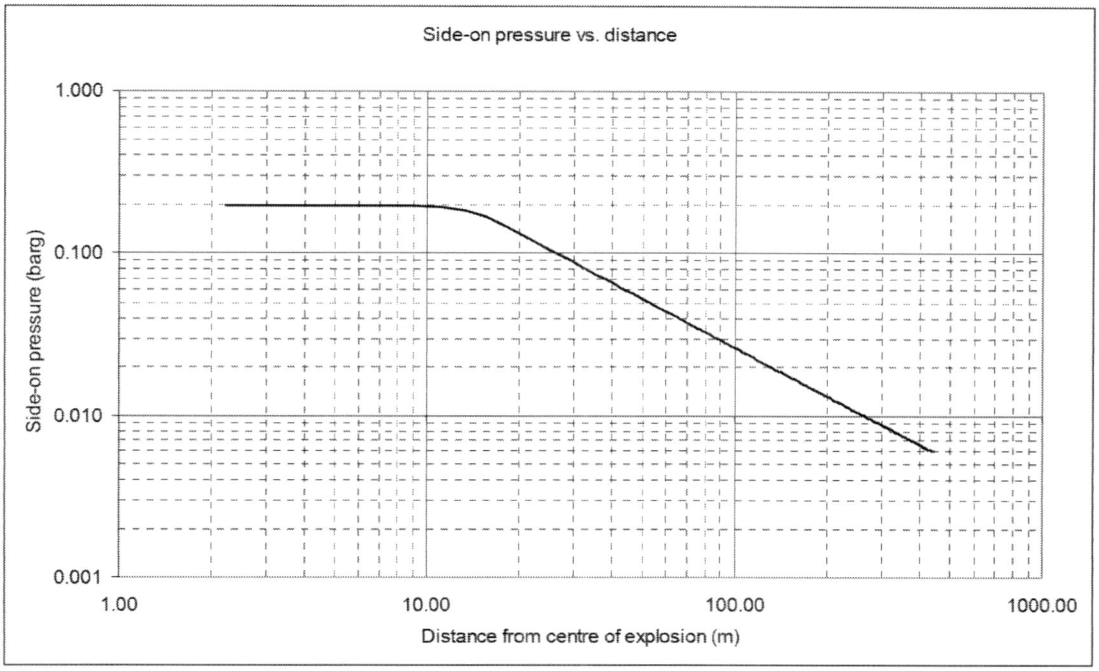

Figure 11.1: Blast Loads at varying Distances from the Center of Explosion

The above study when used in conjunction with failure frequency analysis of buildings can yield design loads with respect to distances from the center of explosion. For instance if the design load of a building is 0.20 barg, it should be placed no closer than 2.22 meter from the center of explosion (in this case the source of explosion being the regasification unit). The duration corresponding to an overpressure of 0.20 barg is 110.2 ms.

12. ASSUMPTIONS/SCOPE FOR FUTURE WORK

Congestion in the regasification unit has been added based on engineering judgment. However, blast loads generated will greatly depend on the orientation/placement of the added piping/equipment in the module which in this case has been randomly placed. Hence, the overpressure values mentioned in this paper are highly dependent on the further course of the design of the module, making it highly imperative for an explosion pressure analysis when the design of the module is more complete in nature.

13. ACKNOWLEDGEMENTS

We appreciate the consent given to us by our esteemed clients from Moss Maritime to publish the contents of the paper .Also the paper would have been an impossible task without the contributions and support from senior consultants in Scandpower especially Joar Dalheim(Manager, Norway) ,Jorgen Burtnes(Principal Consultant, Houston) and Hans Jorgen Nordstrand (President , Houston)

14. REFERENCES

[1] Hans Y.S. Han ,Jung Han Lee, Yong Soo Kim, Design Development of FSRU from LNG Carrier and FSRU Construction Experiences, OTC , Houston, Texas, 6-9 May 2002.

[2["Gas Explosion Handbook" (www.gexcon.com)

[3] van den Berg: "The Multi-Energy Method-A Framework for Vapor Cloud Explosion Blast Prediction", Journal of Hazardous Materials, 12.1985.

[4] Brigadeau , Alexander , Scandpower AS: "Details on Dispersion Simulations", October 2006.

[5] OTO Report: "Explosion Pressures Evaluation in Early Project Phase", (Health & Safety Executive) 1999 048.

Process Plant, Pre-commissioning, Commissioning and Start-Up

Safdar Rashid Toor
P.O.BOX # 30738
Street # 4
Doha, Qatar.
Contact # 00974 431 6682, 00974 540 2450
Email:safdarrashid@hotmail.com

ABSTRACT

Process Plant Commissioning is the first job of many start-ups and it is potentially the most hazardous as we are to prove the ability of the plant to operate. There are always unknown factors and more care and the attention are required into the first startup than into subsequent shutdowns and startups. Experience and expertise gained during the first start-up is used again by plant operators and maintenance personnel for future plant shutdown and startups.
In this paper, the emphasis is placed on the key activities involved in pre-commissioning, commissioning, startups and effective management of the startup. Furthermore, some benefits of effective management and disadvantages of poorly managed start ups are elaborated.

INTRODUCTION

Effectively managed startup will bring in significant revenue while poorly managed startup will reduce the Return on Investment for a new plant. Delay in startup mean delays in production of the finished product stream that would bring in revenue and when such period is extended due to difficult startups, it adds to the cost of delivering the project, which could affect the cash flow as was predicted in the economic viability.
Broader strategic impact is the inability to supply customers and markets as promised, creating opportunities for competitors to establish themselves. Experience from start-ups suggests that experienced organizations with well-planned startups can startup a plant and do the job in a reasonably short time while poorly managed startups can take an unreasonable amount of time for the startup.

Main causes of delays are[1]:

- 75% due to equipment failure
- 20% from inadequate equipment
- 5% from process failure

Here are three foundation stones of successful startups

- Planning, planning and more planning;
- Technical competence;
- Good leadership and direction.

Process plant startup is different in every situation as every process plant has different plant startup procedures. Some things are however commonly applicable and remain the same for all the plants. That is, safety, environment, etc. as startup requirements.

The broad objectives of the process plant startup must meet three fundamental criterions:

- The project must be completed on time;
- The project must be accomplished within the budgeted cost;
- The project must meet the prescribed quality requirements;

Safety

This criterion can be represented graphically as below, by well-known project triangle.
Fig # 1

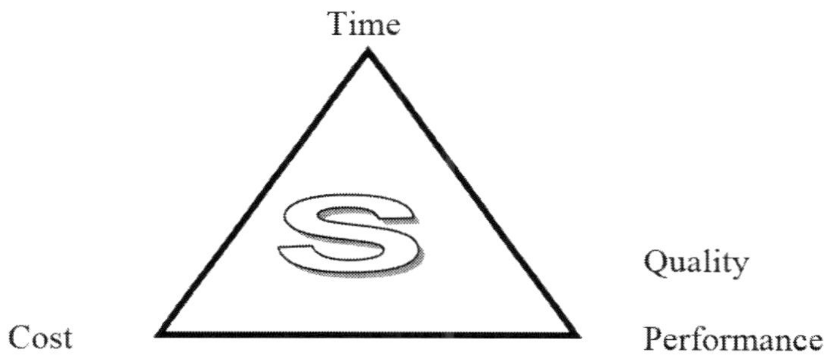

Startup of the process plant depends on the cost, time, quality and safety. The project may be late going into service due to the quality requirements in terms of the comfort that may have to be sacrificed, but under no circumstances can safety be compromised.

Fig #2

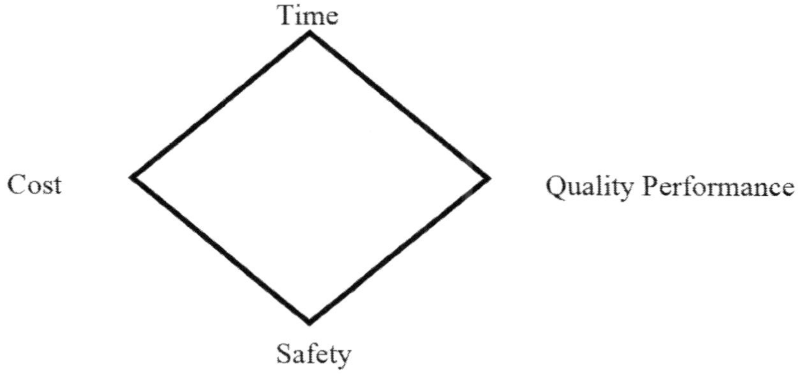

Time

To complete the project in time is like playing a tennis match. If the match is started late and you want to finish on time to avoid the extra expenditure, you must try to finish early then the quality of the match will definitely not be up to that level which the spectators are expecting (quality will be not there).

We can estimate the startup time by following equation:

$$Time = a(0.15 + b + c + d + Ne) \qquad (1)$$

Where

a = Construction time
b = Process Factor (0.15 for radically new process, 0.05 for relatively new and -0.01 for familiar process)
c= Equipment factor (0.15 for radically new, 0.08 for vey new, 0.05 for relatively new and -0.01 for familiar equipment)
D = Labour factor (0.15 for labour in very short supply, 0.05 for labour somewhat scarce and -0.01 for surplus labour).
N= Number of process units involved.
E= Dependency factor (0.25 for interdependent process units, 0.10 for moderately dependent and -0.02 for independent plants.

Its mean to give a quality project, we should not compromise with money and time.

Cost

To complete a housing complex in a local community may exceed the budget due to financing by government, the fixed budget may be frozen due to quality restraints or increased prices, hence forth one may be required to compromise the quality of housing (by using low quality products). The estimation of cost for any process plant startup can be estimated from the following equation:

$$Cost = A(0.10 + B + C + D + NE) \qquad (2)$$

Where

A= Direct fixed capital cost of project.
B= Process factor (0.05 for radically new process, 0.02 for relatively new and -0.02 for familiar process).
C= Equipment factor (0.07 for radically new, 0.03 for relatively new and -0.03 for familiar equipment).
D= Labour factor (0.04 for labour in very short supply, 0.02 for labour somewhat scarce and -0.01 for surplus labour).
N= number of process units involved.
E= Dependency factor (0.04 for interdependent process units, 0.02 for moderately dependent and -0.02 for independent plants.

Quality (Performance)

Quality and performance can be compared to the manufacturing of arms. Like missiles and fighter jets, we cannot compromise the quality and prices but during war we can compromise the specification to achieve on timely delivery in the field. Similarly, in projects, specifications can be relaxed to get the equipment on time for the startup of process plant with out any adverse effect on quality.

Safety

As manufacturing of airplanes, buses or trains, the safety of the passengers is assured and no compromise on safety of airplane, bus or train can be made. Similarly, a serious accident which may kill or injure people will not only cause anguish among the relatives but, while not necessarily terminating the project, could very well destroy the company. For this reason the 'S' symbol when shown in the middle of the project management triangle gives more emphasis of its importance (fig #1).
During the pre-commissioning and commissioning of any process unit, plant startup has always hidden hazards. These hazards are sometime very dangerous and many workers may die during this phase of plant due to unsafe job activities or improper safety work procedures. This is like when flushing is going on in a pipe rack and simultaneously welding is going on. The resulting incident is electrocution of the worker.
Here in this paper, we will discuss these safety hazards during pre-commissioning, commissioning and startup of process plants.

Definitions

When discussing safety, several words are constantly used, often in the wrong context because they are wrongly assumed to mean the same thing. To eliminate any confusion, the following definitions are given:

Accident: An unplanned, unintentional and unexpected event.
Danger: Exposure to harm or loss.
Hazard: Having the potential to cause harm or loss.
Risk: The possibility of meeting danger or suffering harm or loss.
Safety: To be safe, freedom from risk and danger.
What is Safety: Safety means being aware of the hazards and alert to the risk and danger present in a given place or situation at any point in time and ability to conform to the requirements of the safety function.

Accident Cause Chart

An accident is an unwanted occurrence by unsafe acts or conditions which is responsible for delaying the plant startup.

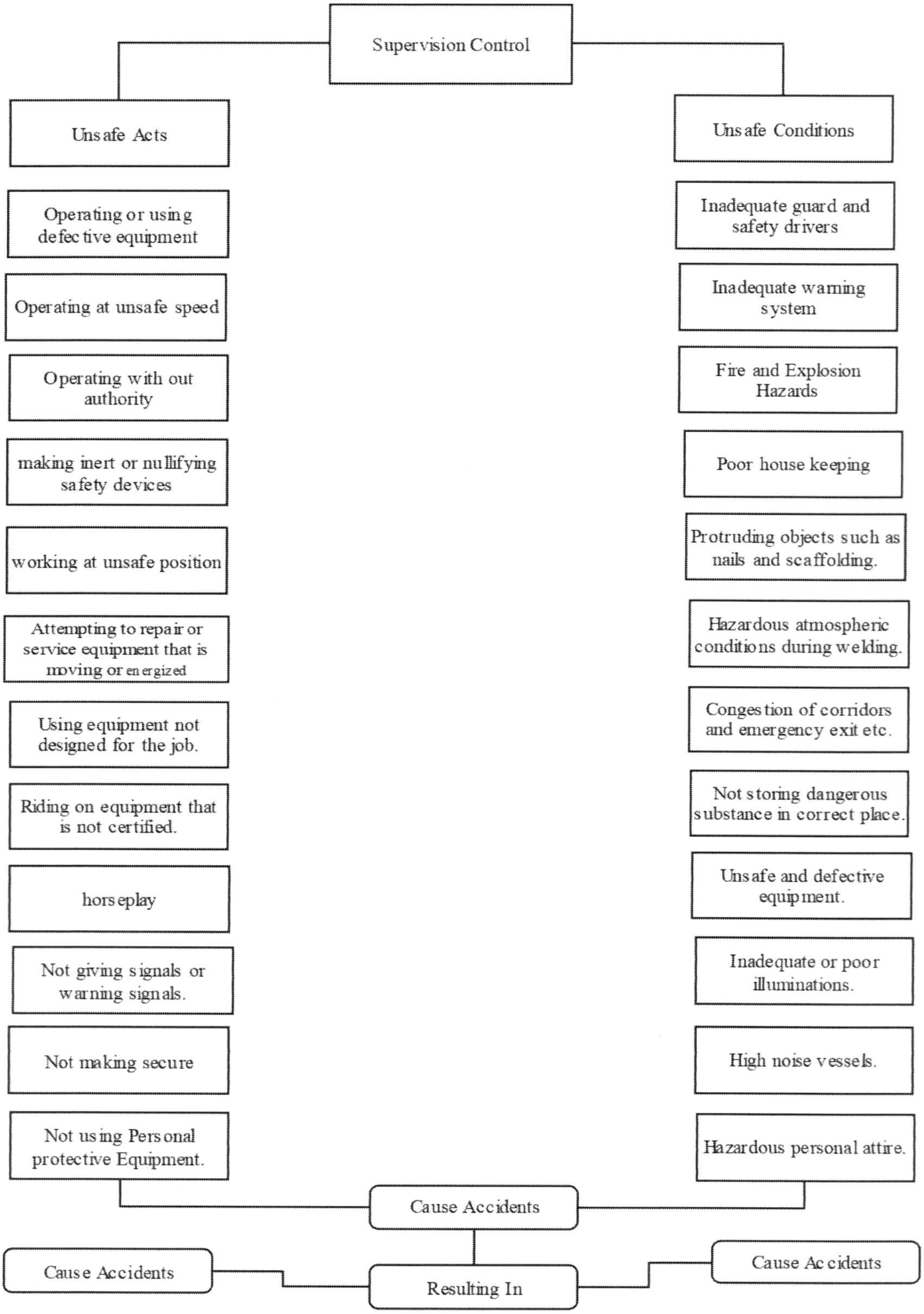
Supervision Control
Unsafe Acts
Operating or using defective equipment
Operating at unsafe speed
Operating with out authority
making inert or nullifying safety devices
working at unsafe position
Attempting to repair or service equipment that is moving or energized
Using equipment not designed for the job.
Riding on equipment that is not certified.
horseplay
Not giving signals or warning signals.
Not making secure
Not using Personal protective Equipment.
Unsafe Conditions
Inadequate guard and safety drivers
Inadequate warning system
Fire and Explosion Hazards
Poor house keeping
Protruding objects such as nails and scaffolding.
Hazardous atmospheric conditions during welding.
Congestion of corridors and emergency exit etc.
Not storing dangerous substance in correct place.
Unsafe and defective equipment.
Inadequate or poor illuminations.
High noise vessels.
Hazardous personal attire.
Cause Accidents
Cause Accidents
Resulting In
Cause Accidents

Before Start-Up, The Following Must Be Completed:

JSA (Job safety analysis)

During the project execution stage a job safety analysis must be carried out and all necessary precautions should be taken to assure the safety of personnel as well as equipment during their transportation. If there is causality or equipment damage then central theme of safety will go resulting in a conflict and tension amongst the effected party relatives and social society and this may delay in plant startup.

Training On and off the job

Training of personnel for the successful startup is very important to ensure that the plant is started up without accidents and incidents. If staff is not well trained then during the startup they can damage the equipment and these factors will also result in delay in process plant startup. This was experienced in the startup of **Ibn-Rushd** (AIFC in Saudi Arabia in 1997-2000) aromatic plant, where the company hired the staff one year before the startup of the plant. They gave the training to staff for almost one year and during the startup time they did not face any problems with the staff or in the startup.

Safety Issues (Commissioning/Startup/Shutdown Issues)

During the pre-commissioning, startup and shutdown time, safety is the core issue of any successful startup or shutdown of the plant. If the procedures are well written and staff is well trained then there is no problem but if staff uses short cuts then such can lead to the case example of **Piper alpha** with very serious consequences. There are many such examples available in addition to this famous one.

The roles and responsibilities of the Commissioning Team

The roles and responsibilities of the commissioning team should be very clear and every one must know about his or her limits. The team leader must be very active and polite, he or she must have a sense of humor and during any emergency must command like a captain. Indeed each successful startup of any plant depends on the team leader and their leadership qualities, how well they manage and train the staff and how well they are doing the planning for successful startup.

Management must hire very good team leaders for successful startup of plant. If the team leaders are not capable then during startup if there is any upset they cannot control the situation and instead of controlling the emergency they will create another emergency for the operational staff.

Technical Inspection and Testing

Inspection of the equipment is an important part of any good startup of the plant and the inspection personnel must be very careful and thorough because this task is critical for the successful startup of the process plant.

Their negligence can delay the plant startup as we saw in many projects due to poor inspection and testing of the equipment resulting in failures during their startup.

Process Plant Machinery

Again, this is the duty of inspection and construction team to properly ensure that the machinery is appropriately installed, all work has been carried out as per procedures and that this has all been passed by quality control personnel. They must confirm before erection of the machinery or equipment, supported by necessary certification that the machinery or part of machine is not damaged, most equipment damage occurs during transportation, loading, and unloading time.

Machinery Failures during Commissioning

Most equipment that fails during startup does so due to improper operator training as mentioned earlier or due to problems with utilities, instrument failure, and loss of connection of instruments (vibration or temperature sensors).

Experience shows that most the equipment fails during startup due to lack of proper training, procedures and planning. Here is list of some critical equipment that are most commonly associated with problems during startup:

- Centrifugal compressors
- Steam turbines
- Large gear speed increaser units
- Electric motors in size categories above 1000HP
- Reciprocating compressors
- Gas Turbines
- Gas expanders

Pre-Commissioning Activities Safety

The most dangerous phase of the process plant startup is pre-commissioning activities. A lot of care should be exercised during this phase. Skilled staff, better planning and training can eliminate or minimize the hazards of pre-commissioning and startup phase. A successful startup requires that the following actions must be performed meticulously well before the project execution and startup. Better planning and a good leadership approach can only start the project on time while a badly managed project will always increase the capital cost. The key to succeed therefore is;

- Adequate overall planning
- Detailed long-term preparation
- Supportive organization
- Detailed operating and emergency procedures
- Established standard and working practices
- Effective personnel
- Trained personnel
- Thorough mechanical completion inspection
- Closely monitored commissioning procedures
- Water trails
- Adequate stores and spare part availability
- Prepared laboratory
- Responsive maintenance support.

Conclusion:

The first point is that no project or plant can be completely free from the risk of not being within the budget and time but we can manage to reduce overhead expenditure by better planning and with good leadership approach. Safety and training of the staff and better planning for the resources is the core values of any plant startup on time.
Therefore, during any successful startup of the process plant, it is necessary to do more and extensive planning and review the planning and progress at every step of the project so that management may know where they are heading for the startup of process plant. One can easily identify the effectively managed project startup benefits and the disadvantages associated with poorly managed projects.
If planning is not good then projects are being delayed in their startup and the result is loss in revenue collection and increasing the project capital cost.

References:

Harvard Technology Middle East 2005.

Prediction of Source Location and Rate of a Chemical Release for Emergency Response: Back Trajectory Approach

Shahryar Khajeh Najafi
Camarillo, California
shah@safersystem.com

Suneetha Burla
Camarillo, California
sburla@safersystem.com

ABSTRACT

With the ever-growing chemical industry, chemical storage and transportation may pose a threat to neighboring communities should there be an accidental release. In order to safeguard the workers and communities, agencies such as the National Institute for Occupational Safety and Health (NIOSH) and the Environmental Protection Agency (EPA) have made it mandatory for all industrial facilities to comply with their listed rules and regulations. The Emergency Planning and Community Right-to-know Act (EPCRA) requires that the public be given access to information on chemicals at individual facilities for emergency planning in case of episodic releases at the site. A large number of state-of-the-art modeling technologies and engineering tools have been developed and are readily available to successfully predict and track the cloud or plume of a chemical during an accidental release using only information on the release scenario and the meteorological data at the time of release. These models provide an assessment of the impact of a release and are essential and extremely useful for emergency planning in order to save lives and resources in the event of a release.

After the events of September 11, 2001, the paradigm has shifted to predicting an intentional release of a toxic chemical, which has potentially high impact if the targets are public locations such as football stadiums, shopping malls, and army bases located either domestically or internationally. In such cases, identification of the source location and the rate of release becomes a big challenge for hazmat personnel tasked with responding to the chemical attack. This is in great contrast to situations involving an accidental chemical release where the source of a release is known, or, in most cases, can be quickly identified.

In an attempt to find a solution to this problem, a method is proposed for locating the source of a chemical using only the basic knowledge of meteorological information (wind speed, direction and stability) and two or more concentration measurements of the chemical released. The technique used in this approach is called the reverse corridor or back trajectory method. This method uses the horizontal dispersion information obtained from surface meteorological data at receptor locations to narrow the search domain for the release location, thus reducing the computational cost. The effects of the main parameters of dispersion, such as wind meander and fluctuations in sensor concentration on the sensitivity of the model, are discussed. This model establishes a confidence level for prediction of a release location under varying circumstances.

The data regarding prediction of the source location and its associated confidence level are valuable pieces of information for emergency responders when seconds count.

1. INTRODUCTION

Source characterization is very critical for emergency response when there is an accidental release or leak of a toxic or hazardous chemical in a plant site, when a chemical is being transported to a different location, and when there is a terrorist attack in a public location such as a football stadium, a shopping mall, or a railway station. In order to identify and characterize the source parameters of a toxic chemical release several research groups have developed and tested various models and methods for source location and the release rate estimation based on the downwind concentration measurements (Shah and Gilbert 2003, Padmanathan et al 2004, Sue and Young 2008 and Rao 2007). Such techniques include Back Trajectories and Cluster Analysis, Bayesian approaches, Variational and Adjoint methods, and Genetic Algorithms. These techniques can be primarily categorized into forward techniques which predict source characteristics from the source to the receptor (sensor) and backward models which invert from the receptor to the source.

This paper mainly focuses on investigating the feasibility and accuracy of locating the source of a chemical release using only basic information on the meteorological data and two or more concentration measurements of the chemical released. The reverse corridor or the back trajectory technique is used to narrow the space domain for the source location and a Gaussian model is used to calculate the concentration at points of interest.

Section 2 presents the methodology for constructing back trajectory from horizontal dispersion parameters obtained from surface meteorological data. The criteria for identifying the possible zone for a release location are discussed. Finally, a method is proposed for locating the release source by searching for the global minimum error between the measured and predicted concentrations.

Section 3 describes the experimental set-up used for current study. The results predicted by the model under several test conditions are presented in section 4. First, the conventional straight-line Gaussian diffusion model is tested under constant meteorological conditions. Considering the non-stationary diffusion conditions, the model is then tested by introducing some random noise to the measured data. The Gaussian model is used for forward and backward calculation of concentration measurements. Section 4.1 includes testing of the model without considering any fluctuation at the measuring points, section 4.2 describes the effect of sensor position on the model sensitivity and section 4.3 presents the results when the model is tested with different levels of noise due to non-stationary nature of meteorological data. Finally in section 5, the conclusions of the work and the usefulness and robustness of the current model are summarized.

2. METHODOLOGY

2.1 Constructing the Back Trajectory

Back trajectory formulation is based on surface meteorological data. Figure 1 represents the methodology for constructing the corridor. The horizontal plume dispersion parameter σ_y, based on ten minute averaging, is obtained from the equation of the form

$$\sigma_y = ax^b \tag{1}$$

Where x is the distance, and a, b are the constants determined by the Pasquill stability condition. The values of a are 0.4, 0.36, 0.34, 0.32, 0.32 and 0.31 for stability classes ranging from A through F respectively. The corresponding values of b are 0.91, 0.86, 0.82, 0.78, 0.75 and 0.71. The horizontal stability class is determined by the azimuthal component of wind direction. The standard deviation of the wind direction fluctuations, based on ten minute running average, is used to derive the horizontal stability. Tables related to the standard deviation of azimuthal fluctuation of wind direction to atmospheric stability are published by the EPA.

The reverse corridor is a wedge originating from the sensor location pointing in the mean wind direction as measured at the meteorological tower. If the wind is blowing from the west, the reverse corridor is drawn from the east. To calculate the wedge angle, a downward distance of x meter is assumed from the receptor against the wind direction (OA). A line segment of length $2.14\sigma_y$, where σ_y is the horizontal dispersion parameter found from equation (1) using an appropriate stability curve, is drawn perpendicular to OA at A. The wedge is then constructed by connecting the other end of the line segment to the sensor location. The factor of 2.14 sets the cloud width at which the concentration falls to about one percent of the centerline value.

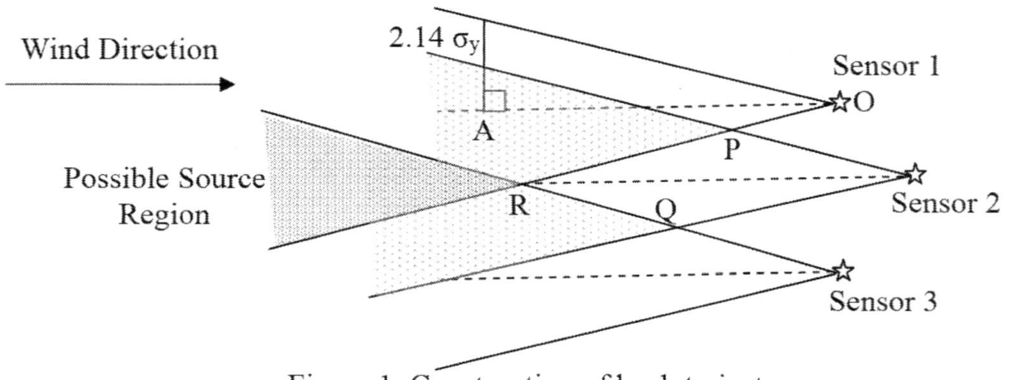

Figure 1. Construction of back trajectory

2.2 Narrowing down the Search Zone

The number of deployed sensors and their distribution play an important role in targeting the location of a chemical release. Rather than searching for an infinite space upwind of the sensors, one can utilize the sensors' distribution to narrow down the search zone.

Assume a case with three sensors: sensor 1, sensor 2 and sensor 3, as shown in Figure 1. A reverse corridor is drawn from each sensor as described in the previous section. The wedge drawn for each sensor encompasses the area where the release could have occurred. The intersection of two wedges provides a region where any release initiated within that region would impact the two sensors. The wedges drawn from each sensor intersect with each other at points P for sensors 1 and 2, Q for sensors 2 and 3, and R for sensors 1 and 3 as shown in the Figure 1. The shaded region formed by the intersection of all of the wedges from sensors 1, 2, and 3 is called the common wedge, which would contain the most probable area of the release source location.

2.3 Locating the Source

A mesh or grid that contains square cells with user definable dimensions is overlaid onto the common wedge as shown in Figure 2. The nodes on each cell define the locus of points used as possible source locations. The algorithm for source location is based on two loops, an outer loop where the rate is incremented, and an inner loop where the domain for possible source location is swept from point R up to a predefined distance in the direction opposite to the wind.

For each trial rate and trial source location, a concentration value is predicted by the Gaussian model for each impacted sensor. Actual measured concentration at each sensor is compared against the model prediction, and the error between the two values is recorded. The minimum recorded value obtained from the inner loop is labeled a local minimum. For each trial rate the procedure is repeated until the absolute minimum, called global minimum, is achieved. This is a very intensive computing process. In order to simplify the process, only the ground level release is studied and the Gaussian model was selected due to its convenience and rapid response. As an alternative, any other model can also be used at increased computational costs.

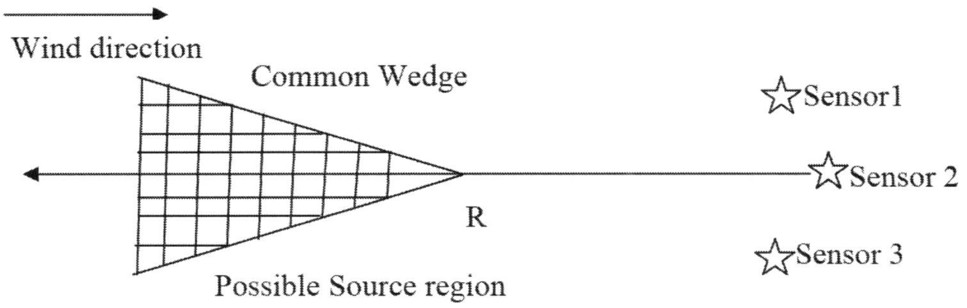

Figure 2. Source Location by back trajectory

3. EXPERIMENTAL SET-UP

To test the feasibility and accuracy of the source location model, simulated releases are carried out using the Gaussian plume model. This is done with the understanding that using the Gaussian model in both forward simulation and the reverse method biases the result in favor of the reverse method. A Cartesian co-ordinate system (x, y) (z=0, for ground release), where x represents the downwind distance and y represents the cross-wind distance from the source location, is used with the x-axis oriented in the mean wind direction and the y-axis in the horizontal cross-wind direction.

The forward Gaussian model is used to compute the concentration of a chemical release at each sensor position from a known location and for a known release rate. This concentration value is called as "measured" concentration. A wide range of possible values for the release rates are chosen, ranging from 0.01 through 10 kg/s, with an increment of 0.01 Kg/s. The search domain is limited to a distance of 10 km from the common wedge apex. A total of 1000 different release rates are considered for the test. For each rate, the entire possible source domain is swept to find the local average minimum error between the predicted and the measured values for all sensors. The absolute global minimum obtained from the simulation of 1000 rate changes yields the most possible source locations and their corresponding release rates.

Three sensors S_1, S_2 and S_3 are located within the plume at a distance (x_1, y_1), (x_2, y_2) and (x_3, y_3) respectively from the source O (0, 0), as shown in the Figure 3, such that they are impacted by the chemical.

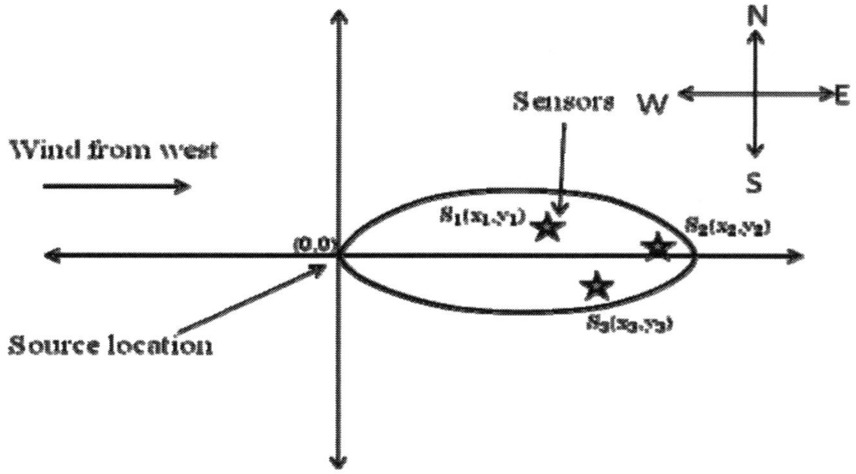

Figure 3. Plume/Sensor impact

The concentrations at each sensor location, $C_1(x_1, y_1)$, $C_2(x_2, y_2)$ and $C_3(x_3, y_3)$ can be calculated using the Gaussian form of plume equation:

$$\langle C\rangle(x, y, z) = \frac{Q_m}{2\pi\sigma_y\sigma_z u}\exp\left(-\frac{y^2}{2\sigma_y^2}\right)\times\left\{\exp\left[-\frac{(z-H_r)^2}{2\sigma_z^2}\right]+\exp\left[-\frac{(z+H_r)^2}{2\sigma_z^2}\right]\right\} \quad (2)$$

$\langle C\rangle(x, y, z)$ = Average concentration (15 min)
Q_m = Release rate (Kg/s)
σ_y,σ_z = Dispersion coefficients = f (stability class, downwind distance)
u = Wind speed (m/s)
y, z = Coordinates (m)
H_r = Release height (m)
For the case in study, z = 0 and H = 0, hence the equation (2) can be simplified to

$$(C)(x, y, z) = \frac{Q_m}{2\pi\sigma_y\sigma_z u}\exp\left(-\frac{y^2}{2\sigma_y^2}\right)$$

(3)

4. TESTING MODEL VALIDITY

In order to test the validity of the model we present some numerical calculations and compare them with the model prediction. The model is tested for the following conditions:

1) Wind speed (1.5m/s, 3m/s and 6m/s)
2) Wind direction from west
3) Stability Classes 1 through 6 (A through F)
4) Different levels of concentration fluctuations (40% and 80% noisy data)

4.1 Case Study (1): Ideal case scenario with no sensor data fluctuations

Three sensors located at (100, 21), (300, 49) and (500, 50) from the actual source location are chosen and the model is tested for the release rates of 0.01, 0.1, 1.0 and 10.0 Kg/s.

This is an ideal case where we do not expect any inherent fluctuations in the sensor data due to turbulence and the sensor data is free of any noise. Under these conditions, the concentration measured by the sensors is the same as that calculated using the forward Gaussian model. This test enables us to evaluate the model alone because several sources of error are eliminated from consideration, knowing that the sensor data is exact. The predicted source locations for various stability classes from A through F, for wind blowing from the west at a speed of 1.5m/s, are plotted as shown in Figure 4 and the results are also tabulated in Table 8 in the Appendix.

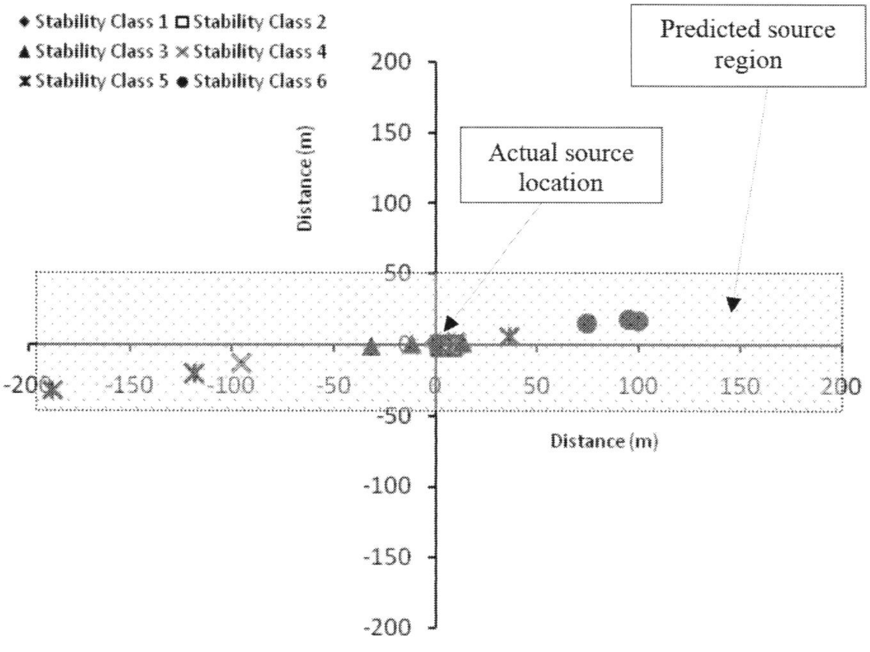

Figure 4. Comparison of the model predicted source location against the actual source location for different stability classes and release rates at wind speed of 1.5m/s

The data points for each stability class correspond to four different release rates viz. 0.01, 0.1, 1.0 and 10.0 kg/s. From Figure 4, it can be seen that the model predictions for the source location values are very close to the actual source location (0, 0). For a given release rate, the accuracy with which the model predicts the source location decreases with the increase in stability class. For lower stability classes, the predicted source locations are concentrated around the origin (0, 0) (actual source location) and as we move to higher stability classes, the predicted source locations are more dispersed and farther from the actual source location. This is especially true in cases with low release rates of 0.01Kg/s and 0.1 Kg/s and under highly stable conditions (5 and 6), where the predicted source location values are far from the actual source location values.

The release rates predicted by the model are also compared against the actual release rates for various stability classes in Table 1. It can be seen that, except for the higher stability classes (5 and 6), the release rates predicted by the model match well with the actual release rates.

Table 1. Comparison of the model predicted release rates against the actual release rates for different stability classes at wind speed of 1.5m/s

Actual Source location	Release rate (Kg/s)	Predicted rate (Kg/s)					
		Stability Class					
		1	2	3	4	5	6
(0,0)	0.01	0.01	0.01	0.01	0.01	0.04	0.01
(0,0)	0.1	0.1	0.1	0.11	0.16	0.25	0.01
(0,0)	1.0	1.0	0.98	0.96	0.94	0.98	0.11
(0,0)	10.0	10.1	9.8	9.6	9.8	7.6	1.6

Similar results are obtained when the model is tested for the wind speeds of 3m/s and 6 m/s blowing from the west. The results for the model predictions for the source location at a wind speed of 3m/s and 6m/s are plotted in Figure 5 and Figure 6 respectively. For higher stability classes (5 and 6) and at low release rates (0.01 and 0.1 Kg/s), the predicted source location values are more dispersed and farther from the actual source location for higher wind speeds of 3m/s and 6m/s. Nonetheless, the predicted source location values are still at a radius of 200m from the actual source, which is a reasonable distance limit to locate the source. A box is drawn on the graph, enclosing all the data points for source location predictions under all stability conditions. Thus the model is tested under several conditions of different wind directions, wind speeds and other sensor locations (sensor locations chosen such that they are neither too saturated nor negligibly impacted), and for all the conditions it has been observed that the source can be located within the area enclosed by the box. Hence, without loss of generality, it can be concluded that under ideal conditions of no sensor fluctuations, the model can predict the source location within a 0-50m radius from the actual source location for a wide range of wind speeds (1.5 to 6m/s), except under conditions of very high stability classes(5 an 6) and very low release rates (0.01 and 0.1Kg/s), in which the source location can be predicted within 200m radius from the actual source. The values for the predicted source locations for various stability classes ranging from A through F, with wind blowing from the west at a speed of 1.5m/s, 3m/s and 6m/s are also tabulated in Table 8, Table 9, and Table 10, respectively, in the Appendix.

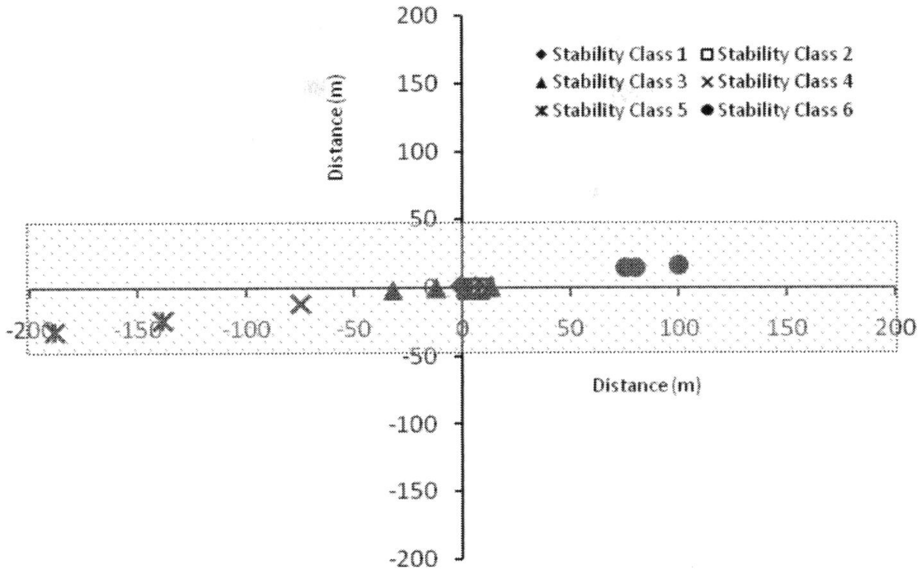

Figure 5. Comparison of the model predicted source location against the actual source location for different stability classes and release rates at wind speed of 3m/s

The predicted release rates for wind speeds 3m/s and 6m/s are tabulated in Table 2 and Table 3 respectively. The results for release rate predictions at wind speeds 3m/s and 6m/s are very similar to the results at wind speed 1.5m/s. Except for the higher stability classes (5 and 6), the model predictions for release rates match well with the actual release rates for all the stability classes.

Table 2. Comparison of the model predicted release rates against the actual release rates for different stability classes at wind speed of 3m/s

Actual Source location	Release rate (Kg/s)	Predicted rate (Kg/s)					
		Stability Class					
		1	2	3	4	5	6
(0,0)	0.01	0.01	0.01	0.01	0.01	0.04	0.01
(0,0)	0.1	0.1	0.1	0.11	0.15	0.29	0.01
(0,0)	1.0	1.0	0.98	0.96	0.94	0.98	0.11
(0,0)	10.0	10.1	9.8	9.6	9.8	9.8	1.6

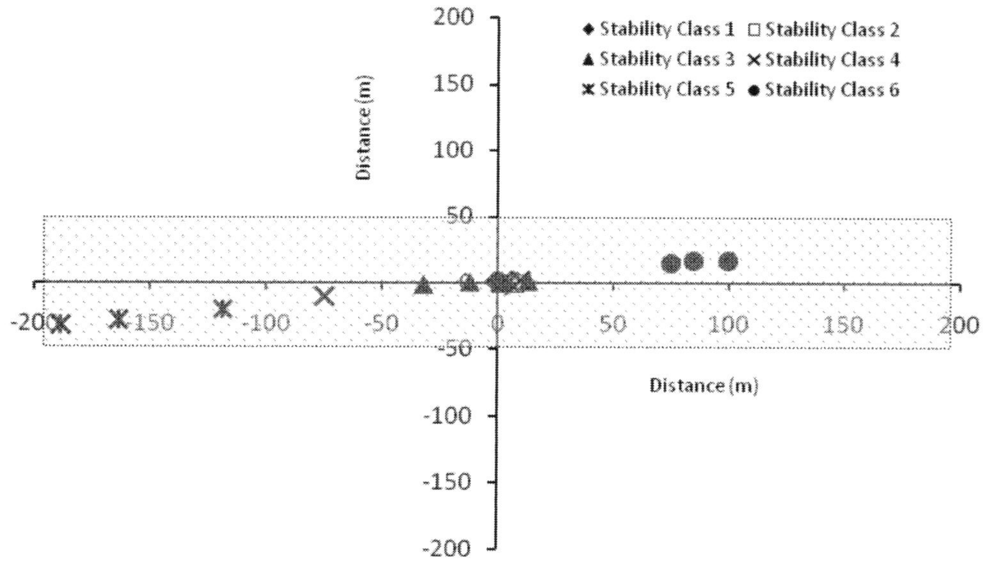

Figure 6. Comparison of the model predicted source location against the actual source location for different stability classes and release rates at wind speed of 6m/s

Similar results for the source location and the release rate predictions at all three different wind speeds (1.5m/s, 3m/s and 6m/s) indicates that the model predictions are independent of the wind speeds, in the range 1.5-6m/s.

Table 3. Comparison of the model predicted release rates against actual release rates for different stability classes at wind speed of 6m/s

Actual Source location	Release rate (Kg/s)	Predicted rate (Kg/s)					
		Stability Class					
		1	2	3	4	5	6
(0,0)	0.01	0.01	0.01	0.01	0.01	0.04	0.01
(0,0)	0.1	0.1	0.1	0.11	0.15	0.34	0.01
(0,0)	1.0	0.99	0.98	0.96	0.94	2.5	0.12
(0,0)	10.0	10.1	9.8	9.6	9.4	9.8	0.98

4.2 Case Study (2): Sensitivity to sensor location

To test the model's sensitivity to the sensor location, different locations other than those mentioned for the previous case study are selected viz. (45, 5), (70, -7) and (100, 11) for the three sensors in the study. A wind speed of 3m/s is randomly selected because we have already seen from previous test results that the model predictions are not very sensitive to the wind speed. The test results are as shown in the following Table 4.

Table 4. Comparison of the model predicted source location against the actual source location for different stability classes and release rates at wind speed of 3m/s

Actual Source location	Release rate (Kg/s)	Predicted Source Location (m, m)					
		Stability Class					
		1	2	3	4	5	6
(0,0)	0.01	(1,-1)	(-1,0)	(2,0)	(13,0)	(-10,0)	(-315,-14)
(0,0)	0.1	(1,-1)	(-1,0)	(-8,0)	(13,0)	(-10,0)	(-45,0)
(0,0)	1.0	(1,-1)	(4,0)	(2,0)	(13,0)	(-10,0)	(-45,0)
(0,0)	10.0	(1,0)	(-1,0)	(2,0)	(7,0)	(0,0)	(-45,0)

From Table 4, it can be seen that the model predictions for the source location using the current set of sensor locations are very different from the model prediction results generated using a different set of sensor locations in the previous Case study (1),especially for cases under high stability classes of 5 and 6 at very low release rates. The predicted source location is about 300m away from the actual source under stability class 6 at very low release rate of 0.01 kg/s.

Table 5. Comparison of the model predicted release rates against the actual release rates for different stability classes at wind speed of 3m/s

Actual Source location	Release rate (Kg/s)	Predicted rate (Kg/s)					
		Stability Class					
		1	2	3	4	5	6
(0,0)	0.01	0.01	0.01	0.01	0.01	0.01	0.01
(0,0)	0.1	0.1	0.1	0.12	0.1	0.1	0.03
(0,0)	1.0	0.98	0.98	0.96	0.94	0.99	0.39
(0,0)	10.0	9.8	9.8	9.6	9.6	9.6	5.8

Similarly, from Table 5 it can be seen that the model predicted values for the release rates under higher stability classes 5 and 6 have changed significantly. This is possible when the concentration values of the released chemical at the sensor location are beyond the measurable concentration range of the sensors. This indicates that the model predictions are highly sensitive to the sensor positions in the field. Hence, it is highly important to take multiple sensor readings at various random locations and care should be taken when selecting the sensors for study. The sensors are chosen such that they are neither too saturated nor negligibly impacted. Before making the decision about the final predicted source location and release rate, the model results for various sensors located at multiple locations are weighed against each other and the readings from negligibly impacted and overly saturated sensors are eliminated from consideration.

4.3 Case Study (3): Sensitivity to sensor data

The same sensor locations, as mentioned for case study 4.1, are selected viz. (100, 21), (300, 49) and (500, 50) for the three sensors in this study.

In this case, some noise is introduced into the sensor data in order to take into account the fluctuations in the sensor data due to turbulence and wind meander. The concentration data measured at the sensor locations using the Gaussian equation are corrupted by adding different levels of noise. Two cases, one with 40% noise in the concentration data and the other with 80% noise, are examined. To study the effect of 40% noise in the concentration data, for each sensor a random concentration value between $0.6C_m$ and $1.4C_m$ is chosen, where C_m is the measured concentration calculated using a forward Gaussian model for the corresponding sensor, and then the simulation was run. This accounts for cases when both the concentration is overestimated and is underestimated due to meteorological data fluctuations. Thus for this case, several random concentration values within the mentioned limit are created for every sensor under consideration. These tests allow us to evaluate the sensitivity of the model to different concentration fluctuations that are inevitable in the real world sensor data.

With the introduction of 40% noise in the sensor data, it can be seen that the model predictions for the source location values are very far from the actual source location (0, 0), for stability classes 5 and 6 at very low release rates of 0.01kg/s and 0.1kg/s. These results when compared with the ideal case study 4.1, with no concentration fluctuations and with all the other conditions the same, show that the model predictions are not significantly affected by the introduction of 40% noisy data.

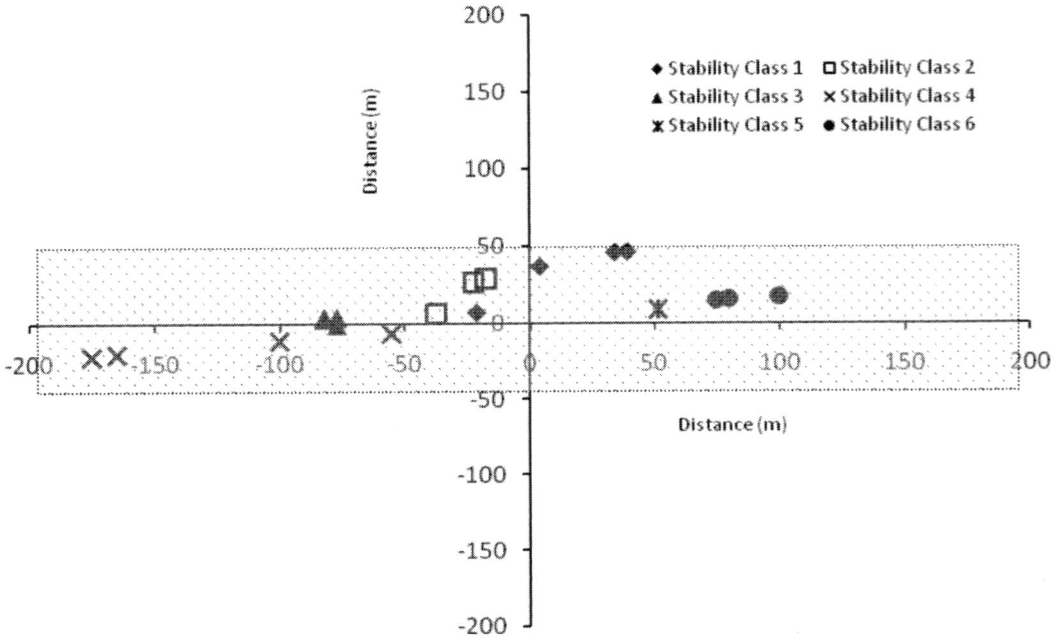

Figure 7. Comparison of the model predicted source location against the actual source location for different stability classes and release rates at wind speed of 3m/s for 40% noisy data

Table 6 shows the predicted release rates with the introduction of 40% noise into the sensor concentration readings. These results when compared against the data with no sensor fluctuations show that the predicted release rates are very different from the actual release rates and the accuracy of the release rate prediction does not follow any trend with the stability class.

Table 6. Comparison of the model predicted release rates against the actual release rates for different stability classes at wind speed of 3m/s for 40% noisy data

Actual Source location	Release rate (Kg/s)	Predicted rate (Kg/s)					
		Stability Class					
		1	2	3	4	5	6
(0,0)	0.01	0.01	0.01	0.01	0.01	0.07	9.8
(0,0)	0.1	0.04	0.04	0.08	0.11	0.24	0.04
(0,0)	1.0	0.62	0.63	0.97	1.6	10.9	0.56
(0,0)	10.0	5	5.2	6.6	11	11	8.1

Similarly, for a case with 80% noise in the concentration data, a random concentration value between $0.2C_m$ and $1.8C_m$ is chosen, and then the simulation is run. The predicted source locations and the release rates are tabulated for different levels of noise. With the increase in the noise in the sensor data to 80%, similar results were obtained for the source location and the release rate predictions.

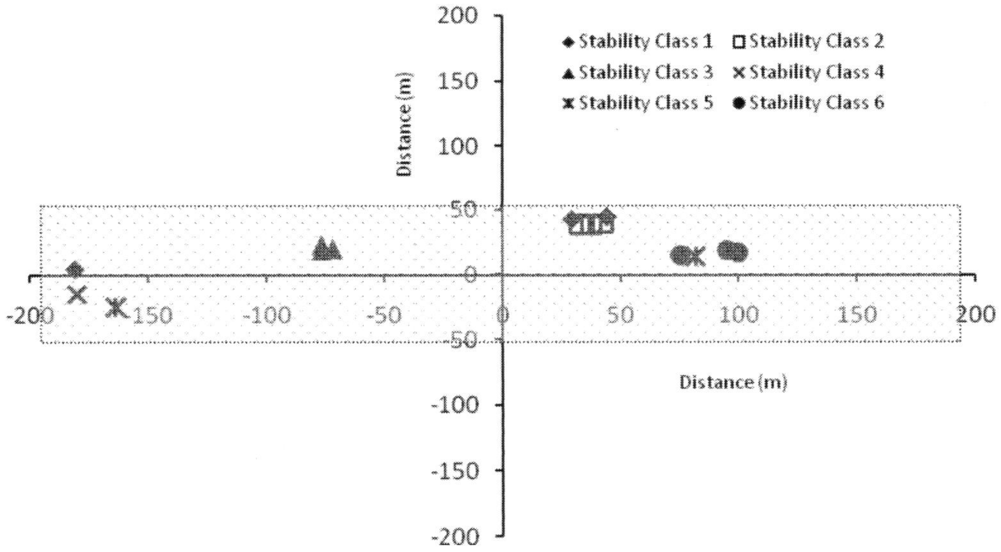

Figure 8.Comparison of the model predicted source location against the actual source location for different stability classes and release rates at wind speed of 3m/s for 80% noisy data

Table 7 shows the predicted release rates with the introduction of 80% noise in the sensor concentration readings. These results match well with the predicted release rate data with no sensor fluctuations. The results when compared with those with the 40% noisy data show that the release rate predictions with the introduction of 80% noise are much better than those with 40% noisy data.

Table 7. Comparison of the model predicted release rates against the actual release rates for different stability classes at wind speed of 3m/s for 80% noisy data

Actual Source location	Release rate (Kg/s)	Predicted rate (Kg/s)					
		Stability Class					
		1	2	3	4	5	6
(0,0)	0.01	0.02	0.01	0.02	0.05	10.9	0.95
(0,0)	0.1	0.3	0.36	0.66	5.8	11	0.04
(0,0)	1.0	0.87	0.99	1.3	1.5	1.3	0.59
(0,0)	10.0	10.4	10.5	9.9	11	10.8	7.3

Nevertheless, it can be concluded that even with a noisy sensor data, the model predicts the source parameters with a decent level of confidence, except under high stability conditions (5 and 6) and at low release rates (0.01 and 0.1Kg/s).

The values for the predicted source locations for various stability classes from A through F for 40% and 80% noisy data are also tabulated in Table 11 and Table 12 respectively in the Appendix.

5. CONCLUSION

In this paper, we explored the robustness of a relatively simple back trajectory model for constructing the reverse corridors in order to trace the source characteristics such as the release location and the release rate of a toxic chemical. Two different cases were studied in order to test the validity of the model. The first test involved an ideal case scenario where exact concentrations were provided using the forward Gaussian model. The second case was such that fluctuation due to wind meander and atmospheric turbulence at measuring points are taken into consideration. The latter was simulated by adding random noise to the measured data. In both cases, the model results match well with the actual data, in their ability to locate the source and the release rate of a chemical release.

The model was tested for different wind speeds, viz. 1.5m/s, 3m/s and 6 m/s, and the results showed that the model predictions are independent of wind speed for a given stability class and they match well with the actual data. The model was also tested for varying stability classes, A through F (1 through 6), and the results for source location showed that the accuracy of the model prediction decreases with the increase in stability class. This is particularly true under higher stability classes 5 and 6 at very low release rates.

Different prediction results for two different sets of sensor positions indicate that the model is sensitive to the chosen sensor location. The results for the source location and the release rates, when the model was simulated with the noisy sensor data, viz. 40% and 80%, matched very well with the true data (except under highly stable conditions at low release rates), indicating that the back trajectory method coupled with a simple Gaussian plume model is capable of providing a reasonable prediction of source variables.

6. REFERENCES

1. Khajeh Najafi , Shahryar, and Gilbert, Ernest J., *Use of real-time measurement for estimating release rate,* Institute of Chemical Engineers Symposium Series, American Institute of Chemical Engineers, 2003
2. Kathirgamanathan, Padmanathan, McKibbin, Robert, and McLachlan Robert I., *Source release-rate estimation of atmospheric pollution from a non-steady point source at a known location,* Netherlands: Environmental Modeling and Assessment, 2004

3. Haupt, Sue Ellen, and Young, George S., *Paradigms for source characterization*, 15th Joint Conference on the applications of air pollution meteorology with the A&WMA, New Orleans: American Meteorological Society, 2008
4. Rao, Shankar K., *Source estimation methods for atmospheric dispersion*, Atmospheric Environment, Science Direct, 2007

7. APPENDIX

Table 8.Comparison of the model predicted source location against the actual source location for different stability classes and release rates at wind speed of 1.5m/s

Actual Source location	Release rate (Kg/s)	Predicted Source Location (m, m)					
		Stability Class					
		1	2	3	4	5	6
(0,0)	0.01	(-1,1)	(2,0)	(-12,0)	(5,1)	(-190,-32)	(100,17)
(0,0)	0.1	(-1,1)	(2,0)	(-32,-2)	(-95,-14)	(-120,-20)	(75,15)
(0,0)	1.0	(-1,1)	(8,-1)	(13,1)	(10,1)	(1,0)	(75,15)
(0,0)	10.0	(-1,0)	(8,-1)	(13,1)	(5,1)	(36,6)	(95,17)

Table 9.Comparison of the model predicted source location against the actual source location for different stability classes and release rates at wind speed of 3m/s

Actual Source location	Release rate (Kg/s)	Predicted Source Location (m, m)					
		Stability Class					
		1	2	3	4	5	6
(0,0)	0.01	(-1,1)	(2,0)	(-12,0)	(5,1)	(-190,-32)	(100,17)
(0,0)	0.1	(-1,1)	(2,0)	(-32,-2)	(-75,-10)	(-140,-24)	(75,15)
(0,0)	1.0	(-1,1)	(8,-1)	(13,1)	(10,1)	(1,0)	(75,15)
(0,0)	10.0	(-1,0)	(8,-1)	(13,1)	(5,1)	(1,0)	(80,15)

Table 10.Comparison of the model predicted source location against the actual source location for different stability classes and release rates at wind speed of 6m/s

Actual Source location	Release rate (Kg/s)	Predicted Source Location (m, m)					
		Stability Class					
		1	2	3	4	5	6
(0,0)	0.01	(-1,1)	(2,0)	(-12,0)	(5,1)	(-190,-32)	(100,17)
(0,0)	0.1	(-1,1)	(2,0)	(-32,-2)	(-75,-10)	(-164,-28)	(75,15)
(0,0)	1.0	(4,-2)	(8,-1)	(13,1)	(10,1)	(-120,-20)	(75,15)
(0,0)	10.0	(-1,0)	(8,-1)	(13,1)	(10,1)	(1,0)	(85,17)

Table 11.Comparison of the model predicted source location against the actual source location for different stability classes and release rates at wind speed of 3m/s for 40% noisy data

Actual Source location	Release rate (Kg/s)	Predicted Source Location (m, m)					
		Stability Class					
		1	2	3	4	5	6
(0,0)	0.01	(-21,7)	(-37,6)	(-77,-1)	(-100,-10)	(-230,-36)	(100,17)
(0,0)	0.1	(4,36)	(-20,30)	(-82,3)	(-175,-22)	(-243,-40)	(75,15)
(0,0)	1.0	(34,45)	(-23,26)	(-82,3)	(-165,-20)	(-210,-33)	(75,15)
(0,0)	10.0	(40,46)	(-23,26)	(-77,4)	(-55,-7)	(50,9)	(80,15)

Table 12.Comparison of the model predicted source location against the actual source location for different stability classes and release rates at wind speed of 3m/s for 80% noisy data

Actual Source location	Release rate (Kg/s)	Predicted Source Location (m, m)					
		Stability Class					
		1	2	3	4	5	6
(0,0)	0.01	(-180,4)	(-230,-7)	(-380,-20)	(-122,2)	(-285,-35)	(100,17)
(0,0)	0.1	(45,45)	(32,39)	(-77,23)	(-300,44)	(-265,-26)	(75,15)
(0,0)	1.0	(30,43)	(38,38)	(-77,20)	(-12,-1)	(-245,-24)	(75,15)
(0,0)	10.0	(45,45)	(43,40)	(-72,20)	(-2,0)	(-180,-15)	(95,15)

2007 NPRA National Safety Conference
May 2-3, 2007
Woodlands, Texas

PASSIVE DEVICE TECHNOLOGY THE TREND AWAY FROM RELIANCE UPON VALVING SYSTEMS TO ADDRESS LINE FAILURE

Andrew Abrams, Thomas Steinbach, Joseph Abrams
Smart Hose Technologies, 2536 South 59th Street, P.O. Box 116828, Philadelphia, Pennsylvania, 19142. 215-730-9000

Abstract

Since the landmark incident in Sanford, North Carolina to the more recent tragedy in Riverview, Michigan, reliance upon excess flow valve technology to stop the potentially catastrophic consequences of hose and line failure has been eroding. Through legislation requiring "passive device" technology to formal warnings against this reliance issued by all of the major regulatory agencies in the United States to special permits that provide regulatory relief for users of passive device technologies, a growing trend has commenced towards eliminating this over-reliance upon excess flow valves in the event of transfer line failure. The basis of our paper is to identify this trend, explain its genesis and provide an understanding of the benefits to safety, cost and regulatory compliance through the growing trend and regulatorily encouraged use of passive devices for line and hose failures.

Introduction

For more than a decade, safety concerns have been proliferating throughout industry – none more so than the liquefied petroleum gas (LPG) industry. Incidents have spawned greater concerns over "weak links" in the safety chain of LPG transfer systems. Incidents have caused countless millions of dollars and claimed many lives. On September 8, 1996, in Sanford, North Carolina, during delivery of propane to a bulk storage facility by an MC 331 bulk transport, more than 35,000 gallons of propane were released. The discharge hose separated from its hose coupling at the delivery end of the hose. Most of the transport's 9,800 gallons of propane and more than 30,000 gallons from the storage tanks were released. If this quantity of released propane had ignited, local authorities estimated that about 125 emergency response personnel could have been seriously injured or killed.

On June 3, 1997, in Caro, Michigan, while unloading propane into a storage tank at an industrial facility, the transport delivery hose ruptured. The ensuing fire and a series of explosions seriously burned the driver, destroyed four vehicles and extensively damaged the facility. Initial estimates of property damage were at least $2.0 million.

The history of major accidents in the transportation of anhydrous ammonia is similar to that involving the transportation of liquefied petroleum gases. Pulmonary injuries are more significant with ammonia while fire damage is more significant with liquefied petroleum gases. An example of a major accident involving the release of ammonia is an incident that occurred, in Houston, Texas. The driver of an MC 331 transport lost control while negotiating an interstate exit ramp. The cargo tank motor vehicle released its entire cargo of 7500 gallons of anhydrous ammonia. The driver was killed in the crash. Five other persons were killed and 78 others were hospitalized, all due to inhalation of ammonia. Another 100 persons were treated for less severe injuries. Favorable wind conditions prevented the vapor cloud from reaching a nearby elementary school.

These incidents as well as many more in other petrochemical industries all pointed to a fundamental flaw in the fluid transfer system - over reliance on the excess flow valve. This recognition spawned a landmark Rulemaking issued by the United States DOT-RSPA in which they declared formally that excess flow valves are not reliable to address hose failures and new technology must be developed and identified to provide a greater degree of safety and reliability. Thus the term, **passive technology** was defined as shutdown equipment that has the capability to shut off the flow of product without human intervention within 20 seconds of an unintentional release caused by complete separation of a delivery hose. This definition and requirement thereby changed the safety landscape. This rulemaking marked the first, of several important, steps that have since been taken to require alternative means to address the consequences of transfer line failure by relying upon systems other than excess flow valves.

Excess Flow Valve Technology and The related problems of Over Reliance

An excess flow valve is, typically, a metallic device inserted into the interior piping of a tank car, just below the valve(s) used to load and unload the car. In the event that the valves are sheared off in a railroad accident, there will be a sudden rush of product out the opening thus created. With nothing to impede the flow of fluid product, the excess flow valve will move toward the opening and seat, thus sealing off the opening. In response to concerns that the then-current regulatory provision for excess flow valves might be ambiguous, in 1985 the DOT published a notice of proposed rulemaking to amend the tank car specifications by adopting what is now the contemporary standard. Proponents of the clarification stated that tank-mounted excess flow valves are not intended to substitute for adequate excess flow equipment in plant loading systems. ``The only use of such valves is for protection against loss of lading due to shearing of external closure during transit." \1\

The hazardous materials regulations (HMR) are quite clear that excess flow valves are limited in purpose and scope: An excess flow valve as referred to in this specification, is a device which closes automatically against the outward flow of the contents of the tank in case the external closure valve is broken off or removed during transit \2\

Excess flow valves, by their nature, must encounter a high-volume, surging flow of product to be activated. If that were not the case, they might function in unintended situations, such as when a tank car is being unloaded with the aid of a strong pump. As designed, essentially any apparatus attached to the outside of the external closure valve will create sufficient internal friction (whether hose or pipe) that the flow of product will not be sufficient to activate the excess flow valve. Excess flow valves, by both design and regulation, are intended to function only when the external closure valve is sheared, broken off, or otherwise removed during transit. These devices may also function as a back-up flow control device during tank car loading or unloading activities.

Despite these clear and obvious functional limitations, in the absence of newer passive device technology, industry has placed great reliance on these types of valves to address a scenario that they are not designed for – line/hose failure. (See NTSB Report in which, inter alia, six of nine companies surveyed rely on tank car excess flow valves as a method of stopping or limiting a leak in the transfer equipment.).

In the face of this limitation and over reliance, what would the regulatory bodies do when the next major incident occurs?

Rulemaking HM-225 and 49 CFR 173.315

New regulations **HM-225A** in effect since July 1, 1999, imposed new requirements on operators of cargo tanks used to transport liquefied compressed gases and new procedures for unloading liquefied compressed gases from cargo tanks. The new regulations seek to prevent unintentional releases during unloading of liquefied compressed gases, assure prompt identification of unintentional releases, and reduce the consequences of unintentional releases. Specific provisions include: New inspection, maintenance, and testing requirements for cargo tank discharge systems. Revised unloading requirements for liquefied compressed gases, including revised attendance requirements. New requirements for emergency discharge control equipment, such as passive systems that will shut down unloading without human intervention and remote control devices that enable an attendant to stop the unloading process at a distance from the vehicle.

The new requirements for emergency discharge control equipment on cargo tanks used to transport liquefied compressed gases are keyed to the degree of risk associated with the transportation of specific liquefied compressed gases. The regulation **49 CFR 173.315** specifies two types of emergency discharge control equipment - **Passive Shut-**

down Equipment. Passive shutdown equipment must shut off the flow of product without human intervention within 20 seconds of an unintentional release caused by complete separation of a delivery hose.

This landmark legislation marked the first time that any regulatory body not only identified the risks associated with excess flow valve reliance but further regulated against their use in lieu of other technology. Since this regulation went into effect several systems have been developed and no recorded incidents have occurred.

These same standards were adopted in Canada. **Safety standard CSA B620** has been revised to introduce off-truck remote or **passive** emergency discharge control systems for highway tanks or portable tanks during the unloading of liquefied compressed gases. This applies to highway tanks and portable tanks of the type TC 331 and TC 51, respectively that are used to transport for example, liquefied petroleum gases (LPG) or anhydrous ammonia, the most common of these gases transported. This revision applies equally to compressed gases with a poison inhalation, refrigerant gases and other compressed gases that pose similar risks. It does not apply to Class 2.2 without a subsidiary risk, for example, gases like argon, helium, etc.

The standard calls for new highway tanks or portable tanks to be equipped with these emergency discharge control systems by January 1, 2003. The standard also includes a five-year retrofit period for the installation of these off-truck remote or **passive** systems on existing tanks consistent with the five-year pressure re-inspection date. This retrofit period will begin January 1, 2005, and will end December 31, 2010. The emergency shutdown systems are also subject to test procedures. Appendix D introduces acceptable test procedures to test the off-truck emergency shutdown systems.

The remote capability introduced in CSA B620 enables operators attending the unloading operation to shut off the flow of product when away from the motor vehicle during delivery. These remote systems are associated with TC 331 tanks that are in metered delivery service, more commonly referred to as Bobtail units. The **passive** emergency discharge control equipment (sometimes referred to as a **Smart hose**) is designed to shut down unloading operations without human intervention. These **passive** systems are found on larger trailer type units.

Post-Rulemaking Atofina Accident and The NTSB's Conclusions

About 3:45 a.m. on July 14, 2001, at the **ATOFINA** plant in Riverview, Michigan, a pipe attached to a fitting on the unloading line of a railroad tank car fractured and separated, causing the release of methyl mercaptan, a poisonous and flammable gas. About 4:09 a.m., shortly after the Riverview Fire Department chief arrived on scene, the methyl mercaptan ignited, engulfing the tank car in flames and sending a fireball about 200 feet into the air. Fire damage to cargo transfer hoses on an adjacent tank car resulted in the release of chlorine, a poisonous gas that is also an oxidizer. The fire was extinguished about 9:30 a.m. Three plant employees were killed in the accident. There were several other injuries; most of the injured were treated for respiratory symptoms and released. About 2,000 residents were evacuated from their homes for about 10 hours. Two tank cars, railroad track, and plant equipment (including hoses and fittings) were damaged in the fire. /3/

The Safety Board determined that the probable cause of the accident was a fractured cargo transfer pipe that resulted from (1) the failure of **ATOFINA** to adequately inspect and maintain its cargo transfer equipment, and (2) inadequate Federal oversight of unloading operations involving hazardous materials. Contributing to the accident was **ATOFINA**'s reliance on a tank car excess flow valve to close in the event of a leak from cargo transfer equipment and the company's failure to require appropriate safety equipment for employees involved in tank car loading and unloading operations.

While **ATOFINA** could not provide data or records to confirm whether or when the transfer pipe that failed in this accident had last been inspected under the mechanical integrity program, the erosion-corrosion that was found within the failed pipe indicated that the program had clearly not been effective. The Safety Board concluded that **ATOFINA**'s failure to implement effective procedures for inspection and maintenance of its unloading pipes and fittings allowed the transfer pipe in this accident to gradually deteriorate and ultimately fail.

Both OSHA and the Environmental Protection Agency (EPA) required ATOFINA to develop and document safety plans for the Riverview facility that included safeguards intended to reduce the risk and consequences of catastrophic releases of hazardous materials. ATOFINA's risk management plan (mandated by the EPA) and process hazard analysis (mandated by OSHA) included an accident scenario that involved the failure of a flexible hose on the unloading apparatus for a methyl mercaptan tank car - a scenario similar to this accident. Under both plans, ATOFINA indicated that the release of methyl mercaptan would be stopped by the automatic closure of the tank car's excess flow valve. Further, **ATOFINA**'s risk management plans explicitly noted that excess flow valves on the tank car would activate in the event of a pipeline or unloading hose rupture. **However, when the transfer pipe/hose failed on July 14, 2001, the excess flow valve on the tank car did not close and stop the release of the methyl mercaptan.**

Calculations made by Safety Board engineers and parties to the investigation indicated that the flow rate of methyl mercaptan through the broken transfer piping was insufficient to cause the excess flow valve to close. Excess flow valves are designed to close and stop the release of product from the tank car in the event a tank car valve or fitting is broken or sheared off during transit. Attaching cargo transfer apparatus to a tank car can change product release rates and flow rate characteristics and can prevent the excess flow valve from closing in the event of an emergency. As noted by the Chlorine

Institute in its *Chlorine Manual Pamphlet 57 and* by the Safety Board in its investigation of a July 30, 1983, accident at the Formosa Plastics plant in Baton Rouge, Louisiana, tank car excess flow valves are not designed to act as an emergency shutoff device during cargo transfer. /4/

To determine whether reliance upon tank car excess flow valves as safety mechanisms during transfer operations is restricted to **ATOFINA** or is a broader problem, Safety Board investigators interviewed a sampling of domestic chemical companies. Interviews with personnel responsible for company safety plans revealed that six of nine companies surveyed rely on tank car excess flow valves as a method of stopping or limiting a leak in the transfer equipment. Only one company reported having remotely operated shutoff valves on the unloading piping just outside the tank car dome. (The other two companies did not respond to the Safety Board's inquiry.) Although the Safety Board's sampling was limited, the results suggest that the **inappropriate use of tank car excess flow valves may be a widespread practice in the chemical industry.**

As a result of its investigation of the Riverview accident, the National Transportation Safety Board therefore made the following safety recommendation to the Occupational Safety and Health Administration: Assist the U.S. Department of Transportation in developing safety requirements that apply to the loading and unloading of railroad tank cars, highway cargo tanks, and other bulk containers that address personal protection requirements, emergency shutdown measures (passive device technology), and the inspection and maintenance of cargo transfer equipment.

Festus, Missouri and the Continued Consequences of Over Reliance on Excess Flow Valves

A chlorine repacking plant receives 4 railcars contained nearly 250,000 lbs of chlorine to be repacked into cylinders and other distribution. After 59 days, a hose wrongfully identified as a chlorine hose catastrophically failed spilling 48,000 pounds of Chlorine over 3 ½ hours. The root cause in addition to the hose failure is that backup systems (ESD'S) as well as Excess Flow Valves (EFV's) failed to work. /5/

DPC Chemical, Festus, Missouri August 2003

Below is the section of the hose that failed after three workers on break heard a loud “pop”, which clearly was this hose failing. The hose was made of stainless steel not suitable for Chlorine uses, although it was identified as hastelloy by the manufacturer, a material that is suitable for Chlorine usage. There is no visual difference between stainless steel and hastelloy, therefore the chlorine repackagers employees relied upon the manufacturers representation that they were indeed shipped the correct hose. Moreover, the facility relied upon the four valves (ESDs and Excess Flow Valves) in case of this precise type of error.

Ruptured chlorine transfer hose with its Teflon inner layer exposed;
when the hose burst it ejected pieces of Teflon as well as other components of the hose.

The illustration below (Figure 1) shows the positions of all valves and hoses as it relates to this incident. Determining the actual positioning of the five ESD valves at tank car station #3 led the United States Chemical Safety Board's investigators to conclude that the chlorine flowed through the partially opened ESD valve #3, to the attached intact liquid CTH line, into the plant side piping, and back through the ruptured hose (see arrows denoting product path). Closing tank car valve A did not stop the release because valve C continued to supply chlorine through the intact liquid CTH line. Even if these events replayed themselves the product flowing through the building because of the failure of the ESD's would have been stopped known passive device technology on the inboard side . The pressure of the product itself would have kept the plunger/valve sealed and limited loss of product to what was in the hose, likely below the RQ (reportable quantity).

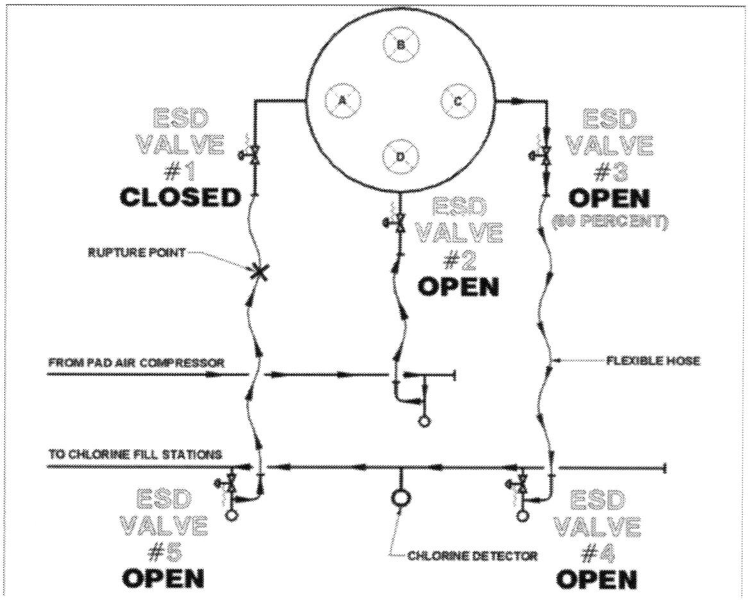

Figure 1

CSB investigators concluded that it is all but impossible to determine all possible routes of escape for hazardous materials or the infinite numbers of LPHC (Low Probability/High Consequence) scenarios. It is however abundantly clear that the hose is the weakest link in the chain and that devices, training and other precautions have proven to be inadequate in combating hose failures. In most reports of hazardous material release incidents we submit that one constant seems to be that employees do not follow regulations, rules and precautions that management, organizations and agencies have mandated and reasonably assume are being followed. Moreover, the valving technology that relies upon excess flow principles has been vastly over relied upon to address these incidents and governmental regulatory bodies have taken notice through he issuance of warnings and landmark Special permits encouraging the use of the passive device system mandated in 49 CFR 173.315.

Regulators Begin their Proliferation of Passive Device Technology Benefits

A Special Permit (previously referred to as an Exemption to Regulation) is temporary regulatory relief from one or more regulations given to a person or class of persons subject to the regulations, or who intend to engage in an activity that would make them subject to the regulations. (b) A Special Permit provides the person or class of persons with relief from the regulations for up to two years, and may be renewed. (c) Special Permits may only be granted from one or more of the requirements contained in the certain sections of the various regulations. To apply for such a Permit, the applying party must identify how they would ensure that they could achieve a level a safety that is **equivalent to, or greater than, the level of safety that would be obtained by complying with the regulation.** Thus the issuance of such a permit indicates that the regulatory body considers the procedure or technology to be of a greater safety value than the regulations promulgate. This is precisely the conclusion that the United States Federal Railroad Administration reached when they issued the landmark Special Permit 12325-SP. (See attached FRA Analysis)

Exemption/Special Permit 12325-N provides that the use of a Smart-Hose Technologies hose assembly will exempt the user from 49 CFR 173-177-Railcar Unloading Attendance Requirements. This regulation normally provides that rail tank car unloading operations must be performed by a reliable person who has been properly instructed in unloading hazardous materials and attended by the unloader until unloading is completed. Section 177.834 requires that a cargo tank motor vehicle be attended by a qualified person (usually the driver) at all times while it is involved in transfer operations to ensure the safe loading or unloading of hazardous materials being loaded or unloaded. The purpose of the attendance requirement in section 174.67 is to ensure the safe loading or unloading of hazardous materials and that, in the event of an emergency; such processes can be quickly halted. The Permit provides that the attendance requirements in the above regulations are relaxed if the rail car operator is using a Smart-Hose hose assembly.

The loading and unloading of tank cars requires the driver to stand and monitor the discharge of gas and fluids from the vehicle. The DOT recognized that this monitoring does not always occur particularly in the cold wintertime when the drivers may retreat from the cold to their vehicles. Air Liquide, a current user of the Smart-Hose passive device technology effectively petitioned the DOT to create a Special Permit 13484 to provide for relaxed attendance requirements for MC 330, MC 331 and 338 cargo tankers for division 2.2 gases.

Warnings are Issued

Nearly two years after the National Transportation Board made its recommendations; all major regulatory bodies in the United States issued warnings. In April of 2004, the United States Environmental Protection Agency issued the following warning: "Notify all facilities that are required to submit risk management plans to the

Environmental Protection Agency that tank car excess flow valves cannot be relied upon to stop leaks that occur during tank car loading and unloading operations and that those companies that have included reliance on such valves in their risk management plans should instead identify and implement other measures that will stop the uncontrolled release of product in the event of a transfer line[hose] failure during tank car loading or unloading."

In, September 2003, the **United States Department of Transportation – Federal Railroad Administration** issued the following warning: FRA is issuing Safety Advisory 2003-02 advising all persons involved in loading and unloading products from railroad tank cars that they cannot rely on internal excess flow valves to stop the flow of product except under the limited conditions for which these valves were designed and installed.

As early as 1980 **OSHA had issued warnings** indicating that in Anhydrous Ammonia service, it is emphasized that the system is not adequately protected without devices designed specifically to protect the system in case of rupture of the applicator tank liquid withdrawal hose. It is strongly stressed that excess-flow valves or other protective devices installed in bulk storage tank feed hose outlet lines are not designed to protect the system in case of an applicator tank withdrawal hose rupture. These devices are required to protect the lines that they serve and are designed specifically for this purpose only. For example, a manufacturer of anhydrous ammonia equipment advised that an excess flow valve installed in a bulk storage tank hose outlet (designed to accommodate a one-inch fill hose that is 15 feet in length) will be spring-loaded to trip should the hose be severed at any point along the 15 foot length of the hose; however, it is doubtful that adequate protection would be provided beyond the 15 foot point. Thus in the case of the applicator tank equipped with the combination valve, the hose outlet excess flow valve provides protection of the system from the bulk tank to the applicator fill/withdrawal valve with doubtful effectiveness in protecting the system beyond this point. It should be noted that the applicator tank involved in the accident was provided with a 3/4 inch liquid withdrawal hose. A reduction in the diameter of the hose line on the applicator tank further reduces the possibility of the bulk tank hose outlet excess flow valve tripping in case of severance of the applicator tank withdrawal hose -- especially in view of the fact that excess flow valves function on the basis of gallons per minute (GPM) flow rates and pressure differentials, i.e., a sudden drop of pressure in the line.

The preceding discussion has been limited to liquid lines and the filling of an applicator tank from a bulk storage tank. However, the same principles apply when filling the applicator tank from a nurse tank. In addition, it is equally important that protective devices be installed in the vapor lines to protect the system in case of vapor line rupture. It should also be noted that the bulk tank that was used to fill the applicator tank involved in the accident was not provided with an excess flow valve on the one-inch hose outlet on the bulk tank.

HM-223 - the Jurisdiction and Perspective Changes

Under the newly enacted Rulemaking (98-4952), otherwise known as HM-223, RSPA clarified the applicability of the Hazardous Materials Regulations (HMR) to specific functions and activities, including hazardous materials loading and unloading operations and storage of hazardous materials during transportation. They also listed in the HMR pre-transportation functions to which the HMR apply. They stated that pre-transportation functions are functions performed to prepare hazardous materials for transportation in commerce by persons who offer a hazardous material for transportation or cause a hazardous material to be transported. They determined that transportation functions are functions performed as part of the actual movement of hazardous materials in commerce, including loading, unloading, and storage of hazardous materials that is incidental to that movement. For purposes of applicability of the HMR, they concluded that ``transportation in commerce" begins when a carrier takes possession of a hazardous material and continues until the carrier delivers the package containing the hazardous material to its destination as indicated on shipping papers or other shipping documentation. Thus, in promulgating this rule, they effectively transferred the regulatory oversight of loading and unloading railcars to the Occupational Safety & Health Administration ("OSHA") and, to a lesser extent, the Environmental Protection Agency.

In light of the National Transportation Board's recommendations to OSHA to include these warnings in their assumption of jurisdiction and related regulator interpretations, will OSHA utilize this opportunity to further the trend away from excess flow valve reliance towards passive device technology? It would appear at first review, the answer is "Yes". In a recent submission seeking a Letter of Interpretation, the following questions were asked and verbally answered in the affirmative with formal comment expected to be issued with 30 days of this submission:

1. Does OSHA have regulations that specify tank car unloading similar to 49 CFR 174.67 (i) (j) and (k) which provide: (i) Tank cars may not be allowed to stand with unloading connections attached after unloading is completed. Throughout the entire period of unloading, and while car is connected to unloading device, the car must be attended by the unloader.(j) If necessary to discontinue unloading a tank car for any reason, all unloading connections must be disconnected. All valves must first be tightly closed, and the closures of all other openings securely applied. (k) As soon as a tank car is completely unloaded, all valves must be made tight by the use of a bar, wrench or other suitable tool, the unloading connections must be removed and all other closures made tight.

2. In the absence of these regulations, will OSHA's General Duty Clause (Section 5a1) which requires each employer to "furnish to each of his employees employment and a place of employment which are free from recognized hazards that are causing or are likely to cause death or serious physical harm" be the defining standard for these procedures?

3. Will the use of Smart-Hose assemblies, *(a passive device, as defined under 49 CFR 173.315, and proven hose technology designed to counteract the hazardous effect of hose rupture or failure during fluid or gaseous transfer operations through its unique, patented and patent pending design that eliminates the potential for disaster through the use of an internal cable connected to specially designed, normally unseated valve "wedges or plungers" located on each end of the cable. In the event of hose separation, stretching to the point of an unsafe condition or coupling-to-hose separation, the valve "wedges or plungers" are released and instantly seat stopping the flow in both directions.)*, in conjunction with the railcar loading and unloading requirements of 49 CFR 174 and the specific requirements of Special Permit 12325 (see below excerpts and complete copy is attached hereto and made a part hereof) satisfy the General Duty provisions of 5a1 thereby making the user compliant with OSHA's railcar loading and unloading regulations and related interpretations?

The direction that has been recommended by the National Transportation Safety Board and promulgated through warnings by the Environmental Protection Agency, the Department of Transportation- Federal Railroad Administration and the US-Department of Labor's Occupational Safety and Health Administration is away from over reliance on excess flow valves towards the regulatorily established Passive Device technology.

Conclusion

Given the numerous, detailed and accurate warnings from manufacturers, advocacy groups, government agencies and from accidents themselves, it is not reasonable for anyone be they industry or agency to be surprised when the next hose bursts and an EFV fails to stop the catastrophic consequences. These warnings are advice to beware to professionals charged with the safety of facilities, communities and individuals and are intended to be actionable.

The "passive device" addresses directly the deficiencies of the excess flow device. These deficiencies are not an indictment of the EFV device, the device behaves exactly as designed when used and installed properly. Rather, where the fault lies is in the properties and attributes that industry affixes to these devices to do a job for which they were not designed. The EFV is designed to trigger at a preset flow rate exceeding normal operation and contemplating the "shearing off" of valves or other catastrophic hardware event. In the event of a transfer line failure, in truth the EFV does not fail it simply works as designed, as the flow through a transfer line is likely within the parameters of normal operation and a rate at which the EFV set to accept.

The "passive device" is designed to shut off the flow of product in both directions in the transfer hose in the event of a failure between the couplings. This device provides safety at flow levels that are common in instances that sometimes are not sufficient to activate the EFV. While the EFV triggers at only high rates a "passive device" will trigger in incidents not only at high and low rates but even when a system is on standby (Festus) making the device not only complimentary but redundant to the EFV. When the next transfer lines fails and the call goes to the responsible party rather than wondering if they have dodged a bullet with an EFV the first question will be, "What did the hose do?"

A proven solution is at hand and industry and government are moving inexorably towards adaptation of passive device technology. The aforementioned warnings are not intended to provide cover to those that ignore them, as they are now so widely known. The fact is that given this knowledge and "best practice" the consequences to those who do take notice and action are likely severe.

Through the use of Passive Device Technology, sanctioned and in some cases mandated, the LPG industry can provide a level of safety for our operators, a level of compliance increasingly mandated and further confidence to our customers that safety does count and we embrace it.

Acknowledgment

We want to thank the various nongovernmental organizations that have supported this effort. We appreciate the contributions by the National Transportation Safety Board, the United States Chemical Safety Board, The United States Department of Transportation, The United States Environmental Protection Agency and The United States Department of Labor – OSHA. We further appreciate the World LP Gas Association for engaging us to research and develop a comprehensive paper on the impact of excess flow valve technology on the LP gas industry.

References

\1\ Docket HM-166W, NPRM at 53 FR at 36418, September 19, 1988; Final Rule adopting the amendment as proposed, 54 FR 38790, September 20, 1989.

\2\ 49 CFR 179.100-13(d).

National Transportation Safety Board, *Hazardous Materials Release From Railroad Tank Car With Subsequent Fire at Riverside, Michigan, July 14, 2001.* Hazardous Materials Accident Report NTSB/HZM-02/01 (Washington, D.C.: NTSB, 2002).

\3\ National Transportation Safety Board, *Safety Recommendation,* Norman Minetta, July 16, 2002.

\4\ The Chlorine Institute, Pamphlet 57, Emergency Shut-Off Systems for Bulk transfer of Chlorine, Edition 3, October 1997 section 3.1.

\5\ The Chemical Safety Board, Investigation Report Chlorine Release, DPC Enterprises, May 2003, incident Festus, Missouri August 14, 2002

2 National Transportation Safety Board, *Vinyl Chloride Monomer Release From a Railroad Tank Car and Fire, Formosa Plastics Corporation Plant, Baton Rouge, Louisiana, July 30, 1983*, Hazardous Material Accident Report NTSB/HZM-85/08 (Washington, D.C.: NTSB, 1985).

Although excess flow valves are routinely used as safety mechanisms in the piping systems of fixed facilities, those excess flow valves are designed and constructed for specific piping systems and the properties of the material flowing through the pipe.

Correspondence dated July 16, 2002, from Marion C. Blakely, Chairman, NTSB, to Allan Rutter, Administrator, FRA, summarizing the Board's investigation, including a public hearing, of this incident.

RISK BASED METHOD TO ESTABLISH INSPECTION INTERVALS FOR PRESSURE RELIEF DEVICES

Philip A. Henry, P.E.
The Equity Engineering Group, Inc.
Shaker Heights, OH

Valerie L. Magyari
The Equity Engineering Group, Inc.
Shaker Heights, OH

ABSTRACT
Properly functioning pressure relief devices (PRDs) are essential to protect plant personnel and equipment, as unexpected overpressure events can cause equipment damage and loss of containment, and result in costly plant shutdowns. PRDs are considered to be the ultimate layer of protection of pressurized systems.

The inspection and testing of relief devices is needed to ensure PRD functionality on demand. But when and how often should PRDs be inspected? Many plants apply inspection intervals to PRDs in general agreement with API 510 and NBIC, without considering historical performance or the potential hazards associated with the relief device application.

Previously, the primary focus of Risk-Based Inspection (RBI) programs has been to create inspection plans that reduce risk and optimize total maintenance and inspection expenses for pressurized static equipment. This approach has been expanded to include a quantitative methodology for PRDs. The PRD RBI technology provides a system approach (includes the damage state of the protected equipment) and evaluates the probability (reliability) and the consequence (criticality) of pressure relief valve failure and sets the appropriate inspection frequency based on risk. Results of recent studies have demonstrated the benefits of the new approach. Compared to a typical API510 approach, the risk based plan can significantly reduce the overall risk in the plant while at the same time a significant reduction in inspection costs can be realized.

A case study of implementation in a refinery process unit illustrates the methodology has been successfully used to establish risk-based inspection intervals for PRDs.

Chemical Safety Database - Application for Flammability Limits of Biogases

Maria Molnarne and Volkmar Schröder

Federal Institute for Materials Research and Testing (BAM)
12200 Berlin, Germany
E-mail: maria.molnarne@bam.de

ABSTRACT

The chemical safety database CHEMSAFE® contains safety characteristics for more than 3000 flammable substances and approximately 1000 mixtures of gases, liquids and dusts. The data is evaluated by experts of two Federal Institutes in Germany, BAM and PTB. The main advantage of CHEMSAFE® is that the data based on own tests of these institutes. The temperature and pressure influence on safety properties are also taken into account. Flammability limits are measured up to 400 °C and 10 MPa. In this paper an application of the database is shown for calculation of flammability limits of biogases. Biogas consists mainly of methane and carbon dioxide, with low levels of hydrogen sulfide and other gases. For the explosion protection in biogas plants, it is necessary to know the flammability limits of the process gases in the air. The composition of biogases from agricultural plants is varying very strongly, so that for each mixture composition the flammability limits should be determined experimentally. This is associated usually with considerable time and effort. A method is presented here which shows that the flammability limits of a biogas with known composition can be assessed with the help of two explosion diagrams. In addition to the methane and carbon dioxide content of the biogas a portion of water vapor is included in the calculation method and the influence of oxygen content in the system is considered.

1. INTRODUCTION

In the recent years biogas has become an important source of renewable energy. During the last three years some serious accidents, accompanied with injuries and fatalities, have happened in biogas production plants in Germany. These agricultural plants must also be controlled to avoid pollution and the risk of explosion. According to the European Directive 1999/92/EC (ATEX 137) the operator of biogas plant in Europe must meet explosion protection measures, such as the prevention of explosive mixtures (primary explosion protection) and ignition sources (secondary explosion protection). For this purpose an explosion protection document must be created. Such a document is based on the safety characteristics of hazardous substances. For biogas plants this safety characteristic is determined by the flammability limits of biogas in air. The biogas consists of mainly methane and carbon dioxide with small percentages of nitrogen, oxygen, hydrogen sulfide and other trace gases.

2. AGRICULTURAL BIOGAS PLANTS

A simple plant, as it can be found in a large number of farms, is shown in Figure 1. The biogas is present in the following operation units: in the digester, the connected gas storage tank and the combined heat and power unit (CHP) with possibly connected emergency flare. For larger plants, which have the permission to feed the gas into the natural gas network, there is a reprocessing plant with gas scrubbing, etc., mostly equipped with a downstream desulfurization unit. Especially in these operating units - with the appropriate vessels, valves, pipes and compressors - a failure may occur that results in the formation of an explosive gas-air mixture.

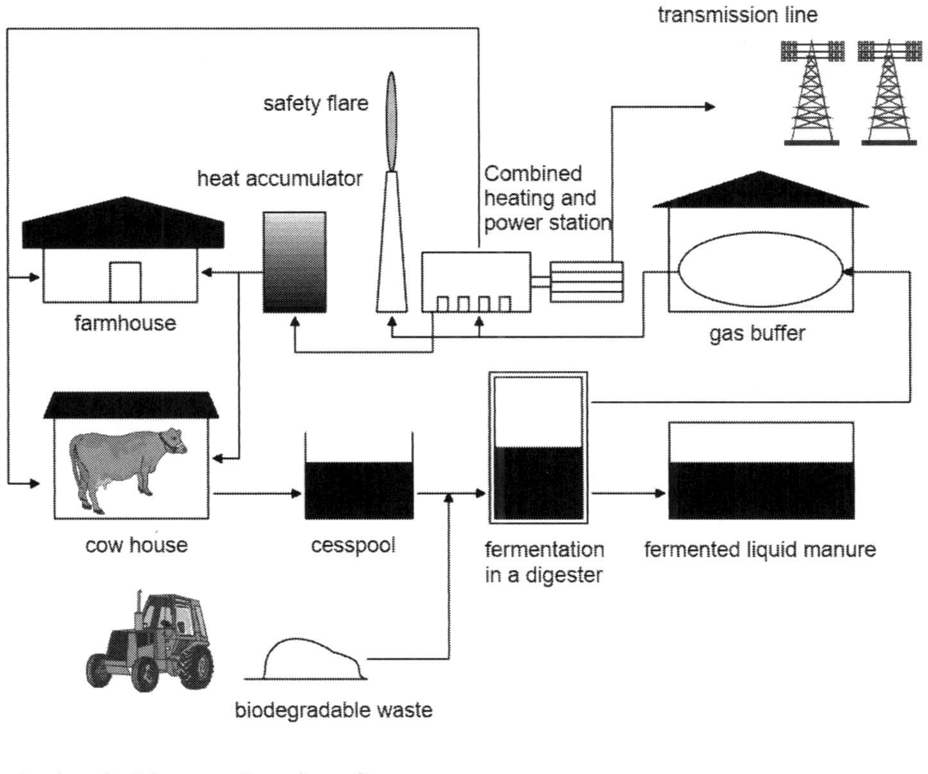

Fig. 1: A simple biogas plant in a farm.

Biogas is a naturally produced gas which is available from decomposition of organic materials in an anaerobic digestion process. The composition varies strongly depending on the introduced organic materials. The raw biogas usually consists of the following components:

- methane (45 mol% to 75 mol%),
- carbon dioxide (25 mol% to 55 mol%),
- water vapor (0 mol% to 12 mol%)
- nitrogen (0 mol% to 5 mol%)
- oxygen (0 mol% to 2 mol%)
- hydrogen sulfide (0.05 mol% to 0.5 mol%), and traces of ammonia, hydrogen and higher hydrocarbons.

The composition is given in mol% which is the same as volume fraction in percent at atmospheric conditions.

In addition, the composition of the process gas depends on the place, the part of the plant, where the gas is located. The biogas in the digester is saturated with water vapor at temperatures of up to 50 °C. On the other hand, the water vapor portion in the gas is significantly lower at ambient temperature. If the gas is fed into the natural gas network, it must be desulfurized, and the proportion of methane has to be increased. For the desulfurization usually some air is added.

The upper (UFL, fuel-rich) and lower flammability limit (LFL, fuel-lean) are referred to the mole fractions (concentration) of the biogas in air at which flame propagation after spark ignition no longer occurs, the last non-ignition point. The range between the lower and upper flammability limits is called as explosion range. In European standards and in this paper the flammability range is called explosion range to emphasize the explosive behavior of the mixtures and set it apart from the flammability of pure combustible substances. The area restricted by the lower and upper flammability limit curves in a fuel-air-inert gas system is called the explosion region. Mixtures with a composition in the explosion region are explosive and dangerous. The flammability limits of pure substances, such as methane, can be taken from data books and safety databases. For a mixture like "biogas" this is not readily available, so that in practice often the flammability limits of methane (4.4 mol% to 17.0 mol%) [1] are applied. At the lower flammability limit this is allowed, because the mixtures are in the "safe" area. At the upper flammability limit, such as by air inlet into biogas, this can lead to dangerous conditions.

It is better, therefore, if the composition is known, to estimate the flammability limits of biogas with the help of so-called explosion diagrams. The explosion ranges can be represented in different types of flammability diagrams. Most frequently diagrams are used with triangular or Cartesian coordinates. Both types of representation allow several possibilities for defining the axis. Unfortunately in the safety literature no uniform definitions can be found. For the present work an axis assignment has been selected, as it is used in the Federal Institute for Materials Research and Testing (BAM). A collection of such flammability diagrams of fuel-inert gas-air mixtures, as measured according to the German standard DIN 51649, is in [2].

3. EXPERIMENTAL DETERMINATION OF THE EXPLOSION REGIONS

The flammability limits, in case of the diagrams used in this paper, have been determined at atmospheric pressure with an apparatus in accordance with DIN 51649 [3]. Since 2004 this standard has been replaced by the DIN EN 1839 [4] tube method. Both standards recommend almost identical test apparatus. The new European standard describes in the tube method a little larger inner diameter for the ignition vessel (80 mm instead of 60 mm). The larger diameter does not affect the results for the mixtures of methane in question here. It was demonstrated by comparing measurements in [5].

In Figure 2 an apparatus can be seen for determination of the flammability limits according to DIN EN 1839 (DIN 51649), as it is used in the BAM. The gas mixture is metered by thermal mass flow controllers at different nominal volume flow rates. For the corresponding gas the mass flow controller must be calibrated by a precision gas meter.

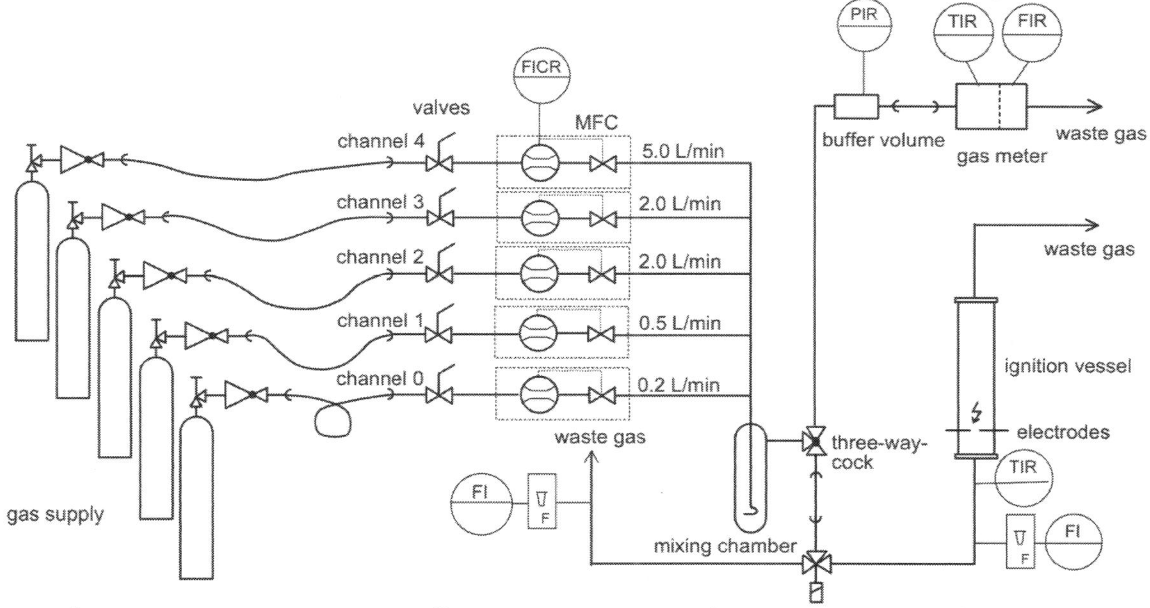

Fig. 2: BAM apparatus according to DIN 51649 and DIN EN 1839 „Tube method“, respectively.

Thereafter the gas is piped through a mixing vessel into the ignition vessel. After purging the ignition vessel the gas is ignited by using high voltage sparks and observed visually whether a flame is detaching from the electrodes and spreading out through the ignition vessel in test mixture. The ignition vessel consists of a vertical glass tube with a length of 300 mm and an inner diameter of 60 mm (DIN) and 80 mm (DIN EN), respectively. The apparatus is tested for functionality by a reference gas facing a test series. With the help of gas chromatography, the composition of the mixtures is randomly tested. The explosion regions of gas mixtures have been determined by varying the percentages of flammable gas, inert gas and air, respectively. The increment in the determination of the flammability limits was 0.2 mol% methane, and for determination of limiting oxygen concentration a step size of 0.5 mol% was used.

4. THE EXPLOSION REGION OF METHANE-CARBON DIOXIDE-AIR MIXTURES

The explosion regions of ternary mixtures can be represented most clearly in the so-called triangular diagrams. Here each side corresponds to a binary mixture and the three corners correspond to the respective pure substances. The composition of a mixture can be read directly from the diagram. The explosion diagram of methane-carbon dioxide-air in Figure 3 is an example of this type of diagrams.

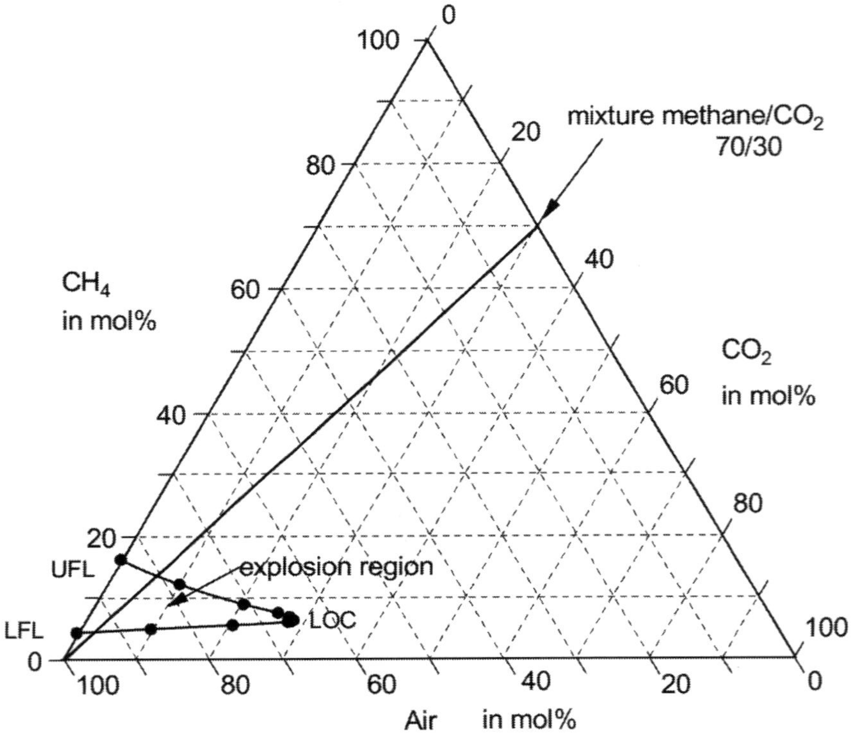

Fig. 3: Explosion region of methane-carbon dioxide-air mixtures [2], measured according to DIN 51649 at 20 °C and atmospheric pressure. The points represent the measured values.

The lower (LFL) and upper flammability limits (UFL) of methane in air are read off on the left side of the diagram. The values are LFL = 4.3 mol% and UFL = 16.3 mol% in this test. The air fraction at the apex of the explosion region is used to determine the limiting oxygen concentration (LOC). The air fraction is found with 65 mol%. This represents an oxygen concentration of 13.6 mol%.

Some basic statements can already be made by using these key data. In case of leakages and the accidental release of biogas in the surrounding atmosphere an explosive mixture can only be formed, if the lower flammability limit of methane is exceeded. Furthermore, it is only possible to obtain an explosive mixture in case of ingress of air into a biogas vessel, if the LOC is exceeded. The LOC of 13.6 mol% refers to a dry biogas, in which the water vapor portion is negligible. For the digesters, in which the biogas has a major portion of water vapor at temperatures up to 50 °C, an LOC of 12.4 mol% can be conservatively used. This value is determined from a methane-water vapor-air explosion diagram.

The straight line shown in Figure 3 corresponds to a biogas with a methane content of 70 mol% and a carbon dioxide content of 30 mol%. By constant addition of air to this mixture the composition of the mixture moves from the right side along the straight line by intersecting the explosion region to the lower left corner which corresponds to pure air. The flammability limits of the biogas (70/30) can be obtained also from the intersection points of the line with the explosion region.

5. CALCULATION OF THE FLAMMABILITY LIMITS OF METHANE-CARBON DIOXIDE-AIR MIXTURES

The lower percentages of other fuel gases which are present in biogas can be added to the methane portion for the calculation procedure. The flammability limits can be determined from the explosion region in Figure 3 for dry biogas with negligible portions of water vapor and oxygen. For better representation of the intersection points with the explosion region a diagram with Cartesian coordinates is often applied. This representation has the advantage that the axes are scalable, and the explosion region can be displayed slightly in optimum size. Moreover, most graphic and spreadsheet programs do not deliver triangle coordinates by default. A disadvantage of this Cartesian type diagrams is, however, that the representation of the axis for the third component is missing. Therefore the portion of air in a mixture can be gained by subtracting the portions of methane and carbon dioxide from 100 mol%. In Figure 4 two biogas compositions with methane-carbon dioxide (70/30) mol% / mol% and (50/50) mol% / mol% are exemplary represented. The flammability limits of biogases are calculated with the help of this diagram by using the mole fractions of methane and carbon dioxide according to equation (1) and (2).

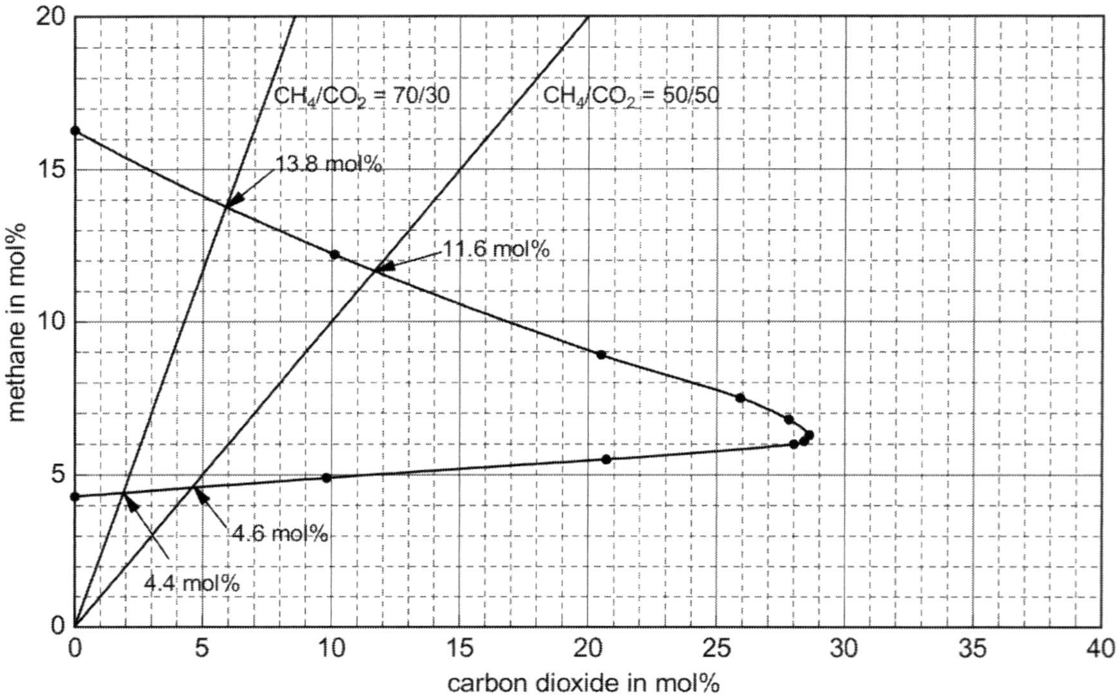

Fig. 4: Explosion region from Figure 3 in Cartesian coordinates.

$$LFL_{Biogas} = \left(1+\frac{x_{CO2}}{x_{CH4}}\right)\cdot LFL_{CH4,CO2} \tag{1}$$

$$UFL_{Biogas} = \left(1+\frac{x_{CO2}}{x_{CH4}}\right)\cdot UFL_{CH4,CO2} \tag{2}$$

x_{CO2}, x_{CH4}	fractions of the biogas components
$LFL_{CH4,\,CO2}$, $UFL_{CH4,\,CO2}$	methane fraction at the intersection point with the limit curves

These equations resulted in the following lower flammability limits

$LFL_{Biogas\ 70/30}$ = (1 + 30/70) 4.4 mol% = 6.3 mol%
$LFL_{Biogas\ 50/50}$ = (1 + 50/50) 4.6 mol% = 9.2 mol%

and the upper flammability limits for the sample mixtures

$UFL_{Biogas\ 70/30}$ = (1 + 30/70) 13.8 mol% = 19.7 mol%
$UFL_{Biogas\ 50/50}$ = (1 + 50/50) 11.6 mol% = 23.2 mol%

If these calculations are carried out for multiple configurations, a diagram can be constructed as shown in Figure 5. From this diagram the flammability limits of the biogas immediately can be read, if the methane content is known.

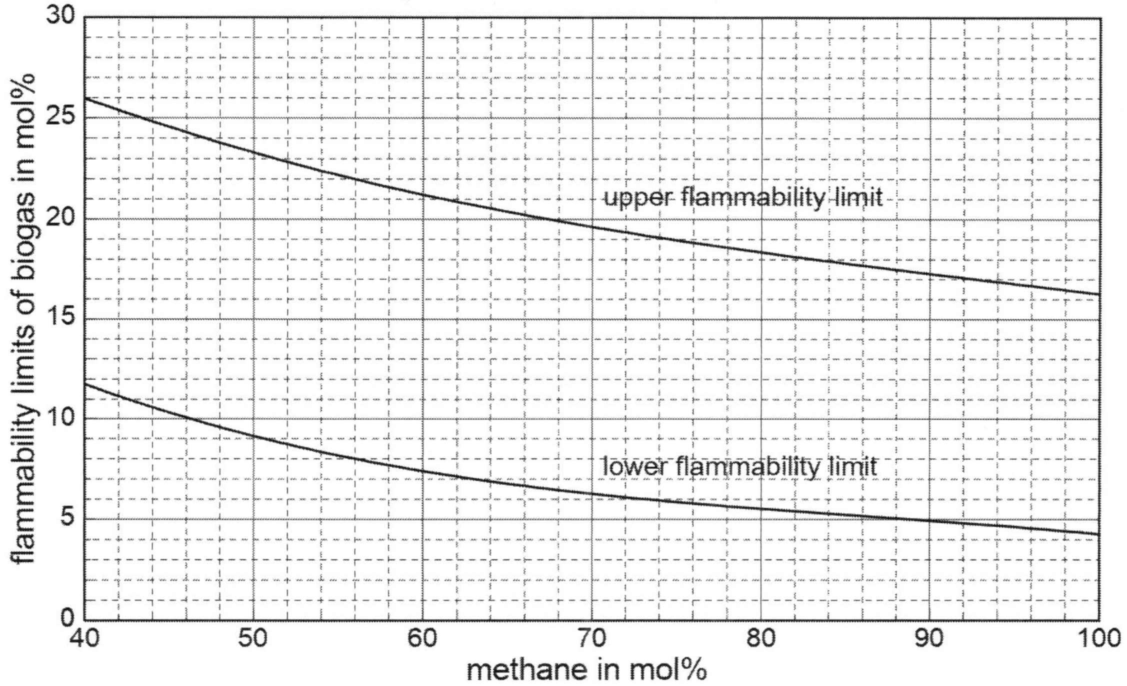

Fig. 5: Flammability limits of dry biogas at known methane content.

6. THE SIGNIFICANCE OF THE FLAMMABILITY LIMITS FOR MONITORING EQUIPMENTS

The first case to be considered is an accidental leakage and release of biogas into the surroundings. The key safety characteristic for this case is, in order to avoid of an explosive biogas-air mixture, the lower flammability limit. It indicates the percentage of biogas above which a biogas-air mixture becomes explosive. For this purpose monitoring equipments (gas warning devices) will be mounted on the units of biogas plants which detect the portion of methane entering into the atmosphere. Here it is important to mention that gas warning devices used in the explosion protection should be tested for functionality by an authorized testing laboratory. The accuracy of the measurement function must be ensured under the operating conditions (temperature, humidity, cross sensitivities).

From Figure 4 and from the calculation examples it is shown that, although the LFL of biogas with different CO_2 portions is quite different, the proportion of methane in the LFL remains nearly constant. For a biogas containing 30 mol% CO_2 the proportion of methane in the LFL is 4.4 mol% and for a 50 mol% biogas it increases only to 4.6 mol% methane (see Figure 4). It also shows that it is acceptable to use the LFL of pure methane as a basis for alarms in gas warning devices. A calculation of the LFL of biogas is therefore in this case not necessary.

The situation is quite different in the upper flammability limit. This case can occur by ingress of air into a part of biogas plants or even during the proper addition of air, such as by desulfurization. Because of an oxygen deficiency the air concentration has the priority in the consideration. It is useful here to apply oxygen sensors for monitoring. The allowable air concentration is calculated from the percentage of methane and carbon dioxide by subtraction from 100 mol%. In our example the biogas with 30 mol% carbon dioxide has an UFL of 19.7 mol% according to equation (2) and with 50 mol% CO_2 an UFL of 23.2 mol%. This corresponds to an air portion of 80.3 mol% and 76.8 mol%, respectively. The permissible oxygen concentration is 16.8 mol% or16.1 mol% calculated from the air concentrations. Using the upper flammability limit of pure methane in air is the oxygen concentration of 17.5 mol%. Using this oxygen percentage the resulting biogas –air mixture can become explosive.

7. CONSIDERATION OF WATER VAPOR IN THE DIGESTER

Before drying and compressing of biogases water vapor is present in different units of the biogas plant in phase equilibrium. The portion of the water vapor in the total gas mixture can be obtained from the temperature dependence of the vapor pressure. According to Table 1 there are significant water vapor contents only at higher temperatures, for example in the digester.

Tab. 1: Vapor pressure of water [6]

Temperature in °C	Water vapor content in mol% at 101.3 kPa
20	2.3
25	3.2
30	4.1
35	5.5
40	7.3
45	9.5
50	12.2

If air enters the digester, the flammability of these mixtures can be conservatively estimated from an explosion diagram of methane-water vapor-air (see Figure 6). This is acceptable, because the water vapor has slightly lower molar heat capacity than carbon dioxide. It does not lead to any explosion hazards, if the carbon dioxide proportion like water vapor portion is considered.

The following example illustrates this approach. The conditions in the reactor after an air access are: T = 50 °C, p = 101.3 kPa, the composition of dry biogas: 70 mol% methane and 30 mol% carbon dioxide. The detected oxygen percentage is 12 mol%.

At 50 °C is a water vapor content of 12.2 mol% according to Table 1.

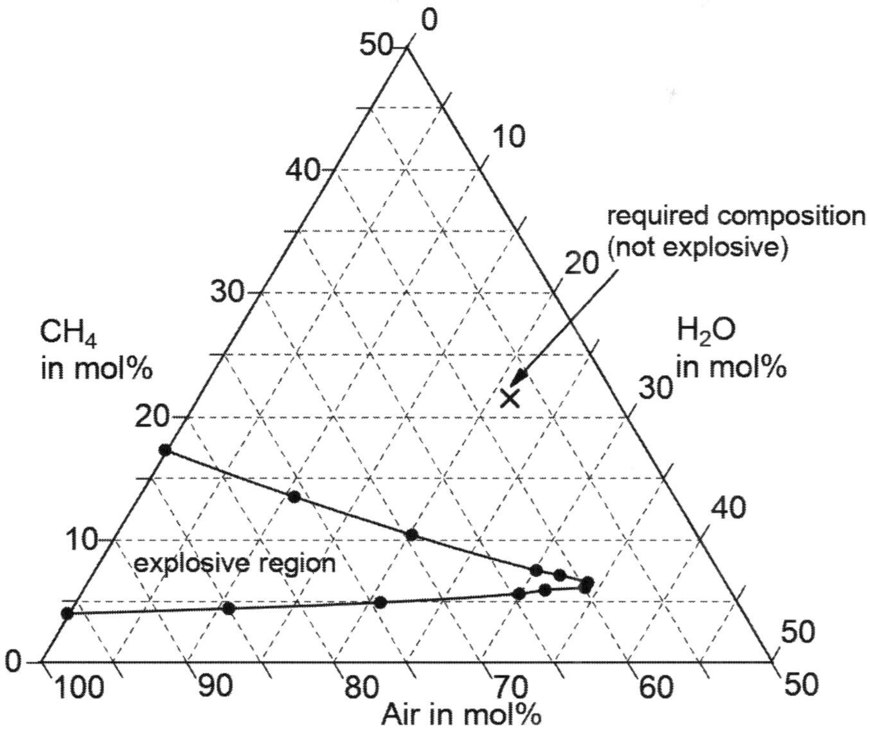

Fig. 6: Explosion region of methane-water vapor-air mixtures [2], measured according to DIN 51649 at 100 °C and atmospheric pressure.

First, the percentage of air is calculated by equation (3) from the measured oxygen content.

$$x_{Air} = x_{O2}/0.21 = 0.571 \triangleq 57.1 \text{ mol\%} \qquad (3)$$

The next step is the calculation of the sum of the inert gases (CO_2 + H_2O) by equation (4).

$$x_{inert} = (1\text{-}x_{Air}\text{-}x_{H20})*0.3 + x_{H20} = 0.214 \triangleq 21.4 \text{ mol\%} \qquad (4)$$

By using these data, the composition of the mixture can be placed into the explosion diagram in Figure 6 and it can be seen that this point is outside of the explosion region. That means that the mixture in our example - in the digester - is not explosive.

8. CONCLUSION

For the indication of the flammability limits of biogas in air it is important to distinguish whether these refer only to the methane content, or to the whole biogas. Both can be found in the safety literature and confusion may lead to dangerous situations.

It is sufficiently safe, to use the lower flammability limit of methane at atmospheric conditions as a warning alarm in biogas plants.

If air is entering into parts of biogas plants, such as by shut downs and start-ups, repairs, etc., the upper flammability limit of biogas should be used to estimate the risks of explosion. The use of the upper flammability limit of pure methane as a reference point for monitoring equipment is not allowed here.

In these cases with the proposed calculation method or from the diagram in Figure 5 of the remaining allowable air content will be determined. Gas warning devices which are applied for measuring the oxygen content can then be set up to these values. For gas alarms the LOC value (see Figure 3) can also be used here, but it overestimates the actual risk.

The influence of temperature on the flammability limits is low. Up to the temperatures of 50 °C, e. g. in the digester or by sunlight, it can be neglected. The influence of higher pressures, such as by compression of biogas in CHP unit may occur, however, must be observed and separately considered. For this purpose explosion diagrams for higher initial pressures are needed.

Avoiding ignition sources as the sole explosion protection measure is usually not enough. Experience has shown that in case of complex systems, usually it is not possible all possible ignition sources to eliminate.

9. REFERENCES

1. DECHEMA, BAM und PTB, *CHEMSAFE® - Database for evaluated Safety Characteristics*, Update 2007, STN International, Karlsruhe (http://www.dechema.de/en/chemsafe.html)
2. Molnarne, M., Schendler, Th. and Schröder, V., *Sicherheitstechnische Kenngrößen Band 2: Explosionsbereiche von Gasgemischen,* Bremerhaven, Wirtschaftsverlag NW - Verlag für neue Wissenschaft GmbH, 2003.
3. DIN 51649: *Bestimmung der Explosionsgrenzen von Gasen und Gasgemischen in Luft,* Berlin, Beuth Verlag, 1986.
4. DIN EN 1839: *Bestimmung der Explosionsgrenzen von Gasen und Dämpfen,* Berlin, Beuth Verlag, 2004.
5. Schröder, V. and Daubitz, R., *Evaluation of Standard Test Methods for the Determination of Explosion Limits of Gases and Vapors*, Proceedings of 11th Int. Symp. of Loss Prevention, Praha, Czech Rep., 2004.
6. Poling, B. E., Prausnitz, J. M. and O'Connel, J. P., *The Properties of Gases and Liquids*, Fifth Edition, New York, McGraw-Hill, 2001.

Development of a Methodology for Toxic and Flammable Gases Sensors Positioning in Oil Platforms

Giordhanne Bruno Carpaneda Gimenes

Mechanical Engineer, CHEMTECH - A Siemens Company
Rio de Janeiro, RJ, Brazil
giordhanne.gimenes@chemtech.com.br

Aristeu da Silveira Neto, PhD.

Federal University of Uberlândia -FEMEC/LTCM
Uberlândia, MG, Brazil
aristeus@mecanica.ufu.br

Solidônio Rodrigues Carvalho, D.Sc.

Federal University of Uberlândia -FEMEC/LTCM
Uberlândia, MG, Brazil
srcarvalho@mecanica.ufu.br

Valter César de Souza, D.Sc.

Consultant, CHEMTECH - A Siemens Company
Rio de Janeiro, RJ, Brazil
valter.souza@chemtech.com.br

Patrícia Carneiro dos Santos

Process Engineer, CHEMTECH - A Siemens Company
Rio de Janeiro, RJ, Brazil
patricia.santos@chemtech.com.br

ABSTRACT

Computational fluid dynamics (CFD) and mechanical systems optimization concepts are currently quite important tools for engineering problems analysis. A typical application of such tools is gas dispersion studies, a theme of great importance for the oil and gas industry. An important safety item for oil and gas production installations (especially oil rigs) is the use of flammable and/or toxic gas sensors to detect gas leaks capable of starting fires, explosions and causing toxic exposure. The

Importance of these sensor rules may be evaluated if one considers all potential costs of compensations, environmental impacts, life loss and profit loss.

This work presents the development of a computational tool that supplies optimum position and number for gas detectors in an oil platform. Associating CFD simulation results with coherent mathematical treatment, this methodology presents improvements in relation to other studies of gas dispersion and sensor positioning. This mathematical treatment supported by this computational tool is responsible for all calculation methods and for the display of graphical results of the locations to place detectors both for toxic (hydrogen sulfide, for instance) and flammable (like methane) gases.

Gas detection projects are usually based on standards and technical notes that do not supply all necessary information for proper sensor positioning. In the same way, although qualitative methodologies are used intensively and are quite accepted, they are not as accurate because, in most cases, the final decision is based on personal analysis and previous experience. Although experience will always remain an important asset in this kind of service, development of such a computational tool as the one described in this paper will reduce project schedules, besides providing more accurate and less intuitive results.

1. INTRODUTION

One of safety items for oil installations is the use of concentration sensors for flammable and/or toxic gases due to high risk proportionated by leaks (fires, explosions, intoxications and others). Consequences of an accident will result in considerable setbacks including compensation, environmental impact and profit stagnation. Furthermore they damage operator's image before investors and society. Pioneering companies in safety studies developed methods for the determination of the positioning of these sensors; however such methodologies are not published because of their strategic value.

Nowadays, gases detection project are usually based on norms and technical notes that don't supply all the necessary information for the correct positioning of the sensors. In the same way, qualitative methodologies are used intensely, however, in spite of being accept; they are not so good because, most of the time, final decision is based on the personal analysis and the engineer's experience.

Due to the mentioned reasons, a new methodology of easy application and in agreement with the norms and specifications in vigor is being developed by CHEMTECH Services of Engineering and Software Ltda. in partnership with the Laboratory of Heat and Mass Transfer and Fluid Dynamics (LTCM) (College of Mechanical Engineering (FEMEC) of the Federal University of Uberlândia (UFU)).

This methodology is based on the use of data resultant from gas dispersion simulations through CFD – Computational Fluid Dynamics. These data are treated with statistical methods aiming to identify the necessary places for the installation of gas detectors. This treatment is executed with the support of a computational tool that is responsible for the process calculation and supplies graphical results of most indicated areas for the installation of detectors.

2. METHODOLOGY

The methodology in development can be used both to toxic gas (sulphidric gas, for instance) and to flammable gas. In the present article, however, it will just be broached the detection of flammable gas, through the example of methane (CH_4). This gas is predominant in the natural gas composition (Table 1) and so it deserves special attention in the studies of gases dispersion and allocation of detectors in oil facilities.

ORIGIN	% VOLUME						Density	Superior Calorific Power (MJ / Nm³)
Country / Field	Methane CH_4	Propane C_3H_8	Ethane C_2H_6	C_4 and Higher	CO_2	N_2		
USA/Panh.	81,8	5,6	3,4	2,2	0,1	6,9	-	42,7
USA/Ashlaw	75,0	24,0	-	-	-	1,0	-	46,7
Canada	88,5	4,3	1,8	1,8	0,6	2,6	-	43,4
Russia	97,8	0,5	0,2	0,1	0,1	1,3	-	39,6
Australia	76,0	4,0	1,0	1,0	16,0	2,0	-	35,0
France	69,2	3,3	1,0	1,1	9,6	0,6	-	36,8
Germany	74,0	0,6	-	-	17,8	7,5	-	29,9
Netherlands	81,2	2,9	0,4	0,2	0,9	14,4	0,640	31,4
Persia	66,0	14,0	10,5	7,0	1,5	1,0	0,870	52,3
North Sea	94,7	3,0	0,5	0,4	0,1	1,3	0,590	38,6
Algeria	76,0	8,0	3,3	4,4	1,9	6,4	-	46,2
Venezuela	78,1	9,9	5,5	4,9	0,4	1,2	0,702	47,7
Argentina	95,0	4,0	-	-	-	1,0	0,578	40,7
Bolivia	90,8	6,1	1,2	0,0	0,5	1,5	0,607	38,8
Chile	90,0	6,6	2,1	0,8	-	-	0,640	45,2
BRAZIL								
Rio de Janeiro	89,44	6,7	2,26	0,46	0,34	0,8	0,623	40,22
Bahia	88,56	9,17	0,42	-	0,65	1,2	0,615	39,25
Alagoas	76,9	10,1	5,8	1,67	1,15	2,02	-	47,7
Rio Grande do Norte	83,48	11	0,41	-	1,95	3,16	0,644	38,54
Espírito Santo	84,8	8,9	3,0	0,9	0,3	1,58	0,664	45,4
Ceará	76,05	8,0	7,0	4,3	1,08	1,53	-	52,4

Table 1: Composition of the natural gas.

Source: www.gasnet.com.br

Methane is an odorless and colorless gas. It has low solubility in water and when dispersed into the air (in certain ratios) it becomes a mixture with high explosive power. Some important physical-chemistry properties for this gas are in the Table 2.

METHANE		
NOMENCLATURE	MOLECULAR FORMULA	MOLECULAR WEIGHT
Methane, natural gas, gas of the swamps	CH_4	16,04
PHYSICAL-CHEMISTRIES PROPERTIES		
Physical state and risk classification: flammable, simple suffocating gas	Color and scent: Colorless of soft and sweetened scent.	
Solubility: 2,86x10^{-3}g/L @ 20ºC	Reactivity: Forms explosive mixies with air and oxidizers agents. Incompatible with halogenous and interhalogenous.	
METHANE FLAMMABILITY		
Lower Flammable Limit: 5 % v/v	Extinguishers agents: CO_2, dry chemical pownder, water vapor for small focuses.	
Upper Flammable Limit: 15 % v/v	Incompatible extinguishing agents: Separately used water can be inefficient. Not to extinguish the flames before eliminating the emptying source, therefore the new ignition can be explosive.	
Toxic products of combustion: He is not toxic, but its partial combustion can generate toxic gases (CO).		
OBSERVATIONS		
It can be flamed by heat, flashes or open flames Cause asphyxia for oxygen reduction. It can provoke giddiness and asphyxia without acknowledgment. In the liquid phase, it can generate burning in the skin and eyes for freezing.		

Table 2: Some properties of methane.

Source: Crowl et al (2002), Cote (2003), Eltschlager et al (2001) e Perry (1997).

According to technical specifications usually employed in the oil and gas industry, alarms in the central control room should go off whenever levels from 20% to 60% of Lower Flammable Limit (LFL) for combustible gases are detected, where LFL is the minimum concentration of gas (% in volume) dispersed in the air that, in contact with an ignition source, is capable to cause combustion of the mixture. In a similar form, the Upper Flammable Limit (UFL) is the maximum gas concentration (% in volume) dispersed in the air that, in contact with an ignition source is capable to cause combustion of the mixture (Figure 1).

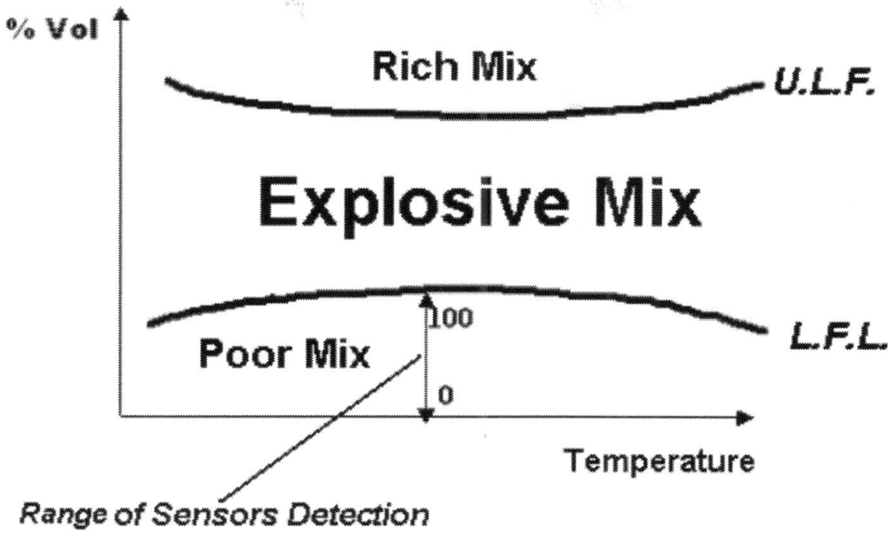

Figure 1: Strips of inflammability of a mixture gas-fuel

Source: Own elaboration

Lower Flammable Limit for methane is 5% in volume in air (Table 2), based on standard conditions of temperature and pressure (20°C and 1 atm). Therefore, the range of detection for this gas will be between 1% in volume (corresponding to 20% of the L.F.L) and 3% in volume (corresponding to 60% of the L.F.L).

In light of this information, three-dimensional gas dispersion simulations are carried out using tools of Computational Fluid Dynamics (CFD) considering geometric aspects, meteorological conditions and gas properties at leak conditions. Scenarios of leaks are simulated with several directions and speeds of wind, including wind calms. Meteorological conditions are important variables in models of gas dispersion (Crown et al, 2002), because gases are dispersed by wind. Atmospheric instability favors dispersion of the gaseous pollutant in the atmosphere. Particularly, in this study developed by the partnership CHEMTECH-UFU, the CFD tool used is the software PHOENICS (Parabolic Hyperbolic Or Elliptic Numerical Integration Code Series) - version 2006, developed by the British company CHAM (Concentration, Heat and Momentum Limited). CHEMTECH is the PHOENIC's licensor in Brazil. More details can be found in the website http://www.cham.co.uk.

In the computational tool in development by CHEMTECH-UFU the user should enter the data referring to leak conditions (probability of leak occurrence in a certain equipment or component) and meteorological aspects (probability of occurrence of each wind direction considered in the simulations). The reference data giving the probabilities of leaks can be extracted from historical reports supplied by the proper operator or from reports published by specialized institutions in health, safety and environment, such as the Health & Safety Executive (HSE), the agency responsible for health and safety regulation in Great Britain. The data for the probabilities of occurrence of winds in the considered directions can be obtained from existing meteorological stations in the area where the plant will be installed. With this information, the probabilities of occurrence of the flow configurations will be calculated. This probability is general and it takes into consideration both the leak occurrence as the wind occurrence and the analyzed direction, synthesizing in a unique value these two variables.

The probabilities calculation, is based on Bayes's Theorem. According to the magazine Ciência Hoje (2006) Thomas Bayes's reasoning is nowadays considered a new form to view the world, as the basis of a true revolution in different fields of the knowledge, from genetics to theology.

Bayes's Theorem is based on one of the basic beginnings of the probability: the probability that an event A occurs simultaneously with an event B, is given by the Equation (1):

$$P(A \cap B) = P(A/B).P(B) \tag{1}$$

Where P(A) and P(B) they are the occurrence probabilities of the events A and B, respectively, and $P(A/B)$ is the probability of occurrence of A considering that B has happened. The development of the Bayes's Theorem is available in specialized literatures in Statistics. The final form of this theorem is given by the Equation (2):

$$P(A/B) = \frac{P(B/A).P(A)}{\left[P(B/A).P(A)\right]+\left[P(B/A^c).P(A^c)\right]} \tag{2}$$

Where A^c is the complementary event of A, also called of *not* A. The bayesian reasoning is not intuitive, being conceptually a 'opinion modifier', in other words, the probabilities are adjusted with the knowledge of new evidences.

After the calculation of the probabilities, the next step is to load the output from CFD simulations (in the case of PHOENICS, extension .phi), separating the values of the variable of interest, in this case, concentration (variable C_1). Based on loaded files, reminding that the number of loaded files is the same to the number of simulations, a new file will be generated (C_2). In this file they will just be printed the computational volumes (Figure 2) that contain concentrations among $20\% \leq C_1 \leq 60\%$ of L.F.L. (for the methane, $1\% \leq C_1 \leq 3\%$ in volume). In the computational volumes out of this range, the value "zero" it will be attributed. At the end of this stage a file will be generated one C_3 file for each file .phi loaded.

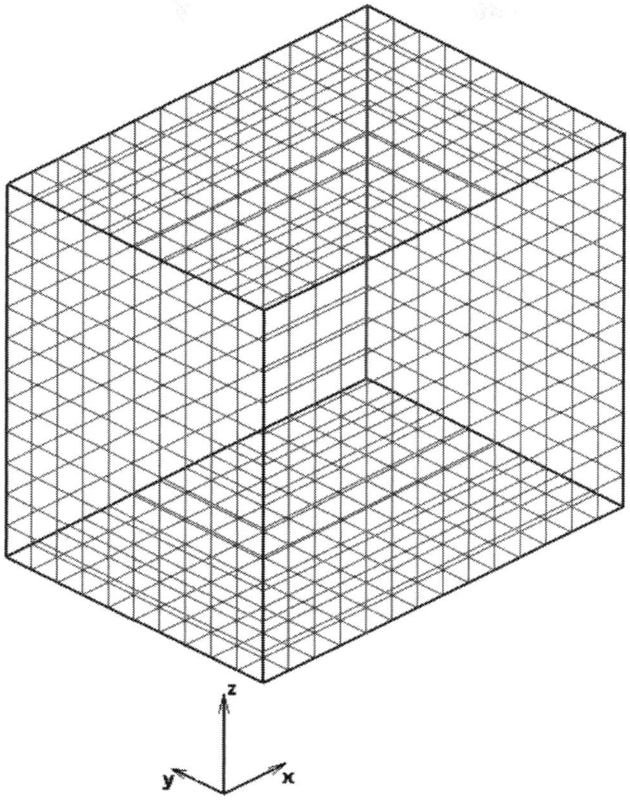

Figure 2: Three-dimensional domain divided in small computational volumes.

Source: Own elaboration.

Soon after, multiply each C_2 file for the respective occurrence probability (resultant of the calculation of the Bayes's Theorem), generating C_3 files. Then, all the files are compared, selecting the highest value for each volume, saving them in an unique file (C_4).

Finally, these values are normalized so that their range varies between 0 (zero) and 1 (one). This final file supplied by the computational tool in development by the partnership CHEMTECH-UFU, may be opened in a graphic software or in the own post-processing module of the CFD tool. This way, it is possible to visualize the field that shows importance index by colors the of the possible areas for allocation of sensors, among 0 (without need) and 1(priority). The Figure 3 summarizes these stages through a schematic flowchart of sequence of files created by this computational tool in development.

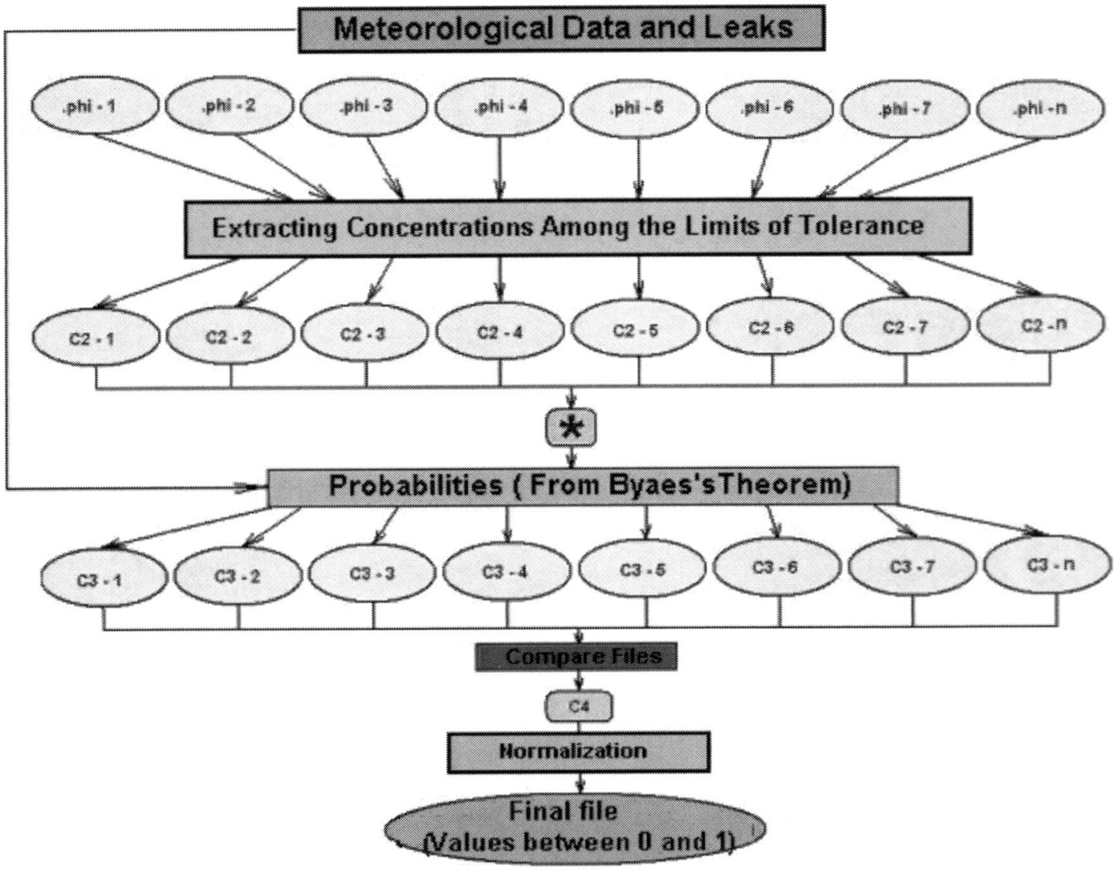

Figure 3: Flowchart of the computational tool.

Source: Own elaboration.

Therefore, at the end of the process, an index will be generated between 0 and 1 indicating probabilities of having gas with concentrations between $20\% \leq C \leq 60\%$ of L.F.L. in the computational volume, considering the probabilities of occurrences of leaks and wind for all the configurations (leak and wind).

It is important to note that is not necessary to know the exact gas percentage in each computational volume. The final file created only considers the existence of gas among $20\% \leq C \leq 60\%$ of L.F.L. this way, to know if the percentage is 21% or 30% adds a little in relation to the needs of the detection system, because any percentage between 20% and 60% of L.F.L. will be detected by the sensors, activating the alarms.

For each computational volume, a number of 0 (zero) to 1 (one) is attributed, that represents in probabilistic terms the classification of importance for each volume in the space potentially receiving a sensor. The methodology offers the advantage of being independent of opinion due to the mathematical treatment explained in this section.

3. RESULTS

Thin computational tool in development was applied to a previous sensor location project executed by CHEMTECH in 2003, in order to validate the proposed methodology. The results obtained in 2003 (through the qualitative methodologies) were compared with the results provided by the new methodology. Figure 4 shows the sensors positioned by CHEMTECH in the earlier study.

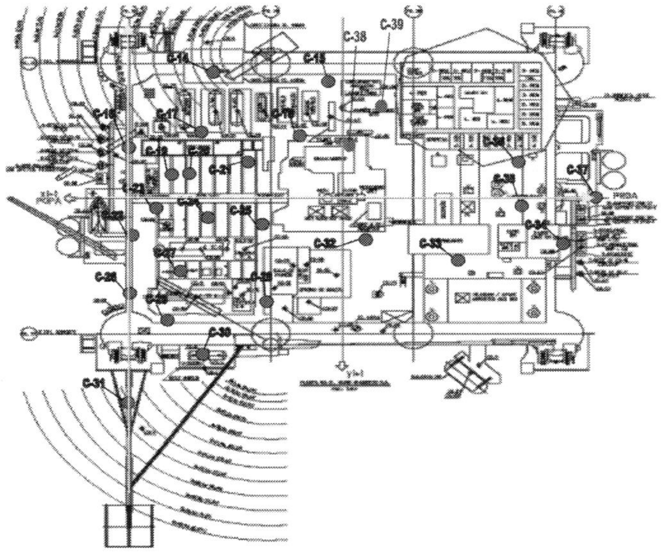

Figure 4: Sensors positioned by CHEMTECH in 2003

Source: Own elaboration

Figure 5 shows the comparison between the original location and the results obtained by the new methodology.

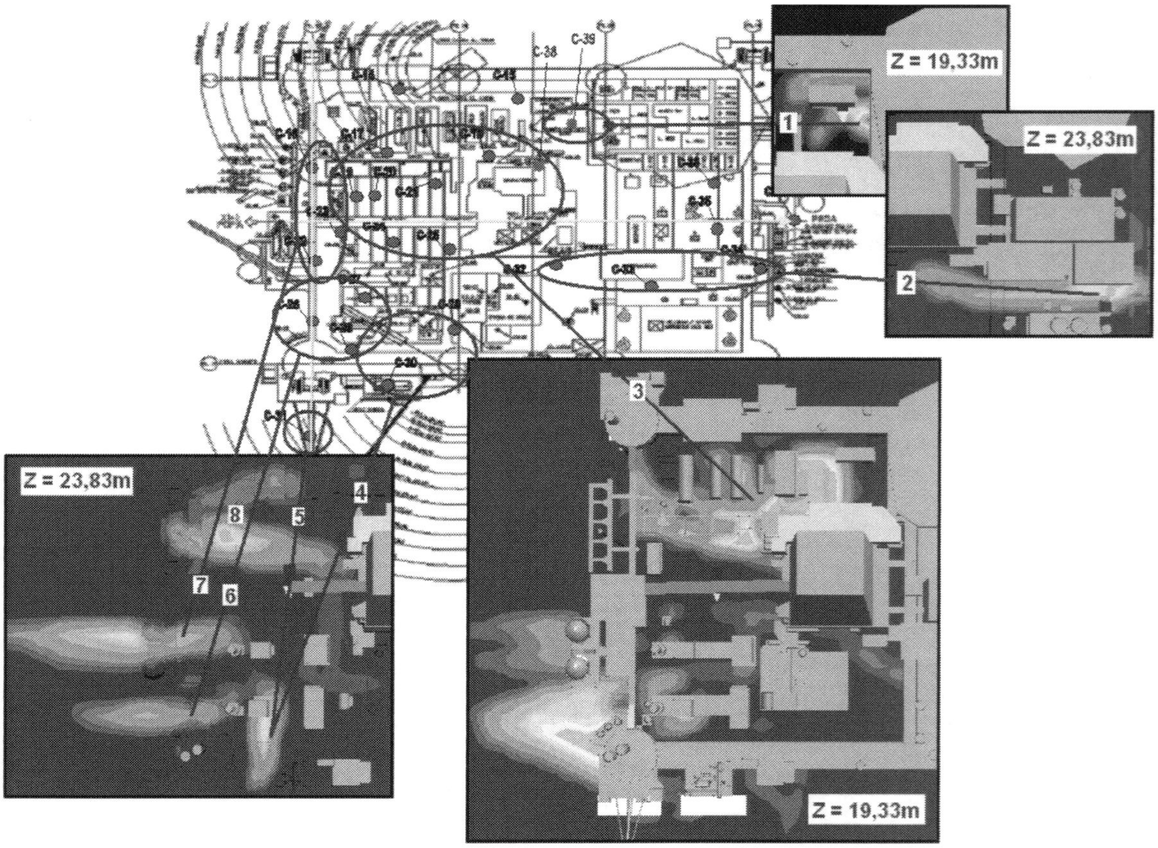

Figure 5: CHEMTECH 2003 versus the new methodology

Source: Own elaboration

The last figure shows that the results provided by the new methodology are similar to the results obtained in 2003. The scale of colors indicate the best regions for the installation of detectors, where "z" is the elevation in meters. With the support of this new tool, it was possible to take into consideration the analysis of many plumes with gas concentration between 20% and 60% of L.F.L. much more quickly. Qualitative methodologies usually analyses only plumes in the threshold gas concentrations for positioning sensors (20% and 60% of L.F.L.) due to many scenarios to be studied.

4. CONCLUSION

This methodology presents improvements in relation to others studies of gas dispersion and positioning of sensors, forming an alliance between CFD simulation results and coherent mathematical treatment. Development of the computational tool described in this document will reduce the time for conception of the projects, besides supplying more necessary and less intuitive results. It is important to remember that, in spite of the great progress that this methodology provides, researches will continue, in order to advance in aspects related to an optimum number of sensors.

5. REFERENCES

1. Ciência Hoje, *Revista de Divulgação Científica da SBPC*, Vol. 38, Julho 2006.

2. Cote, A. E. *Fire Protection Handbook* – Volume II. 19ª Edição. Editora Editora National Fire Protection Association. Estados Unidos, 2003.

3. Crowl, D. A.; Louvar, J. F. *Chemical Process Safety – Fundamentals with Applications.* 2ª Edição. Editora Prentice Hall. Estados Unidos, 2002.

4. Eltschlager, K. K.; Hawkins, J. W.; Ehler, W. C.; Baldassare, F. *Technical Measures for the Investigation and Mitigation of Fugitive Methane Hazards in Areas of Coal Mining.* US Department of Interior - Office of Surface Mining Reclamation and Enforcement. Pittsburgh, PA. Setembro, 2001.

5. HSE, Hid Statistics Report, *Offshore Hydrocarbon Releases Statistics and Analysis, 2002.*

6. Kelsey, A., Hemingway, M.A., Walsh, P.T. and Connolly, S., *Evaluation Of Flammable Gas Detector Networks Based On Experimental Simulations Of Offshore, High Pressure Gas Releases*, Institution of Chemical Engineers, Vol. 80, Part B, March 2002.

7. Perry, R. H. *Perry's Chemical Engineers' Handbook.* 7ª Edição. Editora McGraw-Hill. Estados Unidos, 1997.

8. Pupe, Camila Gonçalves; Soares da Silva, Ályson – *O Uso de CFD na Previsão da Dispersão de Gases em Plataformas Ofsshore*, 2006.

9. Yudkowsky, *An Intuitive Explanation of Bayesian Reasoning, 2003.* http//yudkowsky.net/bayes/bayes.html

A graphical approach for representing hazard scenarios

Rafael Batres

Toyohashi University of Technology
E-mail: rbp@pse.tut.ac.jp

Yukiyasu Shimada

Chemical Safety Research Group, National Institute of Occupational Safety and Health
E-mail: shimada@s.jniosh.go.jp

Tetsuo Fuchino
Tokyo Institute of Technology
E-mail: fuchino@chemeng.titech.ac.jp

ABSTRACT

This paper describes "hazard scenario graph" or HSG which is a visual representation technique for performing hazard identification. HSGs explicitly represent all the intermediate and final consequences in the scenario. HSG is based on ontologies and knowledge representation techniques with which knowledge extraction is possible. The method explicitly represents causality, substances, equipment, physical and chemical transformations, plant operations and personnel, facilitating the integration with other engineering tools such as CAD systems.

1. INTRODUCTION

Several methods for hazard identification have been developed, including checklists, What If Analysis, FMEA, and HAZOP. Among these, HAZOP is probably the most common method for hazard identification of a detail design. Broadly speaking, once a potential hazard scenario is generated, engineers capture the resulting information into two parts: initial causes and consequences. Causes of a scenario normally refer to a single initial cause. However, consequences can be recorded in multiple ways. Sometimes, as a single final consequence, such as an explosions or damage to facilities, but also as several consequences (either final or intermediate) grouped together in a single paragraph. As a result, intermediate parts of the scenario remain in the minds of the team or are hidden in the analysis report (typically in a tabular form). Furthermore, information is often captured in text format which limits the reuse of information offered by CAD and other plant information systems.

This paper describes "hazard scenario graph" or HSG which is a visual representation technique as an aid to hazard identification. HSGs explicitly represent all initial causes, intermediate and final consequences in the scenario. HSG is based on ontologies and knowledge representation techniques with which knowledge extraction is possible. The method explicitly represents causality, substances, equipment, physical and chemical transformations, plant operations and personnel, facilitating the integration with other engineering tools such as CAD systems.

This paper is structured as follows. The HSG notation is presented in section 2. Next, the method for its construction is presented in section 3. The proposed method is illustrated with the hazard analysis of a DAP (Diammonium Phosphate) production process in section 4. Finally, section 5 presents the conclusions and discussion.

2. HAZARD SCENARIO GRAPHS

A Hazard scenario graph (HSGs) is a visual representation technique based on concepts defined in the ISO 15926 standard that specifies definitions for the representation of knowledge about process plants [1].

A hazard scenario graph is composed of the following elements from the ISO 15926 standard:

1. Activities. An activity is a possible individual that brings about change by causing an event. Classes of activities include physicochemical processes, plant operations, and abnormal situations. Examples of activity are "Level increasing" and "Closing feed valve". Activities are shown as square boxes.
2. Events. An event is a possible individual that has zero extent in time, which means that it occurs at an instant in time. Events are shown by ovals.
3. Physical objects. A physical object is a distribution of matter, energy, or both. Examples of physical object are a tank, pump, a stream of material, a person. Participating entities are enclosed by hexagons. For example, tank T-01 is a participating entity in the "level increase in T-01" activity..
4. Causal relations are represented as solid arrows identified by the phrase *cause of event*
5. Temporal relations (beginning and ending) are represented as solid line with a filled circle identified by either the words *beginning* or *ending*
6. The participation relation is represented as a solid line with an empty circle identified by the word *participation*. A participation relation indicates that physical objects are resources, instruments, or performers of an activity

2.1 Causality and time-related information

Causality refers the links between the initial cause and the final consequences of the hazard scenarios passing through a series of intermediate consequences. Causality in HSGs is captured by means of the *cause of event* relation. A cause of event indicates that an event (the caused) is caused by an activity (the causer).

Another important element of hazard scenarios is the precedence-in-time relationships, represented by beginning and ending relations. A beginning relation marks the temporal start of a possible individual (an activity or a physical object). An ending relation marks the end of a possible individual.

Causality in HSG follows the properties described by Shoham [2] some of which are listed here:

1. A cannot cause B if B is the cause of A.
2. A cannot cause itself.
3. Causes cannot succeed their effects in time. If A causes B, A precedes B in time.
4. Activities are temporally bounded. For example, explosions, runaway reactions, mixing operations, all have a beginning and an ending. Similarly, a reactor during a runaway reaction has also a beginning and an end.
5. The property of transitivity [3]. If A causes B and B is the cause of C, then A is also the cause of C.

3. CONSTRUCTING HAZARD SCENARIO GRAPHS

Here we describe the construction of a hazard scenario graph when used together with a HAZOP study of a continuous process. We assume that the HAZOP team has already identified the system boundary, decomposed the system into subsystems or nodes.

Step 1. Select the subystem to be analyzed

Step 2. Select a deviation from the design intent of the subsystem. If the resulting deviation is temporally bounded, the deviation is represented as an *abnormal* activity. In this case, the deviation is identified by a verb or verb phrase. Examples include temperature increase, pressure decrease. If the resulting deviation is not temporally bounded, the deviation is represented as a *normal* activity ending in an event. In this case, the activity is a normal behavior (the intended behavior). For example, a "no flow" deviation is represented as an activity for the specific flow in the intention and event for the cessation of flow linked by the *ending* relation.

Step 3. Identify all possible causes of the deviation. Create a HSG for each cause and represent the cause as an activity. Classify the activity as equipment malfunctions, undesired physicochemical phenomena, abnormal operations, or external events (impacts, earthquakes).

Step 4. For each cause (in a single HSG) identify a possible propagation scenario based on the topology of the plant. An activity causes an event that is the beginning of another activity.

During the construction of the scenario part-whole relations can be used to indicate that activities or events occur during the time boundaries of another activity. For example, an "abnormal" no-flow situation in a pipe can only occur when material is flowing in that pipe.

4. CASE STUDY: HSG OF A DAP PROCESS

This case study is based on the DAP reaction system found in [4]. This is a process in which ammonia and phosphoric acid react to form diammonium phosphate (DAP), a non-hazardous product. The DAP flows from the reactor to an open-top storage tank. Low ratios of phosphoric acid/ammonia result in an off-specification product, but the reaction is safe. If both, ammonia and phosphoric acid increase, the rate of energy release may accelerate, resulting in increase in temperature and pressure and reactor failure. If too much ammonia is fed to the reactor, unreacted ammonia may flow into the DAP storage tank, causing ammonia release and consequently personnel exposure. The schematic is shown in Figure 1.

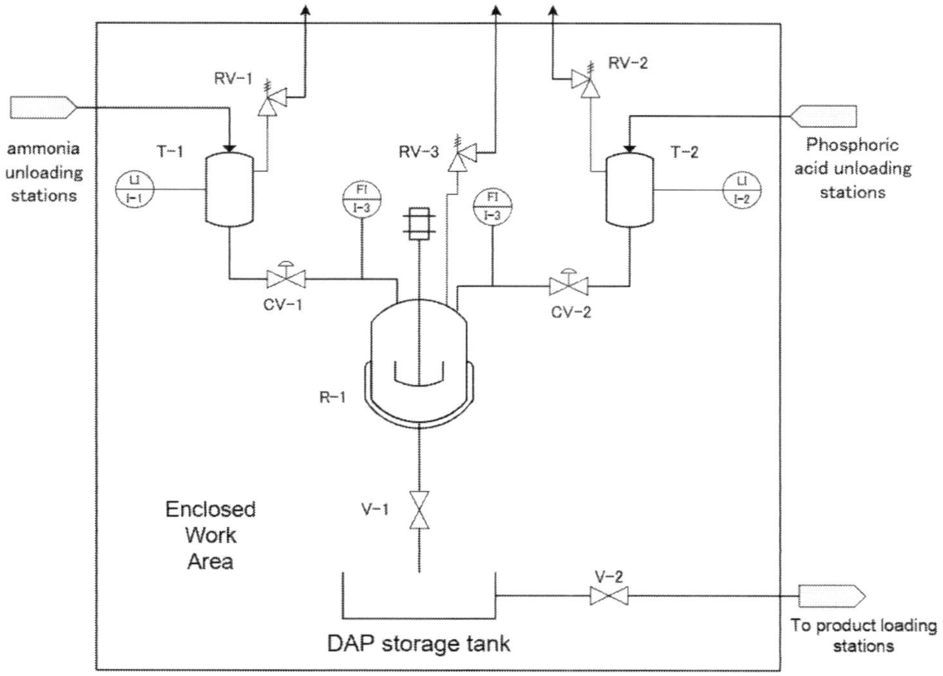

Figure 1. Schematic of the DAP reaction system

Step 1. Subsystem to be analyzed: Phosphoric Acid Feed Line to the DAP Reactor.

Step 2. A deviation is proposed by combining the guideword *no* with the parameter *flow*. The resulting deviation (no flow in Phosphoric Acid Feed Line) is not temporally bounded for which it is represented as an event (that ends the activity of normal flow).

Step 3. Identify all possible causes of the no flow deviation of Step 2. One of such causes is "line leaks". This cause is temporally bounded so it is written as a verb phrase and represented as an activity.

Step 4. Identify possible propagation scenarios for "line leaks". Two of the possible scenarios are shown in Figure 2, leading to "personnel exposure" and "loss of DAP production".

The resulting HSG is shown in Figure 2. Note that the deviation selected in Step 2 is shown explicitly in the diagram, facilitating the process of deducing that one of the consequences is the "loss of DAP production".

The same diagram could be obtained using an alternative approach. The initial cause is obtained by finding equipment malfunction, undesired physicochemical phenomena, and external events. For the *Phosphoric Acid Feed Line*, equipment malfunctions include no flow of ammonia in and out the tank; undesired physicochemical phenomena include corrosion, erosion; abnormal operations include inadequate maintenance. External events include external impacts. Figure 2 shows how the diagram that results from corrosion in the line (an undesired physicochemical phenomenon).

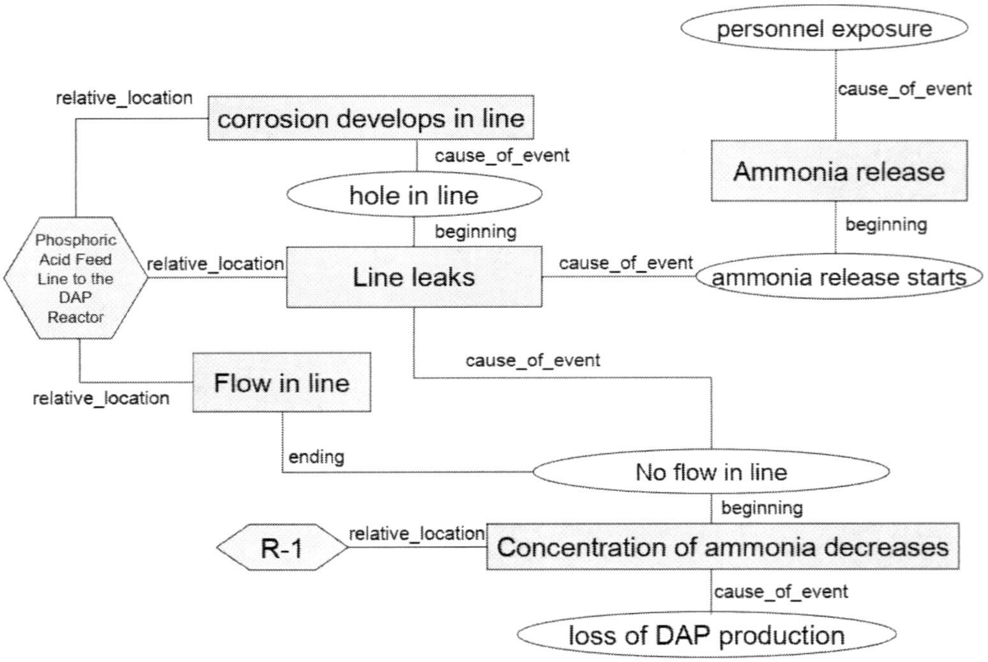

Figure 2. A Hazard Scenario Graph of the DAP reaction system

5. CONCLUSIONS

This paper introduced a graphic approach for representing hazard scenarios, which can be used as together with HAZOP or other hazard identification methods.

One advantage of the HSG representation is that with the use of adequate software tools engineers can search and combine scenarios in an automatic way. For example, if we were interested in finding the causes of ammonia release, the computer environment would combine several scenarios including those that apparently do not lead to ammonia release but lead to corrosion in the line.

6. REFERENCES

1. Batres, R., West, M., Leal, D., Price, D., Katsube, M., Shimada, Y., Fuchino, T. An upper ontology based on ISO 15926. *Comp. & Chem. Eng.*, 31, 5-6, 2007, pp. 519-534
2. Shoham, Y. *Reasoning about Change*, Cambridge, Massachusetts, MIT Press, 1988
3. Findler, N. V., Bickmore, T., *Applied Artificial Intelligence* 10 (1996) 455-487
4. CCPS, *Guidelines for Hazard Evaluation Procedures, with Worked Examples*, AIChE 1992

How to Apply Process Safety Management to Capital Projects

Hunter Shang

3M China Lt., Co.
Shanghai, P.R. China
hshang@mmm.com

ABSTRACT

As a project progresses from beginning to end, how is good Environmental, Health and Safety (EHS) control maintained? When are EHS gate reviews needed during the progress of a capital project? How does one verify and confirm all related requirements have been met before starting the next phase of the project? As a practice example, the "7 Phases and 6 Gates" system integrates project requirements, local laws / governmental requirements, and the 14 elements of the OSHA PSM standard into the overall progress measurement of the capital project, while at the same time, uses an electronic-database to carry out all approval flows, action item tracking and houses all documentation.

1. Background introduction

A capital project normally involves multiple functions within an organization and numerous internal and external requirements. It is difficult to ensure the project progresses smoothly without any EHS and/or operations issues. In the following example, a plant is setting up a new coating production line. The following issues will be considered:

i) Process scope and manufacturing requirements;
ii) Compliance to local laws and regulations;
iii) Hazard and risk evaluation;
iv) Operability;
v) Cost or spending;
vi) Supplier verification and certification;
vii) Contractor management ...

What needs to be done, and when? At the same time, how does it meet related requirements or receive confirmation by each function member? So, a clear flow and a multi-function communication platform is necessary. In this case, based on OSHA PSM standard and local law requirements, we separate the overall project process into seven continuous phases, then, apply thread-up safety tools of MOC, PHA, PSSR, etc. to control project progress with function team gate review. Please see the F-1 chart: *Overall Flowchart of the "7 Phases and 6 Gates" System.*

F-1: Overall Flowchart of the "7 Phases and 6 Gates" System

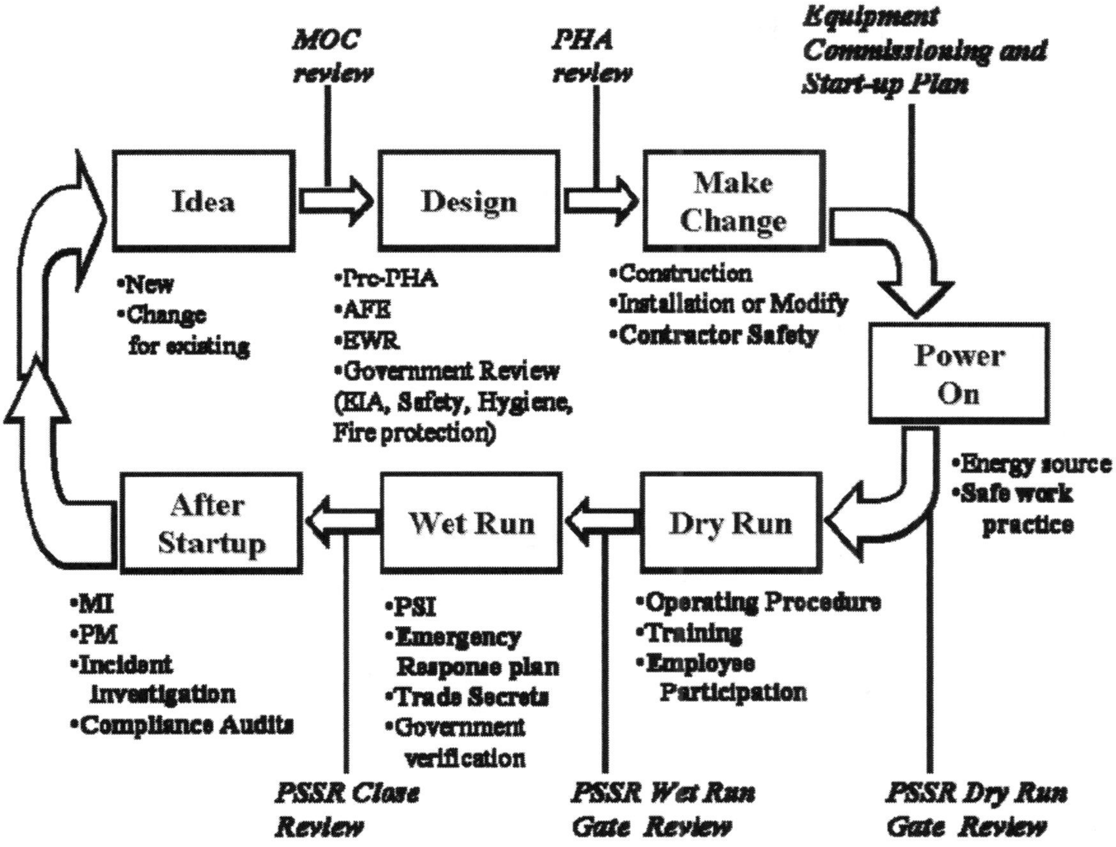

2. "7 Phases and 6 Gates" system Introduction

2.1 Project Phases – "7 Phases"

A capital project usually goes through the following seven phases from first to last. We call it "7 Phases". Please see the F-2 chart: *Project Phases.*

F-2: Project Phases

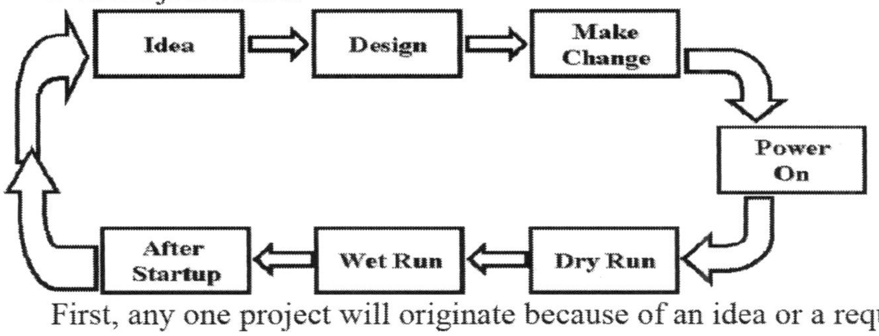

First, any one project will originate because of an idea or a request. It may be a new requisition, or a change for an existing situation. We call the first phase the "Idea" phase.

Then, it needs to be designed based on process requirements or the owner desirability in the light of related regulations. The result is commonly a drawing, a specification or a model, etc. We call the second phase the "Design" phase.

After finishing the formal design, we may need to construct related equipment or facilities by suppliers, and install or modify it on site. We call the third phase the "Make Change" phase.

After basic installation or modification is finished, we attach all kinds of energies; electric, compressed air, water, etc. We call the forth phase the "Power On" phase.

Next, we need try to run the equipment to check if the related functions are right. We call the fifth phase the "Dry Run" phase.

After basic functional testing, we'll try to run the whole equipment or process with materials or loads to check if all overall functions of the process meet the original design requirements. We call the sixth phase the "Wet Run" phase.

Finally, we enter into the normal production or using phase after we verify all functions and related requirements. We call the seventh phase the "After Startup" phase.

Of course, during the "After Startup" phase (normal using phase), we may repeat the entire cycle from the first phase again if we have any more new ideas.

2.2 EHS Gate Review – "6 Gates"

As a project progresses from beginning to end, how is good EHS control maintained? It's a really tough topic! As a core value, EHS should be considered throughout whole project process. In this example, we set up six EHS gate reviews for the whole project process. We call it "6 Gates" review. Please refer to F-1 chart.

The first gate review is the "MOC review" between the "idea" phase and the "design" phase of the project. When we have a new idea, we usually need go through the Management of Change (MOC) system before the formal project design. Generally the goal is a feasibility study of the project. Except business factors, it will focus on big hazards and risks identification of EHS aspects based on related laws / regulations and corporate requirements. So, this is a vital step for the whole project, it is also a basis of other successful gates reviews and EHS systems applications, for example, PHA, PSSR, contractor management, etc.

MOC is one of the 14 elements of the OSHA PSM standard. It should be organized by a multi-functional team, and all results of MOC should be approved by the team and related authorized people before the change implementation. As an actual project communication platform, normally we also discuss and define the future safety control plan and flow of the project in the MOC review meeting.

The second gate review is the "PHA review" before the "make change" phase. The goal of Process Hazard Analysis (PHA) is to complete comprehensive and detailed study to identify potential process hazards/risks and define safe guards and protection action plans based on the actual process design information. PHA is also one of the 14 elements of the OSHA PSM standard, and usually it uses some combination of tools such as: HAZOP, what-if, checklist, FMEA and LOPA etc. Of course, except the formal PHA review, project team will have lots of related work in the "design" phase of the project. For example: Authority for Expenditure (AFE) application; Government review and pre-evaluation for Environmental, Health, and Safety (EHS); Pre-PHA review, etc.

The third gate review is the "Equipment commissioning and start-up plan" review before the "Power On" phase. The goal is to make sure the team has a clear and orderly trial run schedule and no EHS issues happen during testing. During machine/project commissioning normal equipment operations do not exist and a number of potential hazards may be present, and each individual takes responsibility to be informed of such hazards. So, this step will define and inform all related EHS issues of the project trial run phase, meanwhile, energy source control and safe work practice (hot work permit, confine space entering permit, etc.) management will be critical.

At the same time, Pre-Startup Safety Review (PSSR) will be also initiated in this phase. A PSSR should be required before startup of new or modified equipment or facilities. Normally, a PSSR will include an organized checklist of key EHS and equipment requirements. Here we divide the whole PSSR process into three parts ("dry run" phase, "wet run" phase and the "after startup" phase) It is mainly used to determine:

- Whether new and modified facilities have been constructed in accordance with engineering design specifications;
- Safety and training requirements have been met;
- Checks that operating, maintenance, and emergency procedures are adequate and documented.
- Whether all action items (from MOC, PHA, etc.) have been completed

The fourth gate review is the "PSSR Dry Run Gate Review" before the "Dry Run" phase of the project. The goal is to make sure all related EHS issues of the "dry run" phase have been addressed before the project equipment has been started. By getting the approval of the team and related authorized people, the project team can start to dry run the equipment. From the beginning of the "dry run" phase, "Employee Participation" practice will be increased, involving lots of production people and engineers. The "Operating Procedure" will be formed, and the "Training" also need be done.

The fifth gate review is "PSSR Wet Run Gate Review" before the "Wet Run" phase of the project. The goal is to make sure all related EHS issues of the "wet run" phase have been addressed before the project equipment has been started to wet run. At this time, new chemicals or materials will be involved to the production line, increasing the likelihood of potential risks of fire, explosion and toxic hazards. So, this is a vital step of whole PSSR process. In this phase, local government verification (such as: fire protection system certification, explosion-proof equipment certification, etc.) need be confirmed; Process Safety Information (PSI) and Emergency Response Plan (ERP) should be also drafted and employees trained on their use.

The Sixth gate review is "PSSR Close Gate Review" after the "Wet Run" phase of the project. The goal is to make sure all related action items of whole project have been completed. The step is usually executed as a deliverable between the project team and the manufacturing owner. Of course, it also needs the approval of the team and related authorized people. After the equipment startup, the manufacturing team will finally define the procedures and plans of "Prevention Maintenance" (PM), "Mechanical Integrity" (MI), and the "Compliance Audits". "Incident Investigation" also needs to be implemented during the normal production phase.

3. E-Database and Documentation Management

From above "7 Phases and 6 gates" system, new problems will undoubtedly appear. How do we conveniently go through all gate approval flows of all function teams? How can we easily track the status of so many action items? How do we document and keep all related files / drawings for the life of the equipment / process? An Electronic-Database is a good solution which may help resolve those issues.

The following is a main menu of a plant PSM database for your reference:

Main Menu of PSM Database
1. Procedure sub-database
2. MOC sub-database
3. PHA sub-database
4. PSSR sub-database

5. PSI sub-database
6. MI sub-database
7. Incident Investigation sub-database
8. Training manual
9. Contractor safety sub-database
10. Safety Permits sub-database
11. EHS communication conforming sub-database
12. Action items sub-database

Each of the 12 sub-databases listed above are linked to each other in the total PSM database using the overall flow of "7 Phases and 6 gates" system. Each one has its own functions: Creating and editing the request forms, store function, email and approval function, editing action items function, etc. The PSI sub-database may store all projects and product information (including equipment, utilities, processes, chemicals, etc.). The action items sub-database may directly send separate action item to related responsible people, and it will clearly state the following points in a separate email: What's the action? Who is the responsible person? When is the target date? Who is the final approver?

These functions working together help us to integrate all requirements, flows, actions and documents into the same database, improving project implementation efficiency and operability.

4. Acronyms

AFE – Authority for Expenditure
CFR - Code of Federal Regulations
EWR – Engineering Work Request
EIA – Environmental Investigation agency
MI - Mechanical Integrity
MOC - Management of Change
NFPA - National Fire Protection Association
OSHA - Occupational Safety and Health Administration
PHA - Process Hazardous Analysis
PSI – Process Safety Information
P&ID - Piping and Instrument Drawing
PM - Preventive Maintenance
PSM - Process Safety Management
PSSR - Pre-Startup Safety Review

5. REFERENCES

1. Regulations (Standards -29 CFR), *Process Safety Management of highly hazardous chemical. – 1910.119,* Occupational Safety & Health Administration, U.S. Department of Labor, http://www.osha.gov

Process Safety Incident

Hydrogen Compressor Panel Explosion

Date & Time : 24.01.2007 at 2:30 AM

Presentor :
Shantanu Date
Reliance Ind. Ltd , India

Description:

While stopping H2 compressor from the panel small explosion took place.

Both doors of the control panel were thrown out, which struck the operator resulting in minor injury to his head

Hydrogen Compressor Panel

Hydrogen Compressor Panel cable termination Area

Hydrogen Compressor Panel cable termination Area

Possible causes of Hydrogen ingress to the Panel

- Contamination of purge air going to panel by hydrogen .

- Ingress of Hydrogen to panel from atmosphere in case of purge system failure.

- Entry of hydrogen through cables entering the panel.

Probable cause # 1

Possible cause include ingress of hydrogen to local instrument air header through damage to actuator of integral actuator type recycle/decompression valves.

LEL checked at Instrument Air header with hydrogen pressure applied to these valves. No leak observed.

Probable Cause # 2

Probable causes of ingress to panel include

- Ingress from atmosphere in case of Purge failure.
  - Sealing of panel checked and found to be ok.
  - Panel purge system was also found to be working.
  - No ‘ Loss of Pressurisation ’ alarm

Probable Cause # 3

Probable causes of ingress to panel include

- Ingress though cable entry.

Panel LEL was checked by lining up Hydrogen to compressor.

- Observed 40% LEL in panel.
- Observed 100% LEL in Junction box on skid
- Individual inputs were checked and cause was narrowed down to Suction pressure switch diaphragm.

Hydrogen Compressor System

Area specification : Zone 2 Gr II C

- Panel – E Exdep (iia)
- P - Purged with Inst air as media
- E - Enhanced safety through purge interlock which isolates the total power to panel incase of purge failure detection
- d - Purge controller Power supplies protected with Ex-Proof JBs
- (iia) - All field instruments through barriers.

- E/I signals at voltage level of 24VDC not through barriers.

H2 Compressor

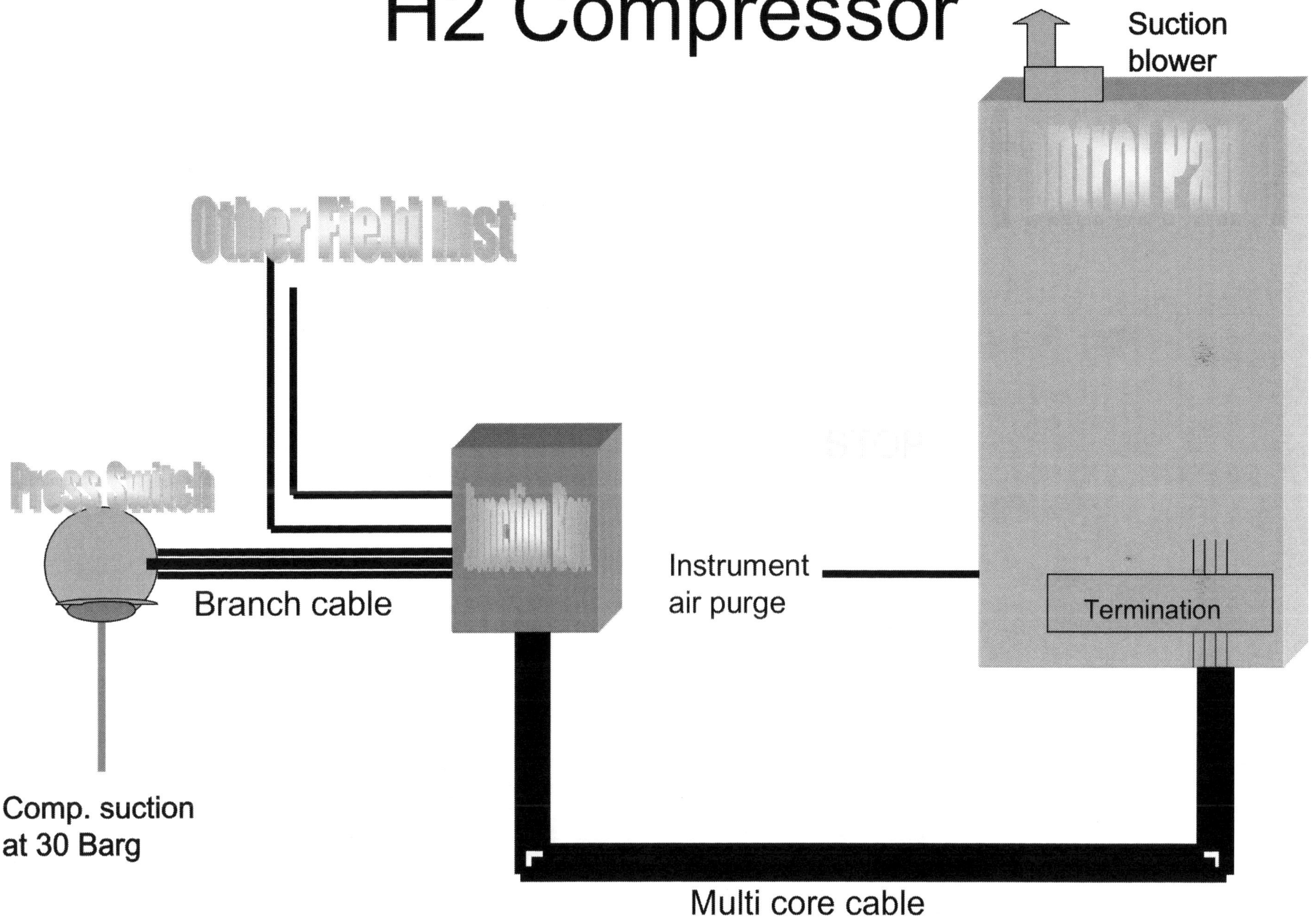

PTA2 H2 Panel Explosion Area

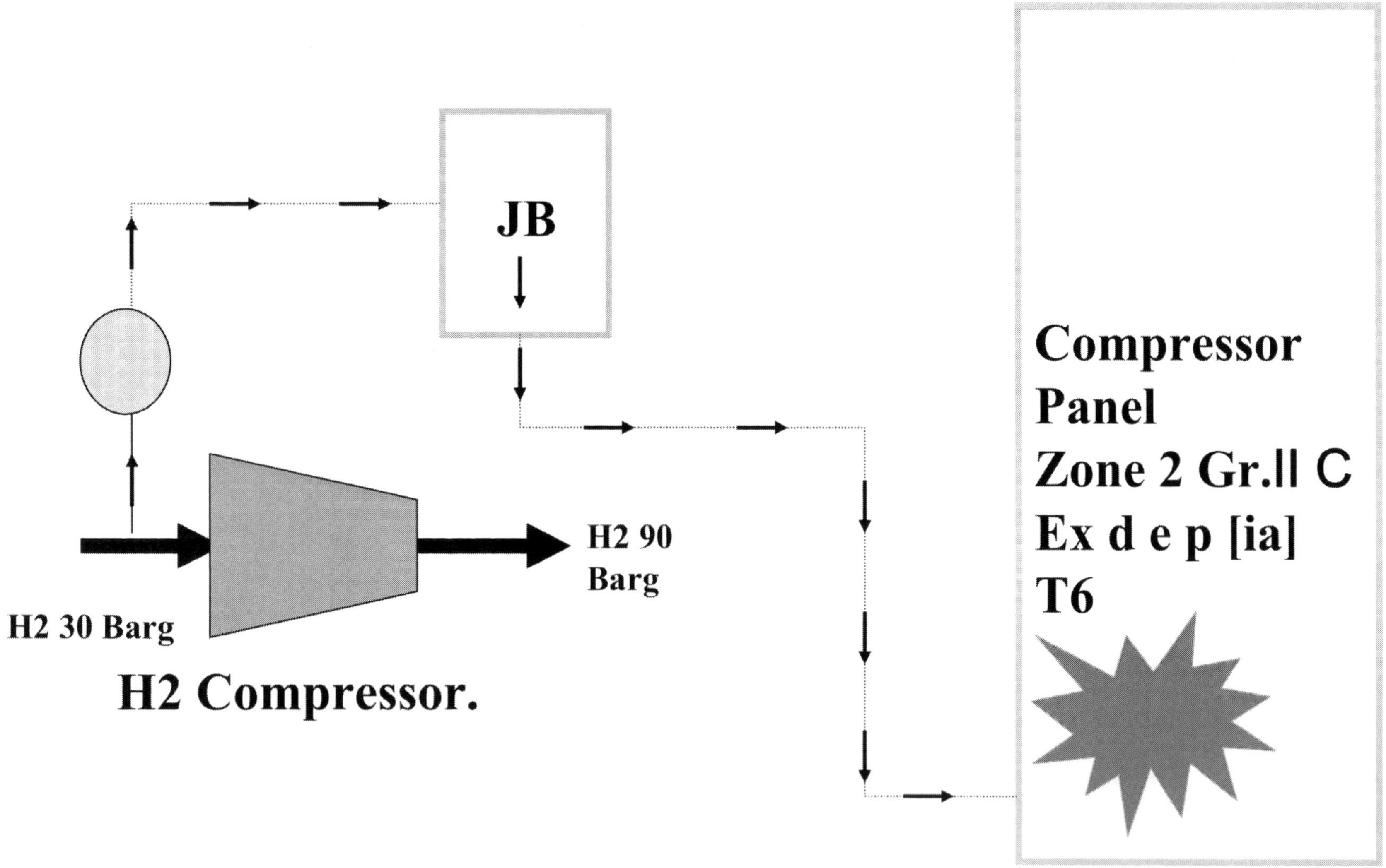

Panel Inside Area classified as safe if pressurized.

Observations :

1. Pressure switch impulse tube contained water
2. Multicore instrument cable from junction box to panel had no seal.

Root Causes :

1. Pressure switch diaphragm got dislocated due to pressurization by water , backed up due to passing NRVs during plant shutdown when compressor was depressurised.
2. Allowed hydrogen ingress into pressure switch Absence of seal in inst. cable allowed hydrogen entry to panel via cable .

Applicable Standards

- ❖ ISA 12.04.01 – 2003 – Electrical apparatus for explosive gas atmosphere – Pressurized Enclosure

- ❖ ISA 12.01.01 – 1999 – Definition & Information pertaining to Electrical instruments in Hazardous Locations.

ISA12.04.01-2003

- **Type px pressurizing**

  pressurization that reduces the classification within the pressurized enclosure from zone 1 to unclassified non-hazardous

- **Type py pressurizing**

  pressurization that reduces the classification within the pressurized enclosure from zone 1 to zone 2.

- **Type pz pressurizing**

  pressurization that reduces the classification within the pressurized enclosure from zone 2 to unclassified non-hazardous.

Standard for Pressurization:ISA 12.04.01
H2 compressor Panel –Type px Pressurization

Table 1 — Determination of protection type

Flammable substance in the containment system	External Zone classification	Enclosure contains ignition-capable apparatus	Enclosure does not contain ignition-capable apparatus
No containment system	1	Type px [a]	Type py
No containment system	2	Type pz	No pressurization required
Gas/vapor	1	Type px [a]	Type py
Gas/vapor	2	Type px (and ignition-capable apparatus is not located in the dilution area)	Type py [b]
Liquid	1	Type px [a] (inert) [c]	Type py
Liquid	2	Type pz (inert) [c]	No pressurization required [d]

Standards For Pressurization.ISA12.04.01
H2 Compressor Panel protection – IP65

Table 2 — Design criteria based upon protection type

Design criteria	Type px	Type py	Type pz with indicator	Type pz with alarm
Degree of enclosure protection according to IEC 60529 or IEC 60034-5	IP4X minimum	IP4X minimum	IP4X minimum	IP3X minimum
Verifying purge period	Requires a timing device and monitoring of pressure and flow	Time and flow marked	Time and flow marked	Time and flow marked
Door/cover requiring a tool to remove	Warning, see 5.3 and 6.2 b) ii)	Warning, see 5.3 (Note 1)	Warning, see 5.3 and 6.2 b) ii)	Warning, see 5.3 and 6.2 b) ii)

Standard for Pressurisation ISA 12.04.01
Safety Provision & Safety devices

Table 3 — Safety devices based upon protection type

Design criteria	Type px	Type py	Type pz
Safety device to detect loss of minimum overpressure	Pressure sensor, see 7.9	Pressure sensor, see 7.9	Indicator or pressure sensor, see 7.9 d)
Safety device(s) to verify purge period	Timing device, pressure sensor, and flow sensor at outlet, see 7.6	Time and flow marked, see 7.7 c)	Time and flow marked, see 7.7 c)
Safety device for a door or cover that requires a tool to open	Warning, see 6.2 b)	Warning, see 6.2 b)	No requirement
Safety device for a door or cover that does not require a tool to open	Interlock, see 7.12 (No internal hot parts)	Warning, see 6.2 b) (No internal hot parts)	Warning, see 6.2 b)
Safety device for hot internal parts when there is a containment system (see Clause 15)	Alarm and stop flow of flammable substance	Not applicable for protection type since internal hot parts not permitted	Alarm (normal release not permitted)

ISA 12.01.01 –1999 Wiring Methods

Seals are installed in conduit and cable systems to

- confine an explosion occurring in an enclosure or a conduit system to only that enclosure or that portion of the conduit system.
- minimize the passage of gases, vapors, or liquids and prevent the passage of flames through the conduit or cable system from a classified to an unclassified area, or from one enclosure to another;
- prevent *pressure piling* the buildup of pressure inside conduit systems (ahead of an explosion's flame front) caused by precompression as the explosion travels through the system. Exploding precompressed gases may reach pressures that would exceed the design pressure of the enclosures.

ISA 12.01.01- 1999 Wiring Methods

- Seals and drains shall be installed in each conduit or cable connection to devices installed on process lines that contain flammable fluids and depend upon a single compression seal, diaphragm, or tube (such as a Bourdon tube) as a barrier between the process fluid and the conduit or cable.

  This is to prevent flammable process fluids from entering conduit or cable systems and being transmitted to unclassified areas.

- Typical examples of such devices are solenoid valves and pressure, temperature, and flow switches or transmitters. This requirement applies even in unclassified areas.

ISA 12.01.01 –1999 Wiring Methods

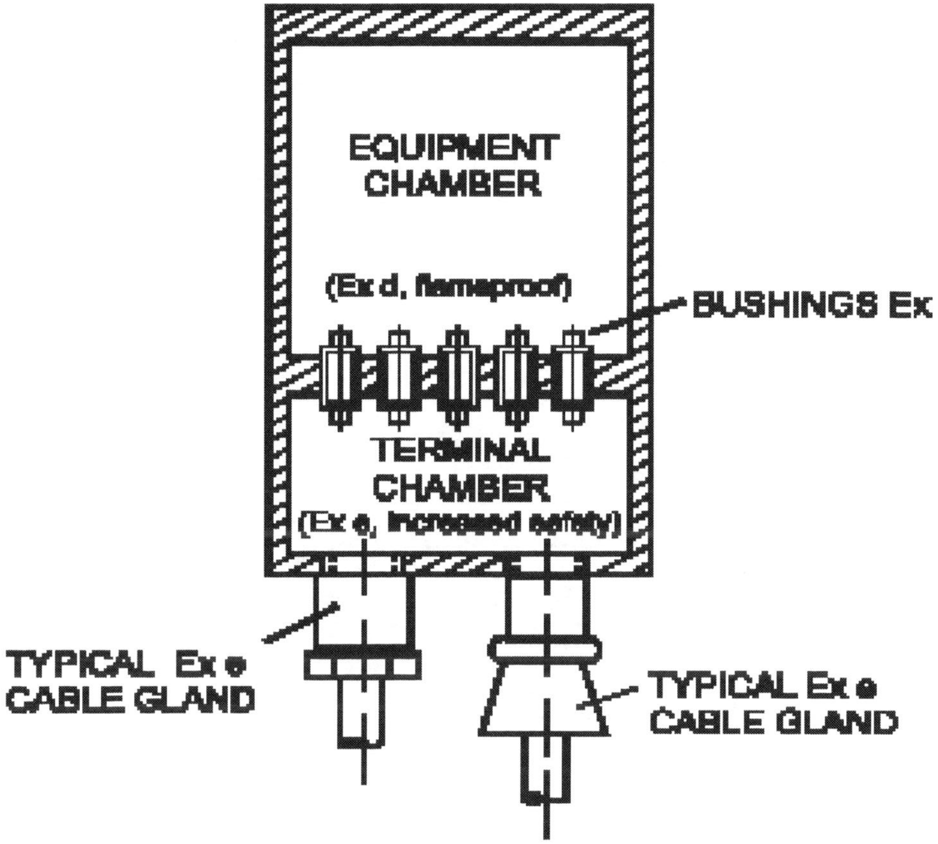

Figure 5 — Cable system (indirect entry)

ISA 12.01.01-1999 Wiring methods

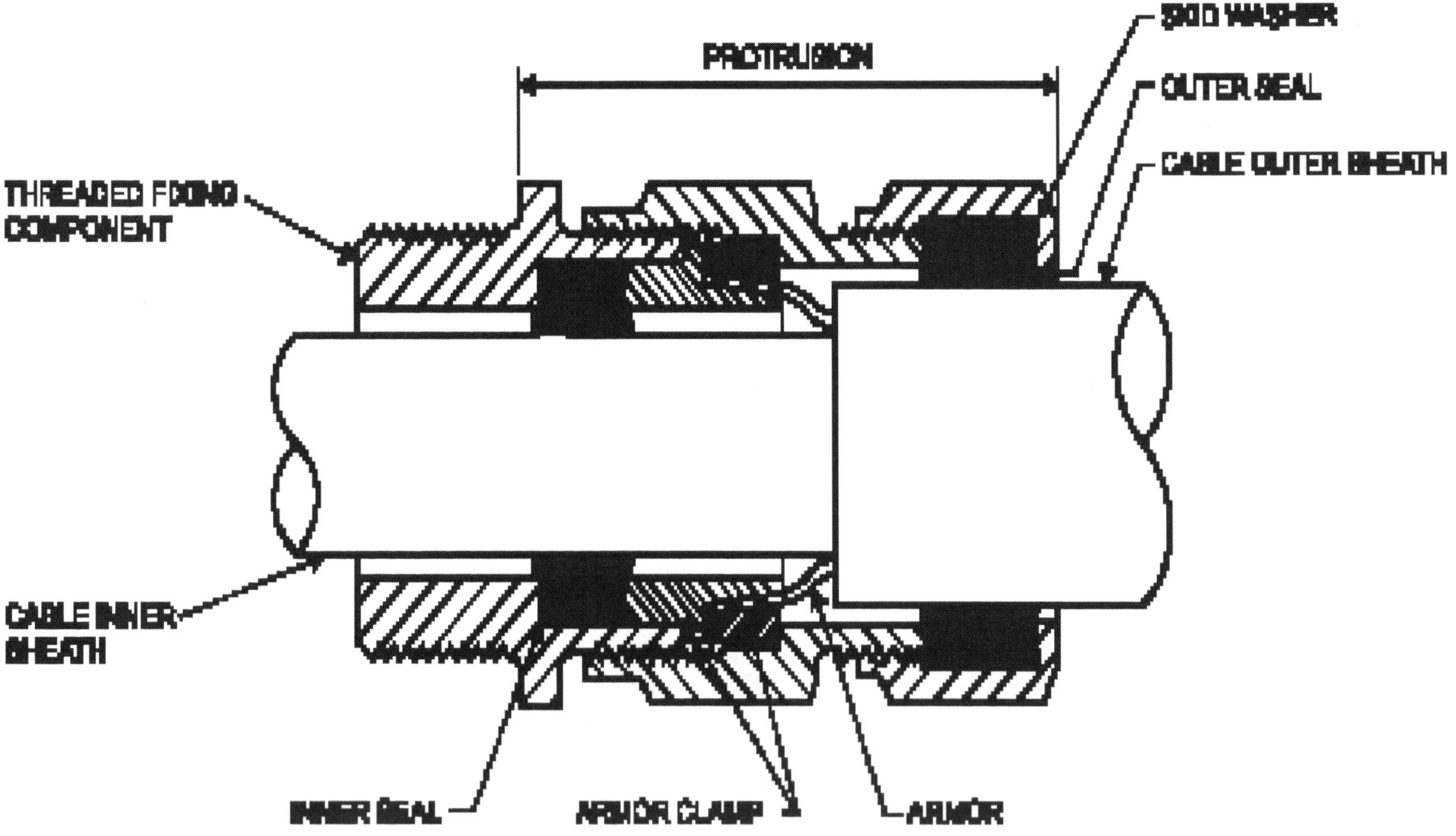

Figure 6 — Cable gland (indirect entry)

ISA 12.01.01 –1999 Wiring Methods

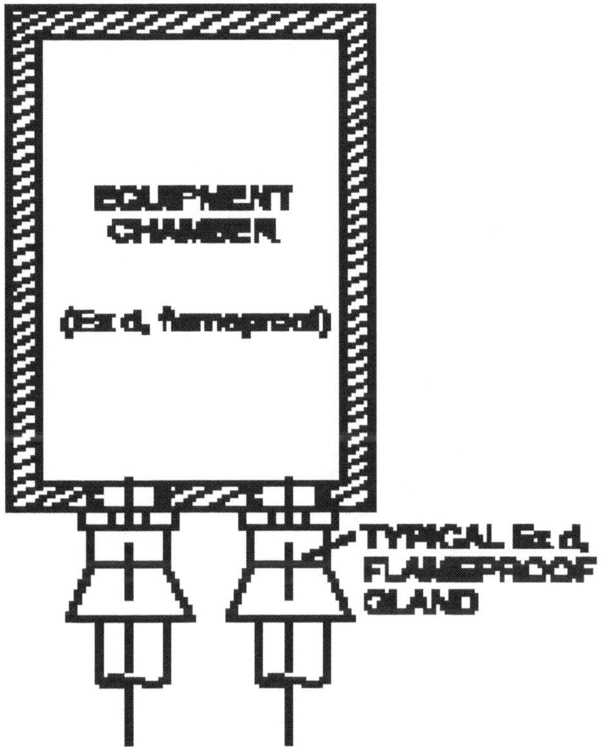

Figure 7 — Cable system (direct entry)

ISA 12.01.01-1999 Wiring Methods
Type of seal to be used at Instrument End

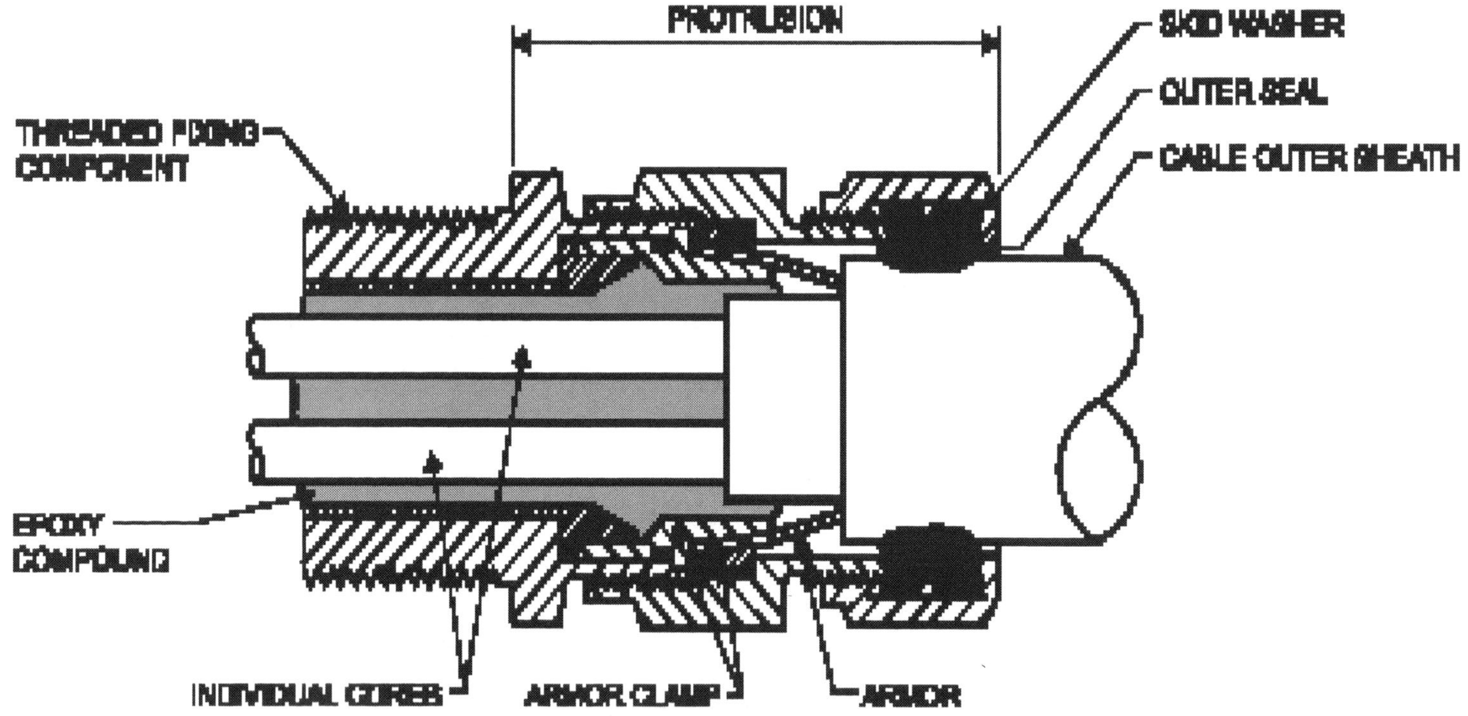

Figure 8 — Cable gland (direct entry)

Systemic Causes Responsible

- There is a practice in the organisation to order cables in bulk. Procurement is done as per applicable codes & standards.
- In this case it was established that though compressor skid was supplied by renowned vendor , supply scope excluded cables as per company's approach.
- Correct cables were procured as per vendor specification & applicable codes

BUT -- - -

During project execution , correct cable was not used. It went undetected as there was no leak in Pressure switch that time.

AFTER 10 YEARS OF OPERATION - -

- Led to unsafe incident when another cause (back up of process water due to passing of Non Return Valves) created leak in pressure switch & combined with unsafe condition already present due to systemic cause .

Case Study of a Hydrogen Explosion in an Electrical Panel

Timothy J. Myers
Exponent, Inc.
Natick, MA
tmyers@exponent.com

Alfonso F. Ibarreta
Exponent, Inc.
Natick, MA
aibarreta@exponent.com

Copyright 2008

Prepared for Presentation at
American Institute of Chemical Engineers
2008 Spring National Meeting
42nd Annual Loss Prevention Symposium
New Orleans, LA
April 7-9, 2008

UNPUBLISHED

AIChE shall not be responsible for statements or opinions contained in papers or printed in its publications

Case Study of a Hydrogen Explosion in an Electrical Panel

Timothy Myers, Ph.D.
Exponent, Inc.
Natick, MA
tmyers@exponent.com

Alfonso Ibarreta, Ph.D.
Exponent, Inc.
Natick, MA
aibarreta@exponent.com

Abstract

A chemical facility generated nitrogen for use in its processes by removing oxygen in air using pressure swing adsorption (PSA). In order to remove residual oxygen in nitrogen from the PSA process, hydrogen was added to the nitrogen and the mixture was then passed through a catalyst bed, converting the remaining oxygen to water. The water was then removed in a drying or dehumidification process yielding high purity nitrogen. During what had become a routine process upset at the facility, a worker opened an electrical panel to disconnect a pneumatic line, which in turn caused a solenoid valve to open and reset the system. As the system reset, an explosion occurred in a neighboring electrical panel and the panel door struck the worker in the head, causing serious injuries. A subsequent investigation revealed that the explosion was caused by the ignition of hydrogen that had migrated into the panel.

As is often the case, multiple factors contributed to the cause of the incident, any one of which could have prevented the incident. In this paper we discuss the root causes and contributing factors of the incident and lessons learned that can prevent similar events from occurring in the future. Issues discussed include: identification of scenarios during process hazard analyses, initial electrical classification of the area, design and maintenance of electrical conduit systems, maintenance of safety shut-off valves, management of change, procedures for working on electrical panels, and response to process upsets.

1. Background

An accidental release of hydrogen caused a small explosion in an electrical panel at a chemical facility, causing the electrical panel door to fly outward and strike a worker in the head. The incident occurred during a valve switching operation at a nitrogen-generating and purification processing facility, injuring one man. Investigation of the incident indentified several factors that contributed to the incident.

2. Process

The process equipment involved in the incident is shown in Figure 1 and consisted of three sections: (1) a pressure swing adsorption unit that separated nitrogen from air, creating crude nitrogen, (2) a catalytic oxidation unit which reacted residual oxygen in the nitrogen with hydrogen, and (3) a drying section that removed water from the nitrogen. A continuous oxygen analyzer monitored the final product quality.

2.1 Pressure Swing Adsorption (PSA) Generator

In the PSA, filtered air is passed through a Carbon Molecular Sieve (CMS) material and oxygen in the air is adsorbed. The CMS is then regenerated in a desorption stage by venting to atmospheric pressure, releasing adsorbed oxygen. Two adsorption vessels are used to alternate between the adsorption and desorption cycles while providing a continuous stream of nitrogen-rich exhaust. The crude nitrogen stream exiting the PSA process typically consists of about 0.5% oxygen content.

2.2 Catalyst Oxidation Unit

After the PSA stage, hydrogen is added to the nitrogen stream and the mixture is then circulated through a catalyst bed, in order to convert the remaining oxygen to water. The purification is a continuous process, with hydrogen being injected at the proper flowrate to maintain a slightly higher than stoichiometric (2 to 1) ratio of hydrogen to oxygen. The ratio is maintained with a ratio controller that adjusts the hydrogen addition rate based upon measurements of oxygen in the raw nitrogen stream and hydrogen in the mixed stream. The hydrogen from a pressurized source passes through a pressure regulator, a solenoid valve, the control valve and then a manual shut-off valve (as shown in Figure 1) before being introduced into the main nitrogen stream. The stream then passes through a mixer and a catalyst bed, where the oxygen and hydrogen are converted to water. This process produces a final nitrogen stream with 5 ppm maximum oxygen content.

2.3 Dryer

Once the excess oxygen is consumed in the catalytic reactor, the generated water is removed first cooling the stream and removing condensed water, and then by passing the stream through a desiccant dryer. Two dryers are present in parallel and flow alternates between the two dryers while one dryer is regenerated by heating. The final product stream is dry nitrogen gas.

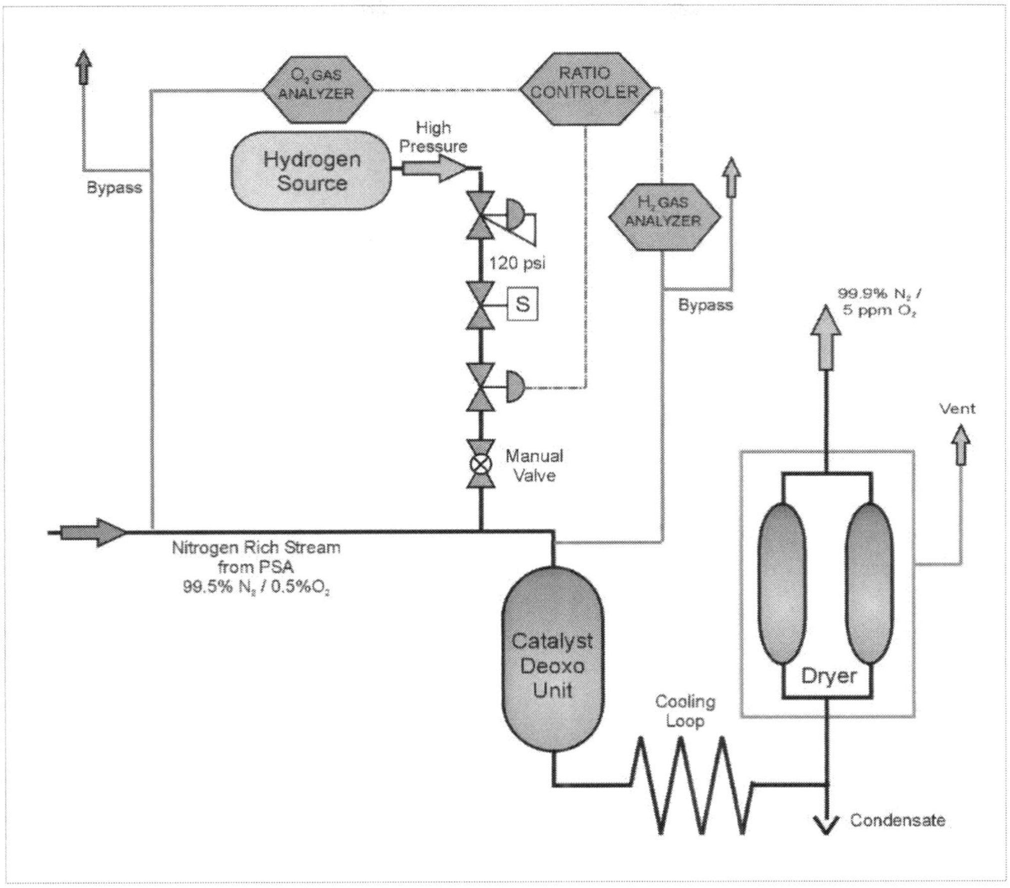

Figure 1: Flow diagram of selected components of nitrogen generation unit

3. Events Preceding the Incident

3.1 Initial Process Hazard Analysis

Prior to the installation of the nitrogen generation process, the facility owner had performed a limited hazard analysis to determine the electrical classification of the area where the nitrogen generation system would be installed. The system was going to be installed in an area with a roof, one wall, and three open sides. Because of the relatively open environment, and the low density of hydrogen, the facility deemed that in the event of a release hydrogen would accumulated near the top portion of the equipment. The facility therefore decided to require gravity ventilation at the high point of the roof, and only classify regions one foot below the roof and higher as Class I, Group B, Division II [1]. All other regions were classified as general purpose, and did not require special electrical equipment. The facility also specified that the hydrogen pipe header should be welded construction and any flanged connections should be at least 8 feet above floor level.

In addition to the facilities requirements, the manufacturer of the hydrogen system used electrical equipment suitable for a Class I environment [1] on the hydrogen line, and placed seals in the electrical conduit between this equipment and other electrical equipment, consistent with practices for a Class I area. Other electrical equipment in the unit, however, was not designed for a Class I environment because of the electrical classification determined by the facility.

3.2 Changes to System Prior to Incident

Prior to the incident, a solenoid shut-off valve on the hydrogen feed line (shown in Figure 1 with an 'S') was replaced. Because it was part of the hydrogen feed flow system, the original solenoid valve was explosion proof and rated for a Class I, Division 1 area. The facility replaced the valve with a general purpose valve from the same manufacturer. During the investigation it was determined that the replacement valve had been installed backwards. Like many solenoid valves, when installed backwards the valve would not properly seal in the closed position, but would leak when the pressure drop across the valve was sufficient to counter the solenoid spring force – approximately 35-40 psi. The hydrogen pressure upstream of the valve was regulated to 120 psi; providing enough pressure to open the valve and allow hydrogen to leak through the system. A hydrogen flow control valve was also present in the system, but as is common with many flow control valves, it was not designed to be leak tight or to serve as a shut-off valve.

3.3 Initial Failure and Operator Response

On the morning of the incident there was a process shut down for four to five hours. The shut down was caused by a failure of a valve to open and release the gas from one of the desiccant drying vessels. This, apparently, was a frequent event at this facility, and operators had devised a process for resetting the system. The system was reset by disconnecting a pneumatic line that provided nitrogen as a pilot to a valve on the dryer vent line. The operators would disconnect the pneumatic line inside a low-voltage control cabinet which would cause the valve to open. After a few minutes of delay, a relay in an adjacent high-voltage cabinet activates a switch and resets the process.

3.4 Final Catastrophic Failure

Immediately after the reset procedure on the day of the incident, a pocket of leaked hydrogen gas ignited in the high voltage electrical cabinet, blowing the cabinet door away from the box. After the explosion, an injured worker was found with severe head injuries due to impact from the door of the high voltage electric cabinet. A photograph of the cabinet, with bulged walls an damaged hinges, is shown on the left in Figure 2. The door to the cabinet, shown to the right in Figure 2, was found approximately 20 feet away where it had impacted a tank. The worker's hard hat and safety glasses had been damaged and knocked off his head. The low voltage electrical cabinet was found open, with the pilot line disconnected, indicating that the worker was in the process of resetting the system using the method commonly used at the facility. Junction box gaskets on electrical conduit near the high voltage electrical box were also bulged outward indicative of internal pressurization.

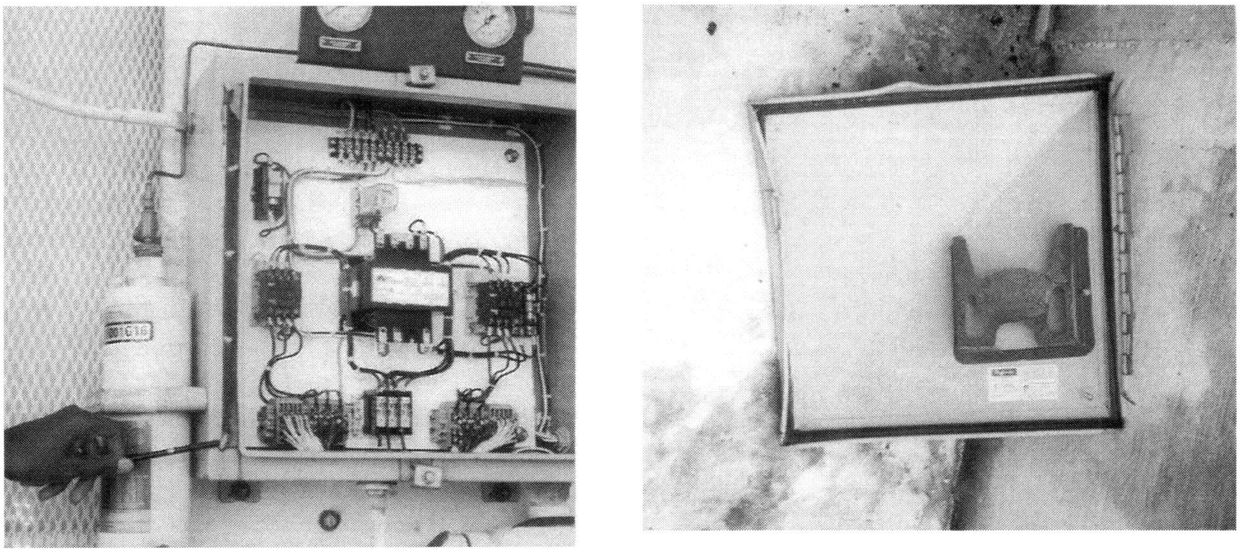

Figure 2: Photographs showing damage to electrical box and door.

4. Incident Investigation

4.1 Source of Energy to Pressurize Cabinet

Investigators originally identified three possible scenarios that could have led to the explosion of the electrical cabinet.

1. Nitrogen Pressurization - Pressurization by process nitrogen by a leak through conduit between the desiccant driers and the high voltage electrical cabinet.
2. High Voltage Electric Arc – A high voltage electric arc in the high voltage cabinet could have led to pressurization of the cabinet.
3. Hydrogen Combustion – Hydrogen migrating into the high voltage electrical cabinet, mixing with oxygen and igniting by a spark from a switch in the electrical cabinet.

The first two scenarios were ruled out. Pressure testing indicated that there was no leak path between the desiccant heater and the electrical panel, and inspection of the electrical cabinet showed no evidence of electric arcing. The third hypothesis could not be ruled out and further testing and analysis were performed to determine whether hydrogen could have migrated into the electrical cabinet

4.2 Source of Hydrogen Leak into Cabinet

Pressure testing was used in an effort to identify leaks through the electrical conduit from the electrically actuated hydrogen shut-off valve and a pressure switch to the high voltage electrical cabinet. The valve and pressure switch were both designed in a manner that did not allow gas to migrate from the hydrogen process side to the electrical conduit. Pressure testing verified that no leak path existed between the hydrogen feed line and electrical conduit. Thus, no direct leakage path from the hydrogen line to the high voltage cabinet was identified.

Other possible hydrogen leak paths existed because the hydrogen shut-off valve was installed backwards and would allow hydrogen to continue to flow through the system even during a shutdown. The system design also enabled an unforeseen leak path of hydrogen through the processed gas exhaust. During a shutdown there was no nitrogen flow through the system. Thus, during the shutdown, portions of the system downstream of the hydrogen addition point that normally handled low concentrations of hydrogen in nitrogen would fill with pure hydrogen. One investigator hypothesized that hydrogen could have been leaking inside the electrical cabinet of a hydrogen analyzer and then migrated through electrical conduit to the high voltage cabinet. This was not tested experimentally, but appears unlikely because gas entered the cabinet at a very low flow rate and pressure, and the cabinet was not air tight so gas leaking in the cabinet would have preferentially leaked out of the cabinet rather than through the tortuous path of the electrical conduit to the high voltage panel.

Another possible leak path into the high voltage electrical cabinet was identified. The drying system was a pre-assembled system purchased from a third-party. Since the dryer was intended for nitrogen service, its manufacturer configured it to use the dried process stream as the pilot gas for solenoid valves associated with the dryer. During the shutdown, eventually the desiccant drier would have filled with hydrogen. Thus, when the pilot line was disconnected to reset the system, it would have released hydrogen and not nitrogen.

Because, this connection was frequently disconnected and reconnected, it may have been leaking before the worker disconnected it as well. This gas could then leak into the neighboring high voltage electrical panel through electrical conduit where it was ignited. The system had been reconfigured before this potential leak path could be tested by experiments.

Although the exact source of hydrogen entering the high voltage electrical cabinet could not be identified, two possible sources were ultimately identified: a leak through electrical conduit from the hydrogen analyzer, or a leak from the pilot line from the drier.

5. Lessons Learned from Incident

Even though the exact leak source of hydrogen into the high voltage cabinet was not identified, multiple factors were identified that contributed to the cause of this incident and could have caused other incidents. The mitigation of any of these factors could have prevented the final incident. This particular case therefore offers multiple lessons to be learned.

5.1 Failure to Identify the Hazard

The original hazard analysis focused solely on external hydrogen leaks from the hydrogen header that could potentially enter electrical equipment. It did not consider possible leak paths that would allow hydrogen to migrate from process piping or vessels to electrical cabinets.

Additionally, the hazard analysis failed to identify upset conditions where high concentrations of hydrogen could enter the downstream portions of the process and potentially migrate into electrical equipment. During normal operation combustion of hydrogen once mixed with nitrogen was not possible because hydrogen was being added to a nitrogen stream at concentrations of approximately 1%, well below the 4% LFL of hydrogen, and the oxygen concentration was approximately 0.5% well below the 5% minimum oxygen concentration of hydrogen. However, as this incident demonstrated, failure of the hydrogen shut-off valve would allow much higher concentrations of hydrogen to exist during a shutdown. When mixed with air, the hydrogen would be ignitable. Had this been identified, the facility may have taken additional measures to seal or purge electrical conduit and cabinets.

When performing hazard analyses or determining electrical classification of areas, facilities should consider potential upset conditions as well as normal operation and determine whether extra protection is needed to address these upset conditions.

5.2 Incorrect Installation and Maintenance of Equipment

Automatic shutoff of hydrogen to the system relied on proper functioning of a single solenoid valve. During routine maintenance, however, the valve was replaced and installed backwards. Like many solenoid valves, the valve was designed to be installed in one direction so that a higher pressure upstream than downstream would act to close the valve when it was not powered. When installed backwards a higher pressure upstream than downstream acts to open the valve. In this case a pressure drop of 35 to 40 psi was sufficient to overcome the force of the spring in the valve and cause the valve to open.

When replacing or performing maintenance on safety related valves such as a shutoff valve, facilities should consider having a second person verify that the valve is properly installed through either visual inspection, or where possible, through testing.

5.3 Redundancy of Shutoff Valves

This incident also demonstrates that if the hydrogen shutdoff valve failed open due to other reasons, hydrogen would continue to flow into the system during shutdowns. Facilities should consider redundant shutoff valves or double block and bleed combinations for shutoff valves with safety implications.

5.4 Use of Process Stream from Desiccant Dryer for Instrument Pressure

Because the desiccant dryer was designed to treat nitrogen, the dryer had been designed to use the process stream as a pressure source for pilots on solenoid valves. The process stream was then routed through the electrical cabinets. Using an independent source of instrument air, rather than the process stream would have prevented hydrogen from flowing through gas lines within electrical cabinets and being released when the pilot line was disconnected. When using process streams in place of instrument air, facilities should consider the possible composition of the process stream during upset conditions as well as the intended composition. Kletz describes a somewhat different incident where solvent vapors in a drier process stream entered electrical equipment causing an explosion [2].

5.5 Improper Handling of Process Upset

The facility had experienced this same process upset, the failure of a drier exhaust valve to open, on previous occasions. Rather than addressing the failure that was causing the upset, employees had developed an ad hoc procedure for resetting the system. No management of change procedures or hazard analysis were followed to develop a safe procedure for addressing the upset. The manual for the desiccant drier included specific instructions for troubleshooting this upset that did not require disconnecting a line in an electrical box. The manual recommended using a "push-to-test button" on the solenoid valve, and if that did not open the valve, manually disconnecting the pilot line at the valve. Both of these steps would have been performed remote from the electrical panel. The troubleshooting procedures provided by the manufacturer would have also identified the cause of the reoccurring process upset allowing the problem to be resolved and preventing future upsets.

When possible, facilities should consider implementing management of change procedures or performing process hazard analyses before addressing process upset conditions. These hazard analyses should result in specific procedures to use to address routine process upsets. The procedures should also not counter the instructions or recommendations set forth by equipment manufacturers. Additionally, where possible the causes of routine upsets should be addressed to stop or minimize future upset conditions.

5.6 Improper Alarm Management

During this event, several alarms would have been activated. It is known that the reset being performed by workers was a result of a system shutdown alarm. At the time of the incident, the high hydrogen alarm would have been also have been present, indicating that a significant process upset had occurred. The implications of these alarms should have been considered before addressing the upset. Workers should have considered closing manual hydrogen shutoff valves and de-energizing equipment before performing work.

6. Conclusions

As often is the case, this incident required several simultaneous failures of equipment and improper actions by workers. The original process design failed to identify hazards during process upset conditions, a solenoid valve had to be installed backwards, operators had to miss the improper installation, and hydrogen alarms had to have been ignored by operators. Several lessons were learned through the investigation of this case that can prevent future incidents from occurring.

7. References

[1] NFPA 70, National Electrical Code, National Fire Protection Association.

[2] T. Kletz, Learning from Accidents, Third Edition, (2001) pp. 22-31.

Chemically Unstable Gases – Flammability of Ethylene Oxide Mixtures in Sterilization Processes

Volkmar Schroeder and Enis Askar

Federal Institute for Materials Research and Testing (BAM)
Berlin, Germany
E-mail: volkmar.schroeder@bam.de

Aydan Acikalin and Joerg Steinbach

Technical University Berlin
Berlin, Germany

ABSTRACT

For the assessment of explosion hazards by industrial sterilization processes with ethylene oxide (EO), the flammability regions of 3-component systems EO/nitrogen/air, EO/carbon dioxide/air and EO/water vapor/air were determined. The tests were performed at temperatures of 20 °C and 100 °C and at pressures of 40 kPa and 100 kPa in accordance with the standard test method EN 1839-B.

The observed flammability regions are similar in shape and typical for mixtures with ethylene oxide. According to the molecular heat capacities the regions get larger with nitrogen and smaller with carbon dioxide. They become larger with increasing pressure and increasing temperature.

Using experimental data a semi-empirical model was created that allows the calculation of flammability limits of process gases in sterilization processes. Such process gases can consist of EO, nitrogen, carbon dioxide, water vapor and air.

The model is based on the assumption that the adiabatic flame temperatures along the boundary curves of a flammability region have a certain temperature profile that is nearly independent of the type of the inert gas. The adiabatic flame temperatures were calculated by using the "Gaseq" Code.

Using a temperature profile calculated from only one experimental system EO/inert gas/air it is possible to predict the flammability limits of systems with other inert gases or of process gases containing several inert gases.

1. INTRODUCTION

Ethylene oxide (EO) is among the top 3 % of high volume chemicals. EO is both, a flammable and a chemically unstable gas that can be explosive without any amount of oxygen or air. During sterilization procedures the used EO can easily form explosive mixtures. In the past, EO already caused a number of explosions in sterilization industries. One example is the explosion accident happened in Ontario, California, 2004 [1]. Subsequently, producers and users of sterilization plants in Germany required exact knowledge about the flammability limits and explosiveness of process gases created by sterilization processes.

During such sterilization cycles gas mixtures are being created that consist of EO, nitrogen (N_2), carbon dioxide (CO_2), water vapor (H_2O) and air. For safety reasons it is recommended to run the process in such a way that the mixtures do not become explosive. Therefore certain limiting concentrations of EO and air must not be exceeded. The accurate knowledge of the flammability limits under process conditions is a prerequisite for those protection measures.

Both, the upper and lower flammability limit (UFL and LFL, respectively) are understood as the mole fractions (concentration) of EO in the gas mixture at which flame propagation does not occur. They are the last non-ignition points. This European definition is somewhat different to the one in the United States where according to ASTM E 681 the mean value between ignition and non-ignition point is taken. Nevertheless, because of the small step sizes the difference is not very important in practice. The range between the LFL and UFL is the so-called flammability range. If inert gases are present in a 3-component system the term flammability region or explosion region is being used. Inside this region the process gas is explosive and can be ignited by ignition sources.

Flammability regions of 3-component systems are given in [2, 3]. However, most of the studies were performed at atmospheric conditions. For higher temperatures and non atmospheric pressures only a few data is available.

The flammability region of EO mixtures is strongly influenced by the type of the inert gas, the temperature and the pressure. In sterilization processes mixtures with more than one inert gas can be created at different temperatures and pressures. Hence, the flammability limits had to be measured for all these conditions. As the matter of fact, this is not feasible due to the high experimental effort.

Therefore, the goal of this study was the systematic experimental determination of basic data on sterilization processes and the development of a reliable calculation method by means of this data.

2. EXPERIMENTAL

The flammability limits were investigated systematically for the systems EO/N_2/air, EO/H_2O/air and EO/CO_2/air at temperatures of 20 °C and 100 °C and pressures of 40 kPa and 100 kPa. The experimental set-up is schematically shown in Fig. 1.

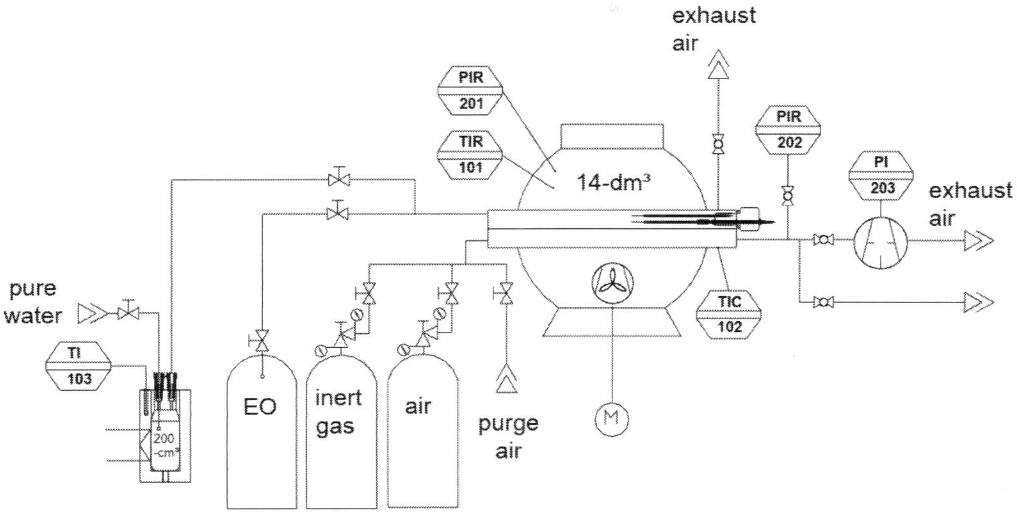

Fig. 1: Experimental set-up according to the standard DIN EN 1839, bomb method.

The ignition tests have been performed following the bomb method of DIN EN 1839 [4]. The ignition vessel used was a heatable spherical steel autoclave with an inner volume of about 14 dm^3. The ignition source was located in the center of the autoclave. It was a 0.12 mm fusing NiCr-wire (exploding wire) soldered across two electrodes at a distance of 5 mm. The ignition energy was controlled by an electronic switching unit that allows the adjustment of the ignition energy between 1 J and 50 J. According to DIN EN 1839 the energy was set between 10 J and 20 J for all tests.

Filling and explosion pressures were measured by piezoresistive pressure gauges (from Keller, type PAA-10, sensitivity 10^{-4} of full scale, precision 0.5%). They were calibrated prior to the experiments. After ignition the pressure-time courses were recorded and stored in a computer.

The temperatures of the autoclave and the gas phase were measured using 0.5 mm NiCr-Ni thermocouples.

The gas mixtures were prepared according to a manometric method on the basis of the partial pressures. Subsequently, the test mixtures were ignited in the ignition vessel. The overpressure caused by the reaction was measured in order to characterize the explosibility of the test mixture. As criterion for a successful ignition served the pressure rising factor F (ratio of explosion pressure p_{ex} and initial pressure p_0) with $F = p_{ex}/p_0 > 1.05$. In case of the lower initial pressure of 40 kPa the criterion was increased to $F > 1.10$ because of a possible influence of the ignition source on the explosion pressure.

The flammability limits were determined by variation of the EO concentration in the mixture. The step size was 0.2 mol% near the LFL and 0.5 mol% near the UFL.

3. TEST RESULTS AND DISCUSSION

The experimentally determined flammability regions (explosion regions) for the investigated systems EO/N_2/air, EO/H_2O/air and EO/CO_2/air at different pressures and temperatures are shown in the following Fig. 2 – 4.

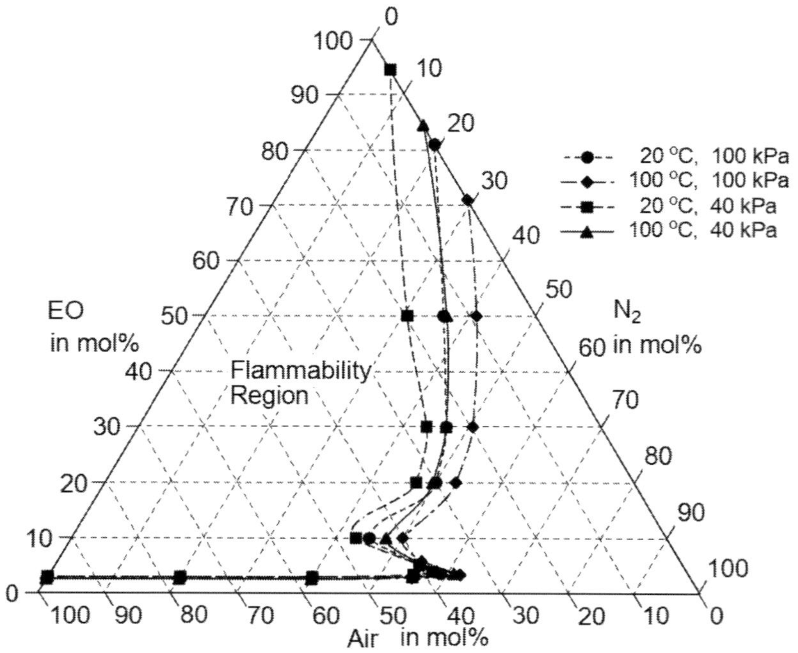

Fig. 2: Experimentally determined flammability regions for the system EO/N_2/air at different temperatures and pressures.

At lower EO fractions, i.e. in the fuel poor range, the run of the boundary lines that enclose the flammability regions is similar to those of other common flammable gases. Here, the oxidation reaction of EO with air dominates the decomposition of EO. At about 10 mol% EO a reversal point can be observed. Approaching the fuel rich range, the decomposition reaction of EO becomes more influential and causes the typical expansion of the flammability region.

With increasing pressures and temperatures a further expansion of the explosion region is observed. The influence is still rather low in the range of LFL. However, in the range of UFL the influence of initial pressure and temperature on the decomposition is particularly strong.

Also different inert gases have an effect on the size of the flammability region. The region decreases with an increasing molar heat capacity of the inert gases in the following order: N_2, H_2O, CO_2. This corresponds to the theory of Burgess and Wheeler [5]. They assumed that a minimum flame temperature that is largely independent of the type of inert gas shall be reached. In case of inert gases with higher molar heat capacities more energy is needed to heat up the system. Therefore more amounts of the flammable gas have to be burnt resulting in a smaller flammability region.

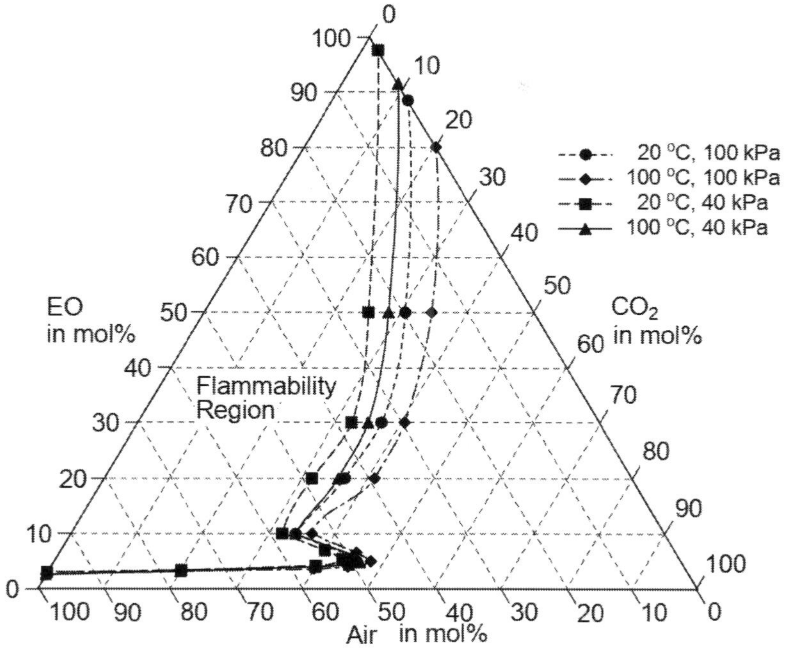

Fig. 3: Experimentally determined flammability regions for the system EO/CO_2/air at different temperatures and pressures.

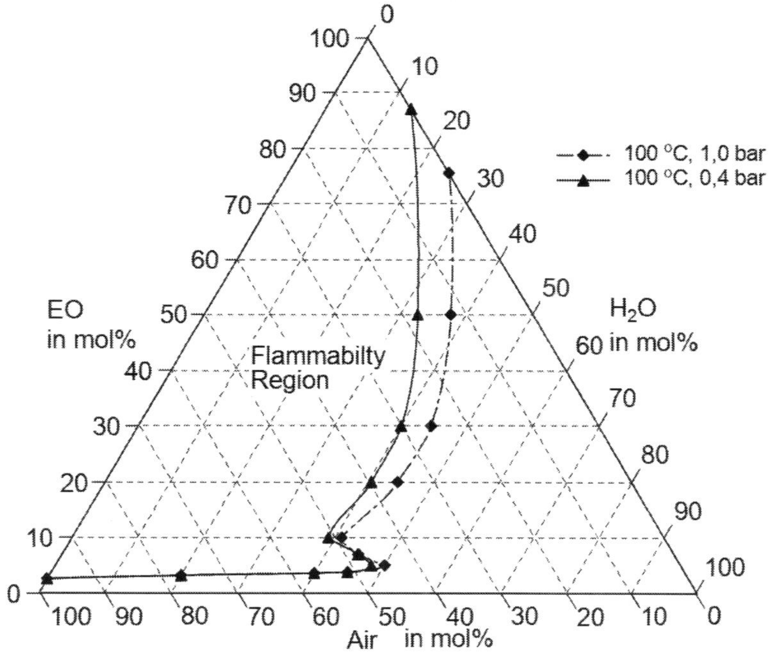

Fig. 4: Experimentally determined flammability regions for the system EO/H_2O/air at different pressures.

4. CALCULATION OF FLAMMABILITY LIMITS

The complete numerical computation of flammability limits is a complex and ambitious task. For many substances not feasible as the complete kinetics of the combustion process and all transport and radiation phenomena had to be taken into account.

Practically, it is much easier to apply semi-empirical methods. Such a model was developed in the 1990s at the University Paderborn, Germany, for the calculation of flammability regions of flammable gas/inert gas/air mixtures [6]. The applied model was based on the assumption of Burgess and Wheeler of "constant minimum flame temperatures". At first the adiabatic equilibrium flame temperatures are calculated for the experimentally determined LFL and UFL. At this stage no inert gases are considered. In the following steps different amounts of inert gases are included and the flammability regions are computed. For this purpose the flame temperature is regarded to be constant.

Using the code "Gaseq" [7], the intention of the present study was to prove that the assumption of Burgess and Wheeler can be used to predict correct flammability regions for EO mixtures, too.

"Gaseq" has a Windows graphical user interface and is easy to run. It allows the computation of equilibrium concentrations and adiabatic temperatures for gas reactions at constant volume or pressure. All calculations were performed for EO and its reaction products by means of NASA thermodynamic data sets [8]. The reaction kinetics were not considered. The solution of the system of equations is based on the minimization of Gibb's energy G according to equation (1):

$$\frac{G}{RT} = \sum_{i=1}^{nSp} \left(\frac{y_i G_i}{RT} + y_i \ln \frac{y_i}{\sum y_i} + y_i \ln p \right) = Min \qquad (1)$$

On the product side the following species were considered: C, CH_4, CO, CO_2, C_2H_4, C_2H_6, C_3H_8, C_2H_4O, CH_3, CH_2O, H, OH, H_2O, O, O_2, CH_3CHO, C_2H_2, C_3H_6, C_4H_8, C_4H_{8CIS}, C_4H_{8TRANS}, N_2, HCN, CN, NH_3, NO, HCO, NH_2, NH, N, CH, CH_2. The formation of carbon black was not taken into account because only small amounts of carbon black were observed during the experiments.

In addition it was necessary to modify the calculation procedure. Especially in those cases where the "inert gases" water and carbon dioxide take part at the equilibrium reactions extremely high amounts of H_2 and CO were calculated. This led to calculated equilibrium temperatures that were much lower compared to the case with nitrogen. Furthermore gas analysis resulted in much lower fractions of H_2 and CO. Therefore the program was modified in such a way that the inert gas fractions were kept constant on the product side. Only the combustion products of EO took part in the equilibrium reactions. Based on this assumption the calculated temperatures at the flammability limits were similar for all types of inert gases following the theory.

Furthermore it was observed that the calculated temperatures along the boundary line of a flammability region are not constant for the whole curve but varied between 1300 K and 1600 K (see Tab. 1). However, the general shape of the temperature profiles found was similar for all three types of inert gases.

Tab. 1: Calculated adiabatic flame temperatures (equilibrium temperatures) for EO/N_2/air mixtures at 20 °C and 100 kPa (see also Fig. 2)

Flammability limits EO in mol%	N_2 in mol%	Flame temperatures in K
3.0	0.0	1369
3.0	20.0	1371
3.0	40.0	1376
3.2	55.0	1437
3.6	59.0	1389
5.2	55.0	1355
10.0	45.0	1485
20.0	50.0	1381
30.0	46.5	1409
50.0	36.0	1478
81.0	19.0	1539

For a better modeling the mean temperature profiles of the three investigated inert gases were taken in order to calculate the flammability regions. Now, by providing arbitrary but specific flame temperatures and using the “Gaseq” code, it was possible to calculate the corresponding EO concentrations (flammability limits) for different amounts of inert gases. It shall be emphasized that this can be done for a gas mixture containing a single inert gas but also for process gases containing more than one inert gas and for different initial temperatures and pressures.

In Fig.5, an experimental explosion region is compared with a calculated one. The differences are very small. A first application of the new calculation method was the verification of the explosiveness of a specific process gas consisting of EO, N_2, CO_2, H_2O and air in sterilization chambers at 100 kPa and 60 °C. Our client also required to know the maximum allowable EO concentration in such a gas mixture. The ignition tests showed that the process gas was not explosive. The flammability limit was found to be 43 mol% EO. Applying the new calculation method the flammability limit was determined to be 42.5 mol%. The calculation method worked very well.

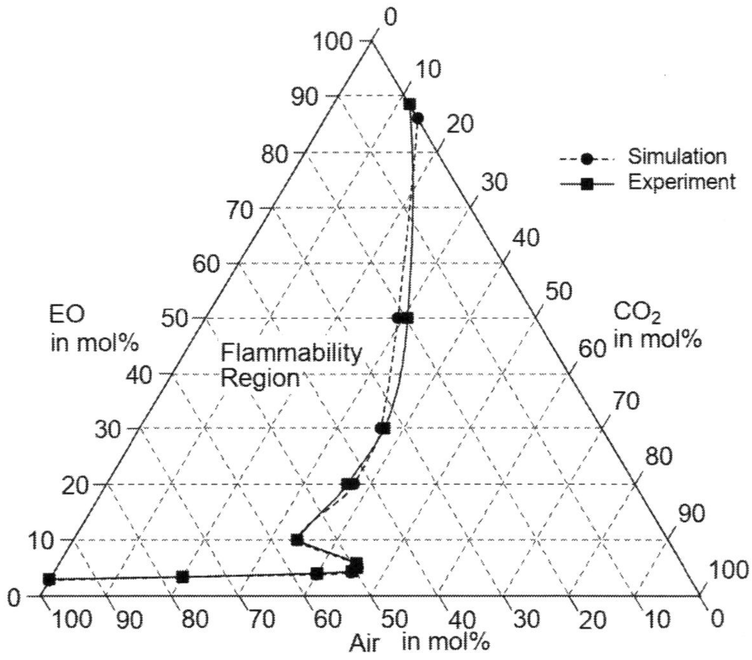

Fig. 5: Comparison between simulation and experiment for the system EO/CO_2/air at 20 °C and 100 kPa

5. REFERENCES

1. U.S. Chemical Safety and Hazard Investigation Board, *Investigation Report No. 2004-11-I-CA, Sterigenics,* 2006
2. Molnarne, M., Schendler, T., Schröder, V., *Sicherheitstechnische Kenngrößen Band 2: Explosionsbereiche von Gasgemischen,* Verlag für neue Wissenschaft GmbH, Bremerhaven 2003
3. DECHEMA, BAM und PTB, *CHEMSAFE® - Database for evaluated Safety Characteristics,* Update 2007, STN Karlsruhe
4. DIN EN 1839: *Bestimmung der Explosionsgrenzen von Gasen und Dämpfen,* Beuth-Verlag, Berlin 2004
5. Burgess, M. J., Wheeler, R. V., *J. Chem. Soc.*, 1911, 99, S. 2013-2030
6. Thiel-Böhm, A.., *Explosionsgrenzen methanhaltiger Brenngasgemische - experimentelle Bestimmung und Simulation; Dissertation,* Universität-GH Paderborn, VDI Verlag, Fortschrittberichte, VDI Reihe 3, Nr. 258, Düsseldorf 1991.
7. Morley, C. "GasEq – A Chemical Equilibrium Program for Windows", online auf http://www.gaseq.co.uk/, Stand: 2007
8. Gordon, S., McBride, B. J., *NASA Reference Publication 1311,* 1994

Effects of Polyethylene Size on Ignition of Electrostatics Discharge

LIU Yi, ZHAO Dong-feng, LU Shuai, ZHANG Lin, YIN Fabo

(College of Chemistry and Chemical Engineering, China University of Petroleum, Dongying 257061, China)

ABSTRACT

Minimum ignition energy of coarse polyethylene granule and screened polyethylene dust is tested with a standard dust explosion testing equipment of Hartmann tube. By comparing the minimum ignition energy and different types of electrostatic discharge energy, the effects of polyethylene size on ignition of electrostatics discharge has been investigated. The results show that, the minimum ignition energy of polyethylene has an exponential relationship with maximum particle size while the particle size is less than 2 mm. No types of electrostatics discharge can ignite the coarse polyethylene granule with average diameter above 2 mm. For the polyethylene dust with diameter below 0.5 mm, electrostatics discharge of corona discharge, brush discharge and cone discharge are excluded from the potential ignition source. Sparks and propagation brush discharge are the possible ignition source. Preventing the occurrence of sparks and propagation brush discharge is an effective measure to defend against the dust explosion of polyethylene from an industrial view.

1. INTRODUCTION

With the high-speed development of petrochemical industry in China, both numbers of polyethylene (PE) production unit and its throughput are all increased. Now the maximum throughput of high-density PE for a single production unit is 200,000 tons per year in China. However, the explosion accident potential has also increased along the process of PE manufacture. Based on our investigation, sixty-eight PE explosion accidents have happened in China during the last several years. The accidents caused a great loss of life and damage to property, which was also a big disturbance to the manufacture and management for the enterprise. The accident investigations showed that most accidents occurred at the silo and attributed to electrostatics discharge.

To assess the hazard of PE electrostatics discharge, the following steps should be included. First, is it possible to form combustible mixture? Second, is it possible for the electrostatics charge to accumulate? Third, what types of electrostatic discharge will occur? Finally, is the energy of electrostatic discharge big enough to ignite the combustible mixtures? During the process of PE manufacture, PE is conveyed with air, and the concentration of PE is above its upper explosive limit, so it is easier for the mixture of PE and air to form combustible mixture. In addition, the volume resistivity of PE is larger than 10^{14} Ω • m, and the electrostatic charge because of friction, collision and segregation could not be eliminated by grounding. Electrostatic would accumulate with a high speed. When the field strength is larger than the breakdown field strength of air, different types of electrostatic discharge will occur. As a result, the key step to evaluate the hazard of electrostatic discharge during the process of PE manufacture is to assess the ignition ability of electrostatic discharge.

Most of previous studies focus on the types of electrostatic discharge, energy of electrostatic discharge and method of evaluation [1-6]. However, more detailed information about the effects of PE size on ignition of electrostatics discharge, which is important to assess the hazard of electrostatic discharge of PE, has not been sufficiently explored.

In this study, the effect of PE size on ignition of electrostatics discharge has been conducted experimentally.

2. EXPERIMENTAL

2.1. EXPERIMENTAL APPARATUS

The experimental system is schematically shown in Fig. 1, which is composed of a Hartmann tube, dust dispersion part, ignition part and control part. To observe the ignition of PE granule conveniently, the upper part of Hartmann tube is made of Plexiglas, which volume is 1.2 L. The dust dispersion part is composed of a gas-supplying unit, a solenoid valve, a gas nozzle and a sample dish. The ignitions part consists of a pair of copper wire electrodes with 0.4 mm diameter. The electrodes were positioned in the middle of the tube, which locates 7.6 cm above the bottom of the tube. The distance between the electrodes tips is approximately 5 mm. The startup time of the ignition time and on-off time of the solenoid valve were set and controlled by the programmable logic controller.

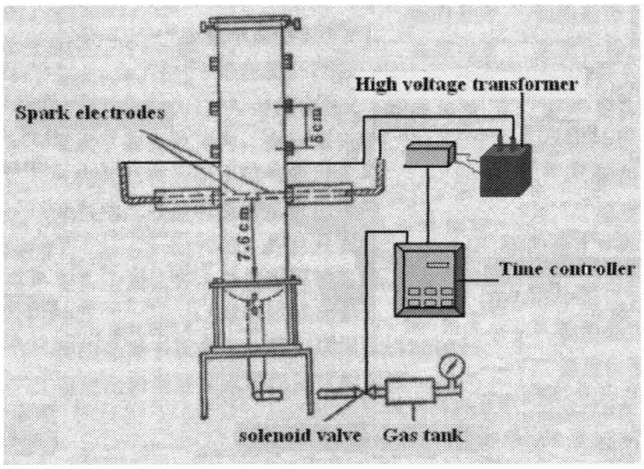

Fig. 1 Experimental system

2.2. EXPERIMENTAL CONDITIONS AND PROCEDURE

The experiments were carried out at atmospheric pressure with the upper end open and the lower end equipped with a dust dispersion part. A specified mass of PE granule weighed by a balance was put on the sample dish evenly. The dust was dispersed into the tube with pressurized air. The PE granule used in the test was made of PetroChina Company Limited, which is a product of high-density PE. The size distribution of the original product is shown in Table 1. Before each experiment, the sample was stored in a drier of 50℃ for 24 h to eliminate the moisture.

Table 1 PE particle size distribution

Particle average diameter/ μm	<2000 μm/ %	<1000 μm/ %	<125 μm/ %	<80 μm/ %
1352	100	28.6	0.5	0.2

The detailed experimental conditions are listed as follows:

Injection time: 100 ms, injection pressure: 0.5 MPa, interval between injection end and ignition: 120 ms.

Experimental procedure is listed as follows:

(1) Inject air into the gas tank until the pressure in the gas tank is 0.5 MPa.

(2) Connect the experiment system and all the other devices as shown in Fig. 1. Open the solenoid valve connected to the gas tank.

(3) Put 1 g PE granule on the sample dish evenly.

(4) Charge the high voltage transformer to a certain value and initiate the control unit to trigger the solenoid valve and ignition part.

(5) Judge whether the PE-air mixture is ignited or not in a range of spark energy. If the PE-air mixture is ignited, and then adjust the voltage and capacitance to lower the spark energy and repeat the procedure from step 1 until the mixture cannot be ignited by 20 times.

2.3. EXPERIMENTAL METHOD

At first, the PE granule was ignited with high voltage circuit discharge, where the voltage and capacitance are 9000 V and 0.1 μF respectively. If the PE granule could not be ignited, the method of the high voltage breakdown and low voltage arc circuit continued to be used. The principle of the high voltage breakdown and low voltage arc circuit is introduced as follows. Break down the spark gap at 10 kV voltage first, where the spark energy is less than 10 mJ and could be neglected, and then the low voltage capacitance discharge in the electric arc between the electrode gaps. The ignition energy was changed by adjusting the voltage and capacitance. Due to the low voltage, the energy consumed on the resistance is small, so the discharge energy can be calculated simply by the difference before and after the electric discharge based on the following formula.

$$E=0.5C\times(U_1^2-U_2^2) \quad (1)$$

Where E is the minimum ignition energy (MIE) of the sample particle, C is the capacitance. U_1 and U_2 are the voltage before and after the electric discharge respectively.

3. RESULTS AND DISSCUSSIONS

3.1. EXPERIMENTAL RESULTS

Based on on-site measurement, most of suspended PE granule diameter is less than 0.5 mm. The MIE of the original PE granule and the particles with diameter below 0.5 mm were tested. The results are shown in Table 2. To compare with results of standard PE dust, the MIE of 75 μm PE dust is also listed in the table [7].

Table 2 MIE of the tested PE particles

Sample	MIE (mJ)
Original PE granules	$>10^7$
PE granule with diameter less than 0.5 mm	656
PE dust with diameter less than 75 μm	10

3.2. DISCUSSIONS AND ANALYSIS

3.2.1 RELATIONSHIP BETWEEN MIE AND PARTICLE SIZE

Base on the experimental results, if the MIE of original PE granule is assumed to 10^7 mJ, the relationship between the MIE and maximum particle size is plotted as shown in Fig. 2. The curve shows that there is a linear relationship approximately between the normal logarithms of PE's MIE and its maximum particle size, where the correlation coefficient is 0.996. As a result, when the particle size of PE is less than 2.0 mm, the MIE has exponential relationship with the maximum particle size.

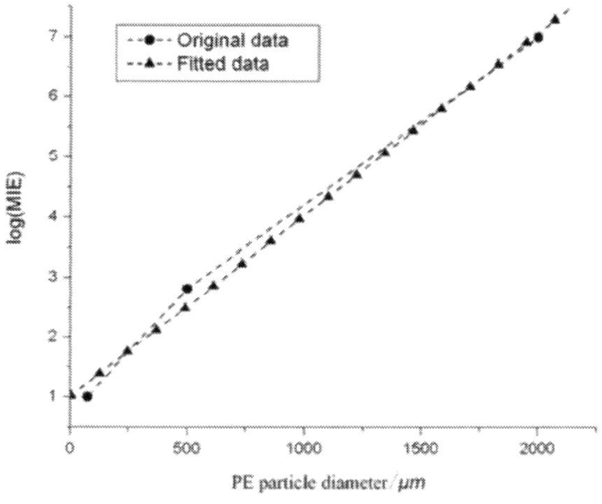

Fig.2 Relationship between MIE and maximum particle diameter of PE

3.2.2 COMPARISON BETWEEN MIE OF PE AND MAXIMUM DISCHARGE ENERGY OF DIFFERENT ELECTROSTATICS DISCHARGE TYPE

During the process of production, transportation and storage, different types of electrostatics discharge will occur while the PE particle collides with pipeline or other devices. Electrostatics discharge type includes corona discharge, brush discharge, cone discharge, spark discharge and propagation brush discharge. Different electrostatics discharge has different maximum discharge energy [8-9]. Corona discharge can just ignite some flammable gases, which has a lower hazard for PE granule. The maximum energy of brush discharge is smaller than 3.6 mJ, which could ignite most flammable gas and some of dust with lower MIE. Cone discharge is also called Maurer discharge, whose maximum discharge energy is 10 mJ, which could ignite most gas and fine dust. Spark discharge generally occurred between two conductors, whose maximum discharge energy is up to 1 J. Propagation brush discharge has maximum discharge energy of 10 J. Almost all the dangerous flammable gas, flammable liquid vapors and the mixture of flammable dust can be ignited by propagation brush discharge.

Comparing the maximum energy of different electrostatics discharge with the MIE results under different particles size, we can conclude that the original PE granule with MIE larger than 10 kJ cannot be ignited by any type of electrostatics discharge. When the maximum particle diameter of PE is less than 0.5 mm, its MIE is larger than the maximum discharge energy of corona discharge, brush discharge and cone discharge, which is smaller than the maximum energy of spark discharge and propagation brush discharge. Under this conditions, much attention has to be paid on spark discharge and propagation brush discharge, and some necessary measures to prevent the two electrostatics discharge has to be taken. When the particles diameter of PE is smaller than 75 μm, its MIE is larger than the energy of corona discharge and close to the maximum discharge energy of brush discharge and cone discharge. In this situation, we should prevent all the electrostatics discharge except corona discharge.

The above analysis confirms that the PE particle size has much effect on the ignition probability. The smaller the diameter of PE is, the more hazards the PE particles are. With the increase of PE particle diameter, the MIE of PE will increase approximately exponentially. Moreover, the hazard of PE dust explosion mainly comes from the suspended dust, so the key index to assess the hazard of on-site electrostatics discharge is the MIE of particles with diameter less than 0.5 mm.

4. CONCLUSIONS

For better understanding the effects of PE size on ignition of electrostatics discharge, the MIE of different diameter PE granules was measured and compared with the maximum electrostatics discharge energy. The conclusions obtained are as follows.

(1) For the PE granule used in the study, the MIE has exponential relationship approximately with the maximum particles diameter of PE when the PE particle diameter is less than 2 mm.

(2) The electrostatics hazard during PE manufacture can be assessed by comparing the MIE of suspended PE particles with the maximum discharge energy of different electrostatics. Some effective measures to prevent PE dust explosion can be taken based on the comparison results.

(3) The smaller diameter has more hazards of PE dust explosion. No kinds of electrostatics can ignite the original PE particles with MIE of approximately 10^7 mJ. When the PE particle diameter is less than 0.5 mm, corona discharge, brush discharge and cone discharge are impossible to ignite the mixture of PE particles and air. Spark discharge and propagation brush discharge should be the possible ignition source. Much attention has to be paid to prevent spark discharge and propagation brush discharge.

5. REFERENCES

1. Dahn, James C., Electrostatic hazard evaluation methods of dust/hybrid mixtures-current & future, *Proceedings of the 4th International symposium on hazards, prevention, and mitigation of industrial explosions*, France: Bourges, 2002.
2. Glor, Martin, Hazards due to electrostatic charging of powders, *Journal of Electrostatics*, 1985, 16(2-3): 175-191.
3. Glor, Martin, Luttgens, G., Maurer, B., and Post, L, Discharges from bulked polymeric granules during the filling of silos-characterization by measurements and influencing factors, *Journal of Electrostatics*, 1989, 23(1): 35-43.
4. Glor, Martin, Ignition hazard due to static electricity in particulate processes, *Powder Technology*, 2003, 135-136(1):223-233.
5. Glor, Martin, Schwenzfeuer, K., Direct ignition tests with brush discharges, *Journal of Electrostatics,* 2005, 63(6-10):463-468.
6. Hoppe, Tom, Jaeger, Norbert, and Terry, John, Safe handling of combustible powers during transportation, charging, discharging and storage, *Journal of Loss Prevention in the Process Industries*, 2000, 13(3-5):253-263.
7. Eckhoff, Rolf R., *Dust explosions in the process industries*, London: Gulf Professional Publishing, 2003.
8. Pratt, Thomas H., *Electrostatic Ignitions of Fires and Explosions*. New York: Center for Chemical Process Safety/AIChE, 2000.
9. Wilson, Norman, Luttgens, G., *Electrostatic Hazards*, Oxford: Buttenuorth-Heinemann, 1997.

148o GE Gasturbine Design in O&G Applications According to IEC62061 Requirements Vs CE Marking

Vincenzo Amato and Elda Casamassima. General Electric O&G, Via Felice Matteucci, Florence, 2 / 50100, Italy

Safety Instrumented Systems are used in gas turbine (GT) systems design to mitigate potential Hazardous situations that can arise from machine deviations. International standard IEC62061, (released in 2005) is machinery sector within the framework of IEC61508.

Starting from 31st December 2005 the IEC62061 came harmonized standard for machinery directive 98/37/CE. All members of UE and EFTA use the machinery directive in order to establish the safety requirements (RES) for machine that will be putted in market.

In this way, the IEC62061 are requested to be satisfied in order to give the presumption of conformity for the loop based on Electrical/Electronic/Programmable Electronic components that are called to ensure safety of machinery. This establishing both a probabilistic approach on safety and, the relative safety integrity of a safety instrumented system to perform its prescribed action over the product life cycle (design, installation, and operation of safety instrumented systems).

This paper describes the application of IEC62061 international standard on GE O&G Gas Turbine for the Safety-Related System (SRS) that are separated from the normal electrical control system. This job allows both ensuring a greater safety factor and faulting independence inside E/E/PE system and satisfies the requirements of machinery directives that, together with others applicable European directives, bring to CE marking.

Domino Effects in the Safety Assessment of Petrochemical Enterprises

Zhao Dongfeng Zhang Bo Wang Wendong

(Center of Environment and Safety Technology, China University of Petroleum (East China), Dongying , China)

Abstract: Petrochemical units are normally evaluated independently for the purpose of optimization, but they could easily be involved in chain accidents which we define as domino effects. Domino effects can cause serious consequences in petrochemical units. However, the existing safety assessment methods analyze equipment separately and ignore the influences of the neighbor units. It is of great urgency and importance that safety assessments take the domino effects into consideration. Based on the domino effect analysis model presented by V.Cozzani, we conceived the framework of domino safety assessment method. Domino safety assessment is a safety assessment method in which domino effects should be analyzed and evaluated. It evaluates the interaction between the surrounding units qualitatively and quantitatively to determine the possibility of happening of a domino effect, so that the results of the safety assessment would be more objective and more close to the facts. Three kinds of factors which could cause domino effects presented by V.Cozzani including thermal radiation, overpressure and explosion missiles have been included in the framework of domino safety assessment method. An industrial example has been studied and it explains the accidents phenomenon well. Domino safety assessment method is a useful supplement to the existing safety assessment method and it has great value for research.

Key words: domino effect domino safety assessment method petrochemical enterprises

1. Introduction

Heinrich describes the occurrence of accidents as a fall of domino chain which contains four elements: P—human error; H—unsafe actions of men or unsafe conditions of machines; D—accidents happen; A—injuries. Once a domino falls, a chain reaction will occur; the dominos behind it will fall one by one. The domino theory intuitively shows the causality of the occurrence of accidents, points out that the accident analysis should be carried out from the phenomena of accidents and should be prevented from the very beginning. However, the disadvantage of the classic domino theory is also obvious. It sticks to the domino chain too absolutely and describes the causes and effects of accidents too roughly. In order to evaluate the falling possibility of the domino chain reaction more accurately, we need to consider the probability of the secondary accident, and then establish an evaluation method which can evaluate the interactional effects of the domino chains.

Petrochemical installations are normally evaluated independently for the purpose of optimization and it is good for the process design. However, due to domino effects, installations would be affected by each other in chain accidents, neighboring equipments or units would be involved. Domino effects can be found everywhere, especially when it happens, the domino chains will fall one by one, and then cause more serious consequences. Nowadays, many countries have introduced the assessment of domino effects to safety assessment, but in China, the safety assessment methods are still focusing on the accidents effects simulation for independent equipments or units and have not considered the effects from the condition of equipments or units which are closely surrounding the assessment units. It means that we have not thought much of the analysis and evaluation of domino effects. So, it's of great importance to bring forward and study domino safety assessment method.

2 .Domino Safety Assessment

Domino safety assessment is a safety assessment method in which domino effects should be analyzed and evaluated. It evaluates the interaction between the surrounding units qualitatively and quantitatively to determine the possibility of occurrence of a domino effect (1). So that the results of the safety assessment would be more objective and more close to the facts. The difference between domino safety assessment and existing safety assessment methods is that domino safety assessment focuses on the influence among all neighboring assessment units and emphasizes on the analysis of the probability of the secondary accidents. Domino safety assessment needs to firstly determine the source of significant hazards which cause accidents which including the potential hazards of installations or equipments. Secondly, determine the assessment range and chose a domino effect assessment method according to the hazards types. Thirdly, carry out an assessment for each evaluation unit and at last, point out the prevention and contingency measures to minimize the influence of domino effects. There are many factors which could result in domino effects including thermal radiation, overpressure, missiles of explosion, toxic substances leakage, and a combination of several or all of these factors could lead to second accidents. In these factors, toxic substances leakage mainly relates to the impact of human beings and environment, and researches have been developed before. Our study mainly focuses on the three factors which could primarily cause domino effects: thermal radiation, overpressure and explosion missiles. [2,3]

2.1. Domino Effects Caused by Thermal Radiation

Thermal radiation produced by fire is a major factor leading to domino effect. Flammable gas or liquid leakage comes out from installations could catch fire or explode when provided with certain ignition energy. The targets of thermal radiation could be people, equipments, facilities, buildings and so on. The detailed assessment can carry out according to the following steps: choose a thermal radiation model, calculate the thermal flux a target received, and then use the probability model to assess the damage. The death probability of human beings and ignition probability of wood due to thermal radiation has already been developed by the existing safety assessment method. So our study focuses primarily on the probability model of different types of equipments damage caused by thermal radiation. Valerio Cozzani developed a probability mode for domino effect caused by thermal radiation. It can assess the equipments damage probability according to the amount of thermal radiation. The models are as follows:

$Y = 12.54 - 1.84\ln(ttf)$, where Y is the damage probit.

For the atmospheric and vertical cylindrical containers,

$$\ln(ttf) = -1.128\ln(I) - 2.667\times10^{-5}v + 9.887;$$

For high pressure containers,

$$\ln(ttf) = -0.947\ln(I) + 8.835v^{0.032},$$

where

ttf is “the time to fail” (s),
I is the thermal flux of the equipment targets (Kw/m^2),
V is the equipment capacity (m^3). [4]

2.2 .Domino Effects Caused by Overpressure

The overpressure theory points out that once the overpressure caused by a blast wave reaches a certain value, it will cause a certain degree of damage to the targets. It is another main factor which could cause domino effects. When explosion occurred in the petrochemical installations, the overpressure can reach to several atmospheric pressures. According to experiments, when two same kinds of explosives with different amount explode, if the ratio of distance from explosion source equals to the cube root of the ratio of the explosive amount, they would produce the same overpressure. According to this model and checking from the blast wave overpressure table of the 1000Kg TNT explosion in the air, we can calculate the overpressure received by the target equipments from the explosion source.

Valerio Cozzani developed the following model for the quantitative assessment of the equipment damage caused by blast wave:

$$Y=K_1+K_2 Ln(\Delta P^0).$$

Where:

Y—equipment failure probit,

ΔP_0 —over pressure (Pa);

K_1, K_2 assumes different values according to four different types of containers,:

for atmospheric pressure vessel, $K_1=-18.96$, $K_2=2.44$;

high Pressure Vessel, $K_1=-42.44$, $K_2=4.38$;

long vertical equipment, $K_1=-28.07$, $K_2=3.26$;

small equipment, $K_1=-17.79$, $K_2=2.18$.

The probit analysis is a wide-used method which could assess the human body reflection to toxic substances and radiations, and it also can be used to assess the damage probability of equipment caused by the thermal radiation, overpressure damage and other factors. The relation of the probability F and the probit Y agrees with the normal distribution: $F=\frac{1}{\sqrt{2\pi}}\int_{-\infty}^{Y-5} e^{-u^2/2}du$. Through the relationship, we can make the conversion between probit and the probability, and make the quantitative assessment. [5,6]

2.3 .Domino Effects Caused by Explosion Missiles

As pressure vessels or storage tanks explode, they release enormous energy. The vessels or tanks could be torn into several pieces or debris and be ejected at high speed. These missiles with high speed or in great quantity have enormous energy when hitting the equipments or people around, could result in serious accidents, which is another major factor that could cause domino effects. Currently the domino effects triggered by the explosion missiles could be taken by the initial quantitative evaluation, which is specifically divided into the following steps: determine the probability distribution of the direction of the fragments' initial projectile and the velocity, the debris path, the probability of collision between fragments and target equipments, determine the probability of damage to equipments. The key point is to establish the initial parameters, which can impact fragments flying projectile. Great study must be undertaken in this field. [7]

3 .Case Study Example

We take a petrochemical plant as an example, the occurrence of a series explosion in a petrochemical plant, which has the thermal radiation, overpressure, explosion missiles can lead to domino effects. This study will focus on the domino effect caused by the overpressure as an example to illustrate the domino effect assessment method. We take a tank explosion which contains 25 tons of nitrobenzene as an example. According to the worst accident evaluation guidelines recommended by the Environmental Protection Agency of the United States, we assume that the one tenth of loss contents combusted and exploded completely. Using the TNT explosion and overpressure model mentioned in part 2.2, rough assessment has been drawn according to the distance listed in table 1. The overpressure varies due to distance is shown in figure 1. Equipments damage probability due to various types of equipments is shown in figure 2. From figure 2 we can see that the probability is 1 or near to 1 at distance of 20 meters from the explosion center. In other words, collapsed and smashed, broken, displacement or other structural damage could happen in a range of 20 meters. As the distance increases, different types of equipments have gradually reduced the probability of damage. This analysis shows that if there are equipments in the range which could cause damage, they will get various degrees of damage, then dangerous contents leak from the damaged equipments, and this contents will explode when meet ignition sources, thereby creates a domino chain of accidents. The analysis by our domino safety assessment method complies very well with the actual accidents.

Table 1 overpressure of a tank explosion (containing 25 tons of nitrobenzene)

Actual distance R/m	14	18	22	30	38	46	54	60
overpressureΔP^0/KPa	197	106	70	39	26	20	16	13

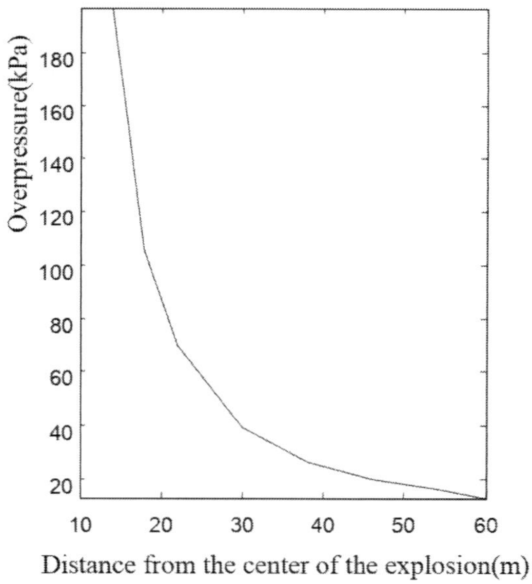

Fig 1 overpressure attenuation due to distance

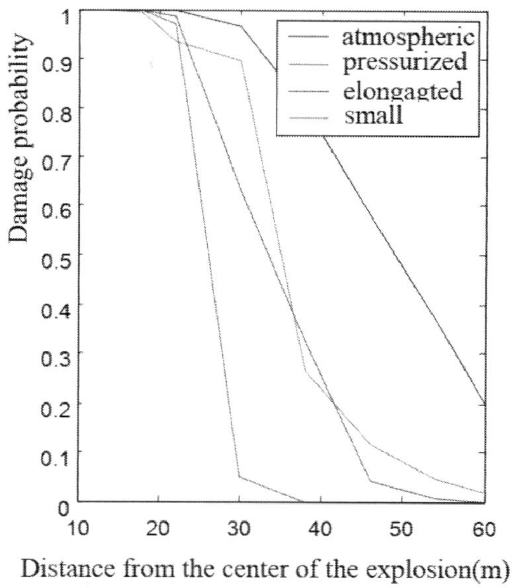

Fig 2 damage probability of different types of equipments

4 .Conclusion

The study presents the concept of domino safety assessment method. Domino safety assessment method could analyze petrochemical accidents more appropriately and it could provide more useful techniques to prevent important domino accidents. Domino evaluation method is a useful supplement for the existing safety assessment methods.

References

1. Khan Faisal I., Abbasi S.A., Estimation of probabilities and likely consequences of a chain of accidents (domino effect) in Manali Industrial Complex. Journal of Cleaner Production, 9 (2001): 493-508
2. Cozzani V., Salzano E., Threshold values for domino effects caused by blast wave interaction with process equipment. Journal of Loss Prevention, 17 (2004): 437-447
3. Cozzani V., Gubinelli G., Salzano E., Escalation thresholds in the assessment of domino accidental events. Journal of Hazardous Materials.A129(2006):1–21.
4. Cozzani V., Gubinelli G., Antonion G.etc, The assessment of risk caused by domino effect in quantitative area risk analysis. Journal of Hazardous Material,A127 (2005): 14–30
5. Cozzani V., Salzano E.,The quantitative assessment of domino effects caused by overpressure Part I Probit Models. Journal of Hazardous Material, A107 (2004): 67-80

6. Cozzani V., Salzano E.,The quantitative assessment of domino effect caused by overpressure Part II Case studies. Journal of Hazardous Material, A107 (2004): 81-94
7. Gubinellia G., Zanellia S., Cozzanib V., A simplified model for the assessment of the impact probability of fragments. Journal of Hazardous. Material, A116 (2004): 175-187

Robust and reliable modelling for a distillation column

Yahya CHETOUANI

Université de Rouen, Département Génie Chimique, Rue Lavoisier, 76821 Mont Saint Aignan Cedex, France. Tel: (0033)235528466, Fax: (0033)235147191, Yahya.Chetouani@univ-rouen.fr

ABSTRACT

The main aim of this paper is to establish a reliable model of a process behavior both for the steady-state and unsteady-state regimes. The use of this accurate model allows distinguishing a normal mode from an abnormal one. Therefore the neural black-box identification by means of a NARX (Nonlinear Auto-Regressive with eXogenous) model has been chosen. It shows the choice and the performance of the neural network. The model is implemented by training a Multi-Layer Perceptron Artificial Neural Network (MLP-ANN) with input-output experimental data. After describing the system architecture, a realistic and complex application as a distillation column is presented in order to illustrate the neural model reliability.

1. INTRODUCTION

In the last few years ever-growing interest has been shown in production quality standards and pollution phenomena in industrial environments. However process development and continuous request for productivity led to an increasing complexity of industrial units. The dynamic nature and the nonlinear behavior of such units pose challenging control system design when products of constant purity are to be recovered.

In chemical process control, the processes typically exhibit nonlinear behavior. The intrinsic highly nonlinear behavior in the industrial process, especially when a chemical reaction is used, poses a major problem for the formulation of good predictions and the design of reliable control systems [1, 2]. Similar problems arise also from the uncertainty for the parameters of the process, such as the reaction rate, activation energy, reaction enthalpy, heat transfer coefficient, and their unpredictable variations.

Most large-scale process models derived from first principles are represented by nonlinear differential–algebraic equation (DAE) systems. Since such models are often computationally too expensive for real-time control. Even if global nonlinear models are available, they may not be appropriate for control and monitoring or fault detection (FD) purpose [3].

On the other hand, when fundamental process knowledge is unavailable or incomplete, or when the resulting model may not be particularly suitable for FD applications, the input/output models identified from plant data may be more useful. Nonlinear autoregressive models have been used [6], as well as models that use a set of polynomial basis functions (e.g. Volterra functions). Also ANNs provide an excellent mathematical tool for dealing with severe nonlinear problems. Another attractive property is self-learning ability. As a result, nonlinear systems can be modeled with a great flexibility. These features allow one to employ artificial neural networks to model complex, unknown and nonlinear dynamic processes.

Distillation is one of the most important separation processes used in many chemical industries. It has extensively been studied [4, 5]. There are essentially two approaches by which nonlinear models can be developed for a distillation column; from first principles by using the process knowledge or empirically from input/output data. The advantages and disadvantages of each approach are well known. In industrial practice, it is not always possible in general to obtain accurate first principles models for high-purity distillation columns.

Most industrial columns are used to separate multi-component mixtures whose constituent elements are often not known completely; the fundamental thermodynamics of multi-component vapor-liquid equilibrium, the physical property data, and other essential constitutive relations required for the successful development of a first principles model are not always available. And even when such knowledge is available, the resulting models usually occur in the form of a very large system of coupled nonlinear ordinary differential equations, and may therefore not always be the most convenient for controller or fault detection (FD) design.

The purpose of this study is to obtain a reduced and reliable model which allows reproducing the dynamics of a nonlinear process as a distillation column. This reliable model enables to reproduce the process dynamics under all operating conditions i.e. steady-state or unsteady state. The present study focuses on the development, and implementation of a NARX neural model for the forecasting of the distillation column dynamics. Experiments were performed in a distillation column and experimental data were used both to define and to validate the model. The performance of this neural model was then evaluated using the performance criteria. Results show that the NARX neural model is representative for the dynamic behavior of this nonlinear process. The experimental set-up, modelling procedure, and prediction results are described in the following sections.

2. EXPERIMENTAL SET-UP

The feed tank (fig. 1) contains a mixture to be separated (Toluene-Methylcyclohexane) with a mass composition at 23 % methylcyclohexane. The operation in continuous mode involves charging the still with the mixture to be separated, bringing the column to equilibrium under total reflux. The product is introduced through the optimal feed tray so that the light components are volatilized, while the heavy part goes down again with the reflux in the column reboiler. The quality of the collected top product of the column depends on the reflux flow rate. The reflux ratio is varied through the magnetic valve by changing the relative quantities of material returning to the column and flowing to product storage. Feed preheating system is constituted by three elements of 250 W each one. In addition it has a low liquid level switch in order to avoid the running if the level is excessively low. The reciprocating feed pump is constituted by a membrane allowing firstly the suction of the mixture and the discharge towards the tank with a flow capacity $F = 4.32$ L.h^{-1}. The column has also a reboiler of 2 liters hold-up capacity, an immersion heater of a power $Q_b = 3.3$ kW and of a level liquid switch sensor which allows the automatic stop of heating if the level is insufficient. The stirring of the mixture in the reboiler is ensured by the boiling mixture. The internal packing is made of Multiknit stainless 316L which enhances the mass transfer between the vapor and liquid phases. In order to approach the adiabatic conditions, heat-insulation made of glass wool is laid around the column. A condenser is placed at the column overhead in order to condense the entire vapor coming out from the column. The cooling medium used in exchangers is water. The heat-transfer area of the total overhead condenser is 0.08 m^2. Moreover the reflux timer (R_t) allows to control the reflux ratio (R_r). It is monitored by the overhead product temperature (T_d). When the required distillate temperature (T_d) is attained, the reflux timer opens. In the opposite case, it remains closed. Distillation supervision control system allows to modify the parameters and to follow their evolution such as the pressure drop (ΔP), the flow or the temperatures at different points of the distillation column. This control system, therefore, must hold product compositions as near the set points as possible. The thermocouples are coupled to a calibrated amplification circuit (4-20 mA, 0–150°C) whose signals are inputted to the computer on-line, which permits the bottom and top temperatures to be obtained. The unit has twelve sensors which measure continuously the temperature throughout the column.

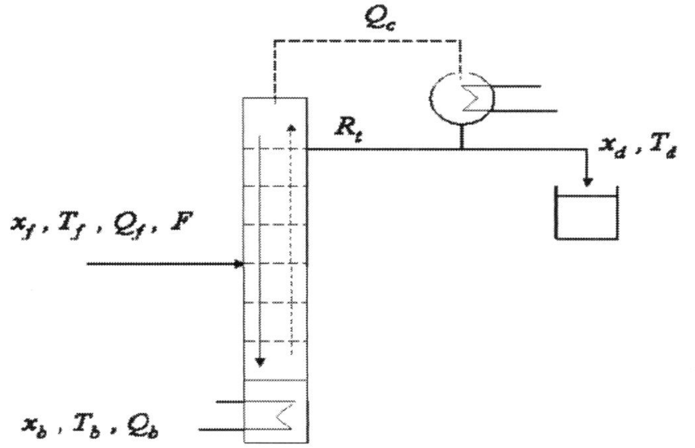

Figure 1. Experimental device: Distillation column

3. IDENTIFICATION OF NARX MODELS

The NARX neural model used to describe accurately the process behavior is the classical MLP-ANNs [7, 8] with one layer of hidden neurons. The NARX model of a finite dimensional system [9] with order (n_y, n_u) and scalar variables y and u is defined by:

$$y(t) = \phi(y(t-1),...,y(t-n_y), \quad u_i(t-n_k),...,u_i(t-n_k-n_u)) \qquad 1 \le i \le m \qquad (1)$$

where $y(k)$ is the Auto-Regressive (AR) variable or system output; $u(k)$ is the eXogenous (X) variable or system input. n_y and n_u are the AR and X orders, respectively. m is the number of the used inputs. n_k is the time delay between u and y. ϕ is a nonlinear function.

The feed-forward MLP-ANN is used in this study and consists in a large number of highly connected nonlinear simple neurons. The log-sigmoid transfer function is used on the hidden layer and a linear transfer function on the output layer. A back-propagation training function for feed-forward networks using momentum and adaptive learning rate techniques is used.

3.1. Calculation of the NN output

Though the applicability of neural networks to solve several nonlinear complex problems has been amply demonstrated, the time taken to train neural networks off-line can be quite excessive [10, 11]. The following steps explain the calculation of the ANNs output based on the input vector.

1. Assign $\hat{w}^T(k)$ to the input vector $x^T(k)$ and apply it to the input units where $\hat{w}^T(k)$ is the regression vector given by the following equation:

$$\hat{w}^T(t) = [y(t-1),...,y(t-n_y),\ \ u(t-1-n_k),...,u(t-n_u-n_k)] \qquad (2)$$

2. Calculate the input to the hidden layer units:

$$net_j^h(k) = \sum_{i=1}^{p} W_{ji}^h(k) x_i(k) + b_j^h \qquad (3)$$

where p is the number of input nodes of the network, i.e. $p = n_y + n_u + n_b$; j is the jth hidden unit; W_{ji}^h is the connection weight between ith input unit and jth hidden unit; b_j^h is the bias term of the jth hidden unit.

3. Calculate the output from a node in the hidden layer:

$$z_j = f_j^h(net_j^h(k)) \qquad (4)$$

where f_j^h is the log-sigmoid transfer function.

4. Calculate the input to the output nodes:

$$net_l^q(k) = \sum_{j=1}^{h} W_{lj}^q(k) z_j(k) \qquad (5)$$

where l is the lth output unit; $W_{lj}^q(k)$ is the connection weight between jth hidden unit and lth output unit.

5. Calculate the outputs from the output nodes:

$$\hat{v}_l(k) = f_l^q(net_l^q(k)) \qquad (6)$$

where f_l^q is the linear activation function defined by:

$$f_l^q(net_l^q(k)) = net_l^q(k) \qquad (7)$$

3.2. Back-propagation training algorithm

The error function E is defined as:

$$E = \frac{1}{2}\sum_{l=1}^{q}(v_l(k) - \hat{v}_l(k))^2 \tag{8}$$

where q is the number of output units and $v_l(k)$ is the lth element of the output vector of the network. Within each time interval from k to $k+1$, the back-propagation (BP) algorithm tries to minimize the error for the output value as defined by E by adjusting the weights of the network connections, i.e. W_{ji}^{h} and W_{lj}^{q}. The BP algorithm uses the following procedure (Eqs.9, 10, 11, 12):

$$W_{ji}^{h}(k+1) = W_{ji}^{h}(k) + \alpha \Delta W_{ji}^{h}(k) - \eta \frac{\partial E}{\partial W_{ji}^{h}(k)} \tag{9}$$

$$W_{lj}^{q}(k+1) = W_{lj}^{q}(k) + \alpha \Delta W_{lj}^{q}(k) - \eta \frac{\partial E}{\partial W_{lj}^{q}(k)} \tag{10}$$

where η and α are the learning rate and the momentum factor, respectively; ΔW_{ji}^{h} and ΔW_{lj}^{q} are the amounts of the previous weight changes; $\partial E/\partial W_{ji}^{h}(k)$ and $\partial E/\partial W_{lj}^{q}(k)$ are given by:

$$\frac{\partial E}{\partial W_{ji}^{h}(k)} = -\left[z_j(k)(1 - z_j(k))x_i(k)\right]\sum_{l=1}^{q}\left[(v_l(k) - \hat{v}_l(k))\hat{v}_l(k)W_{lj}^{h}(k)\right] \tag{11}$$

$$\frac{\partial E}{\partial W_{lj}^{q}(k)} = -(v_l(k) - \hat{v}_l(k))z_j(k) \tag{12}$$

The implementation of the ANN for forecasting is as follows:

1. Initialize the weights using small random values and set the learning rate and momentum factor for the ANN.
2. Apply the input vector given by Eq. 2 to the input units.
3. Calculate the forecast value of the error using the data available at ($k-1$)th sample (Eqs.2, 3 4, 5, 6, 7).
4. Calculate the error between the forecast value and the measured value.
5. Propagate the error backwards to update the weights (Eqs. 9, 10, 11, 12).
6. Go back to step 2.

For weights initialization, the Nguyen-Widrow initialization method [12] is best suited for the use with the sigmoid/linear network which is often used for function approximation.

4. RESULTS AND DISCUSSION

In order to select the optimal number of hidden neurons, tests were performed by varying the number of neurons between 5 and 25 for the temperature (T_d) modelling. The minimal number of inputs is avoided to ensure the model flexibility. Also, the maximum number of inputs is excluded to avoid the over-fitting.

$$T_d(t) = \phi(T_d(t-1), R_t(t-2), \Delta P(t-3), Q_f(t-4), T_f(t-6), F(t-8), T_d(t-1) - \hat{T}_d(t-1))$$

The evolution of the loss function is given in fig. 2 in the training and the validation database. This figure shows that the neurons number in the hidden layer in training can be 7, 10 or 21. However this number can slightly change and becomes 8, 10 or 14 in the test.

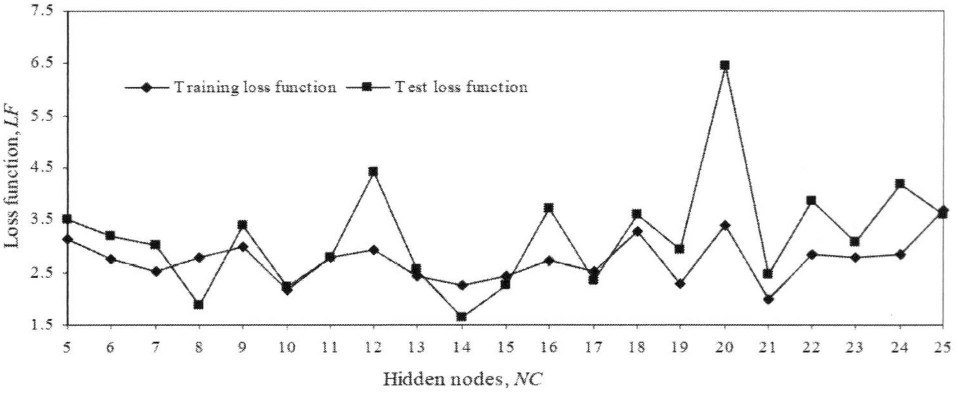

Figure 2. Loss functions

The models which have a structure (7-8-1 and 7-10-1) exhibit the acceptable *LF*. Based on engineering judgment, the model 7-8-1 would be preferred without significant loss of accuracy. Fig. 3 shows the difference between the experimental overhead product temperature and those simulated by the neural model.

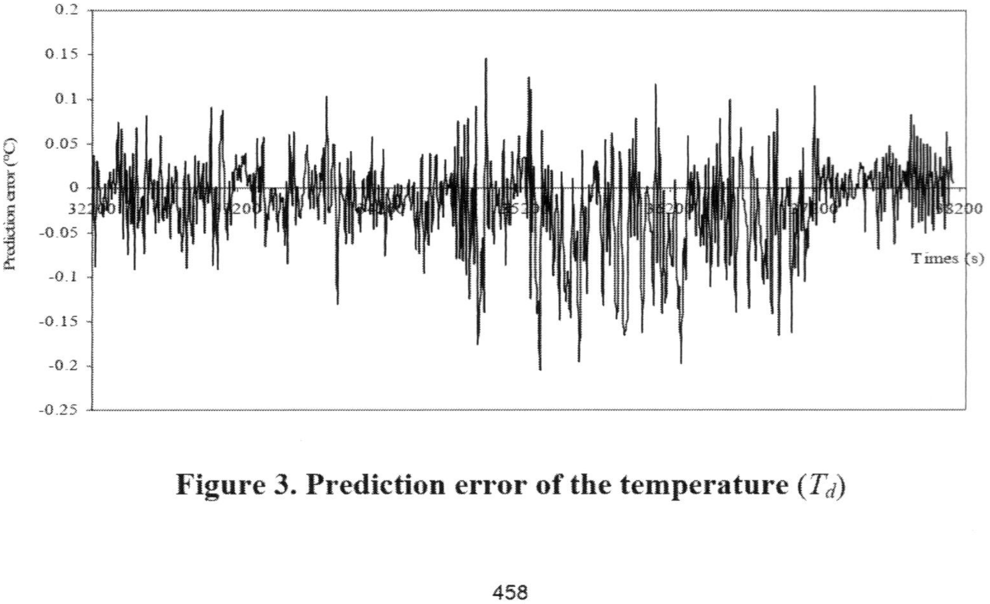

Figure 3. Prediction error of the temperature (T_d)

After analyzing this figure, it emerges that the NARX model 7-8-1 for modelling the temperatures (T_d) ensures satisfactory performances as it is indeed able to correctly identify the dynamics of the distillation column. Therefore, the reduced neural model is considered as reliable one for describing the dynamic behavior of this column. There is a good agreement between the learned neural model and the experiment in the validation phase.

In conclusion the main advantage of the neural approach consists in the natural ability of neural networks in modelling nonlinear dynamics in a fast and simple way and in the possibility to address the process to be modeled as an input-output black-box, with little or no mathematical information on the system.

5. CONCLUSION

This work aims to identify process dynamics by means of a NARX model. The identification of the system dynamics by means of input-output experimental measurements provides a useful solution for the formulation of a reliable model. In this case, the results showed that the model is able to give satisfactory descriptions of the experimental data.

The developed neural models are used in a recursive scheme in order to test their ability to perform the behavior prediction. Finally, the identified neural models will be useful for the detection and the isolation (FDI) of faults which can occur through the process dynamics.

6. REFERENCES

1. Chetouani, Y., *Use of Cumulative Sum (CUSUM) test for detecting abrupt changes in the proccess dynamics*, International Journal of Reliability, Quality and Safety Engineering, 14, 65-80, 2007.
2. Cammarata, L., Fichera, A., Pagano, A., *Neural prediction of combustion instability*, Applied Energy, 72, 513-528, 2002.
3. Huang, Y., Reklaitis, G.V., Venkatasubramanian, V., *Dynamic optimization based fault accommodation*, Computers and Chemical Engineering, 24, 439-444, 2000.
4. Eden, M.R., Koggersbøl, A., Hallager, L., Jørgensen, S.B., *Dynamics and control during start-up of energy integrated distillation column*, Computer Chemical Engineering, 24, 1091–1097, 2000.
5. McAvoy, T.J., *Modeling chemical process systems via neural computation*, IEEE Control Systems Magazine, 3, 24-29, 1990.
6. James, S., Legge, R., Budmann, H., *Comparative study of black-box and hybrid estimation methods in fed-batch fermentation*, Journal of Process Control, 1, 113–121, 2002.
7. Chen, S., Billings, S.A., *Representation of nonlinear systems-The NARMAX model*, International Journal of Control, 49, 1013–1032, 1989.
8. Narendra, K.S., Parthasarathy, K., *Identification and control of dynamical systems using neural networks*, IEEE Trans. on Neural Networks, 1, 4–21, 1990.
9. Ljung, L., *System Identification, Theory for the User,* Prentice Hall, Englewood Cliffs, New Jersey, 1999.
10. Chetouani, Y., *Modelling and prediction of the dynamic behavior in a reactor-exchanger using NARMAX neural structure*, Chemical Engineering Communications Journal, 5, 691-705, 2007.
11. Fung, E.H.K., Wong, Y.K., Ho, H.F., Mignolet, M.P., *Modelling and prediction of machining errors using ARMAX and NARMAX structures*, Applied Mathematical Modelling, 27, 611-627, 2003.
12. Nguyen, D., Widrow, B., *Improving the learning speed of 2-layer neural networks by choosing initial values of the adaptive weights*, International Joint Conference of neural networks, 3, 21-26, 1990.

Specialized Working Procedure

I Made Sukrajaya
BP West Java
Jakarta, Indonesia
sukrajim@bp.com

Yenny Bambang
BP West Java
Jakarta, Indonesia
yenny.bambang@bp.com

ABSTRACT

Everyone knows that standard operating procedures (SOPs) are used for providing clear instructions for safely conduction "critical" or "routine" tasks. But do we forget that non-routine tasks should also have written procedures, called specialized working procedures (SWPs), which should provide clear instructions for safe operations and maintenance?

On closer look, SOP and SWP must be treated the same in terms of purposes, performance requirements and documentation. We should not forget that a majority of risk is due to human factors during non-routine activities and that the task risk assessment or job safety analysis may be insufficient to address all hazard and risk reductions. Risk assessment is used to define the hazard, risk and mitigation plan, but need to be converted to a written procedure to ensure the operator doing the task has clear instructions. By preparing an SWP, it is intended to reduce the frequency of incidents due to human error. An example of SWP is tie-in procedure for plant modification, turn-around operation procedure, non-routine maintenance, etc.

Issues to be addressed when developing an SWP, similar to an SOP; should at minimum include:

- *Overview of work description*
- *Expert personnel that know well the work activity*
- *Responsible person(s)*
- *Tools used for doing all required tasks*
- *Potential Hazard (before and when activities are executed)*
- *Input on Experience Based Hazards from the field*
- *Action steps / accountabilities for key points which must be address if hazards and risks are assessed outside of tolerable limits*
- *Knowledge Test (optional)*

This paper will describe in more detail requirements of SWPs, how to develop, what the key points are which must be addressed, what procedure steps must be considered, the MOC which must be followed-up on and how the SWP should be documented.

1. Introduction

Since 1980s, a number of major industrial accidents have spurred more and more industrial personnel to be aware of the importance of process safety. A major accident is commonly modeled as the iceberg: only a small portion of the iceberg is visible above the waterline, but the tip is supported by much larger and invisible part underneath. Similarly, a major accident is represented by the tip of the iceberg and is caused by many minor and seemingly trivial incidents.

Most of those incidents are usually seen as "easy" tasks; either because they are regular or routine activities, or just "simple" activities such as tightening or loosenening stud nuts. Additionally, some workforce personnel become over-confident with their "used-to-do" activities, such as cable pulling, scaffolding erection, lifting, etc, which they consider these activities as a habit and begin to lose caution during performing the job.

Several practices, such as risk assessment, job safety analysis, work permit and operating procedures are implemented to reduce minor incidents. Although these methods do not overcome the problem completely, they do, with an appropriate approach, reduce the occurrence of minor incidents. Nevertheless, a significant number of cases involve non-routine activities. Hence, a good safety program must not overlook the process of addressing non-routine task.

This paper focuses on discussing the Specialized Working Procedure (SWP) because minor incidents typically occur during performing non-routine activities. It is necessary because of the following considerations:

- there is often no standard procedure for non-routine activities
- there are usually very few person that really understand about the particular non-routine activities
- there is assumption that it is not important to develop standard procedure for non-routine activities, either because the activity is considered trivial or the occurrence is infrequent

Non-routine task can be defined as tasks that are performed apart from the usual work routine. It also refers to an activity which Operating Procedures do not exist or when the Operating Procedure is out-dated or inadequate.

The intentions of the SWP are basically the same as for the Standard Operating Procedure:

- comply with HSE requirements
- reduce process hazards and risk that may occur during procedure preparation until execution of the works. Human error is one primary aspect that needs to be considered.
- establish consistency in developing working procedure
- improving training program to Operator & person that will do the work

2. Hazard Identification and Hazard Analysis

Prior to starting any jobs, routine or non-routine, identifying hazard around the work place is a must. By doing so, the operators are informed of the potential hazard, are encouraged to think several options to perform the work safely, and are aware of the risk behind each options, thus allow them to select the best method to perform the work safely. Hazard identification can be done by walk-around inspection or observation program. A lesson learned from previous similar activity is also a source for identifying the hazard of a task.

After the hazards have been identified, they need to be analyzed. Hazard analysis is a task of analyzing the safety of a performing a job in advance, to ensure that the hazards are fully understood and cautiously tackled and mitigated prior to commencing the job. Commonly, this activity is called Job Safety Analysis (JSA).

Every non-routine task is subject to a thorough job safety analysis. During its preparation, each team member should look carefully at the hazard in the area and determine the best way to eliminate or reduce the risk. Job Safety Analysis should also be reviewed frequently to ensure that it has covered all scope of work and is capable of being followed. Any changes in process, scope of work, technology requires the Job Safety Analysis to be reviewed and changed accordingly.

Although Job Safety Analysis is a good safety program for ensuring safe working environment, it may be insufficient to address all hazard and risk reductions. A majority of workers who did jobs unsafely did so because they believed that it was the safest to do the job. Likewise, some of them did not know how to perform the work, either because the job has never been performed before or because the workers lacked proper training. Another obstacle to safely performing a non-routine job is that frequently there is no experienced person who fully understands and is capable of performing the job safely.

Considering the above explanations, a written procedure for non routine task is required to be developed to ensure adequate information is communicated to personnel involved in performing the work. The written procedure is also intended to provide clear instructions for the operator doing the task. The written procedure is known as Specialized Working Procedure (SWP).

2.1. Human Error Consideration

SWP is intended to provide guidance for consistency hence reducing human error during performing a job. Thus several human error factors should be considered during SWP development and also when the SWP is used as guidance for performing the work on site. The list of human aspect include :

- Human Error for understanding the facility
- Human Error for understanding the hazard and the risk that may occur.
- Human Error due to lack of concern
- Frequency of Error
- Basic approaching to review human error at facility
- Understanding Lesson Learnt from previous experience
- Basic Process of Human Error at Company Based Experience
- Lack of basic knowledge on Risk Based Process Safety

3. What is a Procedure?

A Procedure is a set of steps to be performed to obtain a specified outcome. Generally, a procedure is comprised of action steps, decision steps and repetition steps. Action steps consist of what should be done to perform the job. Usually it is written in imperative sentences and starts with an action verb. The subject of an action steps is always implied of "you". Decision steps are points which a person has to make a choice based on the information available at that time, usually written in yes-no question in an if-then format. Repetition step is a special command or imperative sentence that tells the user to repeat a set of actions and/or decisions until some desired condition is achieved. The repetitive steps must specify exactly which action steps or decision steps to repeat and define the condition that will end the repetition. A procedure also has a starting point defining initial conditions or inputs and a goal-defining output. In summary, a procedure can be described as three elements: starting point, components and goal.

An example of a procedure that most people are familiar with is Standard Operating Procedure. An SOP can be defined as a set of written instructions or steps to allow someone performing a task in a standardized manner, which is the preferred and safest method. SOP is usually intended for a specific task, usually a routine task, and for providing assistance to the operators for maintaining the quality control and ensuring compliance to regulation. An SOP is developed to minimize variance and ensure consistence implementation even if there are temporary or permanent personnel changes.

Similar to SOP, a Specialized Working Procedure is a written instruction for non-routine tasks that allow an operator to perform the task in a standardized manner by the preferred and safest method. Taking the importance of providing clear instruction and ensuring good communication, especially for non-routine work, into consideration, SWP should be treated the same as SOP in terms of purpose, performance, requirement and documentation.

4. How to Develop Specialized Working Procedure

There are a lot of procedures, either Standard Operating Procedure or Specialized Working Procedure, available throughout industry, but few people understand how to develop a procedure, what should be included in the procedures and what a procedure should be look like. This section presents elements which should be included in a procedure. Application in different industries and types of work may necessitate appropriate modification.

A procedure is not just comprised of a set of instructions. Several points are essential in a Specialized Working Procedure document so that the procedure can accomplish its purpose of communicating hazard of non-routine work, and outlining the preferred and safest method, as well as ensuring consistency of performing the non-routine work.

Issues to be addressed when developing an SWP is similar to an SOP, which consist of the following:

- Overview the work description
- Expert personnel that know well the work activity

- Responsible person
- Tolls used for doing all required tasks
- Potential Hazard (before and when activities are executed)
- Input or Experience Based Hazards from the field
- Action steps / accountabilities for key points which must be addressed if hazards and risks are assessed outside of tolerable limits
- Knowledge Test

4.1. Overview of work description

The overview of work description provides an explanation of what objective is to be achieved. It consists of the purpose of developing the SWP, the particular intended user and process and/or regulation standard. Clear boundary definition of work to be covered by the SWP must also be explained in order to provide guidance for the writer to focus and providing clarification for the reader and the user.

Definition of any specialized or unusual terms is also necessary to be described in separate section. The reasons are to familiarize the reader with those terms and to make it easy-to-find when the definitions are needed. Sometimes it can be in the form of a list of glossary provided in the beginning of the document. Which format is chosen depends on its convenience and the writer's preference.

4.2. Expert personnel that know well the work activity

Who should develop the SWP needs to be carefully considered. Usually a SWP is developed by a person who has good knowledge of the task, preferably one who has performed the particular task. Consultation with additional experts is required under some circumstances, for example if there is no experienced person in the organization of this particular task. The name of the preparer of SWP should be clearly stated.

Regardless level of competency, it is encouraged that personnel involved in performing the job also be involved in preparing the SWP. This will help engaging the operators and ensuring that the procedure is to be properly followed.

4.3. Responsible Person

One of the most significant issues to be considered is the person responsible for the performance a particular task. This section should provide complete list of the roles and responsibilities of all personnel for undertaking each step in the SWP. The SWP should specify an appropriate level of responsibility for each of the working team members. Minimum required competencies for performing specific task are encouraged to be defined. By providing clear roles and responsibility of everyone involved, team members are more aware of their task and the limits of their own competence. Additionally, this is also ensures that in the absence of a team member, others should not have increased responsibility or do the work whose competence and knowledge they are unsure of.

It is also important to define the accountability of each task to ensure that each personnel are professionally accountable and carefully consider all decisions they are making.

4.4. Tools used for doing all required task

Any equipments or tools required for performing the entire required task specified in the SWP should be listed. To simplify the list, it is preferable using a general terms for common tools and equipment. This section also includes required certification and inspection to be carried out to all tools and equipments prior to commencing the work. Required personal protective equipment for performing the task is also necessary to be stated.

4.5. Potential Hazards

Prior to writing action steps of the procedure, it is critical to understand about the potential hazard and associated risks of the task. This information can be simply obtained by conducting hazard identification and hazard analysis. Through job safety analysis, preventive action to eliminate or reduce the risk can be identified.

However, not only the hazard before starting the work that needs to be considered. But any potential hazard that may arise during working period should also be assessed and properly mitigated. For example, during hot work activity, there is possibility of potential hydrocarbon release within working location.

For certain activities, such as hot work, the environment where the task is to be performed should be clearly defined, such as required good ventilation, special working location, etc. Each permitted or prohibited area should be specified. Additionally, environmental conditions should also be clearly defined; for example is the work is to be carried out during daylight only or during sunny conditions.

4.6. Input or Experience based Hazard from the field

Risk assessment alone is not considered adequate for identifying all hazards for performing specified tasks. Lesson learned inputs from similar activities are one of the best options for identifying the hazard. Other source is from experience based hazard identification from the field. This can be carried out by having walk-through inspection accompanied by experience person or by simulating the implementation of the procedure in the actual working condition. Through walk-through inspection, the personnel with basic understanding can ask question to enhance its knowledge and confidence for doing the work. By field testing, each action steps can be challenged to be the preferred and safest method.

4.7. Action Steps

Action Steps are clear and simple step by step instructions described in sequential order for performing the particular task in a preferred and safest manner. The instructions should be written as clear as possible to avoid ambiguity and misunderstanding. In case of any two steps to be done at the same time, the sentence should be to differentiate the action to the other.

A SWP should also provide clear guidance for exceptional circumstances if hazards and risks are assessed outside the tolerable limits.

4.8. Reference

A good procedure needs a complete reference section. This section consists of complete list of standard and regulations used for developing the SWP. It also provides source for the reader and user if they need to seek additional information or if they are required to clarify an issue.

4.9. Knowledge Test

A Knowledge Test is an optional feature of a SWP. This test is intended to ensure basic understanding of the involved personnel to grasp and comprehend the potential hazard of the work, the risk associated in the work and key point of the action steps. This test may also be a useful measure of successful hazard communication to all involved personnel.

Besides the above issues, one needs to remember that Specialized Working Procedure is aimed for a specific job and intended to provide guidance for specific users. Hence, it should provide sufficient detail to ensure that the operators with limited experience or knowledge of the procedure can successfully perform the work as described in the procedure without being supervised. To avoid misunderstanding, the procedure should be written in a concise, logical, step-by-step and easy-to-read format.

SWP should be developed based on detail study such as review of process information, lesson learned from previous activity, and observation on how the operator tendency to perform the work.

5. Training

A written procedure, besides ensuring quality and consistency of performing a task, is also intended to reduce the frequency of incidents due to human error. SWP provides a simple set of instructions to perform the work in the safest method and provides role clarification. Having that assistance, confusion over who does what and the actions to do for completing the job can be avoided.

To achieve those objectives, SWP should be prepared as concise and as simple as possible so that a person with basic understanding and limited experience can understand and follow the SWP accurately. Nevertheless, not all operators have adequate basic understanding about a particular task. Thus, training on the task and a review of the procedure is necessary to be carried out prior to commencing the work.

The objective of the training is not simply to familiarize the operator with the procedure or to ensure that they can follow the steps in the procedure, but emphasize understanding of the reason for developing the procedure and to assess where the procedure may not be sufficient. Simulation is one useful concept for an effective training for non-routine jobs.

6. Documentation

SWP should be properly documented and numbered accordingly. Date of preparation and/or date of review should also be precisely stated. SWPs must be readily available at all times. It does not mean that they are physically available, but more that they are up-to-date at all times to ensure it relevancy. So it is necessary to outline updating process for a certain period.

It is also critical to record any deviation from the recommended procedures and what has been done to tackle those deviations. Any changes of standards, regulations and technologies should be followed by reviewed the SWP and updated the SWP accordingly.

Any changes or revision should also be properly communicated to relevant users. For ensuring that every revision is recognized, a section in the procedure should be dedicated for recording any revision: the initiator, date of the revision occurred, the approval authority and details of changes.

7. Study Case

This section describes Hot Work Procedure at Offshore Facilities of BP West Java as one example of Specialized Working Procedure. The purpose of this procedure is to describe the required safeguard when performing hot work, such as welding, cutting, and grinding.

Since 2001, BP West Java limits Hot Work Activities for Offshore facilities. By having issuing this policy, Hot Work Activities should be done at an onshore workshop. Except for special circumstances, for example during total shut down of Turn Around Program, Hot Work Activities are permitted to be performed offshore but with thoroughly consideration and –high-level management approval. As Hot Work Activities considered as risky task and it is infrequently performed, a Specialized Working Procedure is required for performing this work.

Hot Work Procedure of BP West Java comprised of the following:

1. Purpose / Scope
   This section describes the purpose of developing this procedures and the scope of activities covered by this procedure.
2. Definitions
   Several specific terms are explained in this section, as well as providing scope boundary of those terms. For examples, Hot Work is defined as welding, burning, cutting, grinding, hot tapping or work with open flame.
3. General Requirements
   The General Requirements sections illustrate Work Description and Work Location.
4. Key Responsibilities, Inspection and Certifications
   This section describes each the roles and responsibilities of each personnel, for example the Person-in-charge, Issuing Authority, Fire Watch and Personnel Doing the Work. It also contains the required certification, such as Welder Certification and Equipment Certification. Required inspection on welding machine and associated equipment is also necessary to be defined.

5. Procedures
   The procedures are required for the Pre-Plan, Site Preparation and Inspection, including Energy Isolation. This section also contains Permitting Process and required Training.
6. Documents, Tools and References
   Revision of the procedure should always be documented properly, complete with the person initiating revision, date, authority approval and detail revision.

8. Conclusion

Specialized Work Procedures are powerful tools for seizing control of non-routine activities. They define the subtle details that maintain consistent implementation, ensure compliance to regulation, and complete the activity safely. In addition, well-written SWPs act as effective communication tools that contribute to worker understanding and safely job completion.

It is important to remember that SWP should have clear and specific title, reference source, revision dates, approval and detail, intended user, purpose, list of tools and parts, definitions of terms and concepts, safety warnings and cautions, detail and simple step by step instructions for performing the job.

SWP development process also provides significant performance improvements. Development process of a SWP, when properly and fully carried out, allows worker, engineers and managers working together for ensuring safe working condition. As the result, everyone focuses their abilities on doing the best job possible as well as improving working condition.

However, the essential purpose of SWP should be enhanced with assessment of hazard and risk and strong behavior-based safety program, as well as with continuous improvement through observation and training.

9. REFERENCES

1. Stewart, Maurice I, *Risk-Based Operating Procedures and Safe Work Practices*, Indonesia, LDI Training, 1997.
2. Center for Chemical Process Safety (CCPS), Guidelines for Risk Based Process Safety, 2007
3. *BP Indonesia HSE Handbook 6th edition,* Indonesia, 2005

RELEVANCE OF SAFETY, HEALTH AND ENVIRONMENTAL (SHE) FACTORS IN SITING OF CHEMICAL INDUSTRIES

Dr. R. K. Elangovan
E mail : elanrk@yahoo.co.in ; elanrk @ hotmail.com

ABSTRACT

The Siting of chemical industries takes into account of various factors like any other industry, ranging from availability of raw material, land and water to proximity to market and transportation. Even though Safety, Health and Environmental (SHE) factors are very much relevant to Siting of any industry, they attain more significance in siting of chemical industries due to their relevance in operation and survival of the industry.

The consideration of SHE factors in Siting of chemical industries ensures a distance between the factory site and built up area, thereby ensures separation distance between hazards and risks in them from the community apart from extending the benefit of mitigating the effects of any release. The releases from a chemical industry are generally due to the breach of safety, thereby resulting in impact on the health and environment. The releases may be intended like effluents and discharges and unintended like accidental releases, and they may be intermittent or continuous. Therefore, the releases having impact on health and environment are part of any chemical industry. The effects of any release decay rapidly with distance. The separation distance attained by the proper siting tends to create more opportunity for protection there by containing and controlling the effect of releases.

This paper discusses on the relevance and importance of SHE factors in Siting as well as describes a methodology to incorporate SHE factors in site selection of chemical industries.

KEY WORDS

Siting, Accident, Releases, Safety, Health and Environment.

INTRODUCTION

The Siting of chemical industries takes into account of various factors like any other industry, ranging from availability of raw material, land and water to proximity to market and transportation. Even though Safety, Health and Environmental (SHE) factors are very much relevant to Siting of any industry, they attain more significance in siting of chemical industries due to their relevance in operation and survival of the industry. The SHE factors are equally important like business factors in deciding the site selection of the chemical industries. This paper discusses on the relevance of SHE factors and their associated factors in deciding on site selection for chemical industries.

CONCEPTUAL FRAME WORK

RISK

Risk is the likelihood of a specified event occurring within a specified period or in specified circumstances. The different types of risks from chemical factories are as follows:

1) Individual Risk
2) Societal Risk
3) Property Risk

Out of the above, the societal risk attains high significance in plant Siting, since it is the relationship between frequency and the number of people suffering from a specific level of harm in a given population from the realization of specified hazards. The concept of societal risk reflects the likelihood of accidents involving multiple casualties and long- term detriment. Any risk within an area surrounding a dangerous process, as in chemical industries, depends upon its frequency and consequence not only to the individual but also to the number of individuals likely to be involved [6,8] .

COMMUNITY HEALTH

Community Health concerns preventing illness and managing the illness of the community and promoting health. Immunization campaigns, controlling infectious diseases, ensuring safe drinking water, food hygiene, public Hygiene, free from illness or diseases out of industrial activity, ensuring availability of adequate antidotes for the chemicals used in the industry, health centers, first - aid facilities, emergency services like ambulances etc., are some of the activities under community health.

Health Impact Assessment (HIA)

The Health Impact Assessment (HIA) is defined as a combination of procedures or methods by which a policy, program or project may be judged as to the effects it may have on the health of a population. The HIA seeks to predict the health impact of any industrial activity before implementation and it aims at facilitating the avoidance or reduction of negative effects on community health. The HIA should consider both positive and negative effects on human health of any industrial activity and the likely human health effects should be identified to decide on corrective actions and finally aid in the plant siting,

Environmental Impact Assessment (EIA)

The Environmental inventory is a complete description of the environmental as it exists in an area where a particular proposed action, like setting up a chemical factory, is being considered The Environmental Impact Assessment is an attempt to evaluate the consequences of a proposed action on each of the description in the environmental inventors.

The essential steps in an environmental impact assessment are:

- Prediction of the anticipated change in an environmental descriptor
- Determination of the magnitude or scale of the particular change.
- Application of an importance or significance factor to the change [3] .

LEGAL PROVISIONS

Safety

Various Indian Statutes on Safety are as follows:

- The Factories Act 1948
- The Chemical Accidents (Emergency Planning, Preparedness and Response) Rules, 1996
- Manufacture, Storage and Import of Hazardous Chemical Rules, 1989

Health

Various Indian Statutes on Health are as follows:

- The Factories Act 1948
- The Public Liability Insurance Act, 1991
- The Disaster Management Act, 2005

Environment

Various Indian Statutes on Environment are as follows [7]:

- The Environment (Protection) Act 1986
- The Environmental (Protection) Rules, 1986
- The Water (Prevention And Control of Pollution) Act, 1974
- The Air (Prevention & Control of Pollution) Act., 1981
- Environmental impact Assessment of Development Project, Notification Hazardous Wastes (Management and Handling) Rules, 1989
- Manufacture, Storage and Import of Hazardous Chemical Rules, 1989
- Manufacture, Use, Import, Export and Storage of Hazardous Micro-Organisms Scheme of Labelling of Environment Friendly Products (ECO-MARKS)
- Bio-Medical Waste (Management and Handling) Rules, 1998
- Re-Cycle Plastics Manufacture and Usage Rules, 1999
- Coastal Regulation Zone, Rules

- Environment (Siting for Industrial Projects) Rules, 1999
- Noise Pollution (Regulation and Control) Rules, 2000
- Municipal Solid Wastes (Management & Handling) Rules, 2000
- Ozone Depleting Substances (Regulation) Rules, 2000
- Manufacture, Storage and Import of Hazardous Chemical (Amendment) Rules 2000
- The Prevention and Control of Pollution (Uniform Consent Procedure) Rules, 1999.
- The Batteries (Management and Handling) Rules, 2001
- The Disaster Management Act 2005
- National Environment Policy 2006

INTERNATIONAL AGREEMENTS ON ENVIRONMENTAL ISSUES

India is signatory to a number of multilateral environment agreements (MEA) and conventions. An overview of some of the major MEAs and India's obligations under these are presented below:

a) Montreal Protocol on substances that deplete the Ozone Layer (to the Vienna Convention for the Protection of the Ozone Layer), 1987

The Montreal Protocol to the Vienna Convention on Substances that deplete the Ozone Layer came into force in 1989. The protocol set targets for reducing the consumption and production of a range of ozone depleting substances (ODS). India accepted to the Montreal Protocol along with its London Amendment in September 1992. The Ministry of Environment and Forests (MoEF), Government of India, has established an Ozone Cell and a steering committee on the Montreal Protocol to facilitate implementation of the India Country Program, for phasing out ODS production by 2010. The Gazette of India on 19 July 2000 notified rules for regulation of ODS phase- out called the Ozone Depleting Substances (Regulation and Control) Rules, 2000. They were notified under the Environment (Protection) Act, 1986.

b) Basel Convention on Transboundary Movement of Hazardous Wastes, 1989

Basel Convention, which entered into force in 1992, has three key objectives.

- To reduce transboundary movements of hazardous wastes.
- To minimize the creation of such wastes; and
- To prohibit their shipment to countries lacking the capacity to dispose hazardous wastes in an environmentally sound manner.

India ratified the Basel Convention in 1992, shortly after it came into force. The Indian Hazardous Wastes Management Rules Act 1989 encompasses some of the Basel provisions related to the notification of import and export of hazardous waste, illegal traffic, and liability.

c) UN Framework Convention on Climate Change (UNFCCC), 1992

The Primary goals of the UNFCCC were to stabilize greenhouse gas emissions at levels that would prevent dangerous anthropogenic interference with the global climate. The convention embraced the principle of common but differentiated responsibilities which has guided the adoption of a regulatory structure. India signed the agreement in June 1992, which was ratified in November 1993. India has initiated the preparation of its First National

Communication (base year 1994), the potential vulnerability to climate change, adaptation measures and other steps being taken in the country to address climate change.

d) UN Convention on Désertification, 1994

Delegates of the 1992 UN Conference on Environment and Development (UNCED) recommended establishment of an intergovernmental negotiating committee for the elaboration of an international convention to combat desertification in countries experiencing serious drought and / or desertification. The UN General Assembly established such a committee in 1992 that later helped formulation of Convention on Desertification in 1994.

NATURAL FACTORS IN PLANT SITING

The following natural factors need to be considered in plant siting not only because of the hazards associated with them but also their impact on SHE factors from the consequences of such hazards.

a) Earthquake

Earthquake can generate secondary effects such as fire, explosion and toxic release as a result of combined actions of the earthquake wave and environmental conditions. Additional safety factors for considering base shear force which tends to topple building or structures of the chemical plants should be taken into account at the design stage itself.

b) Hurricane / Tornadoes

Hurricane and Tornadoes can induce hazardous chemical releases in chemical industries by high winds.

c) Flooding

Flooding can destabilize the equipments, storage systems, piping, and facilities handling hazardous chemicals there by creating hazardous situations. The chemicals that are highly soluble in water can dissolve in the flood water creating health and environmental hazards.

d) Lightning

The released flammable and explosive chemicals in the gaseous state in the form of vapour cloud and liquid state in the form of spills can find ignition source due to lightning leading to a hazardous situation.

e) Tsunami

A tsunami (pronounced tsoo – nah – mee) is a wave train, or series of waves, generated in a body of water by an impulsive disturbance that vertically displaces the water column. Earthquakes, landslides, volcanic eruptions, explosions, and even the impact of cosmic bodies, such as meteorites, can generate tsunami. Tsunami can savagely attack coastlines, causing devastating property damage and loss of life.

METHOD

The step – by – step procedure in plant siting is given in Figure No. 1. The step- by- step procedure should include details on Safety, Health and Environment to make the site selection process realistic and meaningful [1]. The various general factors relevant to the siting of industries are shown in Table No. 1.

SAFETY FACTORS

The Safety Factors in chemical plant siting are provided in Figure No. 2.

Process Hazard Identification [5]

Various process hazards identification methods used in chemical industries to identify the potential hazards with nature of results are tabled in Table No.2.

Types of Chemical Plant Accidents

The Chemical plant accidents are broadly classified as fires, explosions and toxic releases. The different types of chemical accidents along with their probability and potential for economic loss are provided in Table No. 3. Among the accidents fire accidents is the most common in industries followed by explosion and toxic release. The most damaging type of explosion is an unconfined vapour cloud explosion when a large cloud of volatile and flammable vapour is released and dispersed followed by ignition and explosion. Toxic releases typically results in little damage to capital equipment but personal injures and clean up liabilities are significant[4,8]

The Accident Process

The Accidents in industries follow a three step process and they are:

- Initiation - The event that starts the accident.
- Propagation - The event or events that maintain or expand the accident.
- Termination -The event or events that stop the accident or diminish it in size.

Safety Engineering involves eliminating the initiating events and replacing the propagation steps by termination events[8]

Scenario Identification

The Scenario identification is resorted without regard to importance of initiating event. The unidentified scenario cannot be analyzed and hence it is very much important to identify all probable and worst case scenarios to ensure that the study is complete in all respects.

Accident Consequence

The primary components in consequence analysis are source and dispersion models, fire and explosion models, effect model and mitigating factors. Source and dispersion models provide quantitative information on source rates and dispersion of vapour clouds to some concentration levels. Fire and explosion models convert the information on the cloud for flammable releases into hazard potentials such as thermal radiation and explosion over pressures.

Effect models convert these incident specific results into effects on people and structures. Additional refinement is provided by mitigation factors such as sheltering or evacuation, which tend to reduce the magnitude of potential effects in real accident [10].

HEALTH FACTORS

The Chemical industries can create health hazards to the community by the accidents and chemical releases. The fire can create thermal radiation hazard resulting in `thermal burn injuries and an explosion can create over pressure there by creating hazards by overpressure resulting in injuries and even death. The major health hazard can be created by the exposure to the toxic chemicals utilized and manufactured in the factory. The methodology for incorporating health factors in plant siting is given in Figure No. 3. Health Impact Assessment incorporating Risk Assessment, Exposure Assessment and a quantitative and / or quantitative evaluation of the degree of intake of toxic chemical likely to occur should be carried out to decide on plant Siting. The availability of medical facilities, antidotes, emergency vehicles and ambulances are also some of the health factors relevant to Plant Siting [5].

ENVIRONMENTAL FACTORS

The Methodology in incorporating environmental factors in plant siting for chemical industries is provided in Figure No.4. The Environmental impacts of releases due to accidents as well as process emissions like flue gas, fumes and effluents should be assessed to take actions, which require systematic, reproducible and interdisciplinary approaches. The potential impacts on the physical, biological, cultural and socio-economic aspects of the environment should be thoroughly considered in the EIA study for effective decision making [2, 9]

FINDINGS

The findings of the study are:

- The Safety, Health and Environmental (SHE) Factors are very much relevant in deciding chemical plant siting and equally important like other factors.
- Even though the study considered SHE factors separately, these factors are dependent on each other. Any accident in a chemical industry would affect health and environmental factors and hence a holistic approach must be provided.
- The natural factors need not only affect safety in chemical industry alone. They have the potential to create accidents thus playing a role in complicating health and environmental factors.
- The SHE factors should be considered as a whole instead of considering them individually, for effective decision making on Plant Siting.
- The legal compliances, statutes, codes and standards on SHE factors relevant to Plant Siting should be considered in detail for effective decision making.

CONCLUSION

The Safety, Health and Environmental (SHE) factors are very important like business industries. The consideration of SHE factors in plant Siting ensures the smooth function of the chemical industry and helps in emergency preparedness and legal compliances. The incorporation of SHE factors at the initial stage of a siting study ensures least disturbance / or impact to the society, thereby enhancing effectiveness of all control systems in the chemical industries. The SHE factors need to be integrated with other factors for effective plant Siting.

REFERENCES

1 Apple M, *Plant Layout and Material Handling*, Third Edition, John Wiley & Sons, 1977

2 Bhatia S C , *Managing Industrial Pollution*, Macmillan India Ltd, New Delhi, India, 2003

3 Canter W, *Environmental Impact Assessment*, McGraw – Hill Book Company, New York, 1977

4 Goetsch L, *Occupational Safety and Health*, Prentice – Hall, Englewood Cliffs, New Jersey, Second Edition, 1996

5 Gould J, Glossop M and Loannides A, Review *of Hazard Identification Techniques,* Sheffield: Health and Safety Laboratory, Health and Safety Executive, 2000

6 Illuing .P, *Toxicity and Risk, Context Principles and Practice*, London, Taylor & Francis, 2001

7 Mohanty S K, *Environmental & Pollution Law Manual*, Universal Law Publishing Co. Pvt. Ltd, Third Edition, New Delhi, 2002.

8 Skelton B, *Process safety Analysis An Introduction,* Warwickshire: Institution of Chemical Engineers, 1997

9 Sincero P and Sincero A, *Environmental Engineering; A Design Approach*, Prentice – Hall of India Limited, New Delhi, India, 1999

10 *Guidelines for Chemical Process Quantitative Risk analysis*, Center for Chemical Process Safety, American Institute of Chemical Engineers, Newyork 1989

A. Market
1. Location
2. Population Trends
3. Income trends
4. Consumer characteristics
5. Retail sales trends
6. Industrial Markets
7. Cometition

B. Raw Material
1. Type
2. Location of sources
3. Characteristics
4. Price
5. Terms of sale
6. Availability
7. Storage facilities

C. Labour
1. Cost
2. Attitudes
3. Union situation
4. Supply
5. Location
6. Personnel policies in area
7. Legislation
8. Recruiting
9. Community pattern
10. Relocation

D. Transportation
1. Modes of transportation
2. Methods of operation
3. Handling methods
4. Facilities
5. Costs
6. Freight rates

E. Utilities
1. Power
2. Fuel
3. Water
4. Waste
5. Communication
6. Others

F. Climate
1. Maximum and minimum temperatures
2. Direction of winds
3. Weather changes
4. Humidity
5. Evaluation
6. Climate effects
7. Special weather hazards

G. Federal activities in area

H. Representation in congress

I. Financing
1. Financial aid available
2. Requirements
3. Credit factors
4. Factors effecting loan terms
5. Special inducements – taxes, loans, building

J. Stage government and taxes
1. Structure
2. Financial assistance
3. Government attitudes
4. State regulation and legislation
5. State taxes
6. Future tax Prospects

K. Local government and taxes
1. Structure of government
2. Financial conditions
3. Civil attitudes
4. Local taxes

L. Community Facilities

M. Community appearance

N. The size of community

O. Population trend

P. Community planning and Zoning
1. Planning commission
2. Industrial zoning
3. Building codes
4. Traffic and parking
5. Streets

Q. Others
1. Cost of living
2. Insurance
3. Construction costs
4. Location relative to competition
5. Size and character of other local industries
6. Buying habits

R. Individuals sites
1. Requirements
2. Size and shape of site
3. Geologic considerations
4. Cost of development
5. Location within the community
6. Availability to transportation
7. Topography
8. Price of land
9. Availability of building
10. Personal factors
11. Intangible considerations

Table No: 1.Selected Factors for Consideration in Plant Siting [1]

Ref	Hazard Identification Techniques	Nature of Results
1.	Concept Hazard Analysis	Qualitative
2.	Concept Safety Review	Qualitative
3.	Preliminary Hazards Analysis	Qualitative
4.	Pre-HAZOP (Hazard And Operability Study)	Qualitative
5.	HAZOP	Qualitative
6.	"What if?" Analysis	Qualitative
7.	Check List	Qualitative
8.	Standards / Codes of practice	Qualitative
9.	FMEA (Failure Mode and Effect Analysis)	Qualitative
10.	FMECA (Failure Modes, Effects, and Criticality Analysis)	Qualitative
11.	Critical Examination of System Safety	Qualitative
12.	Cause-Consequences Analysis	Qualitative and Quantitative
13.	Fault Tree Analysis	Qualitative and Quantitative
14.	Structural Reliability Analysis	Qualitative
15.	Reliability Block diagram	Qualitative
16.	Human reliability Analysis	Qualitative
17.	Predictive Human Error Analysis	Semi-qualitative
18.	Vulnerability Assessment	Qualitative
19.	Matrices	Qualitative
20.	GOFA (Goal Oriented Failure Analysis)	Semi-Qualitative
22.	MOSAR (Method Organized Systematic Analysis of Risk)	Semi-Qualitative
22.	Inherent Hazard Analysis	Qualitative

Table No.2 Hazard Identification Methods Applicable to Chemical Industries

Type of Accident	Probability of Occurrence	Potential for fatalities	Potential for Economic Loss
Fire	High	Low	Intermediate
Explosion	Intermediate	Intermediate	High
Toxic Release	Low	High	Low

Table No. 3 Types of Chemical Accidents

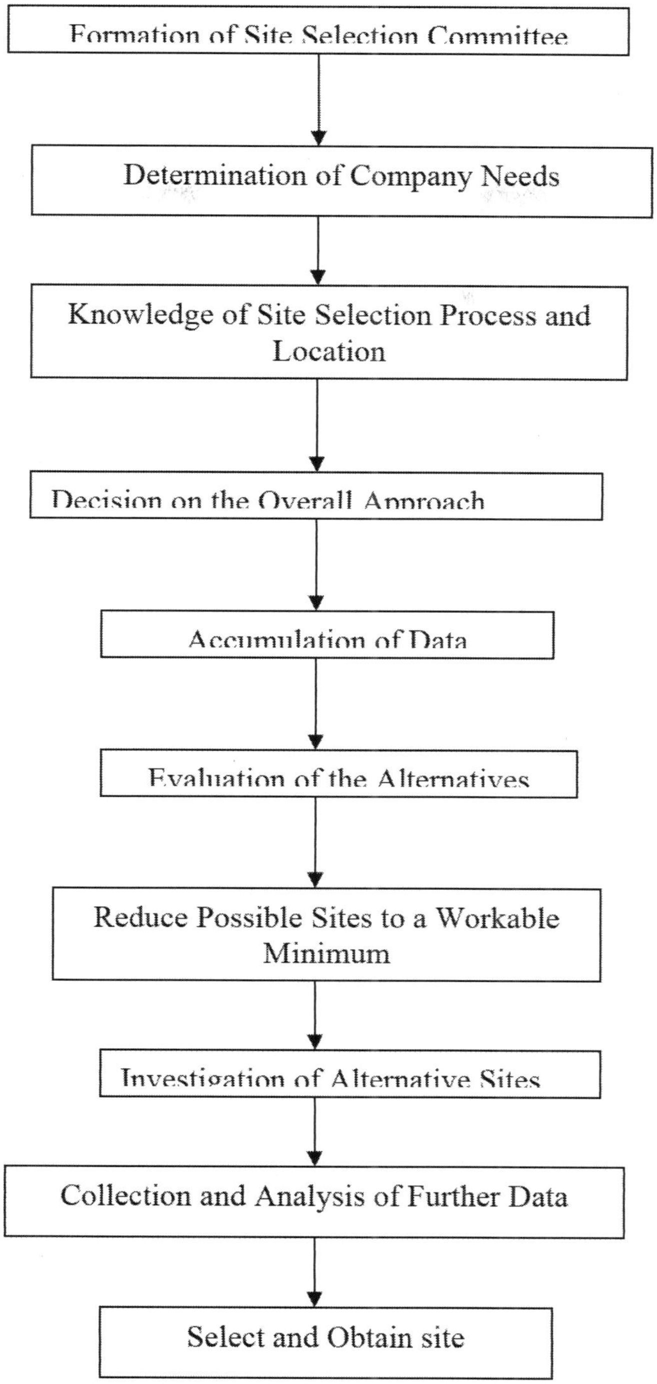

Figure No .1 Step by Step Procedures in Plant Siting.

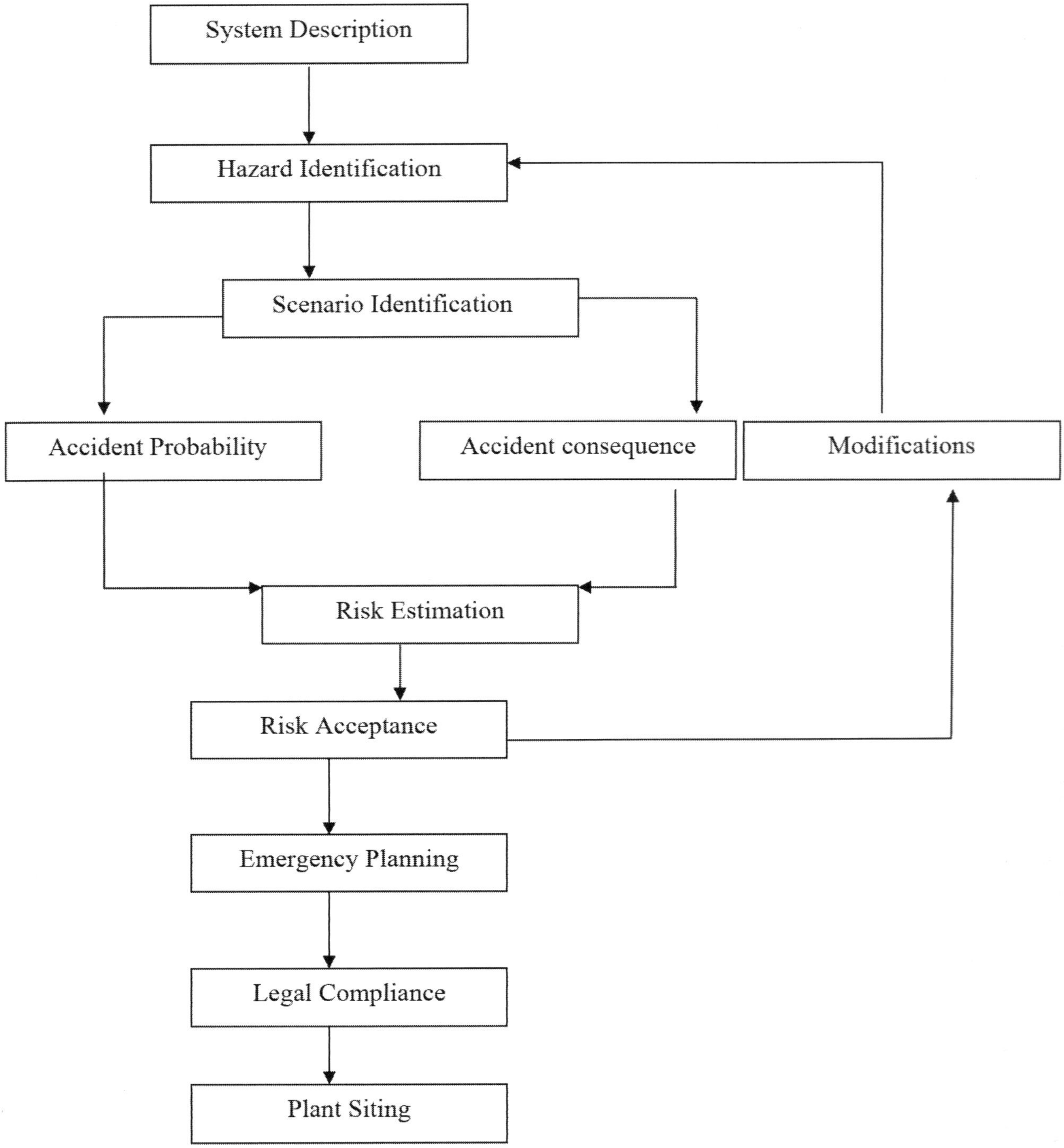

Figure No. 2 Logic of safety Factors in Plant Siting

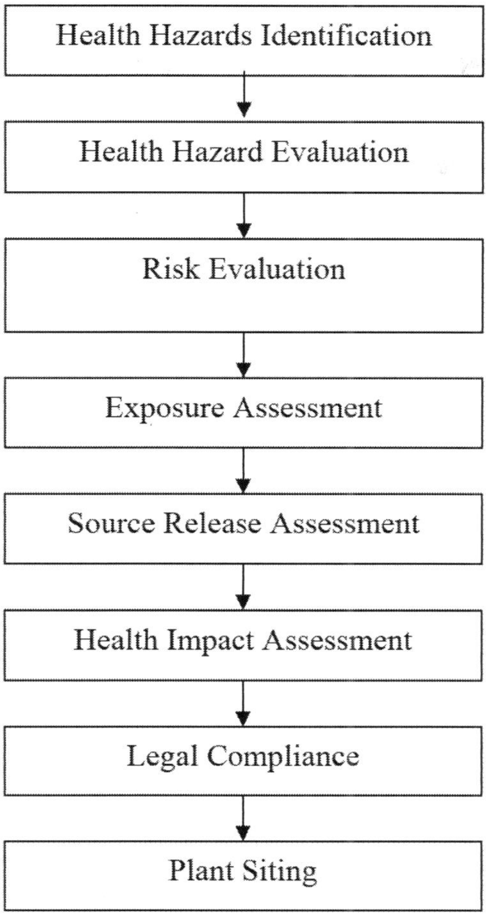

Figure No 3 Health Factors in Plant Siting

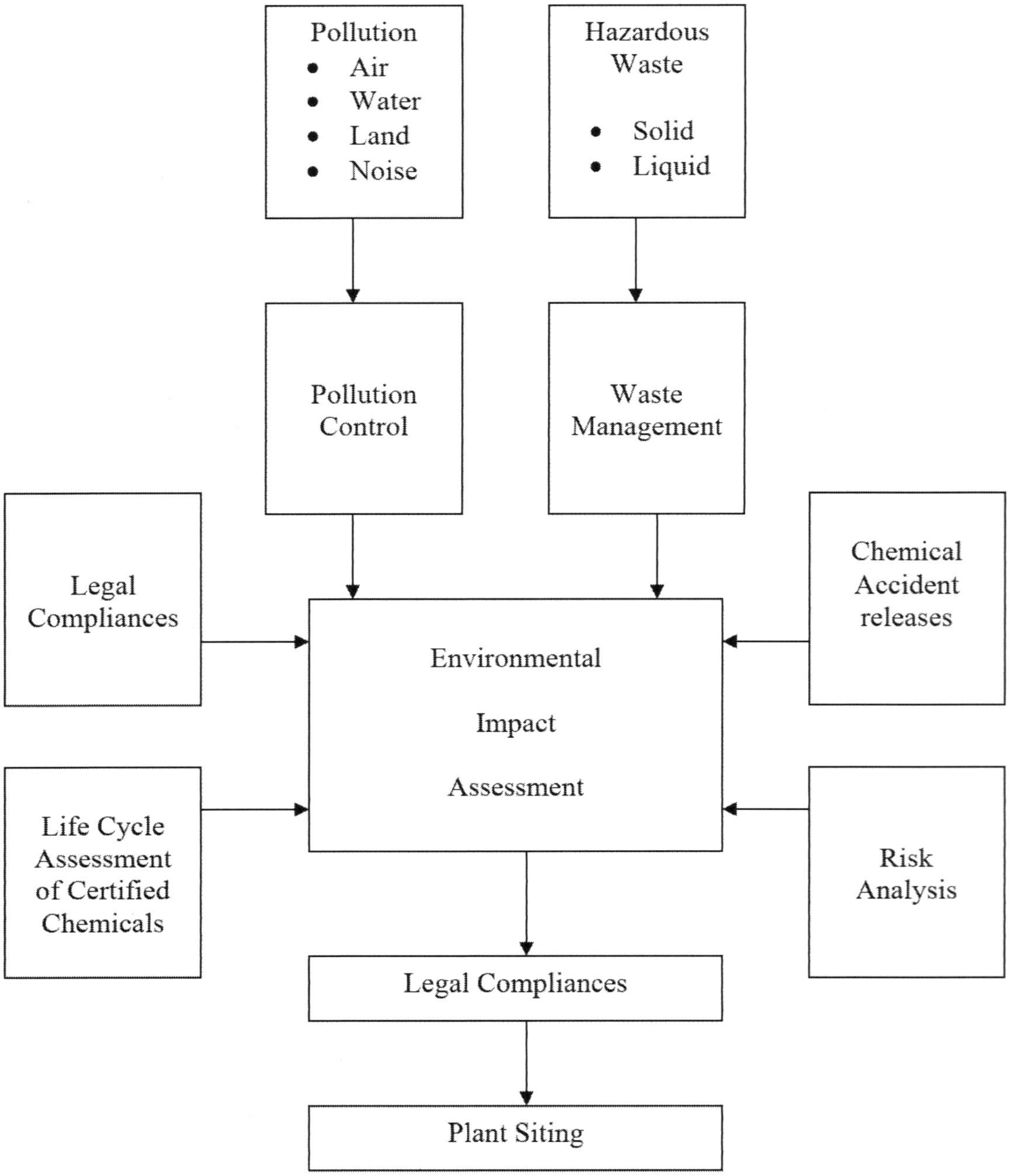

Figure No. 4. Environmental Factors in Plant Siting.

A modeling approach to assess the effectiveness of BLEVE prevention measures

Gabriele LANDUCCI[1], Menso MOLAG[2], Johan REINDERS[2], Alessandro TUGNOLI[3], Valerio COZZANI[3]*

(1) CONPRICI - Dipartimento di Ingegneria Chimica, Chimica Industriale e Scienza dei Materiali Università degli Studi di Pisa Via Diotisalvi n.2, 56126 Pisa (Italy)

(2) Nederlandse Organisatie voor toegepast-natuurwetenschappelijk onderzoek TNO Laan van Westenenk 501 P.O. Box 342 7300 AH Apeldoorn (The Netherlands) Princetonlaan 6 NL-3584 P.O. Box 80015 NL-3508 TA Utrecht

(3) CONPRICI - Dipartimento di Ingegneria Chimica, Mineraria e delle Tecnologie Ambientali Alma Mater Studiorum - Università di Bologna Viale Risorgimento n.2, 40136 Bologna (Italy)

(*) Author to whom correspondence should be addressed.
tel. (+39)-050-511215; fax (+39)-050-511266
e-mail: gabriele.landucci@ing.unipi.it

Submitted for
LOSS PREVENTION SYMPOSIUM

ABSTRACT

Distribution and storage of liquefied pressurised gases is a critical safety issue, often resulting in a very high individual and societal risk, at least in densely populated zones. Several risk assessments pointed out that this is mostly due to the possible occurrence of a Boiling Liquid Expanding Vapour Explosions (BLEVE) of the tank that may be exposed to an intense fire as a consequence of release scenarios. In the present work, a modelling approach is presented for the calculation of fired BLEVE probability. Simplified models were obtained for the estimation of the vessel time to failure with respect to the radiation intensity on the vessel shell. The time to failure was compared to a reference time required for effective mitigation actions and thus related to the escalation probability. The analysis of the effect of protective measures, such thermal insulating coating and pressure relief devices, on the time to failure allowed the identification of the necessary requirements for effective BLEVE prevention.

Keywords: domino effect, escalation probability, vessel time to failure, passive fire protection.

1 Introduction

Domino effect is responsible of severe accidents that took place in the chemical and process industry [1-3]. Several studies pointed out that the more critical step in the quantitative assessment of domino hazards is the availability of reliable models to estimate the possibility of escalation due to the effects of the primary accident [4-7]. In the case of fires caused by the accidental releases of flammable substances, it is well known that secondary events with catastrophic consequences may result as a consequence of flame impingement on equipment and pipes. Moreover, in the case of storages where pressurised liquefied gases (such LPG, chlorine, ammonia, etc.) are present, the escalation may result in a BLEVE, that may have severe consequences due to the extremely high damage distances.

An important benefit to the safety management of possible escalation hazards would come form the availability of an approach for the calculation of the possibility and probability of vessel rupture following external fires, based on simplified correlations able to take into account specific protection factors.

The present study was focused on the development of a simplified methodology for the calculation of the damage probability of process vessels, aimed to the quantitative assessment of domino effect triggered by fire scenarios. The methodology is based on simple analytical functions relating the time to failure of vessels to the radiation intensity. These were validated by an integrated approach, based on:

- available experimental data;
- a commercial finite elements code for complete thermal and mechanical simulations of the behaviour of pressurised vessels exposed to fires;
- a simplified model for vessel failure based on thermal nodes.

Specific correction factors were introduced in order to take into account the effect of protection materials. Damage probability was estimated by a probabilistic function. An approach derived from the Layer of Protection Analysis (LOPA) was used to estimate the probability of effective mitigation on the basis of the calculated time to failure and of site-specific factors. The presence and the delay time for the activation of protection systems was also considered. A fundamental issue in the development of the correlations resulted the presence of thermal protection layers on storage or process vessels. In particular, the possible use of innovative materials for passive protection systems resulted in a high impact on the time to failure. The effect of the protection systems on the time/temperature profile of vessel wall was thus implemented in the simplified threshold correlations for the assessment of time to failure.

2 The approach to the estimation of escalation probability

The use of simplified models, based on a limited number of parameters, is required to limit the computational time required for QRA procedures used in escalation assessment, as they involve the analysis of a high number of accidental scenarios [6,8]. Thus, equipment vulnerability models are usually based on probit functions that relate a "dose" of physical effects to the escalation or damage probability of an equipment item. In the case of radiation as an escalation vector, vessel failure is caused by wall temperature raise and/or by internal pressure increase due to vessel heat-up. This is a rather slow process (time to failure of the vessel is usually of the order of minutes or higher), thus a time lapse exists between the primary event and the secondary events caused by escalation.

Quite clearly, the time to failure (ttf) is a fundamental parameter in the analysis of the domino accidents of equipment exposed to fire: a higher time to failure results in a lower credibility of escalation, since a higher time lapse is available for effective mitigation actions by safety systems or emergency teams (e.g. wall cooling by water deluges). The ttf may thus be compared to a characteristic time for an effective mitigation (tte). The analysis of the effects of protection systems (dumping aimed to depressurization, water curtains by automatic systems, and additional water protection by emergency teams) may be performed by LOPA techniques, and specific damage probability functions may be obtained for the plant under examination. A site-specific or a general evaluation of the tte and its comparison to the ttf may be easily turned out in a probabilistic assessment of the success of mitigation actions. This approach was used to develop specific models relating the radiation intensity on a target vessel to the escalation probability.

3 Estimation of vessel time to failure

Fire may affect a process or storage vessel by one or more of the following modes: i) distant source radiation; ii) full or partial fire engulfment; and iii) jet fire flame impingement. Modelling vessel time to failure in these three situations is extremely difficult given the high complexity of the flame geometries. The wall temperature behaviour in a vessel exposed to an external fire would require a detailed three-dimensional analysis of the thermal flow over the vessels shell, of the thermal gradients in the fluid contained in the vessel and of the effects due to the mixing of the content by natural convection. However, a model based on this approach would require a huge time to run, not justified in a QRA framework due to the uncertainties that usually affect the characterization of the fire scenario. In this specific context, a simplified model able to yield a conservative estimation of the time to failure by a straightforward approach would be more useful.

In the present study, a lumped model was developed for the assessment of vessel time to failure. The model was based on a lumped approach for the calculation of the time and temperature profile, but was improved by an extended validation work, based on an experimental data set for the vessel time to failure, expanded by the use of results obtained from a specific finite elements model implemented for this purpose.

The development of the lumped model was based on thermal nodes modelling. Depending on the fire scenario, this approach attempts to divide the equipment in different zones (or nodes), each of which can be described by a simple set of parameters. The parameters represent physical quantities (e.g. temperature, pressure, thermal conductivity, etc.) averaged over each node. Boundary conditions together with global conservation laws lead to a system of equations which determines the parameters of interest and, in particular, the temperature at each node. This allows the calculation of temperature-time profiles as a function of the radiation mode and intensity on the vessel.

The estimation of these parameters allows the evaluation of the mechanical stress generated in each zone of the vessels and to compare it with the admissible tensile strength of the vessels material (that is strictly dependent on the material temperature). The failure conditions are strictly dependent on the structural design: geometry, material, boundary condition. For the equipment of interest (horizontal cylindrical vessels, vertical cylindrical vessels, etc.) the failure due to an external heat flow is mostly caused by the wall-thinning due to hoop stress and high temperature material degradation. A specific failure criterion was thus established to assess the time to failure of the vessel. This was derived by a direct comparison between parameters representative of the stress field over the vessels shell and of parameters representative of the tensile strength of the material. Thus, the failure criterion requires the detailed analysis of the stress field over the vessel shell and on the boundary structures that may be calculated only by more detailed modeling approaches. This means that with a lumped model only simplified failure criteria may be adopted, based on data fitting of available times to failure for similar vessels and for similar fire scenarios. Further details on the model developed are reported elsewhere [9].

An extended validation of the lumped model was thus necessary to ensure the reliability of the failure criteria implemented. A first set of validation runs was carried out using experimental data available in the literature. Although a significant number of case-studies resulted available, the number of available experiments was not sufficient to carry out an extended validation over the entire field of vessel geometries and of radiation modes and radiation intensities of interest. Thus, a finite elements model (FEM) was developed and validated on the basis of the experimental data. The FEM was used to generate a second data set used to extend the validation of the simplified lumped model.

The FEM allowed the detailed simulation of the thermal and mechanical conditions on vessel shells under fire radiation. Adopting proper failure criteria it was possible to use the knowledge of the temperature and of the stress conditions in each point of the structure to estimate more precisely the ttf. The first step in the simulations was the detailed calculation of the temperatures on the vessel shell as a function of time and radiation mode. A detailed definition of the radiation conditions on vessel shell was possible: Figure 1-(a) shows an example of detailed temperature simulations for a horizontal cylindrical LPG tank engulfed in flames.

The second step of the modeling was the calculation of the stress field as a function of the local temperatures and of the other loads present on the equipment shell (Figure 1-(b)). The calculation of the detailed temperature and stress maps allowed the application of the correct failure conditions and thus a quite accurate calculation of the equipment time to failure, also accounting for the decrease in the allowable stress due to the high temperatures of the vessel walls. The comparison of the results of the simulations carried out with the first set of experimental validation data showed that the predicted ttfs were always conservative and showed a relative error lower than 10%. Thus, the results suggested that the finite elements code could be reasonably used to extend the validation data set available.

The validation of the lumped model carried out using both the experimental and finite elements data for the vessel time to failure pointed out that the lumped model always yields credible and conservative values for the ttf of the vessels. An average relative error lower than 15% is present between the ttf calculated by the lumped model and those obtained by the FEM simulations (Figure 2).

Therefore, the lumped model developed in the present approach may be used with sufficient confidence in the assessment of the ttf of the equipment exposed to fire in the framework of domino quantitative risk assessment, also considering the uncertainties in the characterization of the fire scenario. The ttfs calculated by this simplified approach have a sufficient precision to be compared with the times required for an effective mitigation. The lumped model developed was thus the starting point for the calculation of the escalation probability due to fire and of the threshold values for escalation. As a matter of fact, even if the RADMOD model is characterised by relatively low computational time, its use may require a relevant effort in the analysis of extended areas, also considering that the model, although simplified, requires defining and inputting several parameters of each vessel and radiation mode considered.

The availability of further simplified tools to carry out at least a preliminary assessment of the time to failure is thus useful, in particular to identify the credible domino targets in a complex lay-out. To face this problem, a specific approach was used to define simple analytical functions for the evaluation of the time to failure of the equipment exposed to fire. The more important categories of pressurised equipment involved in domino accidents were identified, and their geometrical characteristics were defined (shape, range of sizes, etc.). Vessel sizes were obtained from typical design data used by engineering companies in the oil and gas sector. The volumes and diameters were derived from vessels typically used for LPG, chlorine, and ammonia pressurized storages, road tankers and tank wagons (Table 1). Cylindrical vessels with horizontal axis and design pressures of 1.5, 2.0 and 2.5 MPa were considered. The design data were verified with respect to section VIII of the ASME codes, and the relief valves were considered to provide the vent area required by API RP 520 standards.

An extended matrix of case studies was thus defined, considering the more credible primary events, the radiation modes and a credible range of radiation values. The lumped model was used to analyze each case study, estimating the time to failure of the selected equipment exposed to a specific fire scenario. A fitting procedure was implemented to obtain analytical functions that directly relate the time to failure of the equipment (ttf) to the value of the radiating heat flow (I):

$$\log_{10}(ttf)=a \log_{10}(I)+b \quad (1)$$

where the a and b coefficients depend on the vessel type and geometry and on the radiation mode. The a and b coefficients may thus be calculated from the fitting of the results of numerical simulations to eq.(1).

First results obtained from the approach are envelope correlations obtained for different vessel categories that evidence the minimum credible value of the ttf as a function of radiation intensity. Figure 3 shows the envelope correlations obtained for vessels fully engulfed in fire.

4 Escalation thresholds and probability

The modeling approach carried out in the present study allowed an extended revision of the escalation thresholds for fire scenarios. The envelop correlations shown in Figure 3 define the minimum credible time to failure for vessels exposed to fire. Thus, if the ttf is compared to a maximum time required for effective mitigation (tte), the threshold values of stationary radiation required for credible escalation may be identified. Table 2 reports the threshold values estimated by this approach.

The Table evidences that the conditions leading to escalation due to fires are highly dependent on the target vessel and on the primary fire scenario. The more the equipment is resistant to the fire, the less credible is the escalation, since more time is available for effective mitigation actions by active safety systems or by the operation of emergency teams. Thus, the probability of damage and escalation may be related to the time to failure of the equipment, comparing the ttf to the tte. These could be estimated implementing a specific analysis of the internal or the external emergency plan of the site. In a simplified approach, a specific probit function may be built to relate the time to failure to the probability of escalation, taking into account the tte:

$$Pr = c - d \ln(ttf) \quad (2)$$

In the absence of site specific data, tte_1 was assumed of 5 min and tte_2 was assumed of 20 min. The following probit function was thus used in the present study to estimate escalation probability:

$$Pr = 9.25 - 1.857 \ln(ttf) \quad (3)$$

where ttf is expressed in minutes.

5 Mitigation by passive protection systems

The correlations and the data reported above are conservative and do not take into account the protection systems of the storage vessels. Even if active systems have an important role in emergency management, active protections, although compulsory, have a limited reliability due to delayed activation, possible damage in the primary event, possible failures of critical components. Therefore, passive protection of pressurized storage tanks is normally adopted to mitigate the effects of external fires. It is thus important to take into account the role of passive protections in the determination of vessel time to failure.

A preliminary analysis of the influence of the thermal insulation layer, performed assuming full engulfment conditions of pressurised vessels, evidenced that the time to failure of insulated vessels (ttf_{pv}) could be obtained simply adding the time to failure of the insulation layer (ttf_p) and the time to failure of the uninsulated vessel (ttf_v) without introducing a relevant error:

$$ttf_{pv} = ttf_p + ttf_v \qquad (4)$$

Thus, the effect of a protection layer on the vessel may be accounted with sufficient precision simply adding a further term to the ttf estimated for the unprotected vessel. In Figure 4 an example of simulation is reported for LPG tanks of different volumes (5, 20, 50 and 100 m^3). The ttfs without the coating (Figure 4-(a)) result much lower in the case of the implementation of a protective layer. In particular, in Figure 4-(b), a 50 mm thick glass wool coating was supposed to be installed on the vessel. Hence, the increment in ttf might allow the possible effective intervention of the emergency measures, to mitigate the fire scenario, avoiding the BLEVE.

6 Conclusions

A simplified approach was developed for the management of the hazard due to external fires involving pressurized storage vessels and the possibility of escalation to BLEVE events. The approach was based on the development of tools for the straightforward assessment of vessel time to failure due to fire radiation in different radiation modes. The estimation of the vessel time to failure allowed the evaluation of the time available for effective mitigation actions and the timing of emergency response in order to prevent the BLEVE. The approach was extended to include the effect of passive protection. The results evidenced that specific and reliable values are needed to characterize the performance of the passive protection systems, since the effective prevention of fired BLEVEs critically depends on the available time to failure.

References

[1] F.P. Lees, *Loss Prevention in the Process Industries*, II Ed., Butterworth-Heinemann, Oxford, 1996.

[2] CCPS, *Guidelines for chemical process quantitative risk analysis*, II Ed., AIChE, New York, 2000.

[3] F.I. Khan and S.A. Abbasi, The world's worst industrial accident of the 1990s what happened and what might have been: A quantitative study, *Proc Safety Prog* 18:3 (2004) 135 - 145.

[4] C. Delvosalle, *A methodology for the identification and evaluation of domino effects*, Rep. CRC/MT/003, Belgian Ministry of Employment and Labour, Bruxelles, 1998.

[5] J. Gledhill and I. Lines, *Development of methods to assess the significance of domino effects from major hazard sites*, CR Report 183, Health and Safety Executive, London, 1998.

[6] F.I. Khan and S.A. Abbasi, Models for domino effect analysis in chemical process industries, *Proc Safety Prog* 17:2 (2004), 107 – 123.

[7] V. Cozzani and S. Zanelli, "An Approach to the Assessment of Domino Accidents Hazard in Quantitative Area Risk Analysis", Proc. *10th Int. Symp. on Loss Prevention and Safety Promotion in the Process Industries*, Elsevier, Amsterdam (2001), pp.1263.

[8] V. Cozzani, G. Gubinelli and E. Salzano, Escalation Thresholds in the Assessment of Domino Accidental Events, *J of Hazardous Materials* 129:1 (2006), 1-21.

[9] G. Gubinelli, *Domino effect in the process industries: quantitative methodologies for the evaluation of consequences*, Ph.D. thesis in Chemical Engineering, University of Pisa, Pisa, 2005.

[10] M.A. Persaud, C.J.Butler, T.A. Roberts, L.C. Shirvill and S. Wright, "Heat-up and failure of Liquefied Petroleum Gas storage vessel exposed to a jet-fire", Proc. *10th Int. Symp. on Loss Prevention in the Process Industries*, Stockholm (2001), pp.1069-1106.

[11] A. M. Birk, D. Poirier and C. Davison, On the response of 500 gal propane tanks to a 25% engulfing fire, *J. Hazardous Materials* 19 (2006), 527-541.

[12] B. Droste and W. Schoen, Full scale fire tests with unprotected and thermal insulated LPG storage tanks, *J of Hazardous Materials* 20 (1988), 41-53.

Nomenclature

a [$s \cdot m^2 \cdot W^{-1}$]	first fitting coefficient for the radiation/time to failure correlation
b [s]	second fitting coefficient for the radiation/time to failure correlation
c [-]	first probit coefficient
d [-]	second probit coefficient

FEM [-]	finite element model
I [$W \cdot m^{-2}$]	radiation intensity
LOPA [-]	layer of protection analysis
Pr [-]	value of the probit function
ttf [s]	time to failure
ttf_p [s]	time to failure of the insulating coating layer
ttf_v [s]	time to failure of the uninsulated vessel
ttf_{pv} [s]	time to failure of the insulated vessel

List of figure captions

Figure 1: (a) Detailed simulation of the shell temperature (°C) calculated for a 100 m^3 LPG tank under radiation conditions due to flame engulfment (110 kW/m^2, 600 s); (b) detailed simulation of the correspondent shell stresses (Pa).

Figure 2: Examples of comparison between ttfs obtained with the lumped model runs and the validation data set. Validation data are both obtained from the FEM model results (labelled with FEM and referred to 25, 50 and 100 m^3 tank volume) and from experimental data [10-12].

Figure 3: Envelop correlation of times to failure obtained for horizontal cylindrical pressurized vessels (1.5 to 2.5 MPa design pressure) under full engulfment conditions.

Figure 4 Comparison between the ttf of pressurized vessels (vessel volume: 5, 20, 50 and 100 m^3) without thermal shielding (a) and with a 50 mm thick glass wool coating (b).

List of table captions

FIGURES

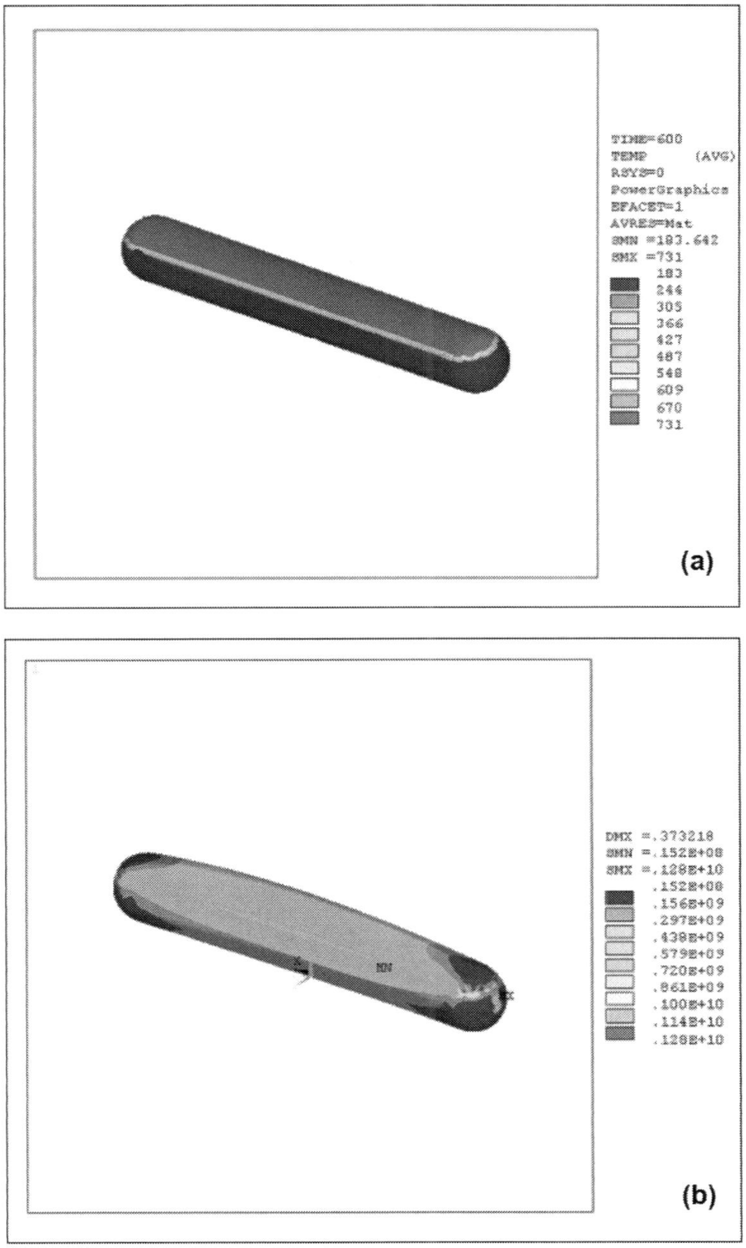

Figure 1: (a) Detailed simulation of the shell temperature (°C) calculated for a 100 m^3 LPG tank under radiation conditions due to flame engulfment (110 kW/m^2, 600 s); (b) detailed simulation of the correspondent shell stresses (Pa).

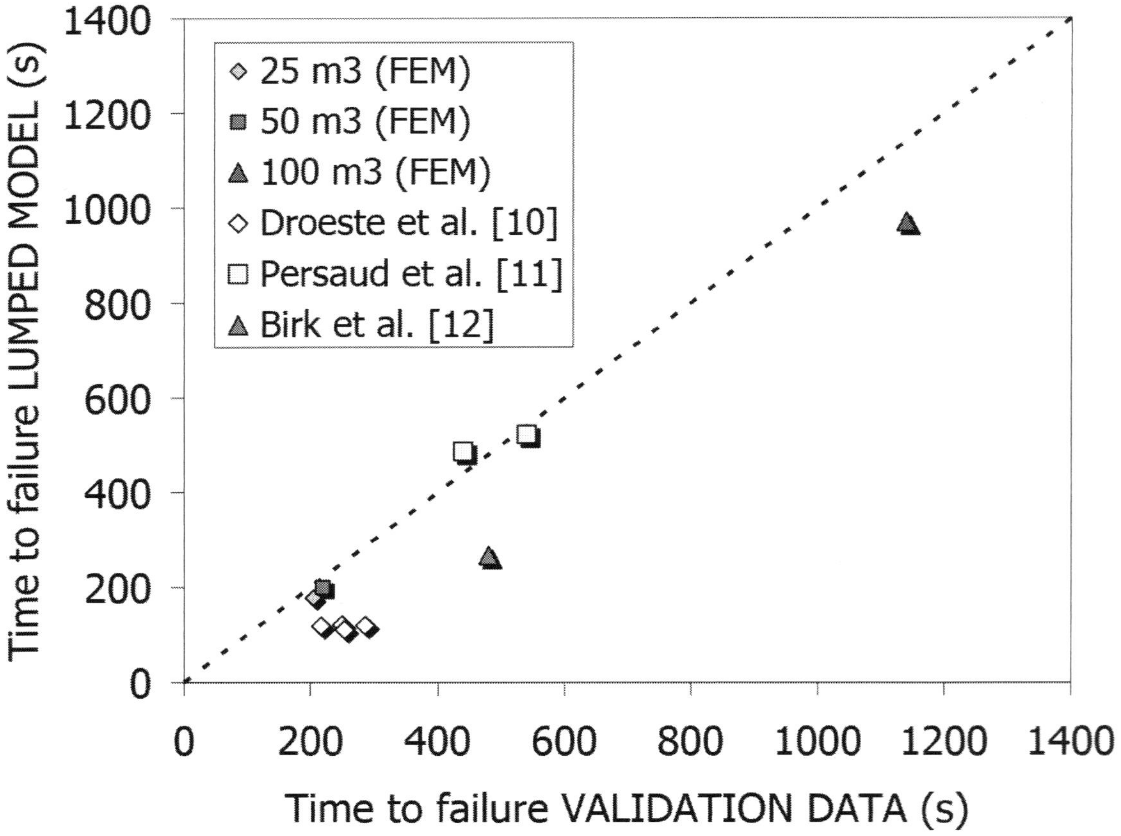

Figure 2: Examples of comparison between ttfs obtained with the lumped model runs and the validation data set. Validation data are both obtained from the FEM model results (labelled with FEM and referred to 25, 50 and 100 m^3 tank volume) and from experimental data [10-12].

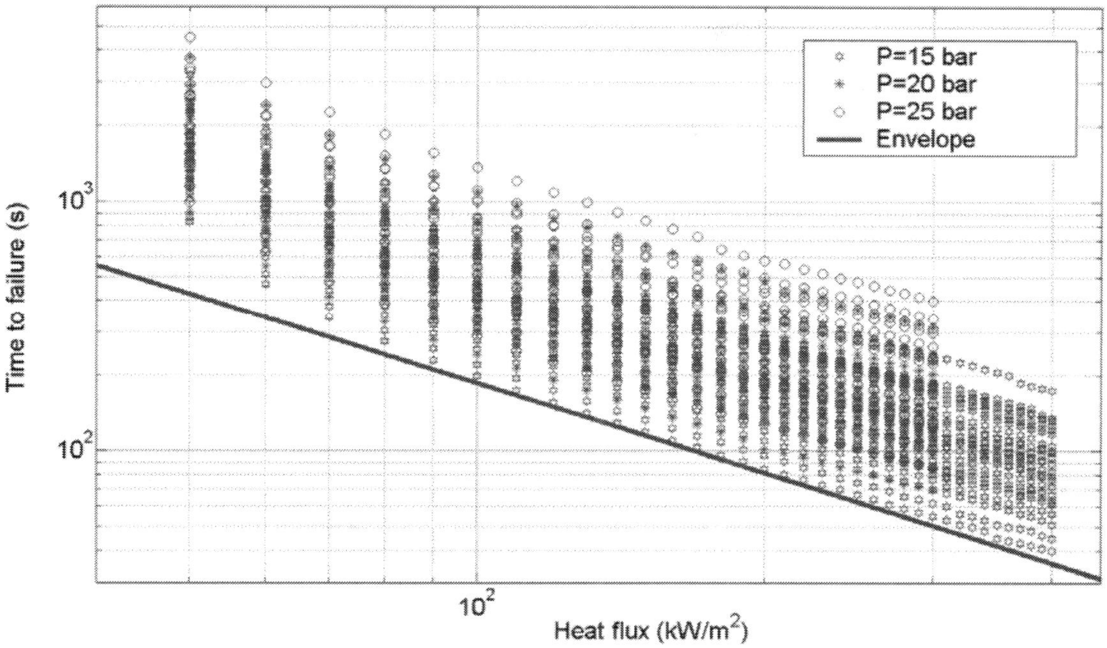

Figure 3: Envelop correlation of times to failure obtained for horizontal cylindrical pressurized vessels (1.5 to 2.5 MPa design pressure) under full engulfment conditions.

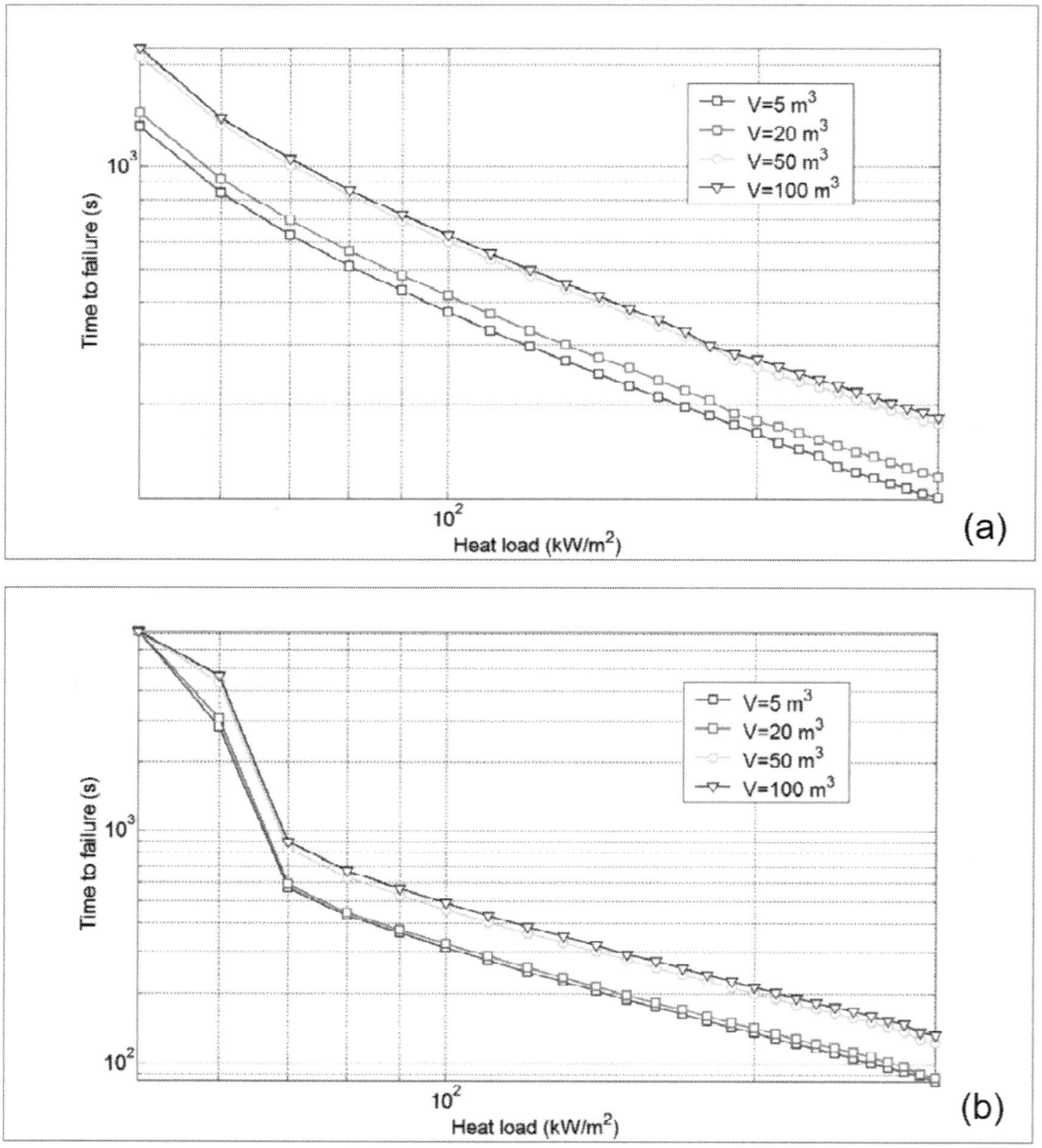

Figure 4 Comparison between the ttf of pressurized vessels (vessel volume: 5, 20, 50 and 100 m^3) without thermal shielding (a) and with a 50 mm thick glass wool coating (b).

TABLES

Table 1: Design data of horizontal pressurized cylindrical vessels

Volume (m^3)	Length (m)	Diameter (m)	Shell thickness (mm)	Area pressure relief valve (m^2)
5	3	1.6	11	0.073
20	7.2	1.9	12	0.164
50	10	2.7	17	0.292
100	18	2.8	18	0.456

Table 2: Estimated threshold values for escalation caused by fire radiation

Time to failure (s)	Radiation Threshold, (kW/m^2)
600	60
900	50
1800	40

LOSS PREVENTION SYMPOSIUM

Fire and Explosion in an Explosives Conditioning Bunker

Timothy J. Myers
Exponent, Inc.
21 Strathmore Road
Natick, MA 01760
tmyers@exponent.com

Peter C. Hinze and Harri K. Kytömaa
Exponent, Inc.
21 Strathmore Road
Natick, MA 01760
phinze@exponent.com and hkytomaa@exponent.com

Copyright 2008

Prepared for Presentation at
American Institute of Chemical Engineers
2008 Spring National Meeting
42nd Annual Loss Prevention Symposium
New Orleans, LA
April 7-9, 2008

UNPUBLISHED

AIChE shall not be responsible for statements or opinions contained
in papers or printed in its publications

Fire and Explosion in an Explosives Conditioning Bunker

Timothy J. Myers
Exponent, Inc.
21 Strathmore Road
Natick, MA 01760
tmyers@exponent.com

Peter C. Hinze and Harri K. Kytömaa
Exponent, Inc.

Abstract

An explosion occurred in a munitions bunker that conditioned an explosive at an elevated temperature, completely destroying the bunker and damaging nearby bunkers. The explosive was used in the production of missile fuel. Investigators believed that the explosion occurred either due to contamination in the explosives or a fire in the bunker initiated by the heating system. Burn damage to combustible materials and witness statements indicated that a fire occurred in the bunker before the explosives detonated. Further investigation determined that the failure of the bunker's heating system initiated a fire that caused the explosives in the bunker to detonate. Information gathered during the investigation revealed that an improperly installed damper partially obstructed the flow of heated air to the bunker. The air heating system was not equipped with over-temperature protection, allowing extremely hot air to enter through ceiling vents near the combustible plywood ceiling of the bunker. The only temperature sensor for the bunker heating system was a thermostat mounted approximately four feet from the floor. A 1/4-scale model of the explosives bunker was constructed to evaluate the thermal air stratification process that occurs inside the bunker when the airflow is deficient. Test results indicated that during low flow rate conditions, hot air entering the bunker stratifies, allowing the wood ceiling to be heated sufficiently to ignite before the thermostat would turn off the heater. The reinforced concrete construction of the incident and adjacent bunkers prevented a chain reaction of explosions in adjacent bunkers.

This incident highlights the fact that careful design and installation is necessary for even relatively standard systems such as HVAC units when handling reactive materials. In this presentation the importance of proper safety interlocks in heating system controls will be discussed.

1. Background

1.1 Incident Description

A commercial explosive was being stored in several adjacent explosives bunkers shown in Figure 1. The bunkers were constructed of reinforced concrete and buried underground. On the day of the incident, witnesses reported seeing smoke exiting exhaust vents on the front of one bunker. The area was evacuated, and eventually the explosives in one bunker detonated, destroying the bunker as shown in Figure 2.

Figure 1: Pre-incident photograph of explosives bunker. Bunker on far right of photograph involved in incident.

No personnel were injured in the incident. Although other bunkers received significant structural damage, the facility design prevented sympathetic detonation of explosives in neighboring bunkers.

Figure 2: Aerial photograph of explosives bunkers after incident. Crater visible in former location of explosive bunker. Neighboring bunkers to the left sustained some damage, but explosives did not detonate.

1.2 Pre-explosion Contents

The configuration of materials present in the bunker was recreated based upon witness statements and inspection of neighboring bunkers and components and is shown in Figure 3 and Figure 4. The bunker interior was 34 ft long, 20 ft wide, and 10 ft, 9 in high (10.4 m x 6.1 m x 3.3 m). Explosive was stored in both fiberboard drums on pallets and in larger aluminum bins. A heating system provided conditioning air through two diffusers present on the centerline of the ceiling of the bunker, with one near the front door of the bunker and one near the rear of the bunker. Exhaust vents were located on either side of the front door of the bunker just above floor level.

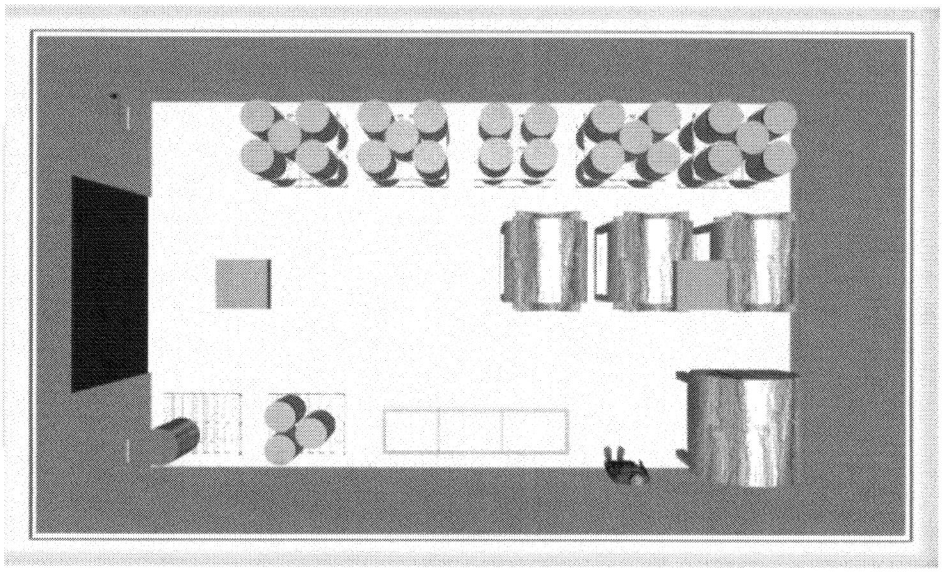

Figure 3: Plan view of bunker interior. Explosives were stored in both drums on pallets and in larger bins shown in the bottom right corner of photograph. Two diffusers located along the centerline of the bunker provided heated conditioning air. Exhaust vents and a door were present on the front of the bunker, shown in the left side of the figure.

Figure 4: View looking from front corner towards back corner of explosives bunker. Two diffusers are visible on the bunker ceiling.

1.3 Initial Investigation

During the initial investigation, many potential causes of the incident were hypothesized. One hypothesis was that the explosive had become contaminated during processing, leading to self-heating and detonation of the explosive. A review of other materials that could have potentially contacted the explosive during handling revealed no incompatible materials. Additionally, witness statements and fire damage to the explosives bunker were consistent with a sustained fire in the bunker before detonation occurred. Witnesses had observed smoke coming from exhaust vents on the front of the bunker before the explosion. As shown in Figure 5, plywood that lined the interior of the concrete bunker walls was consumed by fire in some areas and deeply charred in other areas. The extremely short duration of the detonation could not have caused this fire damage to the bunker interior.

Figure 5: Photograph showing shattering of concrete bunker walls, and fire damage to plywood present on interior of bunker walls.

One potential cause of a fire in the bunker was a heating system that was used to maintain the bunker temperature at approximately 130°F. The remainder of this paper discusses analysis performed on the heating system to determine whether or not it was capable of causing a fire in the explosives bunker.

2. Bunker Heating System

2.1 Layout of Heating System

A schematic of the fresh air heating system for the bunker is shown in Figure 6. The heating system was located on the hillside above the explosives bunker. Fresh air was drawn through filters by a fan, then pushed through three electrical resistance heaters before entering the bunker through two diffuser boxes mounted on the ceiling of the bunker. The air exhausted the bunker through two exhaust ducts located on either side of the front door of the bunker, just above floor level. The system was designed to operate using a thermostat that would call for heat when the temperature dropped below 126°F at a temperature sensor located 4 feet above the ground on the front wall of the bunker. The system would cycle off when the temperature reached 140°F at that same location. Analysis performed after the incident indicated that the exhaust vents on the bunker were undersized and greatly restricted the flow through the heating system, decreasing the flow rate below the intended design value.

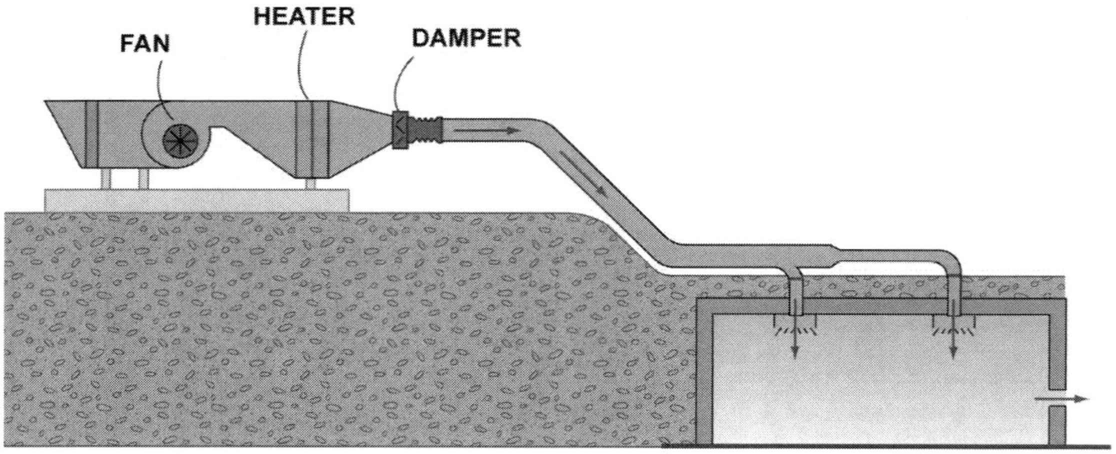

Figure 6: Schematic showing heating system for bunker. Heating system was located on hillside above explosives bunker and heated fresh outside air using electrical resistance heaters. This drawing is not to scale.

2.2 Modification of Heating System – Installation of Damper

During periods when the heating system was not operating, a chimney effect would occur. Driven by buoyancy, warm air would exit the bunker through the ceiling vents and continue out through the heating ducts, and cool air would enter the bunker through the exhaust vents near the floor level of the bunker. This would cool the bunker, causing much more frequent cycling of the heating system than intended.

To prevent this back flow of hot air through the heating system, the facility hired an outside contractor to install a damper downstream of the heating coils. The damper would open when the heating system was calling for heat and close when the system was satisfied, preventing the above chimney effect.

2.2.1 Observation of Damper and Actuator Post Incident

The motor actuator for the damper was found approximately 10% open after the incident. Although the damper was virtually destroyed by the explosion, pieces were recovered and reconstructed. It was discovered that the actuator had been misaligned during installation so that the damper linkage could become stuck in a mostly closed position.

2.3 Heating System as Cause of Fire

With the heating system damper stuck in a partially closed position, air flow rates through the heating system were well below design, causing abnormally hot inlet air temperatures into the bunker. Temperature as a function of heating system flow rate is shown in Figure 7, which demonstrates the theoretical high temperatures present at low flow rates. The installer of the heating system had failed to install a thermocouple downstream of the heating elements that could shutdown the system in the event of abnormal temperatures.

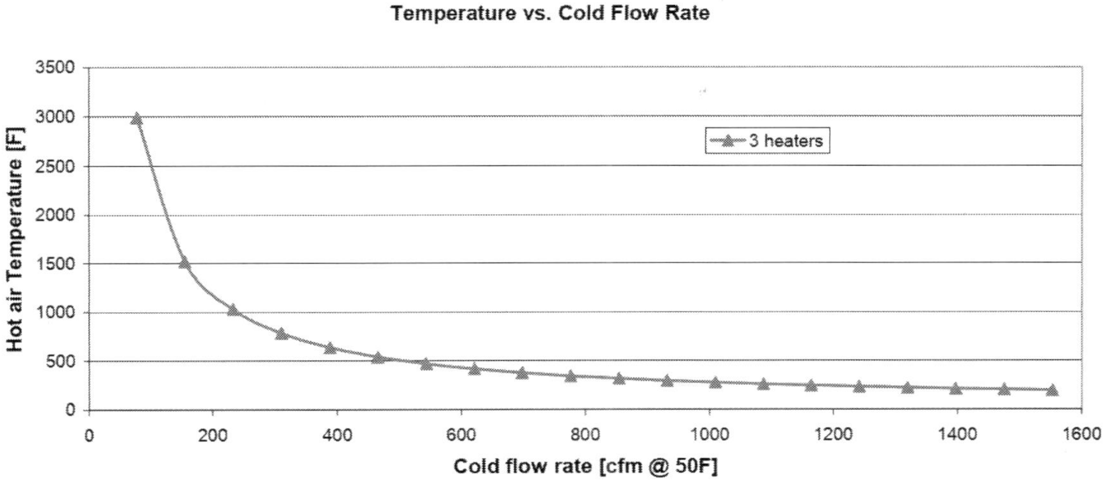

Figure 7: Temperature as a function of heating system flow rate.

If the air entering the bunker is sufficiently hot, it will tend to stably stratify in the bunker creating a layer of hot air against the ceiling over a layer of colder air below. The hot air could potentially heat the wood ceiling of the bunker above the ignition temperature of wood if the air were hot enough. The supply of hot air under this scenario would only go on for a finite duration. The thermostat in the bunker would shut down the heating system when the temperature at the thermostat on the front wall reached 140°F. A scale model of the bunker and heating system was created to determine the maximum temperatures that would occur near the ceiling of the bunker before the heating system thermostat would become satisfied and stop the call for heat. This testing is described in the next section.

3. Scale Model Testing

3.1 Dimensional Analysis

A ¼-scale model of the bunker was built and operated. This model was designed according to the scientifically accepted method of dimensional analysis and its application to model testing [1]. The governing physical equations of natural and forced convective flow and of heat transfer are the momentum equation and energy equation, respectively. Nondimensionalizing these governing equations yields three dimensionless parameters [2]:

$$\Pr = \frac{\nu}{\alpha}, \quad \mathrm{Fr} = \frac{\rho_{hot} u^2}{(\rho_{cold} - \rho_{hot}) g L} = \frac{T_{cold} u^2}{(T_{hot} - T_{cold}) g L}, \quad \mathrm{Re} = \frac{uL}{\nu}$$

The first parameter, Pr, is the Prandtl number and is the ratio of the momentum and thermal diffusivities of air, ν and α. The Prandtl number is a material property of air and varies by less than 1.5% over the temperature range 0°F to 1200°F. Thus, it is essentially identical between the model and full-scale bunkers at all temperatures as the model test was conducted with air.

The second parameter, Fr, is a modified Froude number and is the ratio of inertial forces of the hot air exiting the ceiling diffusers and gravitational buoyancy forces. This group depends on the air velocity, u, size of the bunker, L, gravitational acceleration, g, and hot and cold air densities, ρ_{hot} and ρ_{cold}, (or equivalently for an ideal gas at constant pressure, hot and cold air temperatures, T_{hot} and T_{cold}). This parameter captures the balance between the turbulent mixing created by the diffusers and the competing stabilizing effect associated with buoyancy. Since this experiment seeks to quantify the degree of stratification in the bunker, all variables in the model experiments were selected to match the modified Froude number of the model and the full-scale bunker.

The third parameter, Re, is the Reynolds number and is the ratio of inertia and viscous forces. This group depends on the air velocity of the air entering the bunker, u, the size of the bunker, L, and momentum diffusivity of air, ν. The Reynolds number of both the model and full-scale bunkers are large, indicating that inertia forces dominate viscous forces and the flow is turbulent.

In this type of scaling, one aims to match all dimensionless parameters, but that is not always possible. In this case, it was not possible to match the Reynolds while also matching the Froude number. The importance of the Reynolds number here is to ensure that flow regime out of the vents remains the same, namely turbulent. The geometry of fully turbulent jets exiting the diffusers into the bunker is independent of Reynolds number [3]. While the Reynolds number was smaller than in the actual application, it was maximized by operating the scale model at lower temperatures than the full scale model.

3.2.1 Scaling Test Variables to the ¼-Scale Model

The passage of time does not scale one-to-one between the model and the full-scale processes. The characteristic time to fill the bunker with hot air, τ, is proportional to the length of the bunker, L, divided by the average velocity of hot air flowing into the bunker, u:

$$\tau = L / u$$

The ratio of the fill time of the model, τ_m, to the fill time of the full-scale bunker, τ_{fs}, is given by:

$$\frac{\tau_m}{\tau_{fs}} = \frac{L_m u_{fs}}{L_{fs} u_m}$$

where the subscripts m and fs refer to the model and full-scale bunkers respectively.

When the modified Froude numbers are the same in both the model and full-scale bunkers, the ratio of the average hot velocities is:

$$\frac{u_{fs}}{u_m} = \sqrt{\frac{L_{fs}(T_{fs,hot} - T_{fs,cold})T_{m,cold}}{L_m(T_{m,hot} - T_{m,cold})T_{fs,cold}}}$$

Substituting the ratio of velocities into the expression for the ratios of times shows that, under conditions of equal modified Froude numbers, the ratio of model time to full-scale time is given by:

$$\frac{\tau_m}{\tau_{fs}} = \sqrt{\frac{L_m(T_{fs,hot} - T_{fs,cold})T_{m,cold}}{L_{fs}(T_{m,hot} - T_{m,cold})T_{fs,cold}}}$$

For the ¼-scale model used in experiments, this equation reduces to:

$$\frac{\tau_m}{\tau_{fs}} = \frac{1}{2}\sqrt{\frac{(T_{fs,hot} - T_{fs,cold})T_{m,cold}}{(T_{m,hot} - T_{m,cold})T_{fs,cold}}}$$

The experiments are run with a given initial temperature, which is also designated as the cold temperature, T_{cold}. As a result of the reduced flow, the inlet temperature into the duct reaches a steady hot value, T_{hot}. During the course of a model experiment, temperatures rise in the bunker with time. These temperatures were measured in the model experiments and were scaled to equivalent full-scale temperatures. To this end, a dimensionless temperature difference for both the model and full-scale bunkers is used:

$$\frac{T - T_{cold}}{T_{hot} - T_{cold}}$$

Equating the dimensionless temperature differences of the model and full-scale bunkers yields the expression for the equivalent full-scale temperature in terms of the temperature in the model bunker:

$$T_{fs} = T_{fs,cold} + \frac{T_m - T_{m,cold}}{T_{m,hot} - T_{m,cold}}\left(T_{fs,hot} - T_{fs,cold}\right)$$

3.2 Testing

3.2.1 Scale Model Construction

Photographs of the ¼ scale model are shown in Figure 8 to Figure 11. The interior walls of the model were constructed of non-combustible 1" thick marinate (calcium silicate). Scale models of explosive drums, pallets, and bins were also placed in the interior of the bunker as shown in Figure 9 and Figure 10. A variable speed blower and inline electric resistance heaters were used to control the flow rate and temperature of air into the bunker to match the thermal response of the full scale heating system.

Temperatures within the model bunker were measured by 40 Omega K-Type thermocouples that were recorded by computer through 2 Fluke Hydra data loggers at 8-second time intervals. The locations of the 40 thermocouples included:

- Six thermocouple trees in the bunker interior spanning from the floor to the ceiling.
- Thermocouples on the ceiling of the diffuser box and the metal duct entering the diffuser box.
- Thermocouples mounted on the ceiling
- Thermocouples mounted on the front wall including the location of the bunker thermostat and exhaust air.

Figure 8: Exterior View of ¼ Scale Model

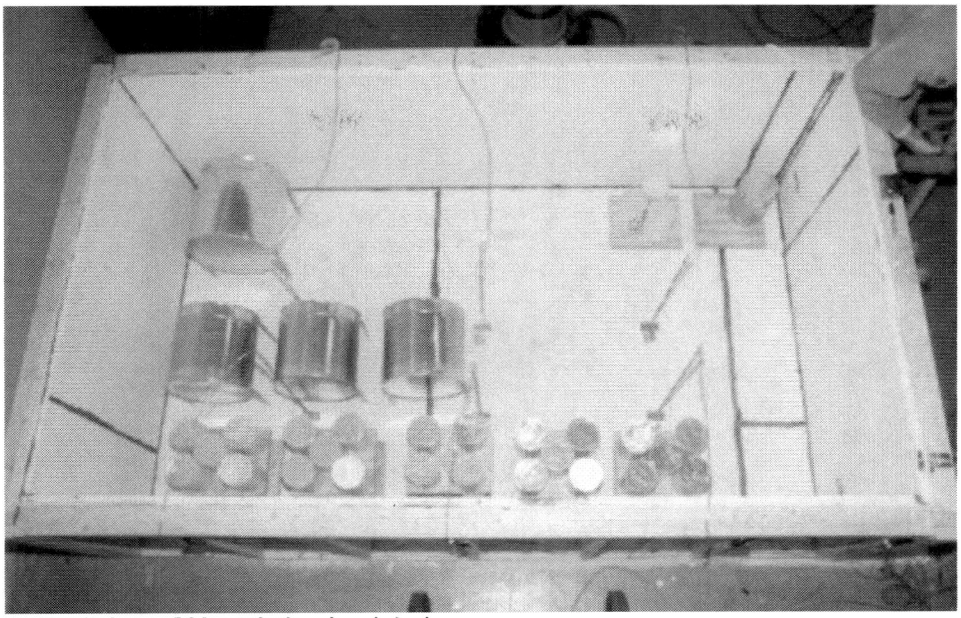

Figure 9: Overhead view of ¼ scale bunker interior.

Figure 10: Photograph of ¼ scale model interior looking towards front. Model ceiling is removed in photograph.

Figure 11: Photograph of ¼ scale model ceiling and hot air diffusers.

Test Results

Test results for two cases are shown below in Figure 12 to Figure 14. All results are shown transformed to the full-scale values. Data in Figure 12 corresponds to a condition where the air flow is approximately 20% of the nominal flow rate with the damper open, yielding an inlet temperature of 785°F. In this case the temperature at the location of the thermostat reaches a shutdown temperature of 140°F after approximately 1 hour. In the scale model the test was allowed to continue to run past this time. After 1 hour, the temperature of the exterior of the rear diffuser hot air duct, which was in contact with wood in the incident bunker was about 590°F. The ceiling temperature within the front diffuser box was about 525°F. Various sources provide a range of non-piloted or autoignition temperature of wood from about 375°F to 550°F [4,5], indicating that wood near the heating ducts in the bunker could ignite.

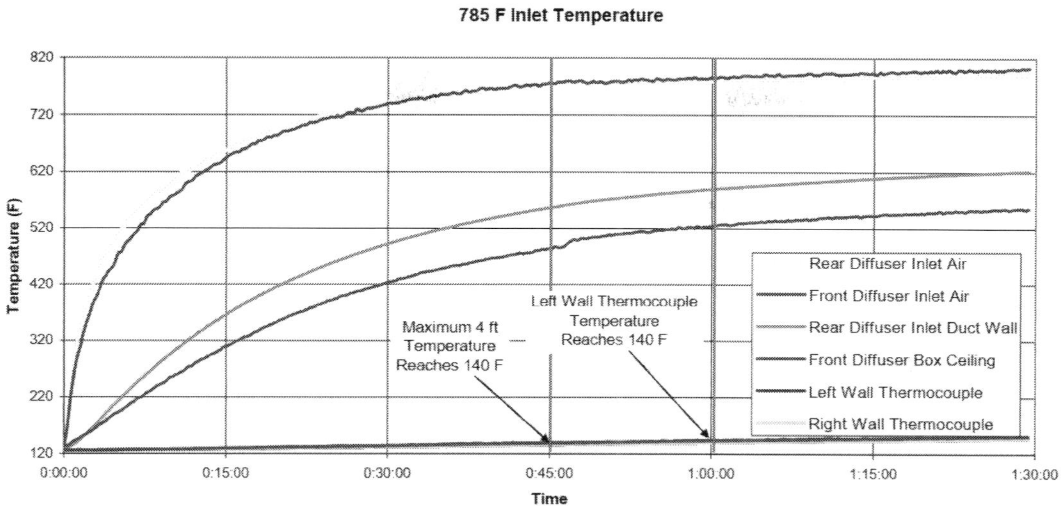

Figure 12: Transient temperatures with reduced airflow and 785°F inlet temperature.

Data in Figure 13 corresponds to an even lower airflow case with an inlet temperature of 1024°F. In this case the temperature at the location of the thermostat reaches a shutdown temperature of 140°F after 16 minutes. At 16 minutes the exterior temperature of the rear diffuser hot air duct was 685°F, and the temperature of the front diffuser box ceiling was 670°F. These temperatures are well above the non-piloted or autoignition temperature of wood.

Figure 14 shows the temperature profile from the floor to the ceiling in the center of the bunker for the 1024°F inlet temperature case. The temperature profiles show that stratification of hot air near the ceiling does occur. Notably the temperature of the ceiling is somewhat colder than the air several inches below it. This is due to heat loss through the ceiling and roof of the model. Analysis shows that the heat transfer loss through the ceiling of the scale model bunker is larger as a fraction of the total heating system power, than in the full-scale model.

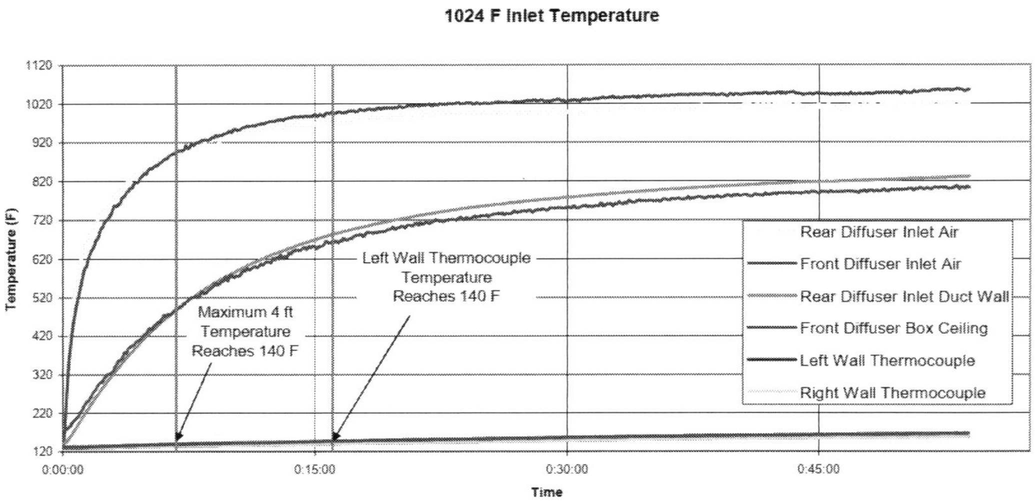

Figure 13: Transient temperatures with reduced airflow and 1024°F inlet temperature.

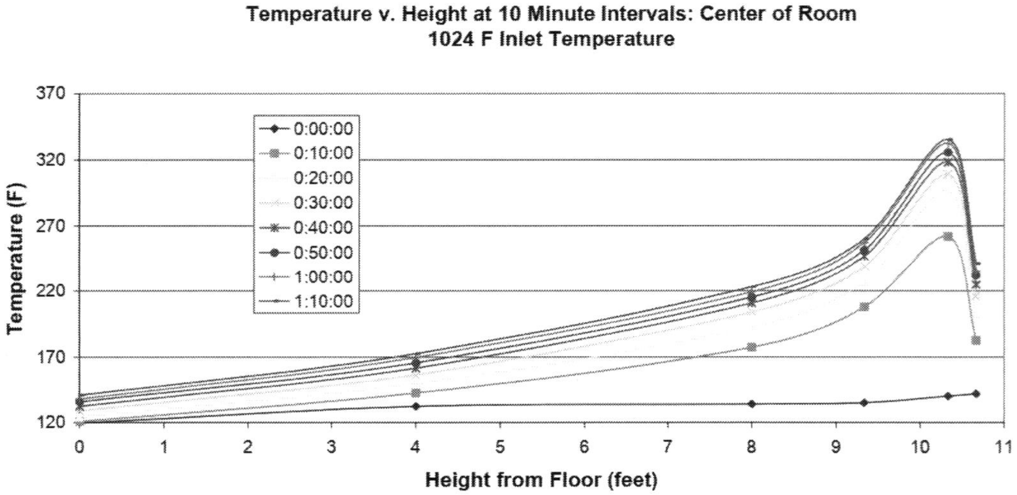

Figure 14: Temperature profile at center of room for 1024°F inlet temperature condition.

4. Lessons Learned

An explosive bunker was destroyed and neighboring bunkers were severely damaged due to a fire than occurred in the bunker. The fire was caused by a damper blocking airflow through a heating system, causing extremely high temperatures. Several specific lessons as well as more general lessons can be learned from this incident.

- Low flow through heating systems can cause excessively high temperatures.
- A single thermostat may have been sufficient to monitor the bunker temperature during normal operating conditions at high airflow rates and lower air temperatures. During upset conditions with low air flow rates and high air temperatures, however,

significant stratification occurred and a single temperature measured near the floor level of the bunker was not representative of the bunker temperature at higher elevations. Facilities should give careful consideration to the location and number of sensors for both normal and abnormal operating conditions.

- The reinforced concrete design and spacing of explosive bunkers was sufficient to prevent the sympathetic detonation of explosives in neighboring explosives bunkers.
- Blockage of airflow through a heating system caused a fire and ensuing detonation. When storing reactive chemicals or explosives, malfunction of even relatively standard equipment such as heating systems can cause significant incidents. Facilities should perform hazard analyses on these simple systems, much as they would for more complicated systems used in the production of reactive chemicals or explosives.
- A thermocouple that was supposed to cause shutdown of the heating system in the event of over temperatures was not properly connected. Facilities should test safety protection systems when equipment is originally installed, modified, and during the life of the equipment to ensure it is properly functioning.
- A damper was installed in the heating system to stop a chimney effect that drew hot air out of the bunker when the heating system was not functioning. This damper introduced a new failure mode into the system that was not identified. Facilities should consider performing hazard analyses whenever systems are modified.

5. References

[1] Avallone, E.A., Ed. and Baumeister II, T. Ed., Marks' Standard Handbook for Mechanical Engineers, Ninth Edition, McGraw-Hill Book Company, 1987.

[2] Incroprera, F.P., and DeWitt, D.P., Introduction to Heat Transfer, Second Edition, John Wiley & Sons, 1990.

[3] White, F.M., Viscous Fluid Flow, Second Edition, McGraw-Hill, Inc., 1991.

[4] O'Halloran, M.R., "Plywood In Hostile Environments, Physical Properties and Applications," APA, Report No. 132, December 1975.

[5] NFPA Fire Protection Handbook, Ninteenth Edition, NFPA, 2003, p. 8-36.

A Series of Small Changes

John C. Wincek, CSP
Process Safety Manager
Croda, Inc.
8 Croda Way
Mill Hall, PA 17751
Phone: 570-748-7796
j.wincek@crodausa.com

Prepared for Presentation at
American Institute of Chemical Engineers
2008 Spring National Meeting
42nd Annual Loss Prevention Symposium
New Orleans, LA
April 6-10, 2008

UNPUBLISHED

AIChE shall not be responsible for statements or opinions contained
in papers or printed in its publications

42nd Annual Loss Prevention Symposium
AIChE 2008 Spring National Meeting
New Orleans, LA
April 6-10, 2008

A Series of Small Changes

By John C. Wincek, CSP
Process Safety Manager
Croda, Inc.
8 Croda Way
Mill Hall, PA 17751
j.wincek@crodausa.com

Abstract

A 6,000-gallon, stirred tank batch reactor was equipped with a one-inch, piston-type sample valve at the lower margin of the straight sidewall. This sample valve was originally installed with a two-inch long pipe nipple in the discharge port of the sample valve. Over a period of several years, recommendations from near miss reports and risk assessments, as well as operator concern with valve plugging resulted in a series of small changes to the sampling setup. Changes included lengthening the discharge pipe nipple, addition of a hand valve with position indication, and installation of heat tracing. These changes, while individually harmless, collectively created a hazard that resulted in the ignition of material as it was being sampled from the reactor.

A 6,000-gallon, stirred tank batch reactor was equipped with a one-inch, piston-type sample valve at the lower margin of the straight sidewall, just above the joint with the bottom dished head. When closed, the piston in the sample valve sits flush with the inside wall of the vessel. This type of valve is necessary to prevent plugging of the valve when high melt point products are produced in the vessel. As originally installed, the sample valve had a short pipe nipple threaded into the discharge port of the sample valve *[Figure 1]*.

A near miss report was submitted, stating that the sample valve discharging at approximately face level of the employee operating the valve created a hazard. Due to the high temperature products that were being sampled, serious burns could result if the product splashed onto the employees face. To minimize the hazard, a one-inch diameter, nine-inches long pipe nipple was threaded into the sample valve outlet, to extend the sample discharge point to the chest level of the operator obtaining the sample. Shortly after installation of the longer pipe nipple, operators discovered the nipple would frequently plug when sampling high melt-point products. When this occurred, the operator would unscrew the pipe nipple from the valve, and either rod it out or replace it with a new pipe.

Several months after this change, during a risk assessment of the vessel, it was recommended that all hand valves on the reactor be fitted with position indicators. These position indicators would be interlocked to hazardous materials charging valves on the reactor to prevent charging flammable and/or toxic materials to the vessel whenever a hand valve was inadvertently left open. For example, if the position indicator on the atmospheric vent valve indicated the valve was open, the automated charge valves for flammable and/or toxic materials would not open. This effort included installing a position indicator on the sample valve. However, because a position indicator was difficult to install on the piston-type sample valve, it was instead decided to install a quarter-turn ball valve on the end of the nine-inch sample valve extension pipe. This ball valve would later be equipped with a position indicator, such that even if the sample valve itself were left open, the operator could be assured that the ball valve was closed before flammable/toxic materials were permitted to the charged. This modification was reviewed and approved using the facility's Management of Change system.

Figure 1: Original Installation

Soon after installing the ball valve and position indicator, (as shown in Figure 2), the operators realized that they could no longer remove the extension nipple, as the position indicator on the ball valve would first have to be removed by an electrician. Episodes of sample line plugging became more frequent due to the addition of the ball valve on the end of the sample extension. Furthermore, plugged sample lines were more difficult to clear, because they could no longer be removed for cleaning. Operators resorted to applying steam, using a hand-held steam hose, to the sample assembly in order to melt the material plugging the line.

A department manager observed an employee unplugging a sample valve using a steam hose, and decided that steam tracing the valve and extension piping would be preferable. A work order was written, and ¼" steam trace tubing, supplied with 150 degree C steam, was installed on the sample valve and extension pipe. The installation of steam tracing was not reviewed using the Management of Change system.

Within a month of the steam tracing installation, while preparing to obtain a sample from the reactor, an operator opened the ball valve on the end of the sampling line, and witnessed a small, pressurized discharge of vapor. He then opened the sample valve to draw a sample into an 8-oz. glass sample jar. Two to three seconds after flow from the sample valve began, a six-inch diameter fireball erupted from the end of the sample extension pipe. The flash fire lasted less than one second, and diminished to a small (1" long) flame. The operator blew out the flame and closed the sample valve.

The ensuing investigation revealed that operators routinely charge a solution of 20 % sodium chlorite in water through the sample valve. 60 minutes before the fire, sodium chlorite had been charged through the sample valve. After the charge, both the sample valve and ball valve had been closed to allow for charging of a hazardous material. The investigation team theorized that residual sodium chlorite solution, trapped between the sample valve and ball valve, was heated by the 150-degree C steam tracing, boiling/evaporating off the water and leaving behind concentrated sodium chlorite. When the operator opened the ball valve, the observed discharge of vapor was the water vapor from the chlorite solution. When the sample valve was subsequently opened, the organic product contacted the concentrated chlorite and ignited, as foreseen by a warning on the sodium chlorite MSDS.

The conditions that permitted this incident to occur developed over a one to two-year period. A perfectly safe sampling arrangement was modified, in part with the "blessing" of an approved MOC, to make the reactor system safer. The MOC reviewers, however, did not realize the modification would increase the frequency of sample line plugging, and that installation of the position indicator would make clearing a plugged sample line more difficult.

Figure 2: Configuration at the time of the incident

An alert department manager noted that the second modification (created by installation of the ball valve and position indicator) created the hazard of melting out a plugged line with a live, open-ended steam hose. The facility had experienced a number of thermal burns, sensitizing the manager to potential burn hazards. The manager asked a subordinate to write a work order to install steam tracing on the sample valve. Neither the manager nor the subordinate thought the change was "worthy" of an MOC. The maintenance mechanics that installed the tracing had been trained to identify work orders that should have an MOC, but did not (perhaps justifiably?) flag this work order as requiring one. While it is easy to blame the lack of a MOC review as the causes of the incident, the four MOC approvers interviewed as part of the incident investigation indicated they did not believe they would have recognized the hazards of the steam tracing in combination with sodium chlorite charging.

The incident investigation team initially considered calling for a more robust review of MOCs. However, they also noted that the hazards of concentrated sodium chlorite were not well known in the facility, and in fact were not noted in the chemical hazards section of the facility's procedures. Further, an incident had occurred in the facility four years prior involving sodium chlorite residue igniting a wooden pallet. This incident, and the hazards of concentrated sodium chlorite, had not been communicated to the employees as the investigation recommended. The investigation team felt that had an MOC been written for this change, and had the hazards of sodium chlorite been well known in the facility, the potential for this incident to occur would have been recognized.

The final corrective actions from the investigation included:

Water-flushing lines and pumps used for sodium chlorite charging.

Water-flushing lines and pumps used for hydrogen peroxide charging (an oxidizer with similar to sodium chlorite).

Investigation of practices used for charging materials, including:

1. Should the procedures specify what specific method/equipment/nozzle to use for each charge?
2. Should the sample valve be used for charging?
3. Are chemical incompatibilities considered when charging multiple raw materials through the same nozzle/lines?
4. Is it acceptable to charge materials to the liquid phase of a product?

Conduct employee training on the hazards of sodium chlorite.

Conduct retraining on the types of small changes that require MOC review.

Combustible Dust Explosion Case History

J. Kelly Thomas, A.J. Pierorazio, D.C. Kirby & M.T. Edel
Baker Engineering and Risk Consultants, Inc.
3330 Oakwell Ct., Suite 100
San Antonio, TX 78218-3024
(210) 824-5960
kthomas@BakerRisk.com

Prepared for Presentation at
American Institute of Chemical Engineers
2008 Spring National Meeting
42nd Annual Loss Prevention Symposium
New Orleans, LA
April 7-9, 2008

UNPUBLISHED

AIChE shall not be responsible for statements or opinions contained
in papers or printed in its publications

Combustible Dust Explosion Case History

J. Kelly Thomas, A.J. Pierorazio, D.C. Kirby & M.T. Edel
Baker Engineering and Risk Consultants, Inc.
3330 Oakwell Ct., Suite 100
San Antonio, TX 78218-3024
(210) 824-5960
kthomas@BakerRisk.com

Abstract

This paper deals with a combustible dust explosion that occurred in 2003. The explosion and subsequent fires resulted in multiple fatalities, widespread injuries and damage to a processing area of the facility. The majority of the damage associated with the explosions was caused by a secondary dust explosion which resulted from the widespread accumulation of hazardous levels of combustible dusts throughout the processing area. The primary combustible dust involved in the explosion was phenolic resin, although similar damage would have been caused by any typical combustible dust allowed to accumulate in a similar manner. Although a number of potential initiating events have been postulated, the most likely initiating event was a gas explosion inside an oven associated with the manufacturing operation.

This event clearly demonstrates that allowing combustible dust to accumulate to hazardous levels within a building, room or similar enclosure creates the risk of a dust explosion. The accumulation of hazardous levels of combustible dust within an enclosure coupled with a competent initiating event can result in a dust explosion. The hazards associated with combustible dusts are widely discussed in the literature and addressed by national codes and standards (e.g., NPFA 654). Nevertheless, combustible dust explosions, such as the one described in this paper, continue to occur. It is therefore important to continue to disseminate information regarding combustible dust explosion hazards and case history information which illustrates that nature of these hazards.

LOSS PREVENTION SYMPOSIUM

42nd Annual Loss Prevention Symposium
New Orleans, LA
April 7-9, 2008

Session 7: Case Histories and lessons Learned

The Importance of Reliable Inertisation and Grounding of Ignition Sensitive Powder Handling Systems

Bill Roper
Ciba China Ltd.
18/F Xin An Building
99 Tian Zhou Road
Caohejing Hi –Tech Park
Shanghai 200233, P.R. China
0086 21 2403 2028
William_leonard.roper@cibasc.com

© 2008 All rights Reserved

Prepared for Presentation at
American Institute of Chemical Engineers
2008 Spring National Meeting
42nd Annual Loss Prevention Symposium
New Orleans, LA
April 7-9, 2008

UNPUBLISHED

AIChE shall not be responsible for statements or opinions contained
in papers or printed in its publications

1. Introduction

This "learning from" will show how ignoring a small defect on a magnetic sieve together with unreliable inerting lead to the eventual increase of oxygen in a powder handling system to the point that an ignition occurred, probably through an electrostatic discharge, resulting in a fire..

The purpose of this paper is:

- To encourage management to establish a system of simple but regular controls of specific safety equipment, using their operation staff, in so doing reduce the probability of incidents.
  To this end a Safety Maintenance and Control systematic (SAMACO) will be presented.
- To bring to management attention the importance of an effective inertisation system which will provide the necessary level of safety for the people and equipment
- To show the consequences of not following the company safety standards recommending the use of a structured incident investigation technique which resulted in an incomplete analysis and a higher level of remaining risk. This highlights managements responsibility to ensure the safety standards, which are there as a provision and a protection for the people, are used

2. The Incident details

Incident Description

At approximately 23:30 on 17th September 2005 an alarm sounded in the milling room on the 4th floor of the powder handling plant C2 and when an operator investigated he found black and melted product residues scattered around various piece of equipment in the room. He smelt the odor of burning but saw no fire. He found a blackened hole in the material compensator above the mill from which smoke had emitted.

2.1 The powder handling equipment being used

The incident happened on the equipment shown on the following diagram.

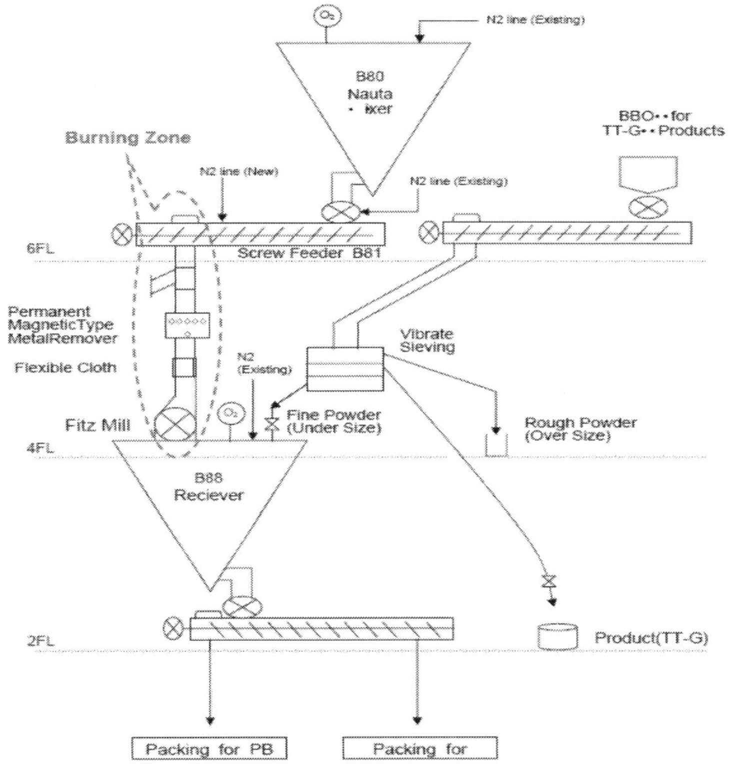

DIAGRAM 1

All the equipment is made of stainless steel apart from the flexible material compensator above the Fitz mill which is made of a nonconductive Poly Propylene

2.2 Description of the process

Product is transported using nitrogen gas as carrier into the Nauta mixer B80. Product from the Nauta mixer B80 is transported via gravity, rotary valve and screw feeder B81 through a permanent magnet metal remover to a Fitz mill. After milling the product passes into the receiver B88 and is ready for packaging.

At the time of the incident the equipment to the right (marked BBO for TTG products in the diagram 1 above) was not being used.

There is a gas balance line between B80 and B88 (not shown on diagram 1. The system is vented to the atmosphere via a dust filter mounted on B80 (not shown on diagram 1)

The atmosphere in this equipment is oxygen reduced to <6%v/v using nitrogen gas (introduced at the receivers B 80 and B88) and the oxygen concentration is monitored by on line analyzers whose sample ports are on receivers B 80 and B88. The oxygen analyzer detection principle is based upon "membrane barrier galvanic method".

2.3 Properties of the product

The product is an organic antioxidant additive used in the plastic industry. The following Fire and Explosion (FIREX) data for this product is known, see table 1.

Table 1

Characteristic	**Result**	**Comment**
Minimum Ignition Energy(MIE) (without induction)	<3mJ	Very low (electrostatically generated discharges could be an effective ignition source)
Powder particle size	<200 microns	Fine
Volume resistivity	$>10^{14}$ ohm.m	Highly insulating
Solvent content	0.08 wt. %	≤0.5 wt. % considered solvent free
Melting point	120 Degrees C	
Burning Class @ 20 and 100 Degrees C	2	Not likely to sustain burning easily
Ignition temperature	410 Degrees C	

3. The incident investigation

Shortly after the incident a statement was made as to the probable causes of the incident and measures taken.
No root cause investigation technique was used an at that time the investigators recorded the following:

- Signs of a fire having taken place in the pipe line between the Fitz mill and the Nauta mixer B80 (see the dotted line on the Diagram 1). They took some photographs which show what they saw(see photo's)

Hole on the flexible cloth

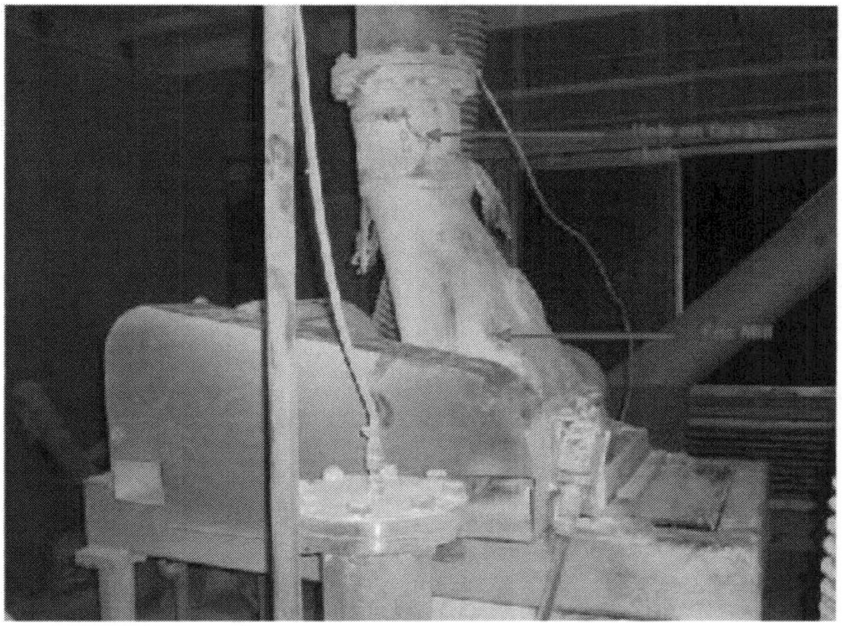

Black stuffs gushed from the hole on flexible cloth

Magnetic metal remover casing and black stuffs

- The maximum Oxygen concentration alarm for the measurement on line at receivers B 80 and B 88 which are set at 6% did not sound.
- The operator heard the Fitz mill alarm sound indicating a trip of the mill
- The Fitz mill had no visible scrape marks or colored spots on the metal indicating heating by friction.

Cutter for milling inside Fitz Mill
(We could not find any spark signs on this cutter)

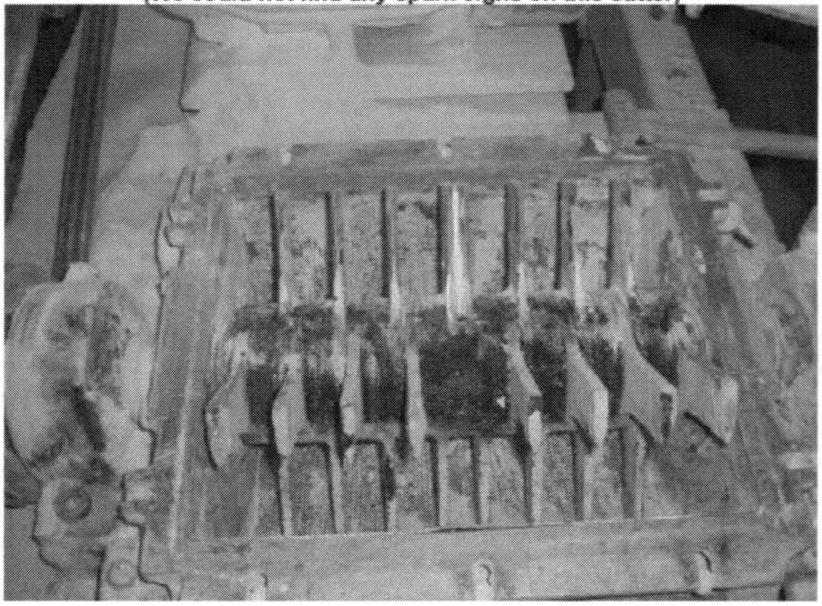

- The screw feeder B 81 did not have signs of friction visible

Black stuffs around screw feeder B81

- The operators had briefly opened the inspection cover on the screw feeder B 81 to check for powder flow
- The permanent magnet metal remover had 2 fasteners on the door of its casing and 1 of these was broken leaving a significant clearance between the door and the casing. This situation had been like this for some time

Black stuffs gushed from magnetic devices

- The resistance to ground for various pieces of equipment was checked and it was found that the magnet bars were insulated from the casing in which they were sitting. A measurement of 20 Mega ohms was recorded for the resistance of the magnet bars to ground.

3.2 *The causes*

The incident investigation isolated the following causes:

1. Air ingress into the system (primarily via the loose magnet door seal and secondly via the inspection cover opening to screw feeder B81) increased the oxygen concentration inside the magnet metal remover to a level >14%(1).
    - Management did not have a system in place to ensure changes(the loose magnet door seal) are corrected immediately

2. Electrostatic build up on a non grounded conductor (the magnet bars) due to the separation process of the powder falling between the screw feeder B81 and the Fitz mill.
    - Management did not have a system in place to ensure the integrity of the grounding of all conductive parts

3. Most probably a spark discharge between the insulated conductive magnet bars and their earthed casing ignited the ignitable dust cloud assumed to have a concentration >Lower Explosion Limit (LEL)

4. The Oxygen meters did not register the dangerously high oxygen concentration at the point where the ignition occurred (this cause was highlighted during the root cause analysis done ca. 2 years later in preparation for this paper)

3.3 *The measures taken*

1. The broken magnet door lid fastener was repaired to prevent air ingress. Inspection of the integrity of the magnet lid sealing system, using a Safety Maintenance and Control (SAMACO) program. If the inspection hatch is lifted on screw feeder B81 then nitrogen gas is purged in automatically at the rate of 100l/min to the screw feeder

2. Grounding of the bar magnet support rail through a grounding wire to earth.

3. Inspection of the integrity of the grounding regularly using a SAMACO program.
4. 2 additional Nitrogen purge points installed: 1 in the lid of the screw feeder B81 (see diagram 1) and the other in the down pipe leading to the magnet metal remover. Both points feed in 100 l/min Nitrogen gas when the Mill is running.

4. The Safety Maintenance and Control (SAMACO) systematic(2)

Equipment installed will only do its job if it works when required. One way to increase the probability of this happening is to continually control the conditions and the integrity of the used equipment and in doing so find any conditions that may cause failure before it actually happens.

Equipment failure prevention begins with the acknowledgment, by people, of the reality that equipment deteriorates over time and the conviction that their progressive action, from housekeeping to preventive maintenance, is the crucial factor for success in slowing down this inevitable wear process.

Safety Maintenance, Service and Control (SAMACO) is that part of preventive Maintenance which is especially concerned with simpler review, upkeep and inspection work not requiring special equipment and normally carried out by the production staff.

The SAMACO is part of preventive maintenance and is to be used for the early detection and elimination of failure, specifically for equipment installed to ensure safety.

4.1 Setting up a SAMACO Program

Implementing an effective SAMACO program should recognize the following 4 steps:

Step 1: Chose the crucial safety equipment you want to control and check

The first step is to set up a catalogue of all equipment and installations needing upkeep, inspection and function checks to ensure safety. Equipment identified from the risk analysis will be one of the inputs here. The compilation of such a catalogue requires a good and comprehensive knowledge of the plant with all its equipment, their function together and the consequences of malfunctions for safety.

In the catalogue for each piece of equipment the following is stated:

-The equipment unit number

And or

-The equipment description

Example of a page in the catalogue:

Equipment number	Name	control interval	Controller	The control instrucitions
R451 - X401	Inner temperature alarm	monthly	Operator	**> Set the temperature limit below the actual inner temperature** The visual temperature alarm blinks. Field 24 of the general
R452 - X401				alarm display lights up and horn sounds
R453 - X401				> Cancel the alarm at the vessel and at the lift
R545 - X401				The visual and audio alarms cease
	Agitator monitor alarm			**> Start the agitator**
R451 - X402			Operator	>Switch the agitator monitor on
R452 - X402				> Turn the agitator off
R453 - X402				The visual agitator alarm blinks. Field 24 of the general alarm display lights up and horn sounds
R454 - X402				> Turn the agitator monitor off
				The visual alarm at the vessel goes out
				> Cancel the alarm at the lift
				The general alarm display light goes out and the horn ceases

Step 2: Define what controls you want to do and how often

The second step involves choosing the exact control/inspection to be carried out on each piece of equipment and their frequency. It may help to ask the question, "how do we ensure that this piece of equipment will always be ready to work"? These controls and schedules will also be

prescribed from the various product risk analysis. Past experience about the equipment reliability, together with help from the equipment suppliers are resources that can be of value here.
In the catalogue for each piece of equipment the following is stated:

-The control/inspection instructions
-How often the control needs to be performed

Example of a control schedule

SAMACO programme		The Control Schedule																Month: Year:
		Week		1		Week		2		Week		3		Week		4		
Equipment number	The Control	Date	Sign	ok	not ok	Date	Sign	ok	not ok	Date	Sign	ok	not ok	Date	Sign	ok	not ok	Remarks
M451-X400	Potential equalisation check													25.1				
M452																		

Step 3: Choose who will do the controlling

The next step is to determine who is going to do the controlling, Controls and simple maintenance tasks can, with the respective training, be entrusted to the operating personnel of a plant. The plant personnel have an interest in the right functioning of the equipment under their control which make them the candidate of choice. Those jobs which cannot be done by plant personnel are delegated to the respective craftsman, some may need to be done by external firms.

In the catalogue for each piece of equipment the following is stated:

-The person/organization who carries out the required control or service

Step 4: Appoint a coordinator who will oversee the program

The next step is to determine who will administrate the control schedules. The program coordinator is responsible for the timely distribution of the control schedules and control instruction. He is also responsible for the filing of the completed control schedules.

4.2 Program effectiveness

Throughout the year additional equipment may be added to the Equipment Catalogue as new needs are identified and as new safety equipment is installed

It is recommended that at least1/year the plant management go over the past results with a view to determining how effective the program has been to reduce failure rates. Redefining inspection intervals, extending or shortening inspection instructions, and adding or subtracting equipment if necessary.

5. Ensuring an effective inertisation system

The following open question remains: Why did the Oxygen meters not warn of an increasing O_2 concentration in the system?

In reviewing the incident for this paper is became that this point had been neglected and the reasons could be a combination of the following:

- Pride- production and EHS thought they had it sewn up using quick and "dirty" methods!
- Production and EHS did not refer to the provision of the internal standard (Accident and Incident Reporting and Investigation)(3) requiring a structured investigation technique for such serious cases
- Misjudgment of the situation by the production and EHS staff

5.1 The hazard scenario:

In addressing the solution to the open point of an ineffective inertisation system the following hazard scenario can be stated:

The sealing of the equipment fails (an active failure)
AND
The Oxygen measurement system does not detect this change (passive failure) leading to the formation of an explosive air powder mixture in the system which is ignited causing an explosion injuring people.

<u>The RISK assessment</u>
Probability assumed to be Low
Severity assumed to be High

Possible reasons for the passive failure could include:

1. The oxygen measurement system is to slow to register changes in concentration
    a. The air sample point is in the wrong place eg. Next to the nitrogen inlet line
    b. The air sampling tube or its filter is blocked or partially blocked
    c. The air sampling tube from the sample point to the detector head is too long
    d. The oxygen measuring technique(electrolytic cell) is not optimal for the requirement
    e. The sample point geometry does not allow for a representative air sample to be taken
    f. Or combinations of the above

5.2. *Verification of the above*

At this point in time the above verification is under way but no results are known.

6. Summary

Simple controls and checks done regularly by the operation staff can greatly aid in ensuring the integrity of equipment as shown by the SAMACO example given

Ensuring compliance with company safety standards no matter how well they are written is a continual management responsibility which if neglected, as in this example, could lead to the same incident being repeated again

In this case management choose to accept the technical and organizational measures in place and take the responsibility for the residual risk until a reliable inertisation system can be found, implemented and verified.

7. References

[1] - Ciba Internal Guideline 16 "Safe Handling of Powders and Dusts", Edition 3 Revison 0 of 2003

[2] - Ciba Internal Guidance Note 11/6 "Safety, Maintenance, Service and Control", of October 1998

[3] - Ciba Internal Guideline 8," Accident and Incident Reporting and Investigation", of March 1998

LOSS PREVENTION SYMPOSIUM

Hydrochloric Tank Explosion

Chandrashekhar Chandwadkar,
National Petrochemical Company,
Nothern Sheikh Bahaei St.,
Mulla Sadra Ave.
Tehran, Iran.
c_chandwadkar@yahoo.com

Ghodaratollah Nasiri
National Petrochemical Company, Iran
nasir@nipc.net

Ali Seyfzadeh
National Petrochemical Company, Iran
hse@nipc.net

Prepared for Presentation at
American Institute of Chemical Engineers
2008 Spring National Meeting
42nd Annual Loss Prevention Symposium
New Orleans, LA
April 7-9, 2008

UNPUBLISHED

AIChE shall not be responsible for statements or opinions contained
in papers or printed in its publications

Hydrochloric Acid Tank Explosion

By
Chandrashekhar Chandwadkar,
National Petrochemical Company,
Nothern Sheikh Bahaei St.,
Mulla Sadra Ave.
Tehran, Iran.
c_chandwadkar@yahoo.com

Ghodaratollah Nasiri
National Petrochemical Company, Iran

Ali Seyfzadeh
National Petrochemical Company, Iran

ABSTRACT

Explosions in Hydrogen Chloride tanks are very rare. However, we have experienced one. The explosion was so powerful that the tank with a capacity 250 KL, 60% filled with 33% HCL, took off from the ground went up to 8-9 meters high in air and landed around 50 meters away. The contents of the tank sprayed all over the nearby area. Though the tank was surrounded by other plants and equipment, fortunately it landed in no-mans land! This averted loss of life and property that would otherwise might have resulted from a cascading effect.

1. Introduction

HCl tanks are generally considered to be safe from fire and explosion hazards; some times there are chances of contamination with flammables. Mainly this depends on the process being followed. Normally there is no problem if there is hydrogen and chlorine combustion and scrubbing with water (i.e. hydrochloric synthesis). However, when hydrochloric gas from the vinyl chloride monomer (VCM) process is used to produce HCl acid one has to be very careful to check the carry-over of flammables. The following case study describes an incident of an explosion in a HCl tank. In spite of nitrogen blanketing on the tank, some of the acts of commission and omission resulted in an explosion.

2. Narration of incident

It was a morning in late November at 6.40 am, near shift change time. Everyone was getting ready to handover the charge to the morning shift personnel and go home. The plant was running smoothly. One of the operators, who was changing his clothes, suddenly heard a loud bang. He just peeped through the window and to his surprise there was no HCl tank! After few moments again he heard loud thud, by this time he was out in open ground and saw a big metal structure in the near by area. It was the HCl tank that he was looking for! The explosion dislodged the HCl tank from its foundation. The tank, having 250 KL capacity, 60% filled with 33% HCl took off from ground went up around 8-9 meters high in air and landed around 50 meters away from its foundation, spraying contents all over near by area! Luckily it was shift change time and no one was outside in plant.

2.1 Photographs of the tank 603B after explosion & tank 603A after implosion:

Tank landed in no-man's-land

View from other side

Foundation of tank

Tank 603 A- Imploded

3. Brief history of earlier incidents:

There was history of minor incidents that should have provided adequate warning of the presence of flammables. In the year 1997 the transfer tank T-603 was emptied to perform welding. While performing welding on the tank a minor explosion was experienced. This was attributed to presence of hydrogen. After two years (i.e. in 1999) again in same tank T-603 another minor explosion was experienced, caused by hammering. This time the vent gas was analyzed and flammables such as acetylene, ethylene, etc were found. This led to the installation of nitrogen blanketing on all HCL tanks.

4. Routes of HCl Production:

4.1 Carbone Lorraine Process:

The SINTACLOR hydrochloric acid synthesis units continuously produce 37% concentrated aqueous solution of hydrochloric acid, from the combustion reaction of chlorine gas and hydrogen and subsequent absorption with de-mineralized water. , The hydrogen and chlorine combine in equal volumes according to an exothermal reaction to form gaseous hydrogen chloride. In practice 5% stoichiometric excess of hydrogen is employed to ensure the complete combustion of chlorine and avoid discharging chlorine waste into the atmosphere and to obtain pure acid.

The remainder of synthesis process involves:

Cooling of the gas formed.

Co-current absorption in the process water

Then cooling of the acid.

Like the combustion, these steps are highly exothermic, which implies the continuous use of cooling water to remove the heat generated by the system. Non absorbed gas is fed to Tail Tower (scrubber) and the weak acid formed is again fed to synthesis unit (Furnace + Absorber) to obtain pure acid.

4.2 HCl gas from VCM plant:

HCl gas produced with the pyrolysis of ethylene dichloride from the Vinyl Chloride plant is directly fed to chlorine line of hydrochloric synthesis unit as shown by the dotted line in fig. The synthesis unit is operated with the burner off and without supply of hydrogen.

4.3 Schematic for both the routes:

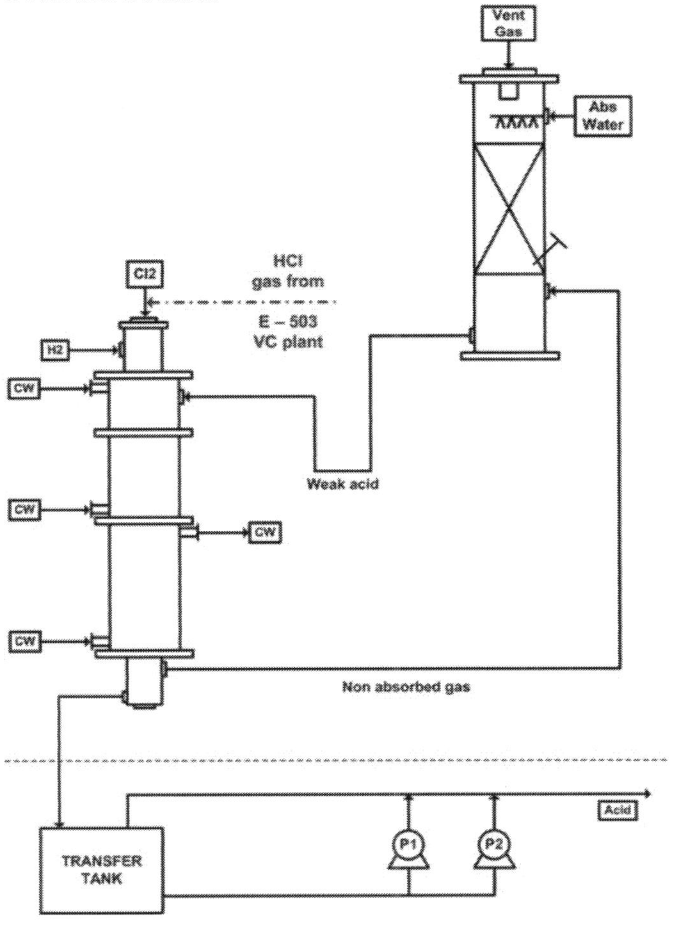

5. Observation & Analysis

- Acid production from SINTACLOR was in progress at the time of incident and just before the incident free chlorine was analyzed and found to be less than 10 ppm, so the possibility of a hydrogen and chlorine reaction is ruled out.
- Rubber lining of tank 605B was redone around 10 days back, so the possibility of hydrogen generation because of acid and tank body reaction is also ruled out.
- There was an interruption in production because of problem with the blower belt (25 Nov. 2006 at around 3 am). The plant was restarted at 2:30 pm on 25 Nov. 2006.
- HCl was being transferred to the tank at the time of the explosion and from the level chart it is clear that rate HCL removal for the five hours before the incident was high which can build static charge in rubber lined tank.
- The contents of the tank T-605 C was transferred in T-605 B. T-605 C contained HCl from the VCM plant gas. VCM plant was being run under capacity. C-502, the HCl Column in the VCM plant was also not functioning to its full capacity, as the plant was nearing annual shut down. This caused more flammable impurities in HCL gas in T-605 C.
- Black spots were observed on the inner surface of tank roof. Black spots imply that hydrocarbons were present.
- Earthing strips were found highly corroded & frailed. They may not be sufficient to take away the static charge produced in rubber lined tank.
- Earlier HCl tank vent gas analysis reports shows high traces of flammable in HCl gas from the VCM stream.
- Internal corrosion in nitrogen line discovered a few days before the implosion in tank T 605C partially/ completely blocked the nitrogen line.
- Common vent line of T-605 B & C was found open as Tank-C, damaged by implosion was being dismantled. This provided the ingress of air in tank, completing the fire triangle.

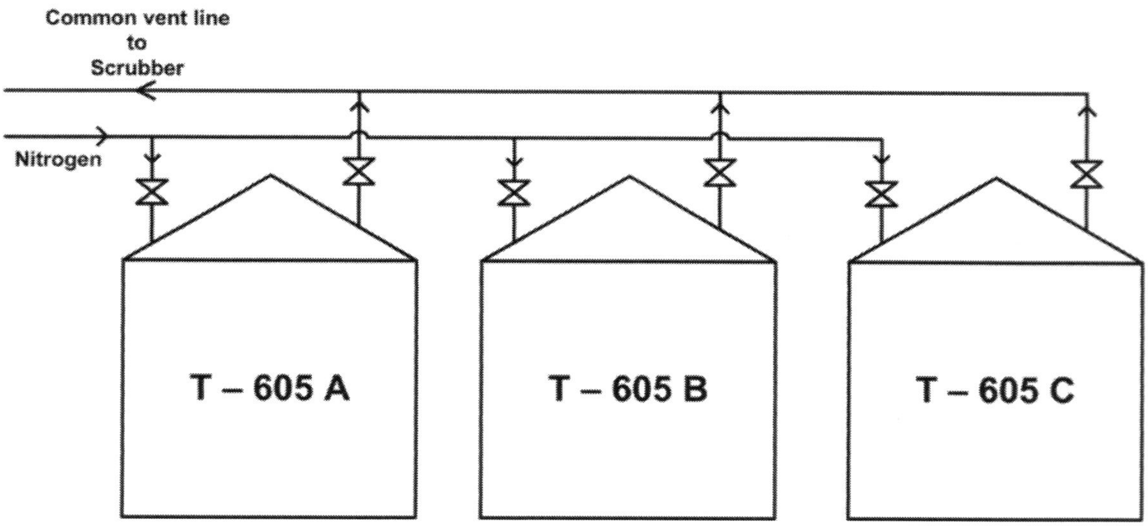

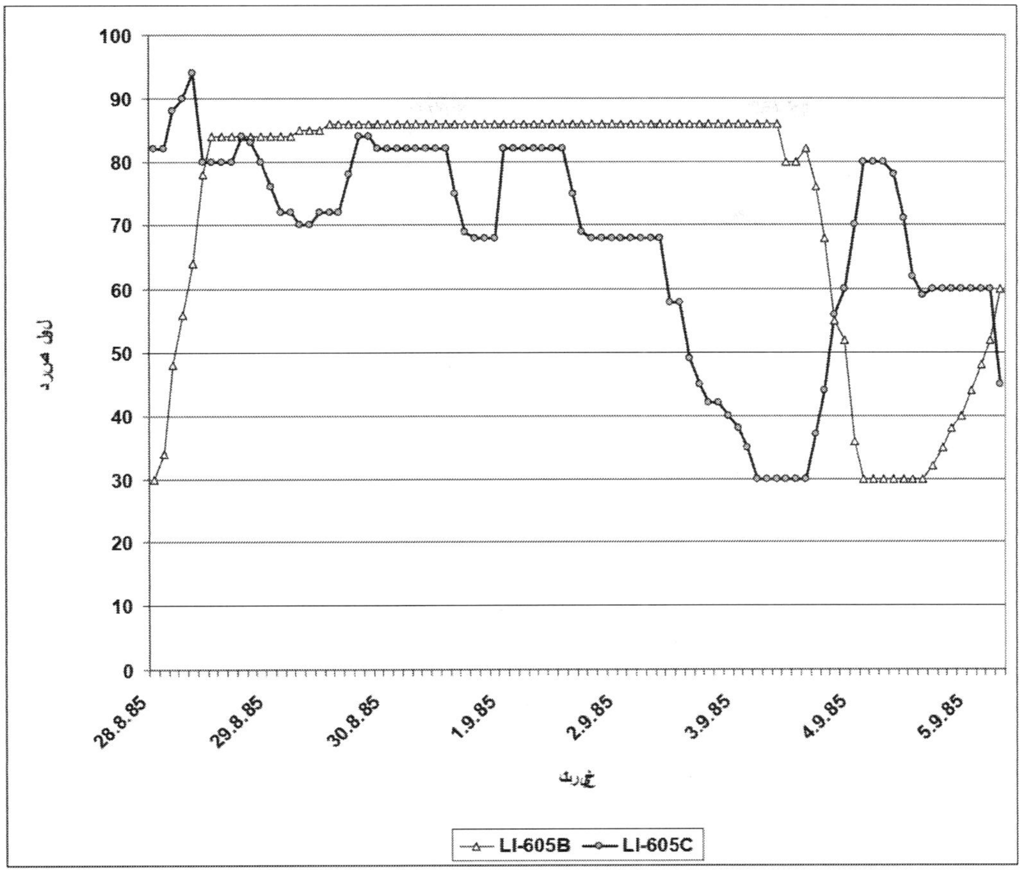

Tank Level Chart

6. Conclusion:

The carryover of flammable gases from the VCM plant HCl gas because of constrains in VCM plant, poor arrangements for static charge removal and the open vent line lead to the massive explosion that resulted in rocketing the 250 KL 60% full HCl tank around 9 meters high in air around 50 meters away from its foundation.

7. Recommendations:

- Limit of flow rate while drawing contents from tank.
- Install online sampling for flammables in VC plant HCl gas stream.
- Require regular sampling of tanks vent gas to find out concentration of flammables.
- Require nitrogen sampling for oxygen content in the tank head space.
- Provide efficient earthing for rubber lined tank.
- Change material of construction of nitrogen line 5 meters from the tanks of concern to FRP.
- While issuing work permit discuss and warn about the residual hazards, in this case the common vent line was opened and kept as it is.

8. References:

[1] Accident Investigation Report NPC.
[2] Documents from 'Carbone Lorraine' process.
[3] Process log sheets.

Synthron Runaway Reaction and Vapor Cloud Explosion

James Lay, PE
U.S. Chemical Safety and Hazard Investigation Board
2175 K Street NW, Suite 400
Washington, DC 20037
Jim.Lay@CSB.gov

Francisco Altamirano, CFEI
Johnnie Banks, CFEI
Jeffrey Wanko, PE, CSP
U.S. Chemical Safety and Hazard Investigation Board
2175 K Street NW, Suite 400
Washington, DC 20037

Prepared for Presentation at
American Institute of Chemical Engineers
2008 Spring National Meeting
42nd Annual Loss Prevention Symposium
New Orleans, LA
April 7-9, 2008

UNPUBLISHED

AIChE shall not be responsible for statements or opinions contained
in papers or printed in its publications

Synthron Runaway Reaction and Vapor Cloud Explosion

James Lay, PE
U.S. Chemical Safety and Hazard Investigation Board
2175 K Street NW, Suite 400
Washington, DC 20037
Jim.Lay@CSB.gov

Francisco Altamirano, CFEI
Johnnie Banks, CFEI
Jeffrey Wanko, PE, CSP
U.S. Chemical Safety and Hazard Investigation Board

Abstract

The CSB investigated a runaway chemical reaction and subsequent vapor cloud explosion and fire that killed one worker and injured 14 (two seriously) at the Synthron, LLC facility in Morganton, North Carolina, on January 31, 2006. The explosion destroyed the facility and damaged structures in the nearby community.

The accident occurred when plant managers increased the batch size for an acrylic additive by 12% and front-loaded all of the additional monomer into the initial batch-mode step in the production process. Subsequent laboratory testing showed that this process change more than doubled the heat release rate in the reactor. The increased thermal load exceeded the capacity of the reactor's condenser, causing a runaway reaction. Faulty bolting practices created a weak point in the pressure envelope, leading to a discharge of flammable solvent vapors into the manufacturing building. The resulting vapor cloud exploded.

The CSB identified numerous factors contributing to the incident, including an inexperienced staff, poor training, a lack of hazard awareness, poor oversight by the multinational parent company, and a lack of engineering safeguards and controls. Lessons learned are described.

1. Introduction

This paper describes a runaway chemical reaction and subsequent vapor cloud explosion and fires that killed one worker and injured 14 (two seriously). The explosion destroyed the facility and damaged structures in the nearby community. The incident occurred at the Synthron, LLC facility in Morganton, North Carolina, on January 31, 2006.

The CSB investigated the incident and published a Case Study in July, 2007 to emphasize the importance of implementing comprehensive safety management practices to control reactive hazards. The Synthron incident was also featured in the CSB's video, Reactive Hazards:

Dangers of Uncontrolled Chemical Reactions, released simultaneously with the Case Study. Both the Case Study and video are available for free download from the CSB website at http://www.csb.gov.

2. Incident Description

2.1 Summary

This incident occurred at Synthron, LLC's Morganton, North Carolina, facility. The company manufactured a variety of powder coatings and paint additives by the solution polymerization of acrylic monomers in a 1,500 gallon reactor.

The company had received an order for slightly more of an additive than the normal size recipe would produce. Plant managers scaled up the recipe to produce the required larger amount of polymer, and added all of the additional monomer needed into the initial charge to the reactor. This more than doubled the rate of energy release in the reactor, exceeding the cooling capacity of the reactor condenser and causing a runaway reaction.

The reactor pressure increased rapidly. Solvent vapors vented from the reactor's manway, forming a flammable cloud inside the building. The vapors found an ignition source, and the resulting explosion and fire killed one worker and injured 14. The blast destroyed the facility (Figure 1) and damaged off-site structures.

The U.S. Chemical Safety Board (CSB) found that the reactor lacked basic safeguards to prevent, detect, and mitigate runaway reactions, and that essential safety management practices were not in place.

Photo courtesy of Morganton Department of Public Safety

Figure 1: Synthron facility after the explosion

2.2 Description

The production department began preparing a 6,080 pound acrylic polymer batch the day before the incident, approximately 12 percent more material than would normally be made in a single batch. The plant superintendent determined the quantities of solvent, monomer, and initiator needed for the batch. The day shift operators then blended the solvents and used some of the blend to prepare the initiator solution. They added the balance to a 1,500 gallon reactor designated as vessel M1 (Figure 2). The second shift operators, in accordance with written instructions, added some of the monomer to the reactor and held back the remainder for use later in the reaction sequence. The day shift arrived on the morning of January 31 and added steam to the reactor jacket to heat the reactor to the temperature specified on the batch sheet, then shut off the steam.

The senior operator took the final step to start the reaction by pumping the initiator solution into the reactor. He then visually checked the flow of condensed solvent through the condenser sight glass to monitor the rate of reaction. While the reaction initially did not proceed as vigorously as he expected, the condensed solvent flow later increased and appeared to be in the normal range.

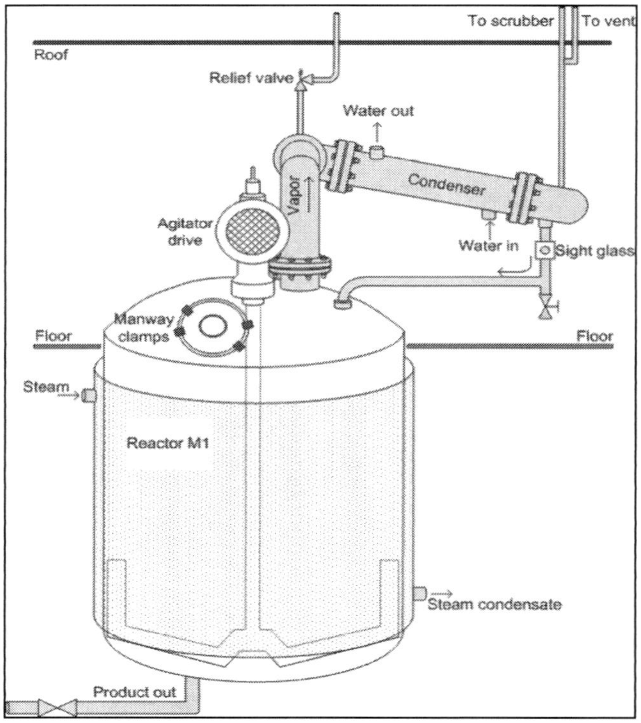

Figure 2: Reactor M1

Several minutes later, the senior operator heard a loud hissing and saw vapor venting from the reactor manway. The irritating vapor forced him out of the building. Three other employees were also forced from the building by the release. Joined by the plant superintendent and the plant manager, the employees gathered outside an upper level doorway (Figure 3). The senior operator re-entered the building wearing a respirator, and was able to start emergency cooling water flow to the reactor jacket. However, the building exploded less than 30 seconds after he exited.

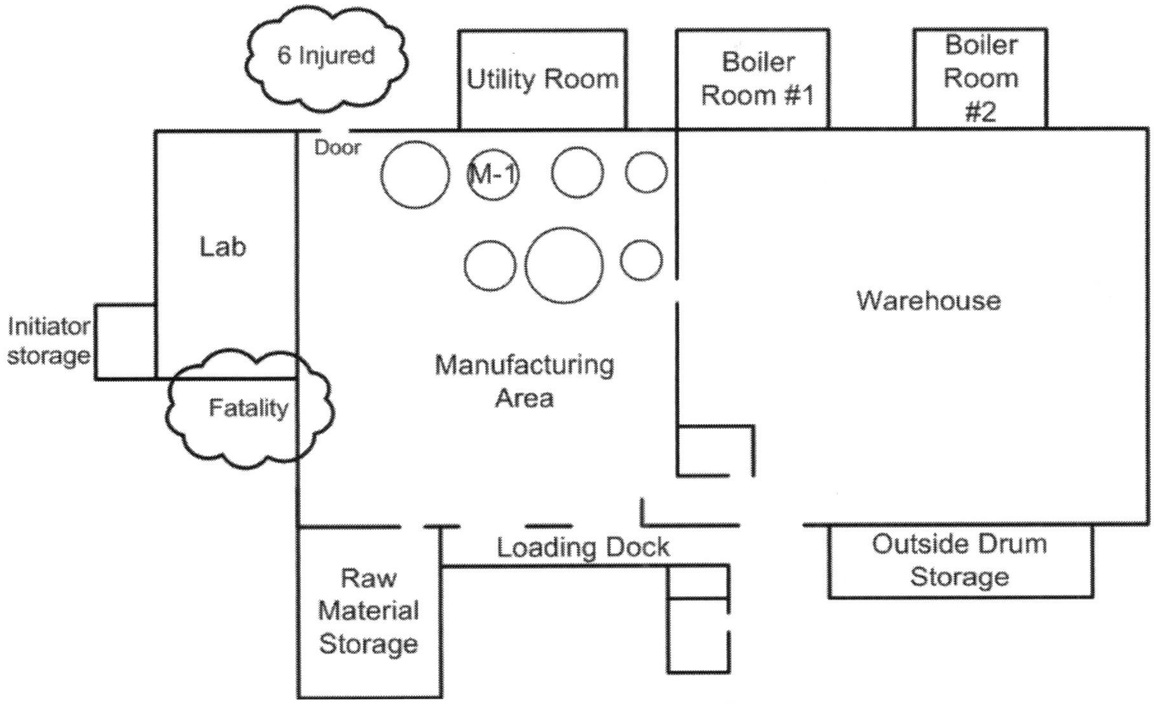

Figure 3: Synthron Facility Layout

The blast knocked down the personnel gathered outside the doorway. All were injured, and one required helicopter transport to hospital. Administrative personnel working in an onsite trailer also suffered minor injuries. The maintenance supervisor was near the lab on the lower level when the explosion occurred (Figure 3). He was severely burned over most of his body and was transported by helicopter to a regional burn center, where he died five days later.

The blast damaged structures in the nearby community. Two church buildings and a house were condemned, and glass was broken up to one-third of a mile from the site. Two citizens driving by the site were slightly injured.

The Environmental Protection Agency (EPA) took control of the site under the CERCLA (Superfund) regulation, remediated contamination from the incident, and eventually razed all of the structures on the property.

2.3 Emergency Response

The Morganton Department of Public Safety responded rapidly and called in mutual aid support from Burke County and nearby municipalities. Employees and Public Safety officers assisted injured employees. The fires following the explosion generated thick smoke, and local residents were asked to shelter-in-place for several hours. The fires were extinguished the next day.

3. COMPANY OPERATIONS

Protex International has owned a controlling interest in Synthron since 1972, and Protex' president is also president of Synthron. Protex, based in Paris, France, is a privately held company with over $100 million a year in sales, and operations in Europe, Asia, and North and South America. In addition to the Synthron facility, Protex operates chemical-related businesses in Massachusetts, New Jersey, and Florida.

Synthron's single U.S. facility was located in Morganton, North Carolina, approximately 70 miles northwest of Charlotte. The 17 Morganton employees included the company's vice president, who was responsible for day-to-day operations. Synthron had 2004 sales of approximately $3 million. Following the explosion, Synthron, LLC filed for bankruptcy under Chapter 7 of the bankruptcy laws.

Synthron produced acrylic polymers, primarily additives for the powder coating and paint industries. The polymers were produced by the free-radical polymerization of acrylic monomers in various flammable solvents. The product being manufactured on the day of the incident, Modarez MFP-BH, was a liquid acrylic polymer.

Polymerization was performed in a 1,500 gallon reactor, rated at 75 psig maximum operating pressure and designated as reactor "M1" (Figure 2). The reactor was located in a manufacturing area adjacent to the warehouse (Figure 3). In a typical reaction sequence, operators added a mixture of solvents and monomers to the reactor. They then injected steam into the reactor's jacket to heat the reaction mixture to a specified temperature, usually the expected mixture boiling point. The steam was turned off and initiator solution metered into the reactor to start polymerization.

The heat given off by the reaction boiled the solvent and monomer mixture, causing hot vapors to flow to an overhead water-cooled heat exchanger where they were totally condensed. Liquid solvent drained from the bottom of the condenser back into the reactor. The outlet of the condenser vented to the atmosphere through a small pipe, keeping the pressure in the reactor very near atmospheric under normal operation. In an emergency, water could be manually directed through the reactor's jacket to increase cooling.

A complete manufacturing cycle could include several reaction steps. The finished liquid polymer was stripped of solvent, cooled, and packaged into drums for shipment to customers.

4. Incident Analysis

The CSB investigators determined that the key factors leading to this incident included

- a lack of hazard recognition,
- poorly documented process safety information,
- ineffective control of product recipe changes,
- a lack of automatic safeguards to prevent or mitigate the effects of loss of control over the reaction,
- improper manway bolting practices,

- poor operator training,
- inadequate emergency plans drills, and
- inadequate corporate oversight of process safety.

4.1 Hazard Awareness

When performing reactive chemistry, companies should maintain a high degree of awareness of the hazards involved. Synthron combined monomers and reaction initiators in the presence of flammable solvent to produce polymer products, but failed to identify the hazards associated with this type of chemistry.

Texts such as Loss Prevention in the Process Industries (Mannan, 2005) and Bretherick's Chemical Reaction Hazards (Urban, 2000) describe appropriate means for characterizing industrial chemical reactions. These include determining the heat generation rate as a function of temperature, the available heat removal capacity, and the potential for excessive monomer or initiator accumulation in the reactor. Failure to control these critical characteristics can lead to severe upsets, including runaway reactions.

Companies can detect hazards by asking "What can go wrong?" to identify the potential consequences of inappropriate mixing, recipe changes, greater- and lower-than-normal heating, lack of mixing, etc. The answers can be used to develop improvements or changes that can prevent reactive upsets or mitigate their consequences. Synthron had not identified the hazards of its reactive chemical operations. No PHAs had been conducted to address "What could go wrong?" during reactor operations.

Furthermore, most of the management and operations personnel at Synthron had been on the job for less than a year, in some cases much less (Table 1), and lacked previous polymer manufacturing experience.1 In addition, Synthron's training program was informal and did not include reactive hazards training.

Table 1: Management and operations personnel tenure at Synthron

Employee	Time at Synthron	Polymer Experience?
Manager	9 months	No
Superintendent	8 months	No
Vice President	5 months	No
2nd Shift Op-1	3 months	No
2nd Shift Opt-2	3 months	No
Chemist	3 weeks	No

Personnel, including site managers, were thus poorly prepared to recognize potentially hazardous changes to product recipes, or to respond to an incipient runaway reaction. Additionally,

[1] Synthron's president had replaced the management team in an attempt to increase sales. A variety of causes had simultaneously contributed to high operator turnover. The senior operator had been brought back from retirement to provide a degree of continuity.

Synthron had no chemical or other engineers on staff, and none had been contracted to evaluate the hazards associated with reactive operations at the site.

4.2 Process Safety Information and Training

Synthron had minimal safety information on its polymerization process, even though this was the core of its manufacturing business. Synthron optimized product formulations to meet customer specifications. However, reaction characterization and calorimetry were not performed to establish process equipment performance requirements and operating limits for safe operations.

When scaling-up new products from the laboratory, the previous Synthron plant manager had typically estimated initial production batch sizes based on past experience. He then gradually increased batch quantities until the sight glass in the condensate return line showed that the condenser was close to flooding, or that another performance limit was being approached (e.g., the reactor pressure increased). Polymerization scale-up to their standard batch sizes had historically been done by simple trial-and-error. The condenser's cooling capacity was not documented, nor could the cooling load placed on the condenser be determined with the available instrumentation. As a result, information essential to the safe operation of the reactor was not available. Polymerization reactors can runaway with disastrous consequences if they are not carefully controlled.

Based on interviews the CSB conducted, Synthron employees and managers had little or no understanding of reactive hazards. They had not been trained on,[2] and did not understand, the margin of safety needed or available in their polymerization operations. Furthermore, they had little knowledge of the sensitivity of the reactor to changes in product recipes, batch sizes, or reaction conditions. Synthron's employees were thus unprepared to recognize and respond to the reactive hazards they faced the day of the incident.

4.3 Batch Recipe Changes

In planning the MFP-BH batch, Synthron managers made several changes that greatly increased the heat released by the reaction in reactor M1 and the potential for a runaway reaction. However, the changes were not effectively reviewed and the hazards went unrecognized. Synthron:

- increased the total amount of monomer to be charged to the reactor by 12 percent. The additional monomer, with a Normal Boiling Point Temperature (NBPT) of 147°C (297°F), was placed in the initial reactor charge.
- reduced the amount of aliphatic solvent, with an NBPT of 81°C (178°F), charged to the reactor by 12 percent.
- increased the amount of aromatic solvent, with an NBPT of 111°C (234°F), by 6 percent.

---

[2] Synthron's training program was informal, relying on unstructured on-the-job training. Testimony indicated that reactive hazards were not systematically addressed.

The customer had ordered 12 percent more MFP-BH than a standard batch would produce. To avoid the additional time and effort of running two half-size batches, the superintendent scaled up the recipe to produce the order in a single batch.

MFP-BH was produced in two stages. In the first, an initial charge of acrylic monomer and solvent was loaded into reactor M1, heated, and then reacted by adding an initiator over a brief period. In the second, the remaining monomer and initiator were co-fed into the reactor over an extended period. Unfortunately, the superintendent placed almost all of the additional monomer for the larger batch into the initial charge of chemicals loaded into the reactor.

According to the batch sheet, roughly equal amounts of aromatic and aliphatic solvent should have been added to the reactor. However, there was not enough of the lower boiling temperature aliphatic solvent available in storage. To compensate, the superintendent and manager decided to make up half the shortfall using the higher boiling aromatic solvent, and to run the batch with slightly less total solvent than specified in the recipe.

Together, these changes:

- increased the total amount of monomer in the reactor by 45 percent,
- increased the concentration of monomer by 27 percent, and
- increased the atmospheric boiling point temperature of the mixture by almost 5°C (9°F).[3]

Each of these changes would be expected to increase the rate of heat release in the reactor. When asked to review the changes in the solvent quantities, the plant chemist estimated that the boiling point of the solvent mixture would increase about 1°C (1.8° F). However, the chemist, manager, and superintendent did not recognize or address the potential impact of the increased monomer amount and concentration on the mixture boiling point, reaction rate, or total rate of heat release.

The combined effect of the changes was to increase the maximum heat output from the reaction to at least 2.3 times that of the standard recipe. Figure 4 illustrates the results of reaction calorimetry performed to simulate the early stages of the incident.

The lower curve shows the heat release rate (thermal power) versus time for the standard recipe, while the upper curve shows the heat release rate for the modified recipe.[4] The lower solid horizontal line is the estimated cooling capacity of the reactor condenser on the day of the incident. Once the heating curve exceeded the condenser cooling line, control of the reaction was lost, resulting in a runaway reaction.

---

[3] The CSB measured the boiling points of laboratory mixtures with the same compositions.

[4] Each curve was generated isothermally at the NBPT of the mixture, closely simulating the operation of the reactor up to the time when control was lost. Heat generation rates during the subsequent runway reaction were higher.

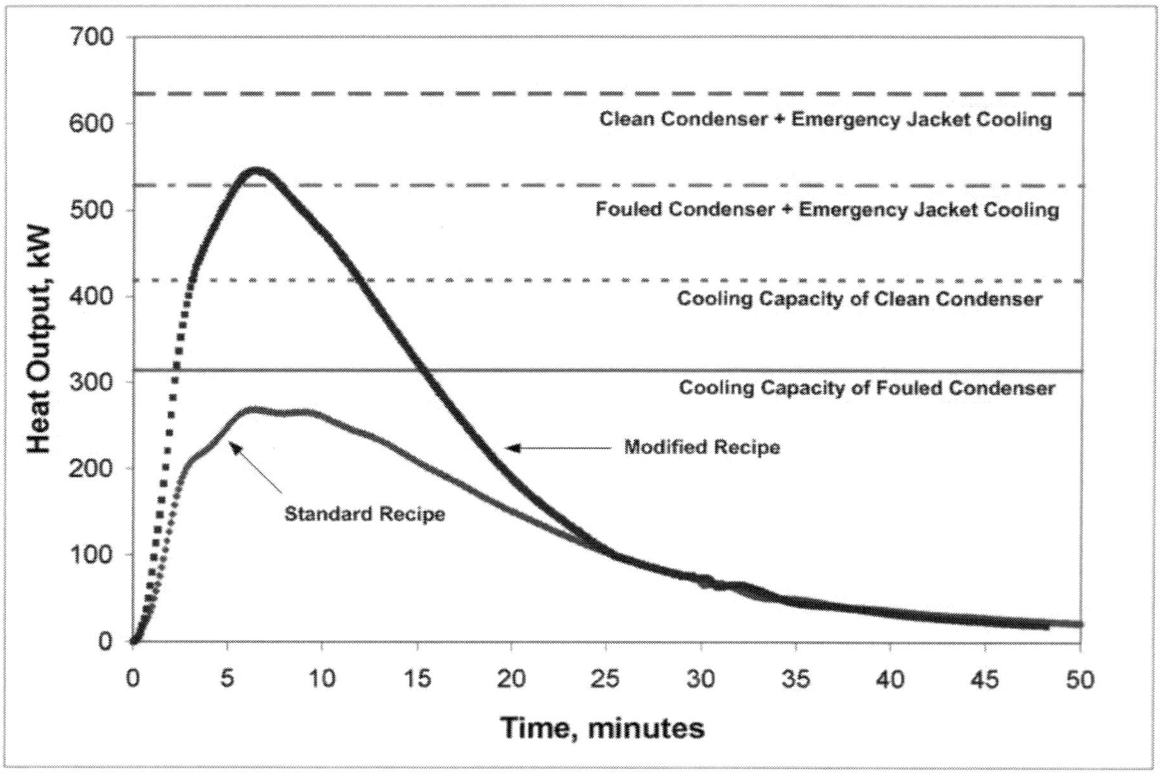

Figure 4: Reactor Energy Release and Cooling Capacity

4.4 Engineering Safeguards

Synthron relied primarily on a procedural safeguard to prevent loss of reactor control: the batch sheet, which was used as an operating procedure at the site. Procedures are essential for safety in chemical processing operation, but are the least reliable form of safeguard for preventing process incidents (CCPS, 2004).

Failures with potentially severe consequences, such as runaway reactions, should have multiple independent safeguards. Examples of safeguards that could have prevented or mitigated this incident, but were not installed at Synthron, include:

- high pressure alarms to notify operators of problems early in the incident when action to control the reaction might still be possible,
- automatic emergency cooling water flow to the reactor jacket,
- automatic shut-off of initiator feed,
- automatic or remotely operated injection of "short stop" solution to stop the polymerization reaction,[5] and
- automatic or remotely operated venting or dumping of the reactor to a safe location.

---

[5] Chemicals, such as phenothiazine, can be injected into a reactor to slow or stop free-radical polymerizations such as the reaction at Synthron.

Good practice is to review the adequacy of safeguards on chemical reactors using a structured method such as Layers of Protection Analysis (CCPS, 2001). Such reviews can help ensure that runaway reactions are prevented or are rapidly and reliably detected and controlled.

4.5 Manway Bolting Practice

Operators opened the reactor manway after every batch cycle to clean the reactor. Long-standing practice at the facility was to then close the manway and secure it using only four of the 18 clamps specified by the manufacturer (Figure 5).[6] The risk posed by this practice had not been recognized because the reactor normally ran at near-atmospheric pressure, for which four clamps were thought to be adequate.

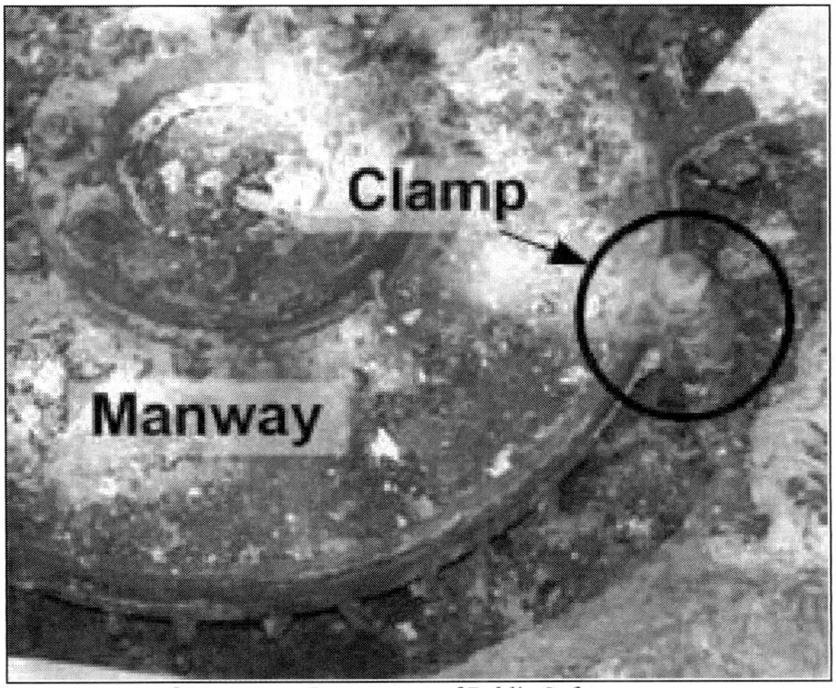

Photo courtesy of Morganton Department of Public Safety

Figure 5. One of four installed manway clamps

The CSB investigators calculated that, with only four clamps installed, the manway began to leak flammable solvent vapors when the reactor pressure reached approximately 23 psig, well below the reactor's maximum allowable working pressure (and likely relief valve set point) of 75 psig.[7] The vapor leak path is clearly visible in Figure 6 (arrow).

---

[6] The reactor manufacturer specified 18 clamps to maintain a tight seal at the reactor maximum working pressure of 75 psig. The clamps were not permanently attached to the manway, and no markings on the manway indicated the required number of clamps.

[7] The reactor M1 relief valve could not be recovered from the debris, and the relief valve records were destroyed in the explosion and subsequent fires. Other relief valves examined were set at the rated operating pressure of the vessels they protected.

Figure 6. Manway gasket and leak path (arrow)

4.6 Condenser Fouling

Synthron failed to establish procedures to maintain the performance of the M1 reactor condenser, an inclined shell-and-tube heat exchanger essential to safe operation of the reactor. In the condenser, solvent vapors flowed inside the tubes and were cooled and condensed by cooling water flowing through the shell (Figure 2). Inspection of the cooling water side of the condenser after the incident revealed that it was badly fouled, likely reducing condenser capacity at least 25 percent (Figure 7).

The condenser's design, with the tubes permanently bonded to plates at both ends of the condenser, made inspecting or cleaning the water side difficult.[8] Furthermore, the company had no program to systematically monitor and control water side fouling.[9] The CSB found no evidence that the water side of the condenser had ever been inspected or cleaned to remove the scale, rust, and sediment that had accumulated during 30 years of service. Synthron's employees lacked the expertise and experience to recognize the risk posed by water side fouling of the condenser.

---

[8] The condenser was a Tubular Exchanger Manufacturer's Association (TEMA) type BEM, fixed tubesheet exchanger with the process fluid on the tube side and cooling water on the shell side. The tubes could not be removed for inspection or cleaning. This design may require periodic chemical cleaning of the shell side to maintain good thermal performance.

[9] Chemical treatment of the cooling water is almost always required to prevent biological growth, fouling, and/or corrosion. The cooling water at Synthron was not treated.

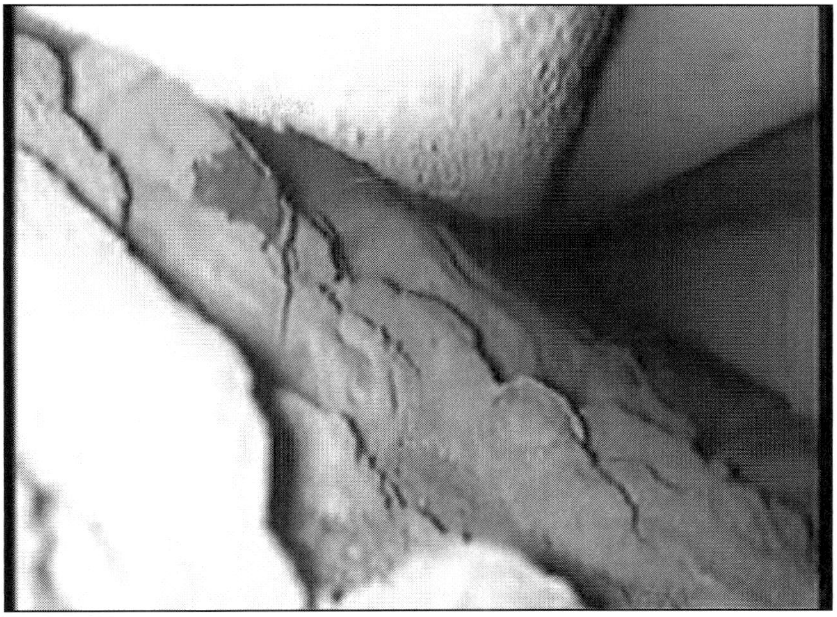

Figure 7: Deposits fouling the water side of the M1 reactor's condenser.

A clean condenser, combined with automatic emergency jacket cooling (Figure 4, upper dashed line), might have prevented the runaway reaction and subsequent explosion.

4.7 Emergency Evacuation Procedures and Training

Effective evacuation plans are important for minimizing injuries and fatalities in chemical emergencies. During this incident, none of the production employees evacuated to a safe location. At the time of the explosion, six employees, including the manager and superintendent, gathered outside a doorway on the upper level while a seventh was on the lower level by the lab. The employee on the lower level was killed, while all the employees outside the doorway were injured, two seriously.

Synthron was unprepared for an emergency; specifically:

- The facility Emergency Action Plan did not list events or describe situations that might necessitate a plant evacuation.
- Operating procedure did not specify employees' actions in the event of a chemical release or loss of reactor control.
- Employees were not trained on the Emergency Action Plan and evacuation drills were not conducted.
- The facility was not equipped with an emergency alarm system.

In this incident, there was sufficient time after the release began to evacuate employees to a safe location. Adequate emergency response planning by Synthron could have prevented the death and serious injuries caused by the explosion.

4.8 Corporate Safety Oversight and Support

Synthron was part of Protex International, a much larger organization. However, the parent company provided little safety oversight or support to Synthron.

While Synthron performed quality control testing and limited product development work in Morganton, the procedures, equipment, and trained personnel needed to characterize reactive hazards were not in place.

However, Protex had the ability to perform detailed reaction characterization work, including reaction calorimetry, at its European facilities. It is good practice (CCPS, 2003) to ensure that reactive safety programs at small facilities are adequately supported by technically qualified resources. Protex did not provide adequate reactive safety support to Synthron.

Furthermore, in the summer and fall of 2005, Synthron's president (also the president of Protex) hired a new vice president, plant manager, and plant superintendent for the Morganton site. While two of these key management employees had degrees in chemistry, they had no previous polymerization experience and were not trained on the parent company's process safety procedures or its testing capabilities. Having little reactive chemistry background or training, they did not recognize the reactive hazards at the Morganton site.

CCPS (1995) stresses that expertise in managing process safety must be ensured when making staffing changes. Synthron's president failed to ensure that the new team he installed at the Morganton site had the requisite training, knowledge, and experience to operate the facility safely.

Finally, Protex did not comprehensively audit or review Synthron's safety program. Such a review should have identified the absence of an effective reactive hazard control program.

5. Regulatory Analysis

Synthron had not implemented a Process Safety Management (PSM) system consistent with the requirements of the Occupational Safety and Health Administration's (OSHA) PSM regulation (29 CFR 1910.119). The site had last been inspected by North Carolina OSHA (NC-OSH) in 1996. Following the explosion, NC-OSH proposed numerous PSM citations, which Synthron contested.

The batch being processed at the time of the explosion, had it been completed, would have contained in excess of the OSHA PSM threshold of 10,000 pounds of flammable liquid. However, only 4,500 pounds of material were in the reactor at the time of the incident, well below the PSM threshold quantity. Nonetheless, a catastrophic incident occurred. Synthron's experience demonstrates that companies working with reactive chemical quantities less than the PSM flammables' threshold still need to consider and guard against potentially catastrophic accidents.

6. The CSB Reactives Study

In its comprehensive 2002 report, "Improving Reactive Hazard Management" (CSB, 2002), the CSB found that reactive incidents are a serious problem in the United States, and that both management system and regulatory improvements are needed to help facilities control reactive hazards. This report studied reactive incidents, causal factors, and preventive measures,[10] and outlined the screening, hazard identification, hazard review, operating procedures, and training needed to prevent reactive incidents.

The report documented 167 serious reactive incidents in the United States between January 1980 and June 2001 that resulted in 108 deaths, hundreds of injuries, and significant public impacts. Ongoing monitoring by the CSB indicates that reactive incidents, such as the Synthron explosion, continue to occur.

The CSB report also found that 70 percent of reactive incidents occurred in the chemical manufacturing industry, with 35 percent due to runaway reactions, such as that which occurred at Synthron.[11] While 42 percent of reactive incidents resulted in fires and explosions, another 37 percent caused toxic emissions. Many reactive incidents occurred at small manufacturing sites such as Synthron.

More than 50 percent of the 167 incidents documented in the CSB report involved chemicals not covered by existing OSHA PSM (29 CFR 1910.119) or EPA Risk Management Program (40 CFR Part 68) regulations. The CSB recommended better coverage of reactive hazards by these regulations. EPA now requires reporting of reactive chemical incidents under RMP reporting rules. OSHA has taken steps to increase industry awareness of reactive hazards, but has not fully implemented the CSB recommendations.

The CSB report identified many valuable sources of good practices for managing reactive hazards. Since the report's publication, additional guidance has become available. The Center for Chemical Process Safety (CCPS) published *Essential Practices for Managing Chemical Reactivity Hazards* (CCPS, 2003), which can be downloaded at no cost from http://info.knovel.com/ccps/ (January 2007). This book lays out the basic steps manufacturers should take to protect against reactive hazards.

Other valuable sources of guidance are available. The EPA's Chemical Emergency Preparedness and Prevention web page http://yosemite.epa.gov/oswer/ceppoweb.nsf/content/ap-book.htm (May 2007) contains useful links. The United Kingdom's Health and Safety Executive also offers an informative web page, http://www.hse.gov.uk/pubns/indg254.htm (May 2007) and the comprehensive booklet, "Designing and Operating Safe Chemical Reaction Processes" (HSE Books, 2000).

---

[10] Available for download at http://www.csb.gov.

[11] Of reactive incidents, 25 percent originate in reactors, with the rest occurring in a wide range of equipment.

7. Conclusion

This incident provides important lessons for manufacturers with operations involving reactive chemistry.

7.1 Identify and Control Reactive Hazards

Manufacturers should take a comprehensive approach, and:

- identify and characterize reactive hazards;
- systematically evaluate what can go wrong, including mis-charging of reagents, loss of cooling, instrument malfunction, and other credible failure scenarios; and
- implement, document, and maintain adequate safeguards against the identified failure scenarios. Multiple, independent safeguards may be needed to reliably ensure the safety of the reactive process.

7.2 Control Change

Chemical manufacturers and others with reactive chemistry operations should control changes to batch recipes, including key operating conditions, such as:

- the quantities, proportions, and sequencing of reactor feeds,
- reaction temperature,
- conditions that could cause initiator or monomer accumulation, and
- conditions that could affect the deactivation of monomer inhibitors or stabilizers.

7.3 Maintain Process Equipment Capability

Manufacturers with reactive chemistry operations should:

- document the performance requirements and capabilities of process equipment, such as the reactor condenser at Synthron; and
- periodically inspect and service process equipment, including the water side of heat exchangers, to maintain appropriate safety margins.

7.4 Train Personnel on Hazards and Procedures

Manufacturers should ensure that worker training includes:

- the nature of the reactive hazards, including process safety margins; and
- operating procedures, including appropriate cautions and warnings, the consequences of deviations, recognition of deviations and abnormal operations, and the appropriate responses to control or mitigate their effects.

7.5 Prepare for Emergencies

Manufacturers should:

- implement an effective emergency plan,
- train employees on the plan,

- install an evacuation alarm system that is audible and/or visible throughout the facility,
- conduct regular exercises to help ensure rapid evacuation to a safe location in an emergency, and
- coordinate their emergency planning with offsite response organizations and the Local Emergency Planning Committee (LEPC).

The CSB made a recommendation to Protex International to establish a program to ensure that reactive hazards at Protex' U.S. facilities are managed in accordance with good industry practices.

8. Disclaimer

This paper has not been independently approved by the Board and is published for general informational purposes only. Any material in the paper that did not originate in a Board-approved report is solely the responsibility of the authors and does not represent an official finding, conclusion, or position of the Board.

9. REFERENCES

[1] Health and Safety Executive (HSE) [Online], *Chemical Reaction Hazards and the Risk of Thermal Runaway,* www.hse.gov.uk/pubns/indg254.htm, London: HSE, (accessed 2007)

[2] HSE, Designing and Operating Safe Chemical Reaction Processes, Rugby, UK: HSE Books, 2000

[3] Center for Chemical Process Safety (CCPS), Guidelines for Hazard Evaluation Procedures - With Worked Examples, 2nd ed, New York: American Institute of Chemical Engineers (AIChE), 1992

[4] CCPS, Plant Guidelines for Technical Management of Chemical Process Safety (Revised Edition), New York: AIChE, 1995

[5] CCPS, Layer of Protection Analysis - Simplified Process Risk Assessment, New York: AIChE, 2001

[6] CCPS, Essential Practices for Managing Chemical Reactivity Hazards, New York: AIChE, 2003

[7] CCPS, Inherently Safer Chemical Processes, A Life Cycle Approach, New York: AIChE, 2004

[8] Mannan, S., Lees' Loss Prevention in the Process Industries, 3rd ed., Burlington, MA: Elsevier Butterworth-Heinemann, 2005

[9] Urban, P. G., Ed., Bretherick's Handbook of Reactive Chemical Hazards, 6th ed., St. Louis, MO: Elsevier Science and Technology Books, 2000

[10] The United States Chemical Safety and Hazard Investigation Board (CSB), Improving Reactive Hazard Management, Washington, DC: CSB, 2002

AUTHOR INDEX

AUTHOR INDEX

American Institute of Chemical Engineers
3 Park Avenue
New York, New York 10016-5991

ISBN: 978-1-60560-210-3